# Lecture Notes in Networks and Systems 899

The series "Lecture Notes in Networks and Systems" publishes the latest developments in Networks and Systems—quickly, informally and with high quality. Original research reported in proceedings and post-proceedings represents the core of LNNS.

Volumes published in LNNS embrace all aspects and subfields of, as well as new challenges in, Networks and Systems.

The series contains proceedings and edited volumes in systems and networks, spanning the areas of Cyber-Physical Systems, Autonomous Systems, Sensor Networks, Control Systems, Energy Systems, Automotive Systems, Biological Systems, Vehicular Networking and Connected Vehicles, Aerospace Systems, Automation, Manufacturing, Smart Grids, Nonlinear Systems, Power Systems, Robotics, Social Systems, Economic Systems and other. Of particular value to both the contributors and the readership are the short publication timeframe and the world-wide distribution and exposure which enable both a wide and rapid dissemination of research output.

The series covers the theory, applications, and perspectives on the state of the art and future developments relevant to systems and networks, decision making, control, complex processes and related areas, as embedded in the fields of interdisciplinary and applied sciences, engineering, computer science, physics, economics, social, and life sciences, as well as the paradigms and methodologies behind them.

Indexed by SCOPUS, INSPEC, WTI Frankfurt eG, zbMATH, SCImago.

All books published in the series are submitted for consideration in Web of Science.

For proposals from Asia please contact Aninda Bose (aninda.bose@springer.com).

Michael E. Auer · Uriel R. Cukierman ·
Eduardo Vendrell Vidal · Edmundo Tovar Caro
Editors

# Towards a Hybrid, Flexible and Socially Engaged Higher Education

## Proceedings of the 26th International Conference on Interactive Collaborative Learning (ICL2023), Volume 1

Springer

*Editors*
Michael E. Auer
CTI Global
Frankfurt, Germany

Uriel R. Cukierman
UTN—FRBA
Mozart, Argentina

Eduardo Vendrell Vidal
DISA
Technical University of Valencia
Valencia, Spain

Edmundo Tovar Caro
UPM, ETSII
Madrid, Spain

ISSN 2367-3370 ISSN 2367-3389 (electronic)
Lecture Notes in Networks and Systems
ISBN 978-3-031-51978-9 ISBN 978-3-031-51979-6 (eBook)
https://doi.org/10.1007/978-3-031-51979-6



This Springer imprint is published by the registered company Springer Nature Switzerland AG
The registered company address is: Gewerbestrasse 11, 6330 Cham, Switzerland

Paper in this product is recyclable.

# Preface

ICL2023 was the 26th edition of the International Conference on Interactive Collaborative Learning and the 52nd edition of the IGIP International Conference on Engineering Pedagogy.

This interdisciplinary conference aims to focus on the exchange of relevant trends and research results as well as the presentation of practical experiences in Interactive Collaborative Learning and Engineering Pedagogy.

ICL2023 took place in Madrid, Spain from 26 to 29 September 2023 and was supported by the Universidad Politécnica de Madrid, InnovaHiEd Academy, the Spanish Conference of Directors and Deans of Informatics Engineering and the Universidad Tecnológica Nacional from Argentina.

This year's theme of the conference was "Towards a Hybrid, Flexible and Socially Engaged Higher Education."

Again, outstanding scientists from around the world accepted the invitation:

## Special Invited Guests

| | |
|---|---|
| Jenna Carpenter | President American Society of Engineering Education—ASEE |
| David Guralnick | President International E-Learning Association—IELA |

## Keynotes

| | |
|---|---|
| Xavier Fouger | Senior Director, Global Academia Programs, Dassault Systèmes |
| Khairiyah Mohd-Yusof | Full Professor in the School of Engineering Education, Purdue University, USA |
| Carlos Delgado Kloos | Rector's Delegate on Digital Microcredentials, UC3M, Spain |
| Antonio Recio Sanroman | Head of Global Learning and Growth Partners, Siemens |
| Miriam Reiner | Director of the VR/AR and Neurocognition Lab at the Technion Institute, Israel |

The following very interesting workshops have been held:

**Workshop/Roundtable on accreditation**
Chair: Eduardo Vendrell Vidal, Universitat Politècnica de València

**Increasing User Engagement in Software Applications for Commercial or Research Purposes**
Mohammad Hajarian and Paloma Diaz, Universidad Carlos III de Madrid, Spain

**Supporting Open Educational Resources Creation, Personalization, Implementation, and Sharing through the Graasp.org Learning Experience Platform and its Associated Open Digital Library**
Denis Gillet, EPFL, Switzerland and Michele Notari, University of Teacher Education, Bern, Switzerland and University of Hong Kong

**Addressing the Engineering Skills Gap: How can industry and educators work together to integrate emerging technologies into student and professional education?**
Chair: Kirsten Williamson, Petrus

**How to Create Virtual Machine-Templates in a Public AND in a Private Cloud Environment**
Michael Dietz, Technische Hochschule Nürnberg, Germany

**The Engineering Classroom: Promoting and Illustrating New Types of Learning and Effective Use of Practical, Evidence-Based Strategies to Support Student Motivation and Academic Success**
Genny Villa, Université de Montréal, Canada and Natalia Rosa Rodriguez Carinthia University of Applied Sciences (Austria)n

**Low-Cost/High-Impact: Success Skills Students Will Actually Use**
Peter J. Shull, Penn State University, United States of America

**From Teaching in the Industrial Age to Teaching in the Digital and AI Age**
Chairs: Carlos Delgado Kloos and Carlos Alario, Universidad Carlos III de Madrid

**Hands-on workshop on developing complex problem-solving skills using Problem Based Learning**
Prof. Dr. Syed Ahmad Helmi Syed Hassan and Dr. Khairiyah Mohd-Yusof

**Open Badges and Micro-credentials: Recognizing Learnings in More Flexible Ways**
Uriel Ruben Cukierman and Juan Maria Palmieri, UTN, Argentine Republic and Eric Rousselle, Open Badge Factory, Finland

**BreakThrough Communication: Interactive Experiential Learning in a Hybrid World**
Susan R. Glaser and Peter A. Glaser, Glaser and Associates, United States of America

**Teach Quantum Computing!**
Chairs: Jose Christen and Maninder Kaur, QURECA

We would like to thank the organizers of the following Special Sessions:

- **Entrepreneurship in Engineering Education 2023 (EiEE´23)**
  Chairs
  Jürgen Jantschgi, Higher College for Engineering Wolfsberg, Austria.
- **AI in learning—a double face Janus (AiL'23)**
  Chair
  Elena Bendíková, Faculty of Education CU, Ružomberok, Slovakia.
- **Digital Education Strategy and Engineering Pedagogy (DESEP)**
  Chair
  Roman Hrmo, DTI University.

Since its beginning, this conference is devoted to new approaches in learning with a focus to collaborative learning and engineering education. We are currently witnessing a significant transformation in the development of education. There are at least three essential and challenging elements of this transformation process that have to be tackled in education:

- the impact of globalization and digitalization on all areas of human life;
- the exponential acceleration of the developments in technology as well as of the global markets and the necessity of flexibility and agility in education;
- the new generation of students, who are always online and don't know to live without Internet; and
- the increasing interdependence between the different sectors of education (secondary and post-secondary education, vocational education).

Therefore, the following main themes have been discussed in detail:

- Collaborative Learning,
- Mobility and Smart Cities,
- New Learning Models and Applications,
- Project-Based Learning,
- Game-Based Education,
- Educational Virtual Environments,
- Computer-Aided Language Learning (CALL),
- Teaching Best Practices,
- Engineering Pedagogy Education,
- Public-Private Partnership and Entrepreneurship Education,
- Research in Engineering Pedagogy,
- Evaluation and Outcomes Assessment,
- Internet of Things and Online Laboratories,
- IT and Knowledge Management in Education,
- Approaches of Online Teaching,
- Virtual and Augmented Learning,
- Mobile Learning Applications,
- Connection between Universities and the Labour Market,
- Further Education for Engineering Educators,
- Educational Virtual Environments.

As submission types have been accepted:

- Full Paper, Short Paper;
- Work in Progress;
- Special Sessions;
- Workshops, Tutorials.

All contributions were subject to a two-step double-blind review. The review process was very competitive. We had to review more than 500 submissions. A team of about 260 reviewers did this terrific job. Our special thanks go to all of them.

Due to the time and conference schedule restrictions, we could finally accept only the best 219 submissions for presentation.

The conference had more than 279 registered participants from 55 countries.

We thank **Sebastian Schreiter** for the technical editing of this proceedings.

ICL2024 will be held in Tallinn, Estonia.

Frankfurt, Germany

Michael E. Auer
ICL General Chair

Madrid, Spain
Mozart, Argentina

Edmundo Tovar Caro
Uriel R. Cukierman
ICL2023 Co-Chairs

# Committees

## General Chair

| | |
|---|---|
| Michael E. Auer | CTI Frankfurt/Main, Germany |

## ICL2023 Conference Chairs

| | |
|---|---|
| Tiia Rüütmann | IGIP President, Tallinn Technical University, Estonia |
| Uriel Cukierman | UTN Buenos Aires, Argentina |
| Edmundo Tovar | Universidad Politécnica de Madrid (UPM), Spain |

## Honorary Advisors

| | |
|---|---|
| Guillermo Cisneros Pérez | Rector, UPM, Spain |
| Stephanie Farell | IFEES President, USA |
| Xavier Fouger | Dassault Systèmes, France |
| Hanno Hortsch | TU Dresden, Germany |
| Hans J. Hoyer | IFEES/GEDC General Secretary |
| Manuel Castro | UNED, Spain |
| Javier Soriano | President of the Spanish Council of Deans of Informatics Engineering—CODDI, Dean ETS de Ingenieros Informáticos, UPM |

## International Chairs

| | |
|---|---|
| Samir El-Seoud | The British University in Egypt (Africa) |
| Xiao-Guang Yue | Wuhan, China (Asia) |
| Alexander Kist | University of Southern Queensland (Australia/Oceania) |
| Alaa Ashmawy | American University Dubai (Middle East) |
| David Guralnick | Kaleidoscope Learning New York, USA (North America) |
| Guillermo Oliveto | Universidad Tecnologica Nacional Argentina (Latin America) |

## Technical Program Chairs

| | |
|---|---|
| Eduardo Vendrell | Universitat Politècnica de València (UPV), Spain |
| Sebastian Schreiter | IAOE France |

## Workshop and Tutorial Chair

| | |
|---|---|
| Valerie Varney | University of Applied Sciences Cologne, Germany |

## Special Sessions Chair

| | |
|---|---|
| Alexander Kist | University of Southern Queensland, Australia |

## Publication Chair

| | |
|---|---|
| Sebastian Schreiter | IAOE France |

## Award Chair

| | |
|---|---|
| Eduardo Vendrell | Universitat Politècnica de València (UPV), Spain |

## Senior Program Committee Members

| | |
|---|---|
| Eleonore Lickl | IGIP Vienna, Austria |
| Andreas Pester | The British University in Egypt |
| Wolfgang Pachatz | Ministry of Education Austria |
| Tatiana Polyakova | Moscow State Technical University, Russia |
| Herwig Rehatschek | Medical University Graz, Austria |
| Cornel Samoila | Romania |
| Thrasyvoulos Tsiatsos | Aristotle University Thessaloniki, Greece |
| Doru Ursutiu | University of Brasov, Romania |
| Axel Zafoschnig | IGIP, Austria |

## Program Committee Members

| | |
|---|---|
| Abdallah Al-Zoubi | Jordan |
| Santi Caballé | Spain |
| Alberto Cardoso | Portugal |
| Dan Centea | Canada |
| Ralph Dreher | Germany |
| Martin Ebner | Austria |
| Christian Guetl | Austria |
| Hants Kipper | Estonia |
| Oleksandr Kupriyanov | Ukraine |
| Despo Ktoridou | Cyprus |
| Jorge Membrillo-Hernández | Mexico |
| Jürgen Mottok | Germany |
| Stavros Nikou | UK |
| Stamatios Papadakis | Greece |
| Rauno Pirinen | Finland |
| Neelakshi Chandrasena Premawardhena | Sri Lanka |
| Teresa Restivo | Portugal |
| Istvan Simonics | Hungary |
| Ivana Simonova | Czech Republic |
| Alexander Soloviev | Russia |
| Matthias Utesch | Germany |
| James Wolfer | USA |

## Local Organizing Committee

| | |
|---|---|
| Alejandro Leo Ramirez | Universidad Politécnica de Madrid, UPM |
| Antonio Molina Marco | Universitat Politècnica de València (UPV), Spain |
| Xavier Molero Prieto | Universitat Politècnica de València (UPV), Spain |
| Bernardo Tabuenca Archilla | Universidad Politécnica de Madrid, UPM |
| Manuel Uche Soria | Universidad Politécnica de Madrid, UPM |

# Contents

## AI and Learning Analytics in Engineering Education

# Collaborative Learning

# Work in Progress: Course Design and E-Learning-Environment for Scientific Competency Development for Bachelor's Degree Students Within the Framework of Self-determination Theory

Doerthe Vieten(✉), Alexandra Reher, and Iris Gross

Bonn Rhein Sieg University of Applied Science, 53757 Sankt Augustin, Germany
doerthe.vieten@h-brs.de

**Abstract.** We present both design and a mixed method evaluation scheme of a semester-long course for scientific literature work. Focus of this contribution lies on the implementation for an e-learning environment with a clearly structured overview as key requirement, reflecting teaching concepts used, that is designed to reduce complexity of usage. Teaching concepts are derived from Self-Determination Theory and are assumed to provide learning conditions to support an optimal motivation profile by focusing on autonomy support, involvement, and structure. Demand-driven student-teacher communication is possible through hybrid communication structures. Approaches to flexibility are reflected by students' own choice of research focus. Situated in 5th semester within a bachelor curriculum, Artificial Neural Networks are used as a thematical framework which easily can be adapted to other subjects. Teaching analysis poll is conducted to obtain timely feedback by students to identify helping and hindering factors for learning, followed by a questionnaire reflecting the acquisition of competences.

**Keywords:** Scientific competency development · e-learning course structure · Mixed method evaluation

## 1 Introduction

Development of scientific working and communication abilities are key requirements for bachelor students, not only for assessment requirements, but also for being able to understand and participate in social aspects of science like scientific communities or publishing. As it is known that some students experience difficulties with scientific writing [1, 2] approaches were developed to embed scientific writing in project courses with subject specific research tasks [3, 4]. Scientific writing is recognized to be an important but also challenging task for students [5], so even if courses in embedded writing are offered there may still be participants who do not finish the course [6]. Research in course dropout states that motivation plays an important role in course completion [7,


M. E. Auer et al. (Eds.): ICL 2023, LNNS 899, pp. 3–11, 2024.
https://doi.org/10.1007/978-3-031-51979-6_1

8], with Self-Determination Theory (SDT) [9] being a widely used research framework [10]. Embedded in a course for scientific working and writing, aims of this paper are to derive criteria from SDT for a motivation promoting course design, second, to implement these criteria at different context levels (content, e-learning environment design, interaction design) and third to verify them by means of standardized feedback.

## 2 Context and Theory

In Bachelor of Computer Science curriculum of Hochschule Bonn-Rhein-Sieg, courses to work with scientific literature are provided for in the 5th semester according to curriculum as described below [11]. Courses are held in small groups of about 16 students and differ by field of interest and methodical approach of the individual lecturer. In each winter semester a variety of different topics (about 10) within the field of Computer Science are offered, giving students the opportunity to enroll in a course according to their own academic preferences. But even though students can choose their topics according to their professional interest and though the course is mandatory there are still students who either don't enroll within the specified semester or choose to drop out of the course.

### 2.1 Theory

Known reasons for course dropout include deficits in motivation [12] or low interest in task at hand [13]. Self-determination theory (SDT) [9] states that motivation comes in different qualities, varying from intrinsic to several forms of extrinsic motivation. Best results for learning settings are shown to be achieved with motivation profiles which are high on the autonomous dimensions and low on external control [14]. Environmental conditions which support an optimal motivation profile include autonomy support, involvement, and structure [14]. Within SDT, Cognitive Evaluation Theory (CET) describes situational context in terms of external forces as deadlines and interpersonal communications in form of instructions which can be perceived as informational or controlling. As a result, intrinsic motivation is affected by the degree of which context is perceived as informational [15]. As for tasks which are not intrinsically motivated, Organismic Integration Theory (OIT) covers the process of integration of these tasks by reflection and experience of coherence with the social environment [15].

### 2.2 Evaluation Results of Preceding Semester

During course of previous semester, one course date was planned and reserved for qualitative feedback. As it was known beforehand that on this date only feedback would take place, only three of 12 participants took part,[1] and even though there was a general agreement with the results among the remaining students on the following course date, these results are best regarded as indications to be investigated further.

[1] Due to a midterm in a parallel course, affected students wisely chose preparation time over feedback. For this semesters course presented in this contribution, we integrated evaluation in course content and expect a higher number of participants.

Focus of previous semesters evaluation was to enquire aspects of the course students regarded as helpful and polled for degree of agreement afterwards. Results are presented in Table 1, all listed results were unanimously seen as helpful.

Even though results are only indicators of course atmosphere, they are in line with theory mentioned above. Importance of structure, as derived from SDT [14], is reflected in its early and double appearance. Informational quality of feedback [15] is reflected in communicating reasons for formalities in science and benefit of adhering to them. Involvement correlates to timeframe for feedback, reachability, and degree of detail in feedback.

**Table 1.** Course feedback during previous semester. Aspects mentioned which were regarded as helpful (in order of appearance)

| *By which aspects of the course do you learn most?* | Agree (%) | Disagree | Abstention |
|---|---|---|---|
| Structure of e-learning environment | 100 | – | – |
| Structure of course | 100 | – | – |
| Feedback as part of grading | 100 | – | – |
| Timeframe for feedback | 100 | – | – |
| Reachability | 100 | – | – |
| Detailed feedback | 100 | – | – |

Aspect not mentioned, but also implemented were autonomy support [14] and experience of coherence with the social environment [15]. Autonomy is special in graded environment, as grading implies rules that have to be followed. To compensate, transparency is provided for every grading relevant aspect of the course. On a content level, students not only enroll in a course according to their own academic preferences but also are required to research and choose their own focus of research. Experience of coherence is provided by using current research articles as examples. Formalities learned are reflected in publications, showing significance of scientific working principles in real life.

### 2.3 Requirements for Course Design

Foremost, e-learning environment should be structured, concise and reflect logical as well as chronological course flow while reducing necessity of user interactions. As environmental conditions should reflect not only structure but also autonomy support [14], second, restrictions in forms of external guidelines or instructions should be communicated in an informational, non-personal manner [15]. Third, reflections about closeness of learnt aspects to reality in communication and working practice may achieve a sense of coherence with the social environment [15].

# 3 Approach

We present a course design oriented along the principles mentioned above as well as a corresponding e-learning course structure based on ILIAS [16] as a solution approach to foster motivation in students and reduce number of deferrals – while still respecting students as responsible for their own learning and working results.

## 3.1 Course Objectives

According to curriculum, course objectives for the students are for their chosen topic to first research academic literature and aggregate the current state of science, second compile their findings in a research article and third present their findings. Competencies to be acquired include professional competencies in the chosen topic, methodical competencies for scientific working and individual competencies for autonomous working. As approaches to flexibility are reflected by students' own choice of research focus [17], in following we concentrate on course planning and methods used.

## 3.2 Differences in This Semester's Student Group

This semester's course is offered out of curricular rotation, being offered for students in need to catch up. Therefore, this semester's learning group differs from last semesters in two important ways: First, as there is only one topic available, students can't choose freely according to their preferences, and topic interest as a source for intrinsic motivation falls away, even though flexibility of choosing one's own research focus remains. Second, all registering students are out of regular curricular timeframe for this course, with semester count ranging from 6 to 18 ($M = 9.13$, $SD = 3.98$), with 76% of them having already deferred from one of the courses offered in previous semester.

## 3.3 Course Planning and Structure

In accordance with the requirements in Sect. 2.3, focus of course design was on structuring course content according to course objectives, and transferal to an e-learning course in a way to reflect structure while reducing all forms of unnecessary complexity.

Course outline for scientific literature work is described in Table 2. As methodical framework, a systematic literature review is performed, with limited number of primary sources. Results are presented at the end of the semester. Feedback is offered every course date and via online communication, to a specified timeframe during the week.

Idea behind course representation in e-learning environment was to transfer planning notes as close as possible, as for compactness of visualization of course content and their sequential connections. Therefore, for implementation it was chosen to present course content in the tabular design shown above (see Table 2). Vertical direction represents thematic progress, while horizontal direction represents consecutive course dates within each thematic phase. Each cell represents a single course date and contains an outline of the content. To our knowledge, as far as we have seen it is not a common practice to present the course structure this way.

**Table 2.** Course outline, structured by phase of work and month.

| Phase of work | Week of Semester, Content | | | |
|---|---|---|---|---|
| **March** | *Pre-Semester*: Course outline, grading scheme and requirement details are provided in e-learning environment | | | **SW1**<br>Introduction |
| **April**<br>Exploration of community research foci | **SW2**<br>Library work and databases<br>HW: Research | **SW3**<br>Abstract usage<br>DL: Result table (graded) | **SW4**<br>Literature Review, Chapter structure<br>HW: Topic choice | **SW5**<br>History and theory of science<br>HW: Methods |
| **May**<br>Research, Data collection and analysis | **SW6**<br>Scientific writing criteria<br>HW: Methods, results | **SW7**<br>Excerpting research articles<br>HW: Result table | **SW8**<br>Literature management, citing<br>HW: Implement pregrade feedback | **SW9**<br>Presentation<br>DL: Final methods and results (graded) |
| **June**<br>Writing of research article | **SW10**<br>Structure of scientific articles<br>HW: Theory, Related works | **SW11**<br>Tables and Diagrams, individual questions<br>HW: Discussion | **SW12**<br>Collaborative working, individual questions<br>HW: Introduction | **SW13**<br>How to present scientific results<br>DL: Presentation, article (graded) |
| **July**<br>Presentation | **SW14**<br>Presentation | | | |

Abbreviations used: SW Week of semester, HW Homework, DL Deadline

Files for course presentations and handouts are linked to their respective occurrence within the outline, and a dropdown area containing homework instructions is added. Benefits of this visual presentation include (a) a simple representation of course structure and relations between course dates, (b) visual compression with reduced need for scrolling, (c) all course material is accessible within the visual context of the respective course date. All course materials are also accessible within a single course folder, named by week of semester and content, resulting in a linear textual representation.

### 3.4 Coherence with the Social Environment

Coherence with social environment is represented in terms of collaborative working tasks, but even more in the aspect of doing research within a database consisting of real scientific communication. Furthermore, as part of content presentation, published literature reviews are discussed as examples, allowing students to substantiate course content by experience.

Group Work [18] is used both in exploration and in research phase. In exploration phase, key focus points of current research are identified by combining individual researched information via shared document, tagging them by research type, subject and focus and combining them in a mind map, working out an overview for key interest points of current research. In research phase, students pair up by interests to do a literature review about one of these focus points, choosing individual articles but using results of all team members for their graded research article.

Cognitive Apprenticeship [19] is used to teach the handling of research articles by example, communicating thoughts and reasons for actions on the process. Explanations can be transferred to individual chosen articles, tried out by students and the process can be reflected on.

## 4 Evaluation

A mixed method approach was chosen to support the exploration of structures that influence learning that would otherwise have remained concealed. [20]. To start with, we conduct a quantitative questionnaire to assess the development of the scientific skills self reportedly. Second, we use the qualitative approach of Teaching analysis poll (TAP) [21] to evaluate the students learning experience.

### 4.1 Questionnaire

In the quantitative part of the methodological section, we run a questionnaire on the acquisition of competences in scientific skills in the literature seminar, carried out at two measurement points in the summer semester of 2023 to ascertain the students' acquisition of competences. The questionnaire was shared as an online survey with the students of the described course via link in the e-learning environment. Answering the questionnaire took about ten minutes. Information on the sample will be described in more detail at the conference.

Questionnaires come in three parts with part one and two being conducted at the first measurement point in the middle of the semester (T1). Part one is carried out through a subjective assessment with a retrospective view, and part two captures the scientific competencies at the actual time of conduct subjectively as well. The aim is to compare the results and discover changes over the first half of the semester. The first part comprises 13 items. Sample items are: *Before the course, I had no knowledge about the structure of a scientific paper* or *Before the course, I had no knowledge about the structure of a Bachelor's thesis.* The second part comprises 16 items. Example items are *My knowledge of evaluating scientific texts has improved as a result of the course* or *My skills in scientific work have improved as a result of the course.*

Lastly, the third part measures the scientific competencies after the course (T2). Therefore, at the end of the semester, the acquisition of competences is surveyed subjectively by reusing items from both parts of T1. The aim is to display the acquisition of competences throughout the semester long course.

### 4.2 TAP

For the qualitative part of the mixed method approach, TAP as an innovative evaluation instrument [21] was conducted in the middle of the semester by an external moderator with a background in didactics in higher education [22], to obtain self-assessment of the students regarding facilitating and impeding factors for learning. TAP was offered voluntarily for students and lecturers.

The course was evaluated with the help of the following three questions relating to the learning process of students, followed by anonymous feedback from the moderator to the lecturer in a confidential one-to-one conversation to support adapting the course to the students' needs within a semester [23]: Which aspects of the course facilitate your learning? Which aspects of the course impede your learning? and What suggestions do you have for improving the impeding points? Hence, the lecturer was provided with qualitative evaluation results on how students perceive the learning process in the described courses [23].

In addition, the TAP questions were slightly adapted by adding the course specific questions *What was helpful and should be kept?* and *What did you wish for and how can this be implemented?* as the lecturers are offered the possibility to add one or two questions according to their individual interests of knowledge to make timely adaptions to the course.

## 5 (Anticipated) Outcomes and Discussion

To this point, only results of TAP are already available. Aspects this semesters' TAP participants ($n = 8$) regarded as helpful are listed in Table 3 below.

**Table 3.** This semester's course feedback, aspects are listed in order of appearance.

| *By which aspects of the course do you learn most?* | Agree (%) | Disagree (%) | Abstention |
|---|---|---|---|
| Own work/need for self-study | 100 | – | – |
| Teammate (group work, discussions) | 50 | 50 | – |
| Course materials (requirements, instructions, ...) | 75 | 25 | – |
| Personalized feedback referring to own solutions | 100 | – | – |
| Structure of approach | 25 | 75 | – |

This semester's student group values work autonomy most, but social aspects are seen ambiguous. Course material and provision of feedback are regarded as helpful. Structure of approach is also mentioned as helpful, though not shared by most students, even though there is a general agreement about the beneficial value of key elements of this structure (material, feedback). In line with this semester's group predispositions (see Sect. 3.2), we assume this to relate to formal aspects like homework and attendance.

Anticipated key outcome of our study is that students increase their scientific skills after the course, namely scientific working, writing, and reading skills. We first anticipate that students report that the practical orientated course helps them to prepare for their bachelor thesis. Secondly, we anticipate that the hybrid structure allows a flexible accessibility of the lecturer which thirdly contributes to an ongoing practice during the semester and therefore finally to a continuous development of scientific skills. All results will be available at conference and included in presentation.

## 6 Conclusion

The methodology should be replicated in further studies, not only because of this semester's peculiarities but also to generate generalizable results by longitudinal study design. Furthermore, a more detailed investigation of reasons for student deferral regardless of curricular demands remains to be done, leading to helpful insights for further adaptations including a more motivating learning structure for these students.

Regarding (anticipated) key outcome of this contribution, due possibility to decide for a research focus according to their own interests, we assume students' scientific competencies to increase as a result of self-determined research. Furthermore, we assume hybrid setting, individual goal orientations and flexible communication between lecturer and class to be regarded as informative and helpful rather than controlling.

Summarizing, the authors recommend supporting students before writing their first major scientific work, as actual scientific working experience can be overwhelming for students with more practically oriented image of scientific work. Based on theory as well as on qualitative data collected, we recommend presenting a detailed structure with clear reference to tasks at hand. Also, we recommend recording the work assignment detailed in writing and to be consistently reachable. This way, we anticipate an increase of both scientific skills and understanding for students completing this course.

## References

1. Moritz, R.E.: Der frühe Vogel - wissenschaftliches Schreiben im akademischen Curriculum. In: Schreibwissenschaft als Disziplin, in JoSch—Journal für Schreibwissenschaft, vol. 11. wbv Publikation (2020)
2. Prätsch, J., Rossig, W.E.: Erstellung und Bewertung wissenschaftlicher Arbeiten. In: Neues Handbuch Hochschullehre, DUZ Medienhaus (2006)
3. Dansereau, D., Carmichael, L., Hotson, B.: Research and teaching: building first-year science writing skills with an embedded writing instruction program. J. Coll. Sci. Teach. **049**(03) (2020)
4. Lee, S.E., Woods, K.J., Tonissen, K. F.: Writing activities embedded in bioscience laboratory courses to change students' attitudes and enhance their scientific writing (2011)
5. Grogan, K.E.: Writing science: what makes scientific writing hard and how to make it easier. Bull. Ecol. Soc. Am. **102**(1) (2021)
6. Davila, Y., Leigh, A., Griffith, N., England, A.: An embedded, flipped and interactive approach to scientific writing. University of Technology Sydney. https://tinyurl.com/mvx3yp4j (2015). Accessed 5 Jul 2023
7. Gökçe, O., Prada, J., Nikolov, N.I., Gu, N., Hahnloser, R.H.R.: Embedding-based scientific literature discovery in a text editor application. In: Proceedings of the 58th Annual Meeting of the Association for Computational Linguistics: System Demonstrations (2020)

8. Zheng, S., Rosson, M.B., Shih, P.C., Carroll, J.M.: Understanding student motivation, behaviors and perceptions in MOOCs. In: Proceedings of the 18th ACM Conference on Computer Supported Cooperative Work & Social Computing, in CSCW'15 (2015)
9. Deci, E.L., Ryan, R.M.: Self-determination theory: A macrotheory of human motivation, development, and health. Can. Psychol. **49**(3) (2008)
10. Badali, M., Hatami, J., Banihashem, S.K., Rahimi, E., Noroozi, O., Eslami, Z.: The role of motivation in MOOCs' retention rates: a systematic literature review. Res. Pract. Technol. Enhanc. Learn. **17**(1), 1–20 (2022)
11. Kursbeschreibung: Literature-Seminar. Course description (in German). https://eva2.inf.h-brs.de/studium/curriculum/2017/matrix/bi/326/de/ (2023). Accessed 26 Mai 2023
12. Fryer, L.K., Ginns, P., Howarth, M., Anderson, C., Ozono, S.: Individual differences and course attendance: Why do students skip class? Educ. Psychol. **38**(4), 470–486 (2018)
13. McMillan, J.: Course change and attrition from higher education (2005)
14. Guay, F., Ratelle, C.F., Chanal, J.: Optimal learning in optimal contexts: the role of self-determination in education. Can. Psychol. **49**(3), 233 (2008)
15. Legault, L.: Self-determination theory. In: Zeigler-Hill, V., Shackelford, T.K., Hrsg. Encyclopedia of Personality and Individual Differences. Springer International (2017)
16. ilias.de. https://www.ilias.de/. Accessed 27 May 2023
17. Li, K.C., Wong, B.Y.Y.: Revisiting the definitions and implementation of flexible learning. In: Li, K.C., Yuen, K.S., Wong, B.T.M., Hrsg. Innovations in Open and Flexible Education, Education Innovation Series. Springer, Singapore (2018)
18. Burdett, J.: Making groups work: university students' perceptions. Int. Educ. J. **4**(3), 177–191 (2003)
19. Collins, A., Brown, J.S., Newman, S.E.: Cognitive apprenticeship: teaching the craft of reading, writing, and mathematics. Technical Report No. 403 (1987)
20. Jick, T.D.: Mixing qualitative and quantitative methods: triangulation in action. Adm. Sci. Q. **24**(4), 602 (1979)
21. Frank, A., Fröhlich, M., Lahm, S.: Zwischenauswertung im Semester: Lehrveranstaltungen gemeinsam verändern. Zeitschrift für Hochschulentwicklung (2011)
22. Penny, A.R., Coe, R.: Effectiveness of consultation on student ratings feedback: a meta-analysis. Rev. Educ. Res. **74**(2), 215–253 (2004)
23. Hurney, C.A., Harris, N.L., Bates Prins, S.C., Kruck, S.E.: The impact of a learner-centered, mid-semester course evaluation on students. J. Fac. Dev. **28**(3), 55–62 (2014)

# Fostering Self-directed Learning in Engineering Undergraduates: A Collaborative Approach

Sherif Welsen(✉)

The University of Nottingham Ningbo China, Ningbo, Zhejiang, China
sherif.welsen@nottingham.edu.cn

**Abstract.** This paper highlights the implementation of a collaborative learning method in a final-year engineering module at the University of Nottingham Ningbo China. To achieve this, students were grouped into pairs based on their preferences, and they worked together throughout the semester on weekly unassessed problems and research assignments. The traditional seminar style was replaced with interactive tutorials held in a teaching lab, where students were presented with engineering design problems and encouraged to develop solutions together. To assess the effectiveness of the collaborative learning approach, a survey consisting of both Likert scale and open-ended questions was administered to 52 students. The aim was to obtain a more comprehensive understanding of the student's perceptions of the collaborative learning approach, how it was implemented in different module exercises, and its impact on their learning experience and skill development. The results indicated that incorporating collaborative learning strategies with technology positively affected how engineering modules were taught and that developing skills such as communication, problem-solving, and critical thinking made the classes more enjoyable and increased student engagement and confidence. Overall, the findings suggest collaborative learning is a valuable pedagogical approach that should be utilized more frequently in engineering education.

**Keywords:** Engineering education · Collaborative learning · Interactive learning · Student engagement

## 1 Introduction

Engineering education has seen essential changes in the past two decades [1]. Different programs of study have been updated to include some group work, especially in modules focusing on project-based learning [2, 3]. Group work benefits undergraduate students by developing essential skills not taught in the classroom, such as teamworking and the development of their communication skills. However, sometimes students face challenges when doing group coursework, impacting their engagement and motivation to study and learn [4, 5].

When students are asked to work in a group to complete an assessed assignment, they sometimes struggle to work together, focusing on the task rather than working as a team. It was found that they are often driven by achieving marks and not learning


M. E. Auer et al. (Eds.): ICL 2023, LNNS 899, pp. 12–19, 2024.
https://doi.org/10.1007/978-3-031-51979-6_2

outcomes [6]. This can be due to different reasons, including how many students are in the group and being in a group they don't like. However, collaborative learning can still be used to inspire non-engaging students during lectures and tutorials. This study aims to empower the next generation of engineers through a self-directed and collaborative learning approach. The study applies collaborative learning and hybrid approaches to deliver a final-year engineering module. The goal is to increase students' interaction in the module and make them more self-directed learners to prepare them for their future post-graduate study and professional career. More importantly, to promote research practices among undergraduate students. The primary research question is: how can paring students in a small group influence their engagement in learning and impact their learning of a final year engineering module? This approach is expected to help students learn widely and deeply. Comprehensive learning results from searching for more topics than those discussed in the classroom. In contrast, deep learning is an outcome of researching issues in detail. The paper is organized as follows: Sect. 2 is a brief literature review on collaborative learning and its application in engineering education. In Sect. 3, the methodology of this study is explained. In Sect. 4, the results from the student survey are presented and discussed, while Sect. 5 concludes.

## 2 Literature Review

In the near future, many universities will likely adopt a hybrid approach that blends face-to-face teaching with virtual classes. This presents a significant task for educators to ensure that all students, regardless of their learning mode, receive a top-notch educational experience. To achieve this, classes and teaching material should be redesigned to accommodate online teaching and integrate interactive activities that foster student engagement, motivation, and teamwork [7]. By utilizing a blend of qualitative and quantitative research methods, a study was conducted to explore the various elements that could impact the satisfaction of learning in flipped education [8]. Out of surveying 171 students, the findings revealed that two critical factors - collaborative learning and the desire for cognition - were instrumental in predicting learning satisfaction. In-depth interviews were also conducted with 12 participants to delve deeper into the collaborative learning process. The outcomes indicated that engaging students in activities that enabled them to become familiar with each other helped enhance their collaborative learning abilities, ultimately promoting innovative and mutual learning skills.

For engineers, working in isolation is not the preferred approach. Instead, they tend to collaborate to tackle complex projects and achieve success. As such, engineering students must cultivate teamwork skills that will equip them to address real-world challenges in the future [9]. By studying undergraduate engineering courses, researchers found that active and collaborative learning methods improved students' design, problem-solving, communication, and group participation skills more effectively than traditional lecture and discussion methods [10]. The study involved 480 students from 17 courses and six schools, with statistically substantial learning improvements observed even after accounting for student pre-course characteristics. A study that employed a collaborative learning approach as a dynamic teaching technique for foreign languages described the pedagogies for teaching English to students who capitalized on the various features available in the online learning environment [11]. The study highlighted the advantages of collaborative learning in enhancing interpersonal, analytical, and communicative competencies through collective undertakings and its vital role in forming a virtual learning society, acquiring foreign language abilities, and attaining linguistic proficiency. Strategic preparation, unambiguous guidelines, and pertinent material can surmount these challenges despite the potential hurdles associated with orchestrating group work online. In a seven-year study, engineering students were trained to present lectures to their peers using project-based and cooperative learning methods [12]. The study tracked daily progress and discovered that this approach boosted students' motivation and skill development while improving their ability to retain knowledge. Additionally, attendance increased, and students became more proactive learners. A group of 42 mechanical engineering students participated in a study that utilized several experiments to differentiate between individual and cooperative learning [13]. The researchers carefully monitored the time spent on task to compare the efficacy of the two approaches. The findings revealed that collaborative learning yielded superior results among students who were granted ample time to develop their skills. To encourage active participation among students in an online engineering education program, a structured approach was introduced. This framework aimed to address the challenges of maintaining student engagement in the online learning environment [14]. The research highlighted various strategies that both teachers and learners could adopt to promote collaborative practices. By taking a proactive approach, these practices could help sustain a constructive and effective learning experience for students pursuing engineering education online. This is evidenced in a review of 62 published articles on flipped learning in engineering education found that students in flipped learning environments outperformed those in traditional classrooms [15]. However, the study highlighted a shortage of qualitative research on the effectiveness of flipped learning in engineering education, and recommended further investigation into whether it can improve students' professional skills, such as interpersonal, self-directed, and lifelong learning skills. The challenges of implementing flipped learning in engineering education were explored in the same article, with studies highlighting issues such as instructor workloads and technical difficulties. There was also a risk of disengagement among students in a flipped classroom [16, 17]. Educators are advised to gradually integrate flipped learning into their modules to mitigate these challenges and communicate the changes to students. Innovative class materials should facilitate collaborative problem-solving among students and their instructors, leading to increased attendance and engagement in lectures [18].

## 3 Methodology

Previous students who attended the module used in this study were surveyed to explore their preference for collaborative learning. Students expressed their preference for collaborating with other partners rather than studying on their own. Additionally, they said they like working in a small group or a pair and, where possible choosing their partner. They also preferred to collaborate on unassessed research tasks than assessed coursework. The existing students were informed to partner in pairs with a collaborator of their choice during the first week of the semester. Students were briefed on the collaborative approach which was then used in the module. A weekly unassessed exercise was created to promote collaboration and was shared with the entire class at the end of each weekly session. Students were asked to work out the weekly exercise with their partners. During the weekly teaching sessions, students were encouraged to take notes of the keywords they wanted to search for and discuss their search with their partners. Students could then bring their questions, if there are any, to the class for further discussion. Students were given some unassessed research assignments based on the scenario of the actual teaching session and debate in the classroom—those small research assignments aimed to further promote collaboration in each pair. Additionally, a few tutorials were moved from the regular classroom to an interactive technology-equipped teaching lab named (WonderLab). In the lab, students were divided into small groups of 2 pairs per group. Students were given small tasks of problem analysis or design. They used the interactive touch displays in the lab for annotation, brainstorming, and discussion among each group first. Then each group shared their idea with the whole class. A survey including 17 Likert of a five-point scale and open questions was shared with final-year engineering students attending the 10-credit module. Completing the student survey was optional, and students were required to fill out a well-informed online consent form before engaging in the survey. The survey was conducted online using MS Forms at the end of the semester. All students attending the module were encouraged to complete the survey by word of mouth, and participants completed the survey voluntarily. The responses were analyzed using Excel and SSPS.

## 4 Results and Discussion

In order to examine the effects of collaborative learning on students' problem-solving and research skills, the researcher asked the participants to share their experiences of completing problem solutions with their partners instead of receiving the entire set of problems and solutions at once. The study results depicted in Fig. 1 indicate that most (75%) participants agreed that discussing the problems with their partners positively impacted their learning experience. In comparison, 88% acknowledged the benefits of information sharing during collaborative learning, followed by individual research. Additionally, all participants strongly agreed (38%) or agreed (62%) that collaborative learning encouraged weekly tutorial assignment discussions. Interestingly, 75% of participants strongly agreed or agreed that writing down detailed solutions for weekly module problems in pairs after discussing them during tutorial sessions was beneficial for their learning. However, 13% of participants did not agree, while the remaining 13% were neutral (see Fig. 1).

To investigate the impact of blending collaborative learning pedagogy with the technology provided in the teaching lab on engineering students' learning experiences, the participants were asked whether they preferred to attend the tutorial in the teaching lab or the regular classroom venue. As their responses demonstrate in Fig. 2, 88% of the participants either strongly agreed or agreed they preferred the interactive teaching lab known as (WonderLab) to better discuss the colored layout design problems learned in this module. The same percentage of the participants (88%) strongly agreed or agreed that the massive touch screens in WonderLab encouraged discussion among each group of two pairs of students. The same percentage of participants either strongly agreed or

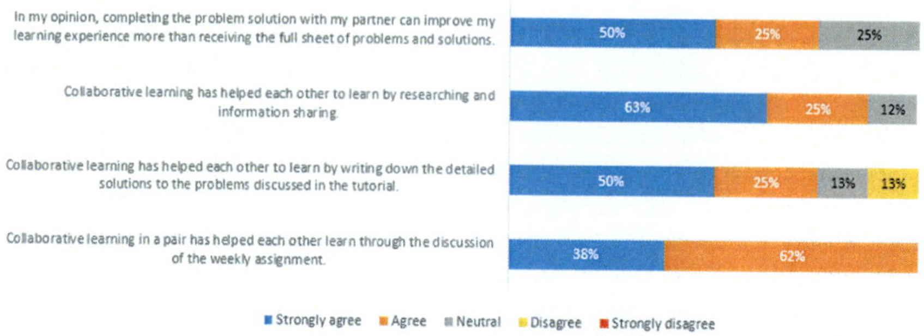


**Fig. 1.** The impact of collaborative learning on problem solving and research skills.

agreed that the touch screens made the interaction enjoyable. 63% of the participants strongly agreed that they prefer to have more tutorials, not just the colored layout design problems, in the lab than in the regular classroom, while 12% agreed. Most participants (75%) strongly agreed that, where possible, they encourage other engineering module convenors to arrange their teaching sessions in the interactive lab, to benefit from the interactive technology and discussion among students.

To evaluate the deployed collaborative learning approach for enhancing students' engagement and learning experience, the participants were asked a group of Likert questions, as illustrated in Fig. 3. 38% of the participants strongly agreed that collaborating in pairs and the interactive discussion increased their confidence, while 50% agreed. The participants strongly agreed and agreed equally that, overall, the collaborative learning approach they used was enjoyable. All the participants strongly agreed (38%) or agreed (62%) that the learning approach improved their learning experience in general. 50% of the participants strongly agreed, and the other 50% agreed that the learning approach promoted effective collaboration and communication in each pair during the entire semester or among the small group of two pairs during the teaching lab sessions. 88% of the participants either strongly agreed or agreed that the approach used in this study effectively helped them to understand the learning material better. Moreover, this approach effectively prepared them for their assessment, including coursework and final exam.

Interestingly, 63% of the participants reported that the approach in this study effectively increased their engagement compared to other group work they did in other previous modules. This is evidenced by students' comments below:

"*The approach implemented in this module felt different than the previous group work. This semester I was able to discuss with classmates and deal with the problem immediately*", "*This semester, it felt easier to communicate with classmates*", "*It gave us more freedom*" and "*This class was absolutely different. The collaborative learning objective was clear and interesting and encouraged me to learn outside the given learning material*". 38% of the participants strongly agreed, and 50% agreed that they recommend using similar methods for learning engineering modules to other students and instructors.

Adopting collaborative learning in teaching the module presented in this study has helped students take ownership of their learning and become more self-directed. This was clear from students' comments:

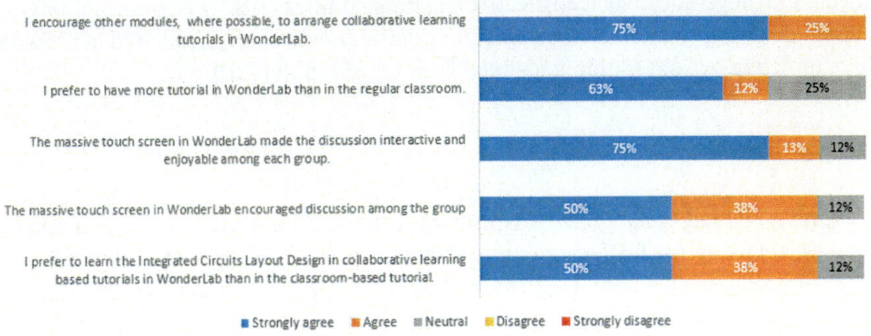


**Fig. 2.** Exploring the effectiveness of collaborative learning in an interactive teaching lab for integrated circuits layout design problems.

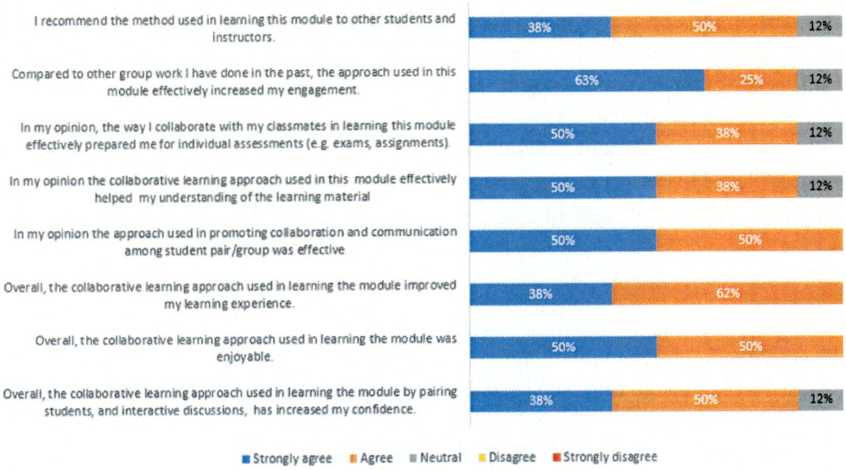


**Fig. 3.** Evaluation of collaborative learning approach for enhanced learning experience and engagement.

"*I can obtain more information than by myself which showed me more ways to study than before*" and "*Made me confident. To be honest, I was a loner before this semester. I studied all by myself, even in the group project, I used to finish my assigned part without any communication with others. Unlike in this semester, I communicated with others frequently*". Students further explained that the learning approach used in this study helped them to develop their skills and commented, "*It helped me to think critically as well as develop my communication skills as in the group we discussed our different ideas, before we brought the discussion to the whole class*", "*I was more encouraged to learn. I asked any question came to my mind freely as our teacher tried his best to get us engaged in the class and communicate with each other*" and "*It made me think in more aspects based on the discussion and comments from my paired student and other classmates*". The findings align with prior research that suggests a correlation between academic autonomy and active student involvement, resulting in heightened student motivation and enhanced creative thinking abilities [19]. This is presumably why students are strongly inclined to attend classroom-based teaching, even when presented with alternative options within a blended learning paradigm [20].

## 5 Conclusion

This study explored collaborative learning to foster self-directed learning and research practices, focusing on final-year engineering students. The approach involved forming small groups and encouraging students to work together on unassessed tasks. The teaching lab also utilized interactive technology to aid problem analysis and design. The study uncovered that students preferred collaborating with peers and working in small groups, which positively impacted their engagement and motivation. Engineering students reported that the collaborative approach facilitated extensive and profound learning, supported their confidence, and improved their understanding of the subject matter. The study's findings emphasized the effectiveness of collaborative learning, particularly when combined with technology in the lab, highlighting the benefits of research and information sharing. Consequently, it is recommended that collaborative learning pedagogies be widely implemented in final-year engineering modules, with students being paired or grouped with a few collaborators. Additionally, incorporating pedagogies like collaborative learning and learning technologies in seminar layouts can promote interaction, self-directed learning, research practice, and skill development.

## References

1. Broo, D.G., Kaynak, O., Sait, S.M.: Rethinking engineering education at the age of industry 5.0. J. Ind. Inf. Integr. **25**, 100311 (2022)
2. Palmer, S., Hall, W.: An evaluation of a project-based learning initiative in engineering education. Eur. J. Eng. Educ. **36**(4), 357–365 (2011)
3. Mills, J.E., Treagust, D.F.: Engineering education—is problem-based or project-based learning the answer. Australas. J. Eng. Educ. **3**(2), 2–16 (2003)
4. Marra, R.M., et al.: Beyond "group work": an integrated approach to support collaboration in engineering education. Int. J. STEM Educ. **3**(1), 1–15 (2016)
5. Chen, J., Kolmos, A., Du, X.: Forms of implementation and challenges of PBL in engineering education: a review of literature. Eur. J. Eng. Educ. **46**(1), 90–115 (2021)
6. Asikainen, H., et al.: The relationship between student learning process, study success and the nature of assessment: a qualitative study. Stud. Educ. Eval. **39**(4), 211–217 (2013)
7. Dwivedi, Y.K., et al.: Impact of COVID-19 pandemic on information management research and practice: transforming education, work and life. Int. J. Inf. Manag. **55**, 102211 (2020)
8. Cheng, F.-F., Wu, C.-S., Su, P.-C.: The impact of collaborative learning and personality on satisfaction in innovative teaching context. Front. Psychol. **12**, 713497 (2021)
9. Gol, O., Nafalski, A.: Collaborative learning in engineering education. UNESCO, International Centre for Engineering Education (2007)
10. Wang, X.W., Zhu, Y.J., Zhang, Y.C.: An empirical study of college students' reading engagement on academic achievement. Front. Psychol. **13**, 1025754 (2022)
11. Sumtsova, O., et al.: Collaborative learning at engineering universities: benefits and challenges. Int. J. Emerg. Technol. Learn. (iJET) **13**(1), 160–177 (2018)

12. Torrijo, F.J., et al.: Combining project based learning and cooperative learning strategies in a geotechnical engineering course. Educ. Sci. **11**(9), 467 (2021)
13. Hsiung, C.M.: The effectiveness of cooperative learning. J. Eng. Educ. **101**(1), 119–137 (2012)
14. Qiu, R.G.: A systemic approach to leveraging student engagement in collaborative learning to improve online engineering education. Int. J. Technol. Enhanc. Learn. **11**(1), 1–19 (2019)
15. Karabulut-Ilgu, A., Jaramillo Cherrez, N., Jahren, C.T.: A systematic review of research on the flipped learning method in engineering education. Br. J. Educ. Technol. **49**(3), 398–411 (2018)
16. Ossman, K.A., Bucks, G.W.: Effect of flipping the classroom on student performance in first-year engineering courses. In: 2014 ASEE Annual Conference & Exposition (2014)
17. Velegol, S.B., Zappe, S.E., Mahoney, E.: The evolution of a flipped classroom: evidence-based recommendations. Adv. Eng. Educ. **4**(3), n3 (2015)
18. Chen, Y., Wang, Y., Chen, N.-S.: Is FLIP enough? Or should we use the FLIPPED model instead? Comput. Educ. **79**, 16–27 (2014)
19. Welsen, S.: Engineering students' engagement and their perspective on compulsory classroom attendance. In: 2022 IEEE IFEES World Engineering Education Forum-Global Engineering Deans Council (WEEF-GEDC). IEEE (2022)
20. Welsen, S.: Impact of blended learning on engineering student attendance post COVID-19. In: 2021 World Engineering Education Forum/Global Engineering Deans Council (WEEF/GEDC). IEEE (2021)

# Collaborative Learning Spaces from Research to Practice: The KAEBUP Platform

Nadia Charalambous[1](✉), Christos Mettouris[2], Ilaria Geddes[1], Evangelia Vanezi[2], George Papadopoulos[2], Constantinos Xenofontos[2], and Marios Kyprianou[2]

[1] Society and Urban Form Lab, Department of Architecture, University of Cyprus, Nicosia, Cyprus
Charalambous.nadia@ucy.ac.cy

[2] Department of Computer Science, University of Cyprus, Nicosia, Cyprus
{mettouris.g.christos,vanezi.evangelia,papadopoulos.george, xenofontos.constantinos,kyprianou.a.marios}@ucy.ac.cy

**Abstract.** In this paper an innovative collaborative online platform, the Research to Practice platform (R2P), developed in the context of the EU funded project Knowledge Alliances for Evidence-Based Urban Practices (KAEBUP), is presented. The platform attempts to eliminate research and institutional barriers in educational urban studies cultures through the development and use of digital resources, structured under Collaborative Learning Activities, a novel concept proposed in a previous research project Emerging Perspectives on Urban Morphology (EPUM), and further developed in this project. The CLAs methodology implements a collaboration approach through the R2P platform, designed to meet the objectives of the higher education institutions (HEIs) in their mission to ensure that architectural and urban design students complete their studies with the skills to enter the professional world on the one hand, and to influence, innovate and support practice through their research work on the other. All participating organisations play an active role in the design of the CLAs and the platform, where learning involves co-construction and co-evolution of knowledge among partners. CLAs offer an innovative way for collaboration between academia, research, and practice/entrepreneurs, aiming at eventually creating and formulating an online community of practice, where the active membership of learners, teachers and practitioners will facilitate an educational social praxis.

**Keywords:** Collaborative learning activities · Blended learning · KAEBUP R2P platform · ICT tools · Urban form studies

## 1 Introduction

### 1.1 Addressing Urban Challenges in Academia and the Profession

There are major challenges faced by contemporary cities and the professionals involved in their design and management, including continued urbanization, increasing migration flows, complex mobility patterns and systems, climate change, ageing populations,


M. E. Auer et al. (Eds.): ICL 2023, LNNS 899, pp. 20–30, 2024.
https://doi.org/10.1007/978-3-031-51979-6_3

health, and social inequalities. This is why urban and planning professionals are now required to have broad knowledge of the variety of issues affecting cities and multidisciplinary skills to address them. It is also why evidence-based approaches to urban design and planning are ever more sought after, to ensure that the challenges are addressed effectively and sustainably, based on sound knowledge, and understanding of the impact of design and planning decisions.

A significant number of HEIs across Europe address the topic of evidence-based urban design and planning, but this still retains a secondary role in curricula compared to traditional methods of teaching design studio. Whilst a small number of successful enterprises in Europe have made research outputs a core element of their practice to address pressing urban challenges including sustainability, mobility, health and social cohesion, there is to date no formal research as to how these companies have formulated their business models and how they effectively integrate research into their work.

Despite the growing demand for evidence-based urban practices by designers and authorities to ensure successful and sustainable results in urban design and governance, there is little awareness in academia with regards to demand for research skills in the profession and the opportunities available for entrepreneurship in this field [1]. Furthermore, the links between businesses and HEIs in many countries remain minimal. There seems to be a 'rupture' at the point where academic research and teaching should 'flow' into the professional world. The purpose of this project is to mend this rupture and create a coherent, seamless, but powerful flow of skills and research into the professional world [2]. Strengthening the links between academia and businesses to mutually benefit from the tools that teaching, and research can provide to enterprises and from the experience of evidence-based practice is a pressing and timely need.

Acknowledging the above challenges, the KAEBUP project's activities focus on developing an entrepreneurial mind-set among students and staff in the fields of architecture and urban design, by training them into how research is used as the basis for professional practice. The purpose of the project arises from the need to innovate teaching practice with a view of understanding what businesses in the field of planning, architecture and urban design require from academia, both in terms of learners' basic and transversal skills, as well as tools, methods, and research findings to apply in the profession. The project addresses three specific needs identified through the needs analysis carried out as preparatory work for the project:

1) strengthening learners' experience of the professional world, and their transversal skills, by working on real-life urban projects;
2) developing students' and staff' understanding of business models for evidence-based urban practices; and
3) co-creating urban knowledge through exchange and involvement of academic and company staff in teaching, research, and practice.

All HEIs involved in this project already run teaching modules which refer to the relationship between research and practice. These may be on how sustainable urban design is developed through research outputs, research-based design studios, methodologies and tools for evaluation and impact assessments. Advancing a co-development of knowledge by involving professional practices, through collaborative, blended-learning approaches, in this teaching will strengthen the HEIs' effectiveness in offering innovative courses addressing contemporary needs through real-life professional projects. It will enable them to test how their research performs when applied to real-life case studies characterized by industry, political and marketing demands. Therefore, it will provide them with the opportunity to explain, clarify and argue how and why their research findings are relevant to design practice and potentially input into ongoing projects. Participating enterprises and NGOs, will enhance their objectives to inform design practice through structured understanding of architecture and urban form, independently assess designs applying research tools for planning evaluations and to adapt design proposals based on changing demands of clients, briefs, and contextual situations, as well as to advance research and practice in the built environment.

The co-development of a critical mass of knowledge about what tools and research findings are needed by contemporary urban practices are proposed to be channeled and elaborated through the online 'research-to-practice' (R2P) platform. The R2P exploits and expands capacities already developed through EPUM's online platform by enhancing its collaborative features and adding a critical mass of knowledge through the input of enterprises and real-life case studies. Student's experience of work-based learning and taking part in transnational collaborative learning activities supports the development of their transversal skills, including digital skills, critical and innovative thinking, inter- and intra-personal skills, global citizenship and media and information literacy. It is important to note here that the R2P platform, although developed in the context of the research findings of contemporary urban practices, it has the capacity, capability and potential to be applied in any research domain that advanced collaborative work practices need to be applied.

The mode of learning which proves to be suitable for such collaborative learning environments is one that facilitates both face-to-face activities, so as to allow institutions and professional practices to work independently, with on-line activities which enable the synchronous or asynchronous collaboration and learning across institutional/professional barriers, in other words, a blended learning approach. The project's learning and training activities thus build upon blended-learning techniques and Open Educational Resources (OER) already developed through the EPUM project by applying the same principles of a mixture of transnational face-to-face and online small teaching activities along with intensive workshops.

## 1.2 Collaborative Blended Learning Approaches in Urban Form Studies

Blended learning, which refers to a learning environment which combines face-to-face instruction with computer-mediated instruction, has gained much popularity in higher education in the past years. It is a term which is endowed with multiple meanings, and it has become apparent through several studies that different models of blending can exist at various levels. The value of blended-learning, however, is not merely about the application of ICT for teaching and learning, "recombining concepts that were previously considered contradictory, such as collaborative-reflection and asynchronous community" [3].

Learners, learning styles, academic programs, subject-matters, disciplines, and institutional frameworks can also become blended. The possibility to combine learning activities which can be carried out at different times and in different places (on-line, in the classroom) combined in interaction with other learning resources, requires specific pedagogical methodologies which take advantage of their collaborative potential and point to the creation of alternative learning environments. Punie [4] highlighted the potential of such learning environments to transcend existing limits, physical, conceptual, and institutional and to place students at the center of the learning, enabling the personalization of learning as well as social interaction at different scales, while being flexible enough to integrate various learning styles, teachers' skills, and curriculums, gradually becoming informal platforms to share expertise and knowledge across organizations. In the OIKONET [5] as well as in the EPUM project, the term Learning Space to address learning outcomes which are in line with those described by Punie has been initiated in the field of housing and urban form studies respectively [6].

The KAEBUP project, further acknowledges the lack of such collaborative, blended learning environments in the field of urban form studies and practices, to explore the potential to link academia, research, and practice, aiming at fostering the co-development of knowledge cutting across institutional and geographical boundaries.

Addressing the pedagogical objectives outlined in the previous section, the project's participants developed, implemented, tested, and are evaluating a blended learning approach which aims at fulfilling a double purpose: to enable participating institutions and professional practices to keep their own program and to facilitate the design and implementation of learning activities in collaboration. Activities carried out in the proposed shared digital platform are integrated with face-to-face activities carried out at the participant institutions through open learning processes (synchronously or asynchronously) as well as in joint hands-on, intensive training workshops, business model workshops, internships, and professional development sessions.

The blended learning approach adopted is supported by the development of the collaborative web-based learning environment, (R2P), structured under specific activities in various thematic areas proposed collaboratively by professors, students as well as practitioners, aiming at breaking down institutional barriers in educational cultures through the development and use of digital learning spaces and resources. These activities are referred to as Collaborative Learning Activities (CLAs) offering an innovative way for collaboration in the education system, by making available resources which are accessible to anyone wanting to access training regardless of their geographical location, educational culture, or ability to travel.

## 1.3 Related Work on Collaborative ICT Platforms

In the context of Computer Supported Collaborative Learning (CSCL), learning takes place through social interaction by using computers. CSCL software systems are collaborative learning environments that utilize technology to facilitate user (teacher to learner and learner to learner) interaction and communication, as well as learning coordination [7, 8]. An effective collaborative learning environment is one that effectively and efficiently supports knowledge sharing in a formed learning group [7]. A number of literature works have contributed towards the development of such learning environments.

In [7] the authors propose using semantic web technologies to build a software tool for knowledge sharing through the usage and management of multimedia annotations in

CSCL. Four annotation categories are supported: definitions (descriptions and explanations), comments, questions, and associations (e.g., links to other resources). Users can annotate web content and link documents and other resources to their annotations. The experimental evaluation showed that students doing collaborative group reading using this system achieved on average much higher scores than students doing group reading using other methods. In addition, the study showed that students in the first group have used collaborative multimedia annotations that contributed to enhanced knowledge sharing.

In case of multidisciplinary learning groups, the authors in [9] proposed an algorithm for composing optimal learning groups in situations where people have different domain backgrounds. The algorithm is integrated in an ontology-based e-learning system that creates self-built educating communities: trainees that participate in the education process gain points through achievements and ultimately become trainers. User profile information is explicitly acquired by having users fill forms. Based on this profile, users are assigned to (or are recommended) learning groups by maximizing the diversity within a group and minimizing the diversity between groups.

In the subject of exchanging data between learner groups, the authors in [10] propose an XML-based procedure using web-services for CSCL data exchange. The data to be transferred from one learner group to another include Moodle forum discussions, online chats, and votes. This allows learner groups to not have to start their discussions from scratch, without any reference of other groups' discussion. The positives of having other learner groups' data available is that the preceding discussions could effectively be used as scaffolding information to participate in the discussions [10]. In addition, the preceding discussions and the information shared provide adequate cognitive workload for the learner to be able to participate in the discussion [10]. A negative point is that the process of actively collecting information and discussion itself is a collaborative learning process in which the user does not participate.

## 2 The Research to Practice Platform (R2P)

The learning platform proposed, titled Research-to-Practice or R2P platform, comprising OER, to support blended-learning activities and co-creation of a critical mass of knowledge, aims to enable and support the project's teaching and learning activities, and to channel and organize research outputs in relation to evidence-based practice and entrepreneurship. As mentioned in the introduction, the platform aims to support effective and efficient higher education systems, to facilitate the exchange, flow, and co-creation of knowledge, to tackle future skills mismatches and to promote excellence in skills development. It aims to address the rupture in the flow of research and skills into the professional world, the lack of concerted effort from the quadruple helix to co-creating urban knowledge and the identified need to strengthen learners' transversal skills.

To this end, following the needs analysis specific to the fields of architecture and urban design, the KAEBUP project developed a Validation Framework for all learning and training activities. ESCO Transversal Skills and Competences relevant to the profession were identified in relation to the learning outcomes specified for all activities. Each skill is then matched to a task and activity that will take place during the project or can be embedded in existing curricula, linking each skill to the task which learners can carry out at a specific time during the project, during the timeline of their modules or in their own time (see following paragraphs). Each participant's profile is associated with a log reporting the completion of each task, which can be reviewed by any participant to enable feedback from across institutions and from peers. The completion of all tasks by a user will 'issue' a transversal skills certificate (a report, see following paragraphs) for architecture and urban design. The outcome is the implementation of an online 'transversal skills module', which can be taken as an elective by students or composed formally in curricula by institutions across Europe either as part of existing courses or, potentially, to be given certification and ECTS credits automatically applied when the module is taken as an elective.

### 2.1 R2P Platform Architecture

To fully support the concept of Collaborative Learning Spaces, allowing the co-construction and co-evolution of knowledge and the active membership of learners and teachers it was decided to exploit the Moodle LMS as the basis for the R2P platform. The selection was based (1) on the wide range of functionalities offered by Moodle, in which all the functionalities needed were included; (2) its popularity among users [11]; (3) it's accessibility as Moodle received WCAG 2.1 Level AA accreditation while accessibility features [12] have been added to the authoring tools so that the content that is produced is as accessible as possible; and (4) it's privacy features [13] that are provided assisting Moodle sites meeting GDPR compliance needs.

The platform, among other functionalities, enables teachers to add Learning Activities in the form of courses in various categories, as depicted in Fig. 1. Each Learning Activity may include many tasks, each of which is assigned with one or more skills, named competencies. To enable the assignment of competencies to tasks, the MOODLE competency framework was used where the "Transversal skills competencies" competency framework was created including 14 skills, as shown in Fig. 2.

When a teacher creates a Learning Activity, they also create the Learning Activity's tasks and then assign each task with skills. During the learning, while the student goes through an activity and performs the requested tasks, a checkbox is provided by the platform's UI (User Interface) that allows the student to declare that the corresponding task is completed (white circle in Activity box in Fig. 1). Upon completion of the task, the student is assigned the corresponding skill(s) of that task. When the student completes all the tasks of the Learning Activity, a report is generated by the customized plugin in the platform upon request (see Fig. 1) that outlines the completed tasks by the student and the gained skills. The report also includes the student's name, the Learning Activity's title and the activity's description. The plugin was developed by the authors as an external plugin and utilizes the R2P database, the Transversal skills competencies" framework and the student's responds through the UI to generate the report.

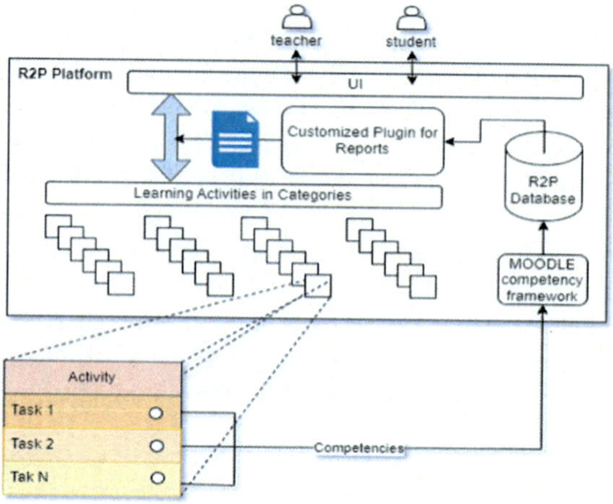


**Fig. 1.** The architecture of the R2P platform

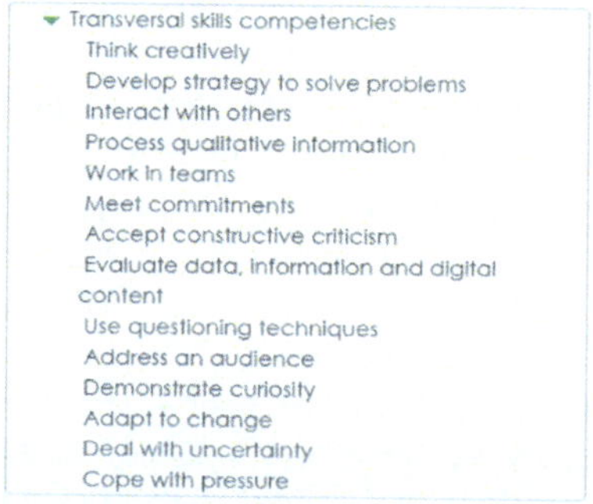


**Fig. 2.** The transversal skills competencies of the R2P platform

The customized plugin for the reports is offered to the teacher as a MOODLE Activity or Resource in the corresponding MOODLE interface for adding Activities or Resources in a Learning Activity (course), as shown by the red rectangle in Fig. 3.

## 2.2 Platform Testing and Evaluation

The first phase of the project focused on the development of the R2P platform's structure and design. Using Moodle Learning Management System (LMS) as the basis for development, the R2P platform was designed, while its functionalities were developed as plugins on top of Moodle LMS. Once a first prototype version was released that included several but not all functionalities of the platform, a hands-on, Business Model Workshop was conducted in the context of the project in Parma, Italy. During the workshop a number of participants used the platform to conduct activities and then evaluated the platform using a questionnaire. The aim was to evaluate the R2P platform in terms of usability and ease of use for the participants of the Parma workshop. The results were overall positive. 75% of Participants stated that they were able to fully understand

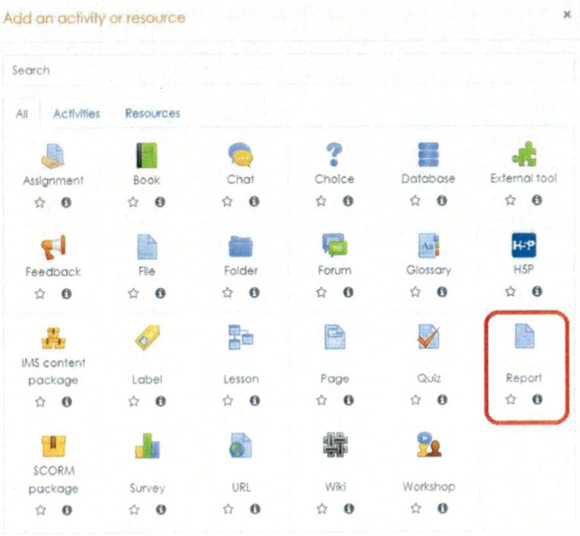


**Fig.3.** MOODLE interface for adding activities or resources (tasks) in a learning activity.

the tasks given to them, while the remaining 25% stated that they understood the tasks given to them. Participants were able to complete the tasks successfully using the R2P platform. Half of the participants stated strongly that the tasks were not difficult at all for them to accomplish using the platform, while the rest stated, "not difficult", with 14% of participants reporting minor difficulties.

The evaluation of the R2P platform provided the development team with valuable feedback as to where improvements need to be made, not only regarding the User Interface (UI) of the platform, but also regarding its functionality. In this context, the development team has undergone improvements for the platform and significantly enhanced its functionality, improving aspects such as communication/collaboration of its users via the platform, uploading of new material where interested users are automatically informed, posting announcements with enhanced and smart notification features (notifying only interested or enrolled to an activity users), automatic report creation and dissemination, and other.

## 2.3 Other Platform Functionalities/Features

The platform offers several functionalities for enabling and supporting teaching and learning activities:

- When someone registers, a corresponding notification goes to the coordinator via email.
- When a teacher uploads material to a Learning Activity, notifications go to all students enrolled in the relevant activity.
- When a user posts in the announcements, notifications to go all users enrolled in the relevant activity (forums).

The platform is accessible to participants upon registration, whereas specific functionalities of the platform remain publicly accessible for long-term exploitation by European institutions. The platform through its functionality communicates and promotes research findings for project dissemination mainly towards the end and following completion of the project.

The platform builds upon the experience developed through the EPUM project and its key features are: 1) ease of accessibility and structured navigation of Open Educational Resources (OER) to enable and support self-learning; 2) transnational interactivity for the purposes of teaching and learning activities (live online lectures, peer-to-peer feedback facilities, electronic submission system for remove assessment, distant-tutoring); 3) intra-personal interactivity enabling learners to access, store and keep track of tasks completed, self-issue template portfolios and skills certificates. Furthermore, the platform has a dedicated space for enterprises to access resources, tools and findings for scoping and testing their potential application in practice. This section of the platform also has an interactive character, by which professionals can request support, information, and feedback from academics to apply the tool, as well as to suggest research tasks to be carried out on their projects and indicating what innovations may be needed to support their work.

## 3 Conclusions - Learning, Co-construction, and Co-evolution of Knowledge through CLAs

The blended learning approach adopted and supported by the digital platform R2P, proves extremely important for the implementation of the KAEBUP project, resulting in the creation of Collaborative Learning Activities among partners throughout Europe, facilitating a community of inquiry which is constituted above and beyond institutional and physical barriers. The CLAs methodology implements a collaboration approach through the R2P platform, designed to meet the objectives of the HEIs in their mission to ensure that architectural and urban design students complete their studies with the skills to enter the professional world on the one hand, and to influence, innovate and support practice through their research work on the other.

The R2P platform strongly supports the methodology proposed, aiming at developing strong links and understanding between academia and practice to design a teaching and training method comprising blended learning techniques and work-based learning through various activities and learning mobilities aimed at:

a) facilitating the involvement of different stakeholders in the design of training courses across EU universities and the development and redefinition of bachelor and master-level programmes;
b) establishing connections between different academic programmes and varied business practices around knowledge co-creation, shared learning, skills development in higher education and long-life learning, competence in defined curricula, and entrepreneurial opportunities;
c) fostering mobility and exchange across institutions to support transversal skills and multi-stakeholder consultation on new curricular topics, and to enhance global citizenship;

d) evaluating and disseminating the project's outcomes to ensure that results are transferable and scalable and the widest possible range of potential users of the project's outputs is reached;
e) setting-up research activities and joint international workshops to address entrepreneurship in the specific fields of architecture, urban design, and planning.

The platform enables the development of OER through the input of different HEIs and professional practices with varied expertise and the implementation of blended-learning techniques, which include international collaborative learning activities, online lectures, remote feedback from peers and teaching staff from international institutions and real-life case studies located in a variety of EU countries. The fundamental benefits of this are the creation of a critical mass of knowledge, informed by organisations and stakeholders in the field of urban practices, with different capabilities and competences, and an effective system to support students' transversal skills development, enabling learners to experience how education, research and practice are linked together and how each can benefit the other.

The digital platform and resources will be interactive not just during the duration of the project when teachers and learners will be able to share data, content, outputs, and feedback, but also in the long term by enabling interested parties to participate through various activities, where learning involves co-construction and co-evolution of knowledge.

## References


1. Chong, G., Brandt, R., Martin, M.: Design Informed: Driving Innovation with Evidence-Based Design, 1st edn. Wiley (2010)
2. Guzzetta, J., Bollens, S.: Urban planners skills and competencies: are we different from other professions? Does context matter? Do we evolve? J. Plan. Educ. Res. **23**(1), 96–106 (2003)
3. Hofmann, J.: Why blended learning hasn't (yet) fulfilled its promises: answers to those questions that keep you up at night. In: Bonk, C.J., Graham, C.R. (eds.) Handbook of Blended Learning: Global Perspectives, Local Designs, pp. 27–40. Pfeiffer, San Francisco, CA (2006)
4. Punie, Y.: Learning spaces: an ICT-enabled model of future learning in the knowledge-based society. Eur. J. Educ. **42**(2), 185–199 (2007)
5. Madrazo, L., Sentieri, C., Charalambous, N.: Applying a blended learning methodology to the study of housing. In Rodrigues Couceiro da Costa, M.J., Roseta, F., Pestana Lages, J., Couceiro da Costa, S. (eds.) Architectural Research Addressing Societal Challenges, vol. 2, pp. 1051–1058. Taylor and Francis Group, CRC Press (2017)
6. Charalambous, N.: Emerging perspectives on urban morphology: collaborative learning activities fostering combined approaches. In: Strappa, G. (ed.) Urban Substrata and City Regeneration. International Seminar on Urban Form Conference Proceedings, Italy (2020)
7. Yang, S.J.H., Zhang, J., Su, A.Y.S., Tsai, J.J.P.: A collaborative multimedia annotation tool for enhancing knowledge sharing in CSCL. Interact. Learn. Environ. **19**(1), 45–62 (2011). https://doi.org/10.1080/10494820.2011.528881
8. Zurita, G., Nussbaum, M.: Computer supported collaborative learning using wirelessly interconnected handheld computers. Comput. Educ. **42**, 289–314 (2004)
9. Dascalu, M.-I., Bodea, C.-N., Lytras, M., Ordoñez de Pablos, P., Burlacu, A.: Improving e-learning communities through optimal composition of multidisciplinary learning groups. Comput. Hum. Behav. **30**, 362–371 (2014), ISSN 0747-5632. https://doi.org/10.1016/j.chb.2013.01.022

10. Tamura, Y., Sumi, K., Yamamuro, T., Maejima, M.: CSCL data structurization and inter-LMS sharing with use of web services. In: Lovrek, I., Howlett, R.J., Jain, L.C. (eds.) Knowledge-Based Intelligent Information and Engineering Systems. KES 2008. Lecture Notes in Computer Science, vol. 5179. Springer, Berlin, Heidelberg (2008)
11. Moodle homepage. https://stats.moodle.org/
12. Moodle homepage. https://docs.moodle.org/dev/Accessibility
13. Moodle homepage. https://docs.moodle.org/400/en/GDPR

# How Many Roads? Critical Thinking and Creativity in Higher Education and Mathematics

Maria M. Nascimento[1,2(✉)] and Paula Catarino[1,2]

[1] University of Trás-os-Montes e Alto Douro, Vila Real, Portugal
mmsn@utad.pt

[2] LabDCT-UTAD/Research Centre on Didactics and Technology in the Education of Trainers, University of Aveiro, Aveiro, Portugal

**Abstract.** We present a decade's work in higher education and teaching subjects in Mathematics in what refers to critical and creative thinking (CCT) at a Portuguese University. Thus, this study aims to summarize and interconnect the contributions of cooperative work among teachers interested in fostering more systematic and intending to foster critical and creative thinking in Portuguese and in parallel with reflecting on their pedagogical practice. The research question: How have CCT practices in the authors' courses changed, driven by their participation in the cooperative work of their research group? The analysis was 3-folded 1) the learning methodologies, 2) the exploratory studies, and 3) the results of the projects. We did a narrative literature review of 17 papers and a book chapter using the three groups analyzing the theme in which (at least) one of the authors appeared. The results revealed that integrating into the cooperative group allowed the two teachers to transform aspects of their practices and contribute to training the students, preparing them for the changes in this century's daily life and labor market. Teaching using active learning strategies does not necessarily promote CCT development, but the change is still in progress. Professional learning communities are drivers for education.

**Keywords:** Critical and creative thinking · Pedagogical practice · Higher education · Mathematics

## 1 Introduction and Context

The authors began their higher education (HE) career at the University of Trás-os-Montes and Alto Douro (UTAD) in the Department of Mathematics at the School of Science and Technology. They have been teaching several disciplines in the area of Mathematics. Their involvement in teaching Mathematics and their students' success remains today.

However, following the Bologna process, we began a path involving critical (CT) and creative thinking about ten years ago. The Bologna process intended that, at the end of the course, the graduate will be able to mobilize the necessary skills to face the challenges of the labor market. So, the study programs must provide the development of


M. E. Auer et al. (Eds.): ICL 2023, LNNS 899, pp. 31–40, 2024.
https://doi.org/10.1007/978-3-031-51979-6_4

entrepreneurship, innovation, creativity, and critical thinking [1]. So, the HE organization model is accomplished through a) moving from teaching based on the transmission of knowledge to teaching based on the development of competencies; b) orientation of the training given towards the specific objectives that must be ensured by the study cycles of the subsystem, university, or polytechnic in which it is inserted; c) determination of the work that the student must develop in each curricular unit - including, namely, when applicable, teaching sessions of a collective nature, personal guidance sessions of a tutorial type, internships, projects, fieldwork, study, and assessment - and their expression in credits according to the European Credit Transfer and Accumulation System, ECTS, based on student work [2]. Following the Bologna Process, student learning should be active and have a strong self-learning component. This self-learning should involve developing analysis, problem-solving, argumentation, group work, and communication skills, reinforcing the potential for lifelong learning [3].

Furthermore, "[a] *student-centred teaching environment fosters students' proactive engagement, which generally translates, in its simplest form, in students asking frequent more challenging questions. Teaching in such settings requires increased confidence, flexibility, and resilience*" [4].

At the same time, the role of teachers should change, leaving their centrality and becoming the moderator/supervisor/coordinator of students. In this way, modifying teaching practices in higher education and "weaning" students from passive audience practices, involving them in the active construction of their knowledge, will be able to enable them to exercise citizenship in a more intervening and participatory way in all aspects, including that of being competent professionals.

To UNESCO, "(…) *professional teaching knowledge has intuitive, practical, and relational dimensions. Collaborative teaching work naturally integrates a dimension of reflection and sharing among peers. Increasingly, this research can be translated into writing, with teachers assuming authorship.*" [5, pp. 89–90].

Therefore, working on critical and creative thinking, and because of current developments in the world, will prepare students to (i) resist the imponderables of their lives with resilience and solid skills, both individually and collectively; (ii) work in jobs that have not yet been created; (iii) make new learnings along their paths; (iv) interact with intelligent machines; (v) meet the complexity of problems that require highly specialized knowledge [3].

For this work, we formulated the research question: How have critical and creative thinking practices in the authors' courses changed, driven by their participation in the cooperative work of their research group? A narrative literature review of 17 papers and a book chapter using three focuses for their analysis is presented: 1) the implemented learning methodologies, 2) the exploratory studies, and 3) the results of the projects we participated in.

## 2 Methodology

Usually, a literature review involves identifying materials for potential inclusion - whether or not a formal literature search is required - to select included materials, synthesize them in textual, tabular, or graphical form, and analyze their contribution or value

[6]. Grant and Booth [6] characterize a literature review by the generic term and from published materials that examine recent or current literature on a range of subjects at various levels and may or may not include an assessment of the quality of the articles. It may or may not be exhaustive and is typically narrative. The analysis of this literature review may be chronological, conceptual, and thematic, among others.

The study of literature published in books and journal articles in the author's interpretation and personal critical analysis [7], in this case, from the authors.

In this work, the literature review was carried out to describe the path of the two authors - the 1st author will be designated by A and the 2nd by B - regarding their critical and creative thinking publications in the last ten years. The narrative of the aspects detected is described.

This work started from the chronological analysis of 17 papers and a book chapter on the theme in which at least one of the authors figured. The final list of references will have an asterisk (*).

Figure 1 summarizes the articles to be analyzed. The analysis of the papers began with their systematic re-reading carried out by the two authors simultaneously to organize the elements of the three following aspects: 1) the implemented learning methodologies, 2) the exploratory studies, and 3) the results of the projects in which we participated.

| Paper n.º [Reference] | Participating author | Year | Original Title [English title] |
|---|---|---|---|
| **1 [8]** | A | 2013 | Insights on web-based peer review: a case study with energy engineering students |
| **2 [9]** | A | 2014 | Estratégias de reforço das capacidades de pensamento crítico: um caso na UTAD [Strategies to reinforce critical thinking skills: a case at UTAD] |
| **3 [10]** | A | 2014 | Come Together: Peer Review with Energy Engineering Students |
| **4 [11]** | A | 2014 | Peer review and critical thinking: Path towards a transversal methodology |
| **5 [12]** | A | 2015 | Learn to cooperate and cooperate to learn: Empowering critical thinking skills through cooperative peer review. |
| **6 [13]** | A | 2015 | Adding value to the learning process by online peer review activities: towards the elaboration of a methodology to promote critical thinking in future engineers |
| **7 [14]** | A | 2016 | Fostering critical thinking through peer review between cooperative learning groups |
| **8 [15]** | A, B | 2016 | Mathematical creativity's understanding in engineering students of a Portuguese university |
| **9 [16]** | A, B | 2016 | Take this Waltz on Creativity- The Engineering Students' Conceptions |
| **10 [17]** | A, B | 2017 | Breaking the habit: engineering students' understanding of mathematical creativity. |
| **11 [18]** | A, B | 2019 | Com Um Brilhozinho nos Olhos: Três experiências promotoras de ensino do pensamento crítico na área da Matemática [With a Twinkle in the Eye: Three experiences promoting the teaching of critical thinking in the area of Mathematics] |
| **12 [19]** | B | 2019 | Cooperative Learning on Promoting Creative Thinking and Mathematical Creativity in Higher Education |
| **13 [20]** | A | 2019 | With a little help from my peers: professional development of higher education teachers to teach critical thinking. |
| **14 [21]** | A | 2019 | Stairway to the Stars: Comparing Health and Tourism Professionals' Views About Critical Thinking |
| **15 [22]** | A | 2019 | Perceptions of Portuguese University Teachers about Critical Thinking Educational Practices |
| **16 [23]** | A | 2020 | The Show Must Go On II: Statistics In (Engineering) Higher Education |
| **17 [24]** | B | 2021 | Teaching Linear Algebra in Engineering Courses Using Critical Thinking |
| **18 [25]** | A, B | 2018 | The Views of Engineering Students on Creativity |

**Fig. 1.** Papers (1–17) and a book chapter (18) in this review.

# 3 Data Collection and Analysis

## 3.1 Implemented Learning Methodologies

Mitchell et al. [26] "define active learning as one-time or ongoing student exercises introduced in the classroom to encourage student thinking and participation to engage students in the learning process."

Of the papers in Fig. 1, 12 (66.7%) involve reports of methodologies implemented in teaching practices by at least one of their authors.

At the beginning of the journey from 2013 to 2016, author A began working cooperatively with other teachers interested in fostering critical and creative thinking more systematically and intentionally at UTAD and, simultaneously, reflecting on their pedagogical practice, resulting in the papers referenced from papers 1 to 7 [8–14]. Methodologies were also addressed in paper 11 involving the two authors. Papers 16 and 17 are works in each author's specific teaching area. From 2013 to 2016, in Papers 1 to 7 [8–14], author A joined the analysis and implementation of a web-based peer review task or tasks at the degree (bachelor) level in a Portuguese Higher Education Institution.

In that time frame, the task was described, developed, and the different results were peer-reviewed and published to disseminate the results. The methodology involves students reviewing and providing constructive feedback on each other's work, facilitating the development of critical thinking skills. Each student or each group of students (in courses with many students) "produced a written document, containing a synthesis and an analysis of the paper (…) using Ennis' six dimensions FRISCO guideline (…) in a Google Drive Doc (digital) template designed by the teacher" [14, p. 35].

Later this task was also referred to by Nascimento et al. in case 3 [18, p. 199]: "The FRISCO guideline allowed students to develop critical thinking skills: identification of reasons, inferences, and credibility of information, among others. The grid was 1st presented in class through an example of the analyses of a newspaper piece".

Paper 11 [18] also refers to other strategies in the classroom, namely: Discovery of the Linear Algebra definitions using images (case 1), explaining the choices in true/false questions in tests (case 1), think-pair-share strategy (cases 2 and 3), class surveys with discussion (cases 2 and 3), and project work (case 3).

Catarino et al. [19] report a quasi-experimental study where the experimental group carried out "(…) *the Trade Questions cooperative method consisted of the analysis by the different cooperative groups of an image related to the contents of Linear Algebra on which they had to elaborate the greatest possible number of questions during four minutes*". In the discussion, the authors mentioned an intervention to promote creative thinking in a Communication and Multimedia degree in a Linear Algebra course. The authors found that cooperative learning activities enhanced experimental group students' creativity.

Active teaching strategies were implemented during the COVID-19 pandemic also by Nascimento and Morais [23]: motivational video reading quizzes, think-pair-share tasks in the theoretical classes, and for the theoretical-practical classes, the use of Padlet as a mural to schedule and follow the pairs/group peer-review activity cycle with feedback from students and the teacher, concept maps, and project work. Catarino and Vasco also reported the think-pair-share strategy [24].

From Paper 1 [8]: "*These lines of research will be carried out by a critical thinking research group which was created in the current school year (2012/2013) by several teachers at the [UTAD]. After the analysis and presentation of previous experiences in the engineering courses (…), their benefits and difficulties, teachers from other scientific areas such as Veterinary, Linguistics, Education, Statistics, Agriculture, and Second Languages teaching showed interest (…). Therefore, this group is working on (…) using a similar global methodology.*"

In Paper 4 [11], based on the description of the path of a group of professors at UTAD (webPACT) in the use online peer review environment, some constraints to the development of communication and critical thinking development of communication and CT skills in students. Some initial adjustments were made to the methodology used between theory and practice, making it completer and more refined. The interventions were carried out in bachelor's and master's courses, including Engineering Degrees, Veterinary Medicine, Basic Education, Communication Sciences, and Mathematics.

### 3.2 Exploratory Studies

Singh writes, "*Exploratory research is the initial research which forms the basis of more conclusive research*" [27, p. 63]. Similarly, it "*allows researchers to explore issues in detail in order to familiarize themselves with the problem or concept to be studied*" [27, pp. 63–64].

Of the papers in Fig. 1, 4 (22.2%) report exploratory studies from both authors regarding students' creativity and mathematical creativity understandings and concepts. Based on surveys done with engineering students, the open-ended questions about their understanding of creativity and mathematical creativity. The answers' content analysis was done and reported in those four papers.

In Paper 8 [15]: "*This exploratory study leaves clues on the connection that needs to be made between mathematical creativity and 'solving problems', maybe it is a way to foster it in Mathematics courses in engineering degrees*" [15, p. 4].

In Paper 9 [16], "(…) *the exploratory analysis of students' definitions* (…) *showed that definitions were affected neither by gender nor by the original area of study, and* (…) *showed the predominance* (…) *of grouped implicit categories (creation, imagination, and originality)*" [16, p. 5].

In Paper 10 [17], the exploratory study also indicated, "[d]*ifferent tasks may encourage mathematical creativity, hence creativity. Following this study for engineering degrees* (…) *in their mathematics courses, for instance, exploring problem-solving and problem-posing in other mathematical subjects such as Statistics*" [17, p. 10].

Finally, in the book chapter [25], the final remarks suggested that "(…) *the use of problem-solving or project-based learning* (…) *would foster students' confidence* (…) *by triggering their curiosity for new approaches toward common problems and challenging their ability to propose and select the most suitable solution*" [25, p. 154].

### 3.3 Project Results

The project was

"(…) *an 'Erasmus + Programme, the 'Critical Thinking Across the European Higher Education Curricula – CRITHINKEDU' project arises from the background and experience of European Higher Education Institutions, business corporations and Non-Governmental Organisations, and their ongoing concern to improve the quality of learning in universities and across different sectors, which converge in a common need on how to better support the development of Critical Thinking (CT) according to labour market needs and social challenges*" [28].

Their main results were:

"*Higher awareness of the need for CT education; Improvement of curricula design and classroom educational practices; Enlargement of academics and researchers' network; Engagement of all institutional levels in supporting CT education; Closer University-Business Cooperation for CT education; Empowerment of teachers' professional development and students' CT; Implementation of activities and teaching strategies that prompt and support CT; Creation of CT scenarios in diverse higher education fields*" [28].

Of the papers in Fig. 1, 7 (38.9%) refer to project results as implemented learning methodologies in Papers 11 [18], 12 [19], 16 [23], and 17 [24] and referred to in paragraph 3.1. The project reports work in CRITHINKEDU resulted in Papers 13 [20], 14 [21], and 15 [22].

The methodologies were "(…) *implemented in a European training course on CT education for university teachers in Rome, Italy. This course aimed to engage participants with CT teaching practices, preparing them with the required pedagogical knowledge and tools suitable for this purpose*" [20, p. 143]. The course was "*drawn upon the proposals of the 'European inventory of critical thinking skills and dispositions for the 21st century' and the 'Preliminary guidelines for quality in critical thinking education'* (…)" [20, p.143]. The course topics for each day are in Fig. 2.

The changes reported in Papers 11 [18], 12 [19], 15 [22], 16 [23], and 17 [24] were done during the Rome course and began their implementation in the 2nd semester of 2017/18 after it and subsequently.

| Course days | Topics for each day | Course days | Topics for each day |
|---|---|---|---|
| Day 1 | What do we want to achieve in our course?<br>1.1.An introduction on the 4C/ID model<br>1.2.Heuristic bias and CT<br>1.3.Characterization of CT: a proposal | Day 3 | How can we support students in CT development?<br>3.1.CT lesson planning, classroom environment, questioning, and cooperative learning techniques<br>3.2.Methods and strategies of Problem-Based Learning (PBL) supporting CT development, creative thinking, and visual literacy<br>3.3.Using Values and Knowledge Education (VAKE) for enhancing CT |
| Day 2 | What do students have to do?<br>2.1.Task analysis<br>2.2.The design of teaching sequences integrating CT<br>2.3.Tips for designing supportive &procedural information to deal with an intuitive mental model | Day 4 | How can we measure the achievement?<br>4.1.Tools to assess CT levels of students |
| | | Day 5 | Are we all ready to go?<br>5.1.Expectations after the CRITHINKEDU course |

**Fig. 2.** CRITHINKEDU course topics for each day (adapted from Paper 13 [20], p.144).

Finally, in Paper 13 [20] discussion: "*What strikes here is that even experienced teachers understand and integrate CT teaching practices in different ways and levels.*

*Thus, professional development and effective change in terms of CT teaching practices development is a complex challenge, slow and time-consuming process*" [20, p. 149].

In the CRITHINKEDU former outcomes, author A questioned (with other project members) different professionals about their views on the critical thinking needs in their areas. In Paper 14 [21]: "*The focus group technique* (...) *was used with a set of open-ended questions* (...). *The participants were professionals of business companies, organizations, and employers from different areas, namely, Health and Tourism*" [21, p. 214]. As a result:

"*In the Health area, the more mentioned CT were Evaluation, Others (...) such as Interpersonal Skills, Establishment of priorities and Communication skills) and Self-regulation; in the Tourism area, the (...) framework CT skill stood out. In the Health area, all the dispositions got more or less the same number of mentions. However, in the Tourism area, Inquisitiveness and Analyticity are highlighted. Finally, CT skills and CT dispositions examples were different in each of these two professional areas*" [21, p. 221].

Finally, in Paper 15 [22]:

"*We carried out semi-structured interviews with five Portuguese university teachers (...). Interviewed teachers (...) were asked to describe their perceptions about CT teaching, namely: how can CT be promoted in HE; what type of interventions, teaching strategies, and evaluation methods are being used to promote CT; and what challenges and limitations teachers have to face nowadays in their CT instruction*" [22, p. 227].

As a result: "*The teachers interviewed emphasize, as for the challenges in CT education, the need to change the students' and teachers' mindset (change of the institutional culture). Also, other challenges and difficulties are mentioned, namely: the lack of institutional support in the promotion of CT; difficulty in implementing activities due to the size of the class (high number of students), organizational conditions (class length) (...)*" [22, p. 236].

Finally, an important finding was referred to:

"(...) *the need to change institutional culture and conditions towards the support of CT educational practices – this will also enable the long-term integration of CT across the curricula and the transferability of skills and dispositions to other contexts. In general, teachers agreed on the importance of being explicit and clear in their CT teaching practice and using authentic situations, dialogue, and active learning strategies to effectively develop students' CT*" [22, p. 223].

## 4 Discussion and Conclusions

Active learning environments are more student-centered in encouraging students to develop skills. In an active learning approach, students take responsibility for the learning, and teachers become facilitators for what happens in and out of the classroom [3, 4, 26]. Several learning methodologies were implemented in the authors' reported teaching practices. These methodologies included web-based peer review tasks, think-pair-share strategy, project work, and discovering mathematical definitions using images. The authors emphasize the importance of cooperative learning activities in enhancing critical and creative thinking skills. They discussed the implementation of these methodologies in different courses and study areas.

Regarding the exploratory studies [27], the authors report their studies about students' understanding and perceptions of creativity and mathematical creativity. They analyzed survey answers from engineering students (and different school years) and found that definitions of creativity were not influenced by gender or the students' original study area. The authors suggest that problem-solving tasks and exploration of problem-posing in various mathematical subjects can encourage creativity and, therefore, mathematical creativity simultaneously.

In the item project results, the authors - author A as a team member and author B as a webPACT member - were involved in an Erasmus + project, "*Critical Thinking Across the European Higher Education Curricula – CRITHINKEDU*" [28], and the project aimed to improve university learning quality by promoting critical thinking skills and dispositions [28]. The authors report changes in higher education, improvement of curricula design, engagement of academics and researchers, university-business cooperation, and empowerment of teachers' professional development and students' critical thinking. They also mention implementing activities and teaching strategies that foster critical and creative thinking and the need for institutional support, among others, since its implementation needs time. The project results in which the authors were involved show that pedagogical approaches positively impact students' CT and creativity take class time and should be developed along all courses of the graduations or masters. Additionally, the findings highlight the value of professional learning communities and teachers' networks in facilitating the learning and teaching change (at least gradually) in higher education, specifically regarding critical and creative thinking at their Portuguese university.

The authors analyzed 17 papers and a book chapter, all being peer-review submissions (available in the references), and reflecting on our work in the last ten years, we did not dare to make either quantitative conclusions or generalizations. Consequently, we acknowledge our work limitations. Since the papers are from the authors' production, it is a biased view, and the sample size is small. Therefore, this work only provides a comprehensive overview of the research and perspectives of the authors engaged in the webPACT cooperative research group. In the future, even with other colleagues, we may broaden the publications to the webPACT to capture the range of relevant research and perspectives in this cooperative research group.

In summary, critical and creative thinking practices in the authors' courses changed, mainly driven by their participation in the cooperative work of their research group. The thematic analysis of this literature review revealed that, throughout this decade, all the implemented work allowed the two teachers to alter (at least some) features of their practices and contribute to more proactively training the students, preparing them for the changes in this century's daily life and labor market. Teaching using active learning strategies does not necessarily promote CT development, but the changes are still in progress. Professional learning communities still are key drivers for education, in our case, higher education.

**Acknowledgments.** This work is financially supported by National Funds through FCT – Fundação para a Ciência e a Tecnologia, IP, under the project UIDB/00194/2020.

## References


1. Melo, A.S.: Enquadramento histórico legal do processo de Bolonha e o seu impacto no sistema de ensino superior português. Revista Observatório **3**(6), 75–141 (2017)
2. Leite, C., Fernandes, P.: Inovação pedagógica: uma resposta às demandas da sala de aula universitária. Perspectiva **29**(02), 507–533 (2011)
3. Conselho Nacional de Educação—CNE, Estado da Educação 2021. CNE, Lisboa (2022)
4. di Benedetti, M., Plumb, S., Beck, S.B.M.: Effective use of peer teaching and self-reflection for the pedagogical training of graduate teaching assistants in engineering. Eur. J. Eng. Educ. **48**(1), 59–74 (2023)
5. UNESCO. Reimagining our Futures Together: A New Social Contract for Education. UN (2022)
6. Grant, M.J., Booth, A.: A typology of reviews: an analysis of 14 review types and associated methodologies. Health Info. Libr. J. **26**(2), 91–108 (2009)
7. Ercole, F.F., Melo, L.S.D., Alcoforado, C.C.: Revisão integrativa versus revisão sistemática. Reme: Revista Mineira de Enfermagem **18**(1), 09–11 (2014)
8. *Dominguez, C., Nascimento, M., Maia, A., Pedrosa, D., Cruz, G.: Insights on web-based peer review: a case study with energy engineering students. In: 2013 1st International Conference of the Portuguese Society for Engineering Education, pp. 1–10. IEEE (2013)
9. *Payan-Carreira, R., Dominguez, C., Nascimento, M.: Estratégias de reforço das capacidades de pensamento crítico: um caso na UTAD. Práticas Educativas: Teorização e Formas de Intervenção, pp. 56–67 (2014)
10. *Dominguez, C., Nascimento, M., Maia, A., Pedrosa, D., Cruz, G.: Come together: peer review with energy engineering students. Int. J. Eng. Pedagog. (iJEP) **4**(5), 34–41 (2014)
11. *Dominguez, C., et al.: Revisão entre pares e pensamento crítico: percurso para uma metodologia transversal. Pensamento Crítico na Educação: Perspetivas atuais no panorama internacional, pp. 357–378. Universidade de Aveiro, Aveiro (2014)
12. *Dominguez, C., Carreira, R.P., Nascimento, M., Morais, M.F.: Learn to cooperate and cooperate to learn: empowering critical thinking skills through cooperative peer review. In: Pensamento crítico na educação: desafios atuais, pp.175–185. UTAD, Vila Real (2015)
13. *Dominguez, C., et al.: Adding value to the learning process by online peer review activities: towards the elaboration of a methodology to promote critical thinking in future engineers. Eur. J. Eng. Educ. **40**(5), 573–591 (2015)
14. *Silva, H., et al.: Fostering critical thinking through peer review between cooperative learning groups. Revista Lusófona de Educação **32**, 31–45 (2016)
15. *Catarino, P., Nascimento, M.M., Morais, E., Campos, H., Vasco, P.: Mathematical creativity's understanding in engineering students of a Portuguese university. In: 2016 2nd International Conference of the Portuguese Society for Engineering Education (CISPEE), pp. 1–5. IEEE (2016)
16. *Catarino, P., Nascimento, M.M., Morais, E., Silva, H., Payan-Carreira, R.: Take this waltz on creativity: The engineering students' conceptions. In: 2016 2nd International Conference of the Portuguese Society for Engineering Education (CISPEE), pp. 1–6. IEEE (2016)
17. *Catarino, P., Nascimento, M.M., Morais, E., Campos, H., Vasco, P.: Breaking the habit: engineering students' understanding of mathematical creativity. Eur. J. Eng. Educ. **44**(4), 449–460 (2017)
18. *Nascimento, M.M., Catarino, P., Morais, E., Vasco, P.: Com Um Brilhozinho nos Olhos: Três experiências promotoras de ensino do pensamento crítico na área da Matemática. Revista Lusófona de Educação **44**, 191–209 (2019)
19. *Lopes, J., Silva, H., Catarino, P., Morais, E., Vasco, P.: Cooperative learning on promoting creative thinking and mathematical creativity in higher education. REICE Revista Iberoamericana Sobre Calidad, Eficacia y Cambio En Educacion**17**(3), 5–22 (2019)

20. *Cruz, G., Nascimento, M.M., Dominguez, C.: With a little help from my peers: professional development of higher education teachers to teach critical thinking. Revista Lusófona de Educação **44**, 141–157 (2019)
21. *Nascimento, M.M., et al.: Stairway to the stars: comparing health and tourism professionals views about critical thinking. In: Tsitouridou, M., Diniz, J.A., Mikropoulos, T. (eds.) Technology and Innovation in Learning, Teaching, and Education. TECH-EDU 2018. Communications in Computer and Information Science, vol. 993, pp. 210–222. Springer, Cham (2019)
22. *Morais, F., et al.: Perceptions of Portuguese university teachers about critical thinking educational practices. In: Tsitouridou, M., Diniz, J.A., Mikropoulos, T. (eds.) Technology and Innovation in Learning, Teaching, and Education. TECH-EDU 2018. Communications in Computer and Information Science, vol. 993, pp. 223–239. Springer, Cham (2019)
23. *Nascimento, M.M., Morais, E.: The show must go on II: statistics in (engineering) higher education. Open Educ. Stud. **2**(1), 214–219 (2020)
24. *Catarino, P., Vasco, P.: Teaching linear algebra in engineering courses using critical thinking. Open Educ. Stud. **3**(1), 76–83 (2021)
25. *Catarino, P., et al.: The views of engineering students on creativity. In: Nascimento, M., Alves, G., Morais, E. (eds.) Contributions to Higher Engineering Education, pp. 135–158. Springer, Singapore (2018)
26. Mitchell, A., Petter, S., Harris, A.L.: Learning by doing: twenty successful active learning exercises for information systems courses. J. Inf. Technol. Educ. Innov. Pract. **16**, 21–46 (2017)
27. Singh, K.: Quantitative Social Research Methods. Sage, New Delhi (2007)
28. CRITHINKEDU Homepage. https://crithinkedu.utad.pt/en/crithinkedu/. Accessed 21 May 2023

# Assessing the Development of Soft Skills Among HEI Students in the VAKEN Process Preliminary Findings from Three Sprint weeks

Christa C. Tigerstedt[1(✉)], Britt Petjärv[2], Karen Malene Elmann Andreasen[3], Mikael Forsström[1], Maira Lescevica[4], Helen Kiis[2], Dalia Karlaitė[5], Vera K. Vestmann Kristjansdottir[6], and Hafdis Björg Hjalmarsdottir[6]

[1] Arcada University of Applied Scienes, Helsinki, Finland
christa.tigerstedt@arcada.fi
[2] TTK University of Applied Sciences, Tallinn, Estonia
[3] UCL University College, Tallinn, Denmark
[4] Vidzeme University of Applied Sciences, Valmiera, Latvia
[5] Vilnius University of Applied Sciences, Vilnius, Lithuania
[6] University of Akureyri, Akureyri, Iceland

**Abstract.** Soft skills are essential for success in the modern workplace, encompassing interpersonal, problem-solving, and communication abilities necessary for effective collaboration. In the VAKEN project, we developed a process to train and assess soft skills within a real-life context, guided by coaches and in collaboration with companies. Our objective was to enhance soft skills during a five-day sprint focused on a company problem. This paper aims to investigate students' perceptions of developing creativity, personal leadership, complex problem solving, and critical thinking through the VAKEN process and its sprints. Specifically, we explore how students perceive the development of soft skills before and after the Sprint Week. Data collection involved self- and peer-reviewed assessments, and daily reflective diaries. The first sprint resulted in enhanced problem-solving skills, while critical thinking improved in the second sprint and creativity in the third. Across all sprints, personal leadership exhibited the least improvement according to student perceptions.

**Keywords:** Soft skills development · Self-assessment · Higher education

## 1 Introduction

The development of soft skills is crucial for success in the modern workplace. Soft skills refer to a range of interpersonal, problem-solving, and communication abilities that are necessary for effective collaboration with others [1, 2]. In the VAKEN project, a collaboration between seven partner institutions from the Nordic-Baltic region, the focus is on a set of soft skills that appears to be emerging from the recent working trends: creativity [3–6], personal leadership [7–12]; complex problem solving [13–16], and critical thinking [17–20].


M. E. Auer et al. (Eds.): ICL 2023, LNNS 899, pp. 41–52, 2024.
https://doi.org/10.1007/978-3-031-51979-6_5

This paper will discuss the development and assessment of soft skills in previous researches and the development and assessment of soft skills as part of the VAKEN process.

The purpose behind the study and this paper is threefold:

1. Development of soft skills is crucial for higher education and the workplace.
2. Deeper research is needed on how to assess the development of soft skills.
3. The findings from the VAKEN project need to be discussed to see how the model, the VAKEN process, can be further used and developed.

The aim of this first paper is to answer the following research question:

How is the development of soft skills perceived among students participating in the VAKEN process before and after an intensive study week based on the Sprint methodology?

The VAKEN approach aims to provide a learning platform to practice and improve four soft skills. Furthermore, the project has developed a process that enables the training and assessment of soft skills in a real-life context with guidance from coaches and in collaboration with companies.

The main goal in this project has been to improve soft skills during a five-day sprint, while working on a company problem in diverse and/or international student teams. This student-centred approach takes the form of intense sprints in which a company challenges students to provide solutions for a complex and real problem. The VAKEN process, consists of two design sprints comprising nine stages, each with smaller exercises targeting various soft skills, and is complemented by the flexible VAKEN toolkit, designed to facilitate the development of these skills.

Creativity is a crucial skill in decision-making and problem-solving, applicable in a variety of fields including science and business [3].

Complex problem solving involves overcoming barriers to reach a desired goal state through multi-step activities [13]. The problem-solving process requires the interaction of cognitive, emotional, personal, and social abilities and knowledge [14].

Personal leadership is the ability to take responsibility for one's life and create a vision for the future [7, 8].

Critical thinking involves analyzing and solving problems and making suitable decisions. It is the ability to comprehend, evaluate various information, and draw conclusions [17].

## 2 Developing and Assessing Soft Skills

Soft skills are essential for both personal and professional success and play a critical role in navigating the rapidly changing worklife. Soft skills are not only necessary for work, but they are also essential for everyday life. Educational institutions must focus on methods and techniques to help students develop these skills [21, 22].

Soft skills have been examined recently in many studies. Hence, scholars seem to somewhat agree that soft skills are crucial and holistic understanding is necessary. These topics need to be taught in specific ways and integrated carefully into curricula. Also, multiple scholars note that assessing this type of learning is hard and measuring the learning is limited if only numbers are used. [23–25].

This chapter reviews and discusses some prior research on soft skills development and learning interventions in higher educational and workplace settings. It also comes with some suggestions regarding how soft skills can be developed and assessed.

## 2.1 Developing and Learning Soft Skills

Research into pedagogy and more specifically design for learning (i.e. didactics) for soft skills in higher education show that the skills are to be learnt over a long period of time and also that it is good to develop them in a real-life context where collaboration with for example industry is one element [26–28]. Problem based learning (PBL) has been one approached which has been proven successful in this case [29]. Soft skills are hard to enhance with so called 'taught' methods, i.e. teamwork is not necessarily learnt by just being part of a team. Teamwork can indeed help the students to see their limitations as well as becoming more aware of the skill called teamwork and because of this start a development process, which in time, will enhance this skill [25].

Biggs [30] stresses both the principle of constructive alignment and that teachers need to engage in learning activities which enhance so called academic skills, like for example critical thinking and problem solving. One aligned mode of teaching which promotes this is problem-based learning [29, 31]. A considerable amount of literature explores whether soft skills interventions work at all [32–35] and if it is possible to teach and develop soft skills efficiently at a university. There have been different initiatives to answer that question, to define methods and best practices for developing, as well as for providing tools for assessing the improvement of the activities carried out [36]. For instance, Almonte, McAfee, Snell and Ahmed [37] demonstrated soft skills intervention courses during a period of five years as an effective way to improve students' knowledge of soft skills and their ability to deploy them.

A project called "Tuning" [38] focused on the presence of generic skills in higher education programs. One of the most important results of the project based on a solid survey is the need for the paradigm shift from a Teaching Based Knowledge (TBK) system to a Competency Based Learning (CBL) and student-centred approach. The project suggests a few methods of developing soft skills: integrating them in the different subjects of a degree; conducting seminars and workshops, devoting one or two weeks at the beginning of each semester to train soft skills, etc. According to Cimatti [21] soft skills are personal resources, they can be activated by the context and by the situation, e.g., simulations or case studies connected to the process and tasks can be particularly helpful. Soft skills are more related to the mindset and the personality of the student [25]. Once soft skills are acquired, they can be transferred from one to another context, then the same learning process can produce results in different surroundings [39]. Furthermore, lectures, small-group learning, and project-based learning all have positive associations with achievement provided by the student-centred instructional elements [27, 40]. Also, one method for learning soft skills might not suit all soft skills learning. I.e. active teaching such as project based and PBL might not enhance all desired soft skills in a certain context [24].

In addition to the above, it is emphasized that the support of facilitators is fundamental. E.g., experts' seminars, coaching and tutoring are effective tools to teach soft skills. Mentors, who can personally support and take care of the individual growth and learning, are considered particularly powerful in the process of development of these competences because they can help students to better understand their attitudes, to develop skills and find their way in the professional life [41]. Soft skills proficiency highly related to the student's personality and behavioural qualities [25]. The teachers' role is of the one of a coach – the student has a more active role (student-centred learning) and gets support throughout the learning process [42].

### 2.2 Assessing the Development of Soft Skills

As such evaluation is the systematic process of collecting, analysing, and interpreting training programs [43–45] to ensure that it is delivered effectively and efficiently [46]. Training evaluation gives comprehensive feedback on the value of the training programs and helps to identify training gaps and even discovers opportunities for improving the overall processes of training programs [47, 48]. To evaluate any training, it is important to choose the right evaluation methodology for the training program. According to many authors, [49–51] the best way to evaluate any change in learning is through assessment or tests. Conducting a pre-test before and a post-test after the training could be applied [29, 52, 53].

Giovannini [54] states that assessment can and must be considered as a learning tool that helps a student to self-regulate the student's own learning process. Here the student centred-learning perspective becomes prominent. The student-centred learning approach can be related to the formative learning, i.e. the importance of having not only start and finish assessment, but also assessment in the middle of the process [26, 55]. As such, the idea to involve students in the assessment comes from the thought that activity and self-reflection are crucial to learning. Self-assessment enhances the student's possibility to see and evaluate or assess the own development [26]. Cinque [56] in turn, argues that soft skills' learning is a willful, intentional, active, conscious, constructive and socially mediated practice that includes a reciprocal intention or action-reflection. Reciprocal can here be understood as a process with two students helping each other in the learning while at the same time achieving a specific learning goal.

Many researchers suggest that soft skills should preferably be assessed with the help of self-assessment and/or peer assessment [31, 57, 59]. There are many reasons for this: first, the assessment and the ability to think about the development of the skill. Second the learning aspect of the self-peer and peer assessment per se. To continue, Kenwright argues: "Assessing an individual's soft skills, however, takes some introspection. For instance, an individual may think of themselves as a "collaborative leader", until colleagues and peers provide feedback (assessors and individuals themselves have a limited view of a person's soft skills)" [25].

Various tools to assess soft skills have been proposed, e.g., questionnaire can be used by students for self-assessment or to do the peer assessment [31, 56]; a portfolio is recommended as a tool to support students in the self-evaluation process, understanding which are the skills one needs to further work on [54]. Evaluation and measuring the learning numerically has not however been considered an efficient tool if not used as a self-assessment. Using only traditional measures can make one miss essential things about the soft-skills proficiency. If one wants to assess students' soft skills it should be done by asking situational and behavioural questions, i.e. soft skill questions and tests that consider an individual's overall personality and characteristics [25, 59]. Kenwright [25] also mentions assessments in the form of reflective material, mentoring sessions or talking with peers as being efficient. More research into efficient assessment of soft skills development and learning is needed [25, 59].

## 3 Method

In this study a survey was chosen as the tool for the students' self-and peer assessment. This means that there is a quantitative approach to the data collected. Surveys involve gathering data from a large group of people with the help of a pre-designed set of questions [60, 61]. The pre- and post-questions were used as a brief self-analysis tool of the four VAKEN soft skills. That gives the participant the opportunity to evaluate the development of their soft skills within a Sprint Week, a vision for the future and implication on which skills they should develop further. A self-assessment questionnaire is easy to administer, and it is often used in assessment of learning and among the best proven methods to assess soft skills learning [42, 58].

Assessment in the VAKEN process is the participants' self-assessment. To be noted, at the end of the Sprint Week participants assess not only themselves but also their team members. This approach will enhance the participants insight into their own skills and give the participant an idea of how they are perceived from others' point of view.

### 3.1 The Assessment Tool Construction and Procedure in VAKEN

The VAKEN pre- and post-questions were aligned with the objectives and tools specific to each stage, reflecting the four defined skills. These questions were formulated based on the Assessment toolbox introduced in the Employ Skills project, developed by UCL and partners [36]. The questions underwent a pilot phase during the VAKEN Sprint Week in Denmark in the fall of 2021. Following the pilot, the questions were refined and reorganized to enhance clarity.

The scale used for the surveys was a 5-point Likert scale, with answers ranging from 1 (strongly disagree) to 5 (strongly agree). The 5-point Likert scale can be interpreted as intervals but Likert scales with fewer responses are restricted to ordinal numbers [62].

The questionnaire consisted of 22 structured questions, with five for creativity, six for critical thinking, five for complex problem-solving, and seven for personal leadership. Participants responded to the same questions before and after the sprint.

### 3.2 Procedure

Data was collected from three sprint events in Kaunas, Akureyri, and Helsinki ($N = 208$). The first VAKEN Sprint Week took place in Kaunas (16–20.5.2022) with thirty students from various countries. Self-assessments were conducted at the beginning and end of the week. The second Sprint Week occurred in Akureyri (31.10–4.11.2022) with thirty-five students from multiple countries. The final Sprint Week was held in Helsinki (12–16.12.2022) with 220 Finnish students.

### 3.3 Method for Analysing Data

Data was analyzed using SPSS and Excel, involving descriptive statistics, paired t-tests, independent samples t-tests, and the Related Samples Wilcoxon Rank test. The non-parametric Wilcoxon Rank test was conducted to validate the results of the paired t-test, considering the potential challenge to normal distribution with Likert scale data. However, since the data consisted of multiple questions with averages calculated, normal distribution was achieved. The results of the Wilcoxon Rank test aligned with those of the paired t-test, confirming statistical significance ($p < 0.05$) in all instances.

## 4 Empirical Analysis and Results

The VAKEN Sprint Week focused on four main soft skills: creativity, personal leadership, critical thinking, and problem-solving. To measure these skills, groups of questions were created within the questionnaire. For example, the skill of creativity was assessed using questions such as "I am curious," "I have an open mind for changes," and "I am good at getting new ideas." The responses for each skill were aggregated, and an average score was calculated. This process was repeated at the beginning and end of the VAKEN Sprint Week, providing researchers with continuous variables representing the level of each skill for each student.

To analyze the improvement of students' soft skills during the first Sprint Week of the VAKEN Sprint in Kaunas, paired t-tests were conducted for each soft skill. As previously mentioned, the soft skill of creativity was assessed using five questions. The mean level of creativity at the beginning of the week was calculated (mean = 3.88, variance = 0.225), serving as the baseline. The same procedure was repeated for the last assessment to determine the students' level of creativity at the end of the Sprint Week (mean = 4.02, variance = 0.299). A paired t-test for two samples was conducted in Excel to evaluate the change in creativity level throughout the week. The null hypothesis, assuming no difference in creativity level between the beginning and end of the week, could not be rejected ($t = 1.147$, one-tail p-value = 0.130). In other words, the students did not show improvement in their creativity level during the VAKEN Sprint Week in Kaunas.

The soft skill of personal leadership underwent the same procedure. At the beginning of the week, the mean was 3.533 with a variance of 0.164. At the end of the week, the mean was 3.483 with a variance of 0.215. A paired t-test for two samples was conducted in Excel to assess any changes in the students' level of personal leadership throughout the week. The results indicated that there was no statistical difference between the beginning and end of the VAKEN Sprint Week in terms of personal leadership ($t = 0.485$, one-tail p-value = 0.316). Thus, the null hypothesis, stating that there was no difference in personal leadership level, could not be rejected.

The same procedure was followed for the soft skill of critical thinking. The mean at the beginning was 3.723, and at the end it was 3.82, with variances of 0.255 and 0.277 respectively. The paired t-test results showed no statistical difference, with a p-value of 0.162. Therefore, the students did not improve their critical thinking during the VAKEN Sprint Week in Kaunas.

Next, the soft skill of problem-solving was analyzed. The mean at the beginning of the week was 3.53, and at the end it was 3.77. The variances were 0.274 and 0.211

respectively. A paired sample t-test was conducted in Excel to assess changes in the students' problem-solving skill. The results indicated a statistical difference ($t = 1.963$, one-tail $p = 0.03$). Thus, the null hypothesis was rejected, demonstrating that there was a significant improvement in the level of problem-solving skill by the end of the VAKEN Sprint week.

The second VAKEN Sprint Week took place in Akureyri, where all students completed a self-assessment survey at the beginning and end of the week. The improvement in creativity was analyzed using the same procedure as before. The mean creativity level at the beginning of the week was 3.81, with a variance of 0.358. By the end of the week, the mean increased to 4.31, with a variance of 0.286. A paired t-test in Excel was conducted, with the null hypothesis stating no difference in creativity levels between the beginning and end of the week, and the alternative hypothesis suggesting a higher level of creativity at the end. The result showed a statistical difference ($t = 4.978$, $p < 0.001$), indicating an improvement in creativity by the end of the week. The same procedure was applied to analyze the soft skills of personal leadership, critical thinking, and problem-solving. The results are presented in Table 1.

**Table 1.** Summary of results from the students' self-assessments in Akureyri.

| Soft skill | Mean start | Var start | Mean end | Var end | t-value | p-value |
|---|---|---|---|---|---|---|
| Creativity | 3.81 | 0.358 | 4.31 | 0.286 | 4.978 | $p < 0.001$ |
| Personal leadership | 3.59 | 0.312 | 3.90 | 0.186 | 3.262 | $p < 0.001$ |
| Critical thinking | 3.42 | 0.24 | 4.03 | 0.358 | 4.850 | $p < 0.001$ |
| Problem solving | 3.39 | 0.361 | 3.88 | 0.342 | 5.112 | $p < 0.001$ |

In the last VAKEN Sprint held in Helsinki, 143 students completed both the initial and final surveys out of a total of 220 participants. Analysis was based on the responses of 143 students who completed both surveys. Similar procedures were followed as in previous sprint calculations. The null hypothesis was rejected for all four soft skills, indicating a statistical difference. In every instance, students demonstrated improvement in their level of soft skills throughout the week. Detailed results can be found in Table 2.

**Table 2.** Summary of results from the students' self-assessments in Helsinki

| Soft skill | Mean start | Var start | Mean end | Var end | t-value | *p*-value |
|---|---|---|---|---|---|---|
| Creativity | 3.46 | 0.494 | 3.82 | 0.428 | 4.78 | $p < 0.001$ |
| Personal leadership | 3.53 | 0.242 | 3.72 | 0.273 | 3.24 | $p < 0.001$ |
| Critical thinking | 3.31 | 0.359 | 3.63 | 0.383 | 4.45 | $p < 0.001$ |
| Problem solving | 3.27 | 0.404 | 3.61 | 0.388 | 4.79 | $p < 0.001$ |

## 5 Discussion

The self-assessment results indicate that students perceive a higher level of soft skills after participating in the VAKEN sprint, aligning with previous research on the effectiveness of this learning design [24, 29, 42].

In the second and third sprints, students consistently rated themselves higher in soft skills after the Sprint Week. Statistical analysis provides strong evidence to support the alternative hypothesis that soft skill levels are higher after the VAKEN Sprint Week. Figure 1 visually demonstrates the perceived enhancement of soft skills, with the starting value representing the mean at the beginning of the week and the end value representing the mean at the end of the Sprint Week.

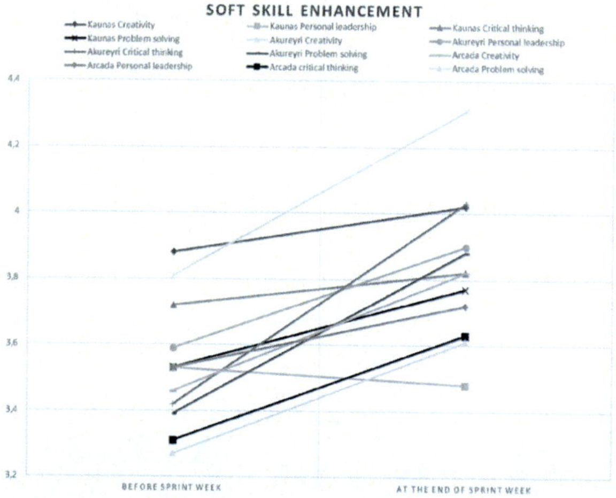


**Fig. 1.** Students perceived level of soft skill, before and after the VAKEN Sprint Week.

The first sprint showed improvement in only one soft skill: problem-solving. Previous research emphasizes the need for pedagogical or design variation to address different learning needs of skills [24]. In the first sprint, students perceived a decrease in their personal leadership skill at the end compared to the beginning. While the results were not statistically significant ($p > 0.05$), it is interesting to compare the first sprint with the latter two. In the last two sprints, there were significant improvements in students' overall level of soft skills.

In the first sprint, students perceived problem-solving as the skill that experienced the most enhancement during the week, aligning with the effectiveness of project-based learning (PBL) in soft skills development [29, 31]. The second sprint highlighted critical thinking, while creativity emerged as the primary skill enhanced in the third sprint. Across all three sprints, personal leadership was perceived as the skill with the least improvement, a finding significant in the latter two sprints. Soft skills are complex, learned over time, and influenced by personality and mindset [25, 28, 37]. This suggests variations in the extent of skill development, potentially influenced by task requirements that may demand varying levels of creativity. The results underscore the need for further analysis of the VAKEN process and its associated tools to enhance students' soft skills. Specifically, there is a call for revision and enhancement of tools targeting personal leadership skills. Existing research also supports the continued development of measurement and assessment methods for soft skills [23, 25].

## 6 Conclusions

The research question we posed in the beginning was: How is the development of soft skills perceived among students participating in the VAKEN process before and after the Sprint Week?

The study demonstrates significant differences in students' perceived levels of soft skills before and after the VAKEN Sprint Week. Participation in the VAKEN sprint weeks generally led to an improvement in students' perception of their soft skill levels. The second and third sprints consistently resulted in statistically significant increases across all four dimensions. However, the first sprint had mixed results, with only problem-solving skills showing significant improvement. This suggests that the tasks and activities assigned during each sprint influence the development of specific skills. Notably, personal leadership showed the least enhancement in all three sprints, highlighting the need for further investigation and refinement of tools for fostering this skill.

Despite the positive impact of the VAKEN sprint weeks, the study has limitations. With only three iterations of the sprint, more data collection and analysis are necessary for increased reliability. Additionally, conducting follow-up assessments beyond the immediate post-sprint period would provide insights into the long-term sustainability and transferability of acquired skills.. This research will contribute to improving the VAKEN sprint weeks in promoting soft skills development, collaboration, and innovation among students.

## References


1. Mitsea, E., Drigas, A., Mantas, P.: Soft skills & metacognition as inclusion amplifiers in the 21st century. Int. J. Online Biomed. Eng. **17**(4) (2021)
2. Kelly, O.: How might The VAKEN project most usefully define "soft skills"? Arcada Working Papers 2/2021, Helsinki (2021)
3. Schulz, B.: The importance of soft skills: education beyond academic knowledge. Nawa J. Lang. Commun. **2**(1), 146–154 (2008)
4. Penman, D.: Mindfulness for Creativity. Piatkus, London (2015)
5. Pölönen, P.: Tulevaisuuden lukujärjestys. Otava, Helsinki (2019)
6. Fernandez-Aráoz, C., Aramaki, K., Roscoe, A.: Harvard business review—from curious to competent. https://hbr.org/2018/09/curiosity#from-curious-to-competent (2018)
7. Ross, S.: A conceptual model for understanding the process of self-leadership development and action-steps to promote personal leadership development. J. Manag. Dev. **33**(4), 299–323 (2014). https://doi.org/10.1108/jmd-11-2012-0147
8. Frese, M., Fay, D.: Personal initiative. In: Wiley Encyclopedia of Management (2015). https://doi.org/10.1002/9781118785317.weom110213
9. Gillard, S.: Soft skills and technical expertise of effective project managers. In: Proceedings of the 2009 InSITE Conference (2009). https://doi.org/10.28945/3378

10. Mastrangelo, A., Eddy, E.R., Lorenzet, S.J.: The importance of personal and professional leadership. Leadership Org. Dev. J. **25**(5), 435–451 (2004). https://doi.org/10.1108/01437730410544755
11. O'Leary, R., Choi, Y., Gerard, C.M.: The skill set of the successful collaborator. Public Admin. Rev. **72**(1), 2, 70–83 (2012)
12. Wesley, S.C, Jackson, V.P., Lee, M.: The perceived importance of core soft skills between retailing and tourism management students, faculty and businesses. Empl. Relat. **39**(1) (2017)
13. French, P.A., Funke J.: Definitions, traditions, and a general framework for understanding complex problem solving. In: Complex Problem Solving: The European Perspective. Lawrence Erlbaum Associates, Inc. (1995)
14. Dörner, D., Funke, J.: Complex problem solving: what it is and what it is not. Front. Psychol. (2017). https://doi.org/10.3389/fpsyg.2017.01153
15. Funke, J.: Complex Problem Solving. (2012).https://doi.org/10.1007/978-1-4419-1428-6_685
16. Dörner, D.: Diagnostik der operativen Intelligenz (Assessment of operative intelligence). Diagnostica **32**(4), 290–308 (1986)
17. Tang, K.N.: Beyond employability: embedding soft skills in higher education. Turk. Online J. Educ. Technol. **18**(2), 1–9 (2019)
18. Kumar, R.R., James, R.: Evaluation of critical thinking in higher education in Oman. Int. J. High. Educ. **4**(3) (2015)
19. Penkauskienė, D., Railienė, A., Cruz, G.: How is critical thinking valued by the labour market? Employer perspectives from different European countries. Stud. High. Educ. **44**(5) (2019). https://doi.org/10.1080/03075079.2019.1586323
20. Halpern, D.: Thought & Knowledge: An Introduction to Critical Thinking, 5th edn. Psychology Press, New York (2014)
21. Cimatti, B.: Definition, development, assessment of soft skills and their role for the quality of organizations and enterprises. Int. J. Qual. Res. **10**(1), 97 (2016)
22. Gray, A.: The 10 skills you need to thrive in the Fourth Industrial Revolution, World Economic Forum. https://www.wefo-rum.org/agenda/2016/01/the-10-skills-you-need-to-thrive-in-the-fourth-in-dustrial-revolution/ (2016). Accessed 22 June 2021
23. Rios, A.J., Ling, A., Pugh, R., Becker, D., Bacall, A.: Identifying critical 21st-century skills for workplace. Educ. Res. **49**(2), 80–89 (2020). https://doi.org/10.3102/0013189X19890600
24. Betti, A., Biderbost, P., García, D.A.: Can active learning techniques simultaneously develop students' hard and soft skills? Evidence from an international relations class. PLoS One **17**(4), e0265408 (2022). https://doi.org/10.1371/journal.pone.0265408
25. Kenwright, B.: The Hard Truth about Soft Skills in Game Development (2022). https://doi.org/10.48550/arXiv.2205.07875
26. Hansén, S.-E., Forsman, L. (eds.): Allmändidaktik. Studentlitteratur, Lund (2017)
27. Hansson, T.: Pedagogik för högskolelärare. Gidlunds förlag, Möklinta (2016)
28. Marton, F., Hounsell, D., Entwistle, N.: HUr vi lär. Norstedts förlag, Stockholm (1986)
29. Aslan, A.: Problem-based learning in live online classes: learning achievement, problem-solving skill, communication skill, and interaction (2021). https://doi.org/10.1016/j.compedu.2021.104237
30. Biggs, J.: What the student does: teaching for enhanced learning. High. Educ. Res. Dev. **18**(1) (1999)
31. Guo, P., Saab, N., Post, L., Admiraal, W.: A review of project-based learning in higher education: Student outcomes and measures. Int. J. Educ. Res. **102** (2020). https://doi.org/10.1016/j.ijer.2020.101586
32. Anthony, S., Garner, B.: Teaching soft skills to business students: an analysis of multiple pedagogical methods. Bus. Prof. Commun. Q. **79**(3), 360–370 (2016)

33. Beard, D.F., Scwieger, D., Surendran, K.: Integrating soft skills assessment through university, college, and programmatic efforts at an AACSB accredited institution. J. Inf. Systems Educ. **19**(2), 229–240 (2008)
34. Joseph, D., Newman, D.A.: Emotional intelligence: an integrative meta-analysis and cascading model. J. Appl. Psychol. **95**(1), 54–78 (2010). https://doi.org/10.1037/a0017286;Source PubMed
35. Matteson, M., Anderson, L., Boydon, C.: Soft skills: a phrase in search of meaning. Libr. Acad. **16**, 71–88 (2016). https://doi.org/10.1353/pla.2016.0009
36. Employ Skills. (2021). Enhancing employability skills in higher education. https://www.ucl.dk/employ-skills
37. Almonte, R., McAfee, H., Snell, T., Ahmed, A.: Measuring the effectiveness of a hybrid pedagogy in a large-scale community college soft skills intervention. Commun. Coll. J. Res. Pract. 1–17 (2021)
38. Tuning Educational Structures in Europe. TUNING Handbook. https://www.unideusto.org/tuningeu/teaching-learning-a-assessment.html (2001–2003)
39. Malykhin, O., Aristova, N., Kalinina, L., Opaliuk, T.: Developing soft skills among potential employees: a theoretical review on best international practices. Postmod. Open. **12**(2), 210–232 (2021). https://doi.org/10.18662/po/12.2/304
40. Schneider, M., Preckel, F.: Variables associated with achievement in higher education: a systematic review of meta-analyses... (2017)
41. Ciappei, C., Pellegrini, M.: Ethical judgment and radical business changes: the role of entrepreneurial perspicacity. J. Bus. Ethics Springer **128**(4), 769–788 (2015)
42. Tigerstedt, C., Fabricius, S.: The skills of the future work life—recognizing and assessing soft skills in higher education. In: Iceri 2019 Proceedings, pp. 7020–7028 (2019)
43. Rachmaniar, R., Yahya, M., Lamada, M.: Evaluation of Learning through Work Practices Industry Program at University with the CIPP Model Approach (2021), License CC BY-SA 4.0. https://doi.org/10.55151/ijeedu.v3i2.55
44. Sunitha, G., Avanija, J.: Program assessment and evaluation for continuous quality improvement. J. Eng. Educ. Transform. **34**(3) (2021)
45. Wanzer, D.L.: What is evaluation? Perspectives of how evaluation differs (or not) from research. Am. J. Eval. **42**(1), 28–46 (2021)
46. Urbancová, H., Vrabcová, P., Hudáková, M., Petrů, G.J.: Effective training evaluation: the role of factors influencing the evaluation of effectiveness of employee training and development. Sustainability **13**(5), 2721 (2021). https://doi.org/10.3390/su13052721
47. Bernardino, G., Curado, C.: Training evaluation: a configurational analysis of success and failure of trainers and trainees. Eur. J. Train. Dev. **44**(4/5), 531–546 (2020)
48. Gakh, D.: Improving structure of the SLMHM on the base of the PDCA cycle. Preprints.org (2021). https://doi.org/10.20944/preprints202104.0490.v2
49. Granberg, C., Palm, T., Palmberg, B.: A case study of a formative assessment practice and the effects on students' self-regulated learning. Stud. Educ. Eval. **68** (2021)
50. McCallum, S., Milner, M.: The effectiveness of formative assessment: student views and staff reflections. Assess. Eval. High. Educ. **46**(1), 1–16 (2021)
51. Supena, I., Darmuki, A., Hariyadi, A.: The influence of 4C (constructive, critical, creativity, collaborative) learning model on students' learning outcomes. Int. J. Instr. **14**(3), 873–892 (2021)
52. Saranya, C., Guduru, R.: Building students' personality for employment readiness through soft-skills training. Randwick Int. Soc. Sci. J. **3**(1), 32–45 (2022). https://doi.org/10.47175/rissj.v3i1.367
53. Hafeez, M.: Impact of teacher's training on interest and academic achievements of students by multiple teaching methods. Pedagog. Res. **6**(3) (2021)

54. Giovannini, M.: Alternatives to development: the contribution of indigenous community enterprises in Chiapas. **28**(7), 1138–1154 (2015).https://doi.org/10.1002/jid.3141
55. Hattie, J.: Visible Learning for Teachers. Maximizing Impact on Learning. Routledge, London (2012)
56. Cinque, M.: "Lost in translation". Soft skills development in European countries. Tuning J High Educ. **3**(2), 389–427 (2015)
57. Wraae, B., Tigerstedt, C.: Self-reflectiveness among entrepreneurship students. Paper presented at 12th European Conference on Innovation and Entrepreneurship, Paris, France (2017)
58. Wraae, B., Tigerstedt, C., Kratzer, J.: Introducing student self-assessment as a new assessment tool in entrepreneurship education. J. High. Educ. Theory Pract. **18**(3) (2018). https://doi.org/10.33423/jhetp.v18i3.563
59. Yang, E., Halpin, P., Handy, D.: Using Psychometric Analysis to Improve Soft-Skill Assessments. Institute of Education Sciences (2022). https://eric.ed.gov/?id=ED624221
60. Bryman, A., Bell, E.: Business Research Methods, 6th edn, 696 p (2022), ISBN: 9780198869443
61. Saunders, M., NK., Lewis, P., Thornhill, A., Bristow, A.: Research Methods for Business Students, 8th edn. Pearson Education (2019)
62. South, L., Saffo, D., Vitek, O., Dunne, C., Borkin, M.A.: Effective use of Likert scales in visualization evaluations: a systematic review. Comput. Graph. Forum **41**, 43–55 (2022). https://doi.org/10.1111/cgf.14521

# Learning-by-Doing as a Method for Teaching the Fundamentals of Light to Physics Educators and Students Online

T. P. Nantsou(✉), E. Kapotis, and G. S. Tombras

Section of Electronic Physics and Systems, Department of Physics, National and Kapodistrian University of Athens, 15784 Athens, Greece
tinanantsou@phys.uoa.gr

**Abstract.** Experimentation in physics and engineering is undeniably vital for improving the understanding of physical laws and engaging students in the scientific process. Following the guidelines established by the American Association of Physics Teachers, the objectives of introducing students to the physics laboratory include teaching them the art of experimentation, fostering analytical and experimental skills, developing conceptual knowledge, establishing the foundations of physics knowledge, and cultivating collaborative learning abilities. This study aims to demonstrate the implementation of a light laboratory designed for K-5 and K-6 students, as well as educators. A part of the laboratory materials was presented to science teachers through a Massive Open Online Course (MOOC). The findings from both laboratories are evaluated based on the comprehension of fundamental laws of light by teachers and students. The educators who participated in the course implemented the experiments in their classrooms and introduced their students to the culture of experimental physics. This research suggests that the educational process of designing an online course on the topic of light can offer original experimental and educational ideas to science teachers, which can be readily tested in classrooms and shared within the broader Physics and Engineering Education community. Specifically, the objective of this study was to assess the feasibility of incorporating innovative learning approaches, such as hands-on experimental teaching using the learning-by-doing method, across essential curricular areas. The students who participated in the online lab gained a deep understanding of the concepts taught in light and optics and were able to successfully apply these concepts.

**Keywords:** STEM education · Online education · Remote laboratory

## 1 Introduction

### 1.1 Education in Physics and Physics Experiments

This study investigates the introduction of scientific methodology and scientific thinking through simple physics experiments in the context of light as an educational method. The interest in this method is intentionally and systematically disseminated in both school education and lifelong learning, as well as in the training of physics educators through Massive Open Online Courses (MOOCs) (Nantsou et al. 2021b).


M. E. Auer et al. (Eds.): ICL 2023, LNNS 899, pp. 53–64, 2024.
https://doi.org/10.1007/978-3-031-51979-6_6

The Learning-by-doing method (LBD) (Nantsou et al. 2020) is an educational approach that offers a comprehensive and in-depth study of scientific concepts through practical, simple experiments in physics. The physics laboratory provides an excellent environment for active learning. Moreover, physics experiments play a crucial role in enhancing conceptual understanding of physical laws and principles, as well as in discovering new physical models that explain observed natural phenomena (Coccia 2020; Nantsou and Tombras 2022a, b). The nature of science is best understood through scientific inquiry itself (National Research Council 2008; Moeed 2013). Directly designing and observing natural phenomena in a real, physical, dynamic, and variable environment during experiments expose students to the uncertainty and ambiguity often encountered in scientific exploration. In designing the lab methodology, we followed the guidelines for teaching physics through experiments established by AAPT (American Association of Physics Teachers) (AAPT 1998). In accordance with AAPT's guidelines, the objectives of an introduction to the physics laboratory include teaching students the art of experimentation, fostering analytical and experimental skills, developing conceptual knowledge, recognizing the foundations of physics knowledge, and cultivating collaborative learning skills. Our aim was to create an environment that promotes critical and analytical thinking, problem-solving abilities while giving students firsthand experience of scientific methodologies.

### 1.2 Hands-on Physics Education in Greece

Despite physics experiments, even some with simple materials, being included in the textbooks of the 5th and 6th grades of primary school and the 1st grade of Greek gymnasium, actual demonstrative experimentation is rarely utilized by teachers in the classroom. Primary school teachers and physics teachers in gymnasium are not trained in the recommended experimental teaching methods outlined in the textbooks, and instead opt for traditional teacher-centered approaches, predominantly through lectures (Nantsou et al. 2021a, b). It is worth noting that the physics lesson in primary school is taught by teachers, many of whom lack experience in physics labs due to a lack of training (Nantsou 2023). In the Greek gymnasium, the subject is often taught by scientists from other disciplines (such as geologists, chemists, mathematicians) who may not possess laboratory knowledge or the necessary experience to effectively teach physics using the recommended experimental approach. Consequently, the teaching of physics in the Greek education system primarily relies on traditional, teacher-centered methods, which are largely influenced by the socio-economic reality of the country.

### 1.3 Hands-on Online Education

Individuals from Greece have shown a high participation rate in online courses worldwide, highlighting the need for the development of online courses in the Greek language. While there is a wide range of online educational materials available, including resources from the Ministry of Education and science educators' personal websites, participants often lack opportunities for interaction with both their peers and the instructor. The online course presented in this research (Mathesis - FORTHE) (Nantsou and Tombras 2022b) focuses on teaching light through practical, real physics experiments using everyday materials. The proposed teaching method encourages a comprehensive exploration of the fundamental principles and laws outlined in the school curriculum. The primary objective of this course is to enhance the skills of science educators at all education levels, enabling them to teach physics concepts in an engaging and enjoyable manner, ultimately facilitating a better understanding of the natural world for both teachers and their students.

## 2 The Online Course in Light

### 2.1 Description of the Online Experiments on Light and Optics

The online course comprised of 20 hands-on experiments focusing on light and optics, utilizing readily available materials. The thematic units covered the fundamental laws of light. The selected experiments were designed to be conducted using simple materials and tools, commonly found in homes, schools, or offices. Materials such as mobile phones, light sources, milk, water, pencils, window glass, mirrors, candles, laser beams, and more were utilized. The objective was to inspire students to explore physics within the familiar setting of their own homes, whether through cooking or engaging in activities in their kitchens. The online course aimed to teach basic concepts including the nature of light, colors, spectrum of light, reflection of light, refraction (Fresnel's law), diffraction, scattering, Rayleigh scattering, absorption, geometrical optics, lenses, mirrors, and the practical application of lenses (such as in camera obscura, telescopes, microscopes, periscopes, kaleidoscopes). Furthermore, everyday phenomena related to lenses, such as the function of the eye, were covered (Fig. 1).

**Fig. 1.** Hands-on experiments on light and optics.

The videos in the online lesson had a short duration and incorporated theoretical elements to serve as viewing models. Students who actively participated in the online lesson watched all the videos showcasing light experiments and successfully completed the tests for each specific thematic unit. These tests required participants to engage in experimentation in order to answer the final test for completing the unit, aiming for a deeper understanding of the experiment they had just observed.

The thematic units presented in the online lesson are: a. Reflection (N = 8) (reflection on a surface, total internal reflection, optical fibers, kaleidoscope, periscope) b. Lens-refraction (N = 6) (refraction in water, optical illusions, lenses, camera obscura, function of the eye) c. Light Spectrum (N = 6) (rainbow using a cd/water, spectrum, colors of the sky (blue, red), moving image, Newton's disc).

## 2.2 Description of Experiments on Light and Optics for K-5 and K-6 Students

The lessons on Light and Optics were conducted both in person and remotely, involving K-5 and K-6 students. The experimental themes of the online class were: a. Nature of light, linear propagation of light, reflection, transparent objects, optical illusions b. Refraction in water, optical illusions, spectrum, colors, lenses, vision, camera obscura, function of the eye. In the remote lab, students watched the experiments in videos and received parallel explanations from the teacher. They then conducted the experiments themselves at home, with their cameras open, while the physics teacher observed the learning-by-doing experimental process (some photos from the online lab are shown in Fig. 2). The students conducted the experiment simultaneously with the teacher and sent the photos at the end of the lesson along with their homework. After completing the online class, a discussion was held to analyze the results and draw conclusions from the experimental process. The total duration of the online lab was 45 min. As homework, students were assigned a worksheet to complete and submit online to the teacher. The lessons primarily focused on hands-on experiments, allowing students to explore additional materials and perform extra experiments beyond the suggested procedures. For instance, during the study of reflection, they had the opportunity to try reflecting light on various surfaces inside their homes (Fig. 2).

**Fig. 2.** Selective experiments of the online lab.

The main objective of the remote lab was to help students discover that physics is omnipresent, even within their own rooms. Through activities like cooking, constructing, testing kitchen utensils, and exploring different sources of light available at home, they realized that the laws of physics apply universally. Some of the experiments on light were also conducted in-person in the school's physics laboratory. The assessment tests were also administered in-person at the school. Although there were assignments and tests that were sent remotely, they were not used as data in this research because their reliability could not be ensured. It was difficult to verify whether students received assistance from adults or other resources at home to answer the tests and complete the homework. These assignments were examined during research and were found to be of a very high level that did not correspond to the teaching level of the course; therefore, they were deemed unusable for research purposes.

## 3 Research Questions

In order to rigorously assess and evaluate our proposal, we meticulously devised a methodological approach and formulated research questions within the scope of our study.

- Is the utilization of an online hands-on laboratory a suitable approach for studying the fundamental principles of light?
- How do the observations obtained from hands-on physics experiments relate to the lived experiences of students?
- Is the utilization of a Learning-by-Doing method suitable for teaching the concept of light through simple hands-on experiments in a school laboratory?

In the online class, students participated in the lab with an aim to learn the basic physics of light. This assessment is based on the student's comprehension of fundamental physics laws as well as their performance in the hands-on lab.

## 4 Methodology

According to the international scholarly literature, it has been observed that the traditional methods of teaching, such as lectures and textbooks, often make it challenging for students to grasp the basic concepts of light (Nantsou 2023). Therefore, it becomes crucial to foster children's interest in physics and natural sciences from an early age by demonstrating the excitement of science and its relevance to their daily lives (Elliniadou and Sofianopoulou 2023). The Learning-by-Doing (LBD) method offers an alternative approach that is based on experiential learning. It promotes active and cooperative learning within the context of relevant and interesting goals for students (Nantsou et al. 2020). To support students and educators in developing their experimental skills, a light laboratory was established, utilizing affordable materials. The primary objective of the light laboratory was to teach students the principles of light and optics through engaging simple activities and physics experiments. The focus was less on theoretical references and more on practical experiments using simple materials. The laboratory comprised 20 experiments, demonstrations, and applications related to light. This work will provide a detailed presentation of the light experiments, simple demonstrations, and constructions conducted in the laboratory.

### 4.1 Participants

The research presented in this study followed a single-case study design, incorporating both quantitative and qualitative research methods. The participants involved in the research included both students and science educators. The students primarily engaged in remote experimentation, but some experiments were also conducted in the school laboratory. The format of the laboratory activities was established considering the strict health and safety regulations implemented during the pandemic, which mandated a significant period of distance education in Greece.

**Quantitative data**. The quantitative data provided here are based on anonymous pre- and post-tests performed by 93 K-5 and K-6 students, as well as an anonymous Students Experience Survey completed by 253 educators.

**Qualitative data**. The qualitative data provided in this work come from 33 personal semi-structured and 13 detailed interviews of science educators.

### 4.2 Research Method

**Research Tools**. Data collection for this study employed a range of techniques to gather comprehensive information and insights. These techniques included the use of worksheets, tests, anonymous evaluations administered before and after participation in both in-person and remote labs, as well as the examination of the homework assigned to students upon completion of the experiments, both in-person and remotely.

**Analysis**. The analysis of questionnaires and tests aimed primarily to study the level of understanding of the natural laws and principles of light and optics, as well as their applications in everyday life. The statistical analysis was conducted using the IBM SPSS Statistics 26.0 software program.

## 5 Results

### 5.1 Students' Results

The initial study was centered on K-5 students, constituting the first experimental group, whereas the subsequent study focused on K-6 students, comprising the second experimental group. All participating students successfully completed both the pre-test and the post-test assessments. The results were graded on a scale from 0 to 10. During experimentation, we recorded the level of understanding concerning the principles and phenomena at hand. In the table below, one can observe the difficulties encountered by students in specific light experiments. Initially, students had to explain the experiment they conducted and interpret the results of their LBD experimentation. The majority of students understood basic phenomena such as reflection, refraction, lenses (Table 1).

**Table 1.** Challenges encountered in the online laboratory

| Physics topic | Correct % | Incorrect % |
|---|---|---|
| Refraction | 80 | 20 |
| Transparent objects | 70 | 30 |
| Optical illusions | 50 | 50 |
| Reflection | 90 | 10 |
| Lenses | 90 | 10 |
| Camera obscura | 30 | 70 |
| Function of eye | 80 | 20 |

All participating students successfully completed the laboratory sessions and subsequent tests. Moreover, a paired-samples t-test was employed to investigate potential differences in students' scores before and after the tests. The results obtained from the pre-test (M = 7.2978, SD = 1.70613, M = 7.3787, SD = 1.46450) and the post-test (M = 9.2522, SD = 1.09661, M = 9.1277, SD = 0.89314) demonstrated a significant improvement in students' understanding of physics through the light online laboratory: $t(45) = -13.415$, $p < 0.001$, $t(46) = -11.750$, $p < 0.001$ (Table 2).

**Table 2.** Results of the online school lab

| Test | N | Mean | Stud. Dev |
|---|---|---|---|
| Pre-test K-5 | 46 | 7.2978 | 1.70613 |
| Post-test K-5 | 46 | 9.2522 | 1.09661 |
| Pre-test K-6 | 47 | 7.3787 | 1.46450 |
| Post-test K-6 | 47 | 9.1277 | 0.89314 |

The light lab was favorably received by the students, as indicated in Table 3, owing to the following reasons:

- It facilitated their introduction to the fundamental principles of light through engaging hands-on activities and physics experiments.
- It acquainted them with rudimentary constructions and hands-on engineering toys.

The outcomes of the pre- and post-tests demonstrate that the lab significantly enhanced students' comprehension of the fundamental principles and concepts of light and optics (Table 2).

**Teachers' Results**

Teachers, particularly those in primary schools, often face challenges when it comes to teaching physics and STEM subjects that involve laboratory skills (Table 5).

**Table 3.** Evaluation of the school lab

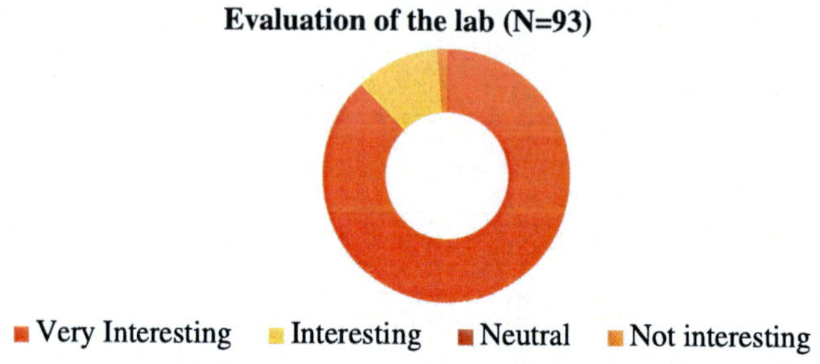


**Table 4.** The education level of the educators who successfully completed the course.

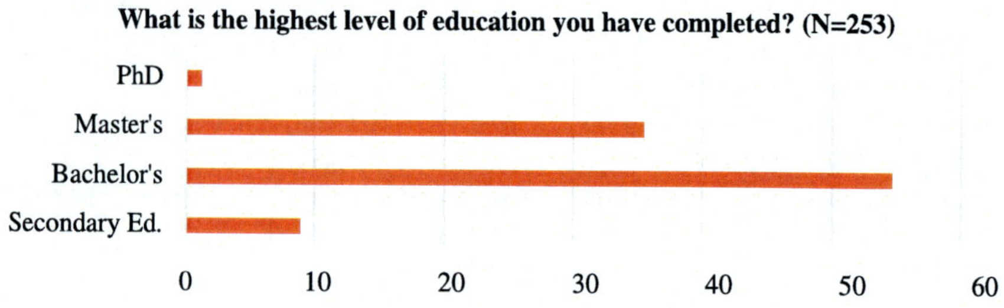


**Table 5.** Specialization of the educators.

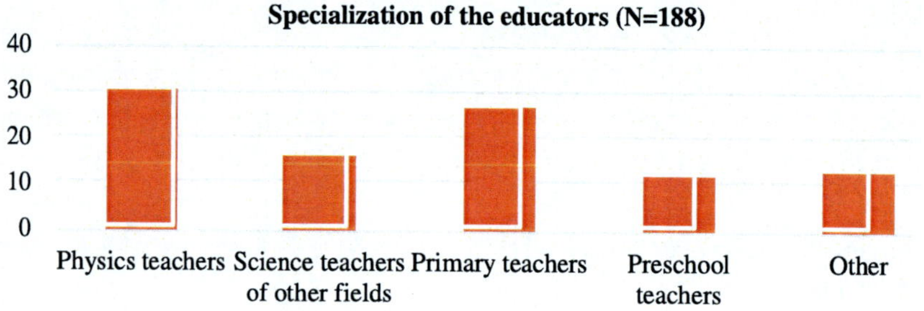


The level of the educators who participated in the online training was found to be very high, as documented in Tables 4 and 5.

47% of the educators conducted all the experiments (Table 6). According to the research, simple experiments helped the participants to understand the laws of physics better (Table 7). They also guided them to share the view that hands-on experiments can be taught at all ages (Table 8). The educators conducted the experiments that were interesting (28%) (Table 6). The suggested physics experiments with everyday materials have a qualitative, rather than quantitative, nature. Their goal is to observe natural phenomena and understand them (Table 7), without necessarily leading to the proof of physical laws through experimental measurements and calculations. The suggested experiments can be performed in any classroom (Table 8), without requiring a laboratory or special facilities, and can also be carried out safely at home by the students themselves.

**Table 6.** I only conducted the experiments that interested me.

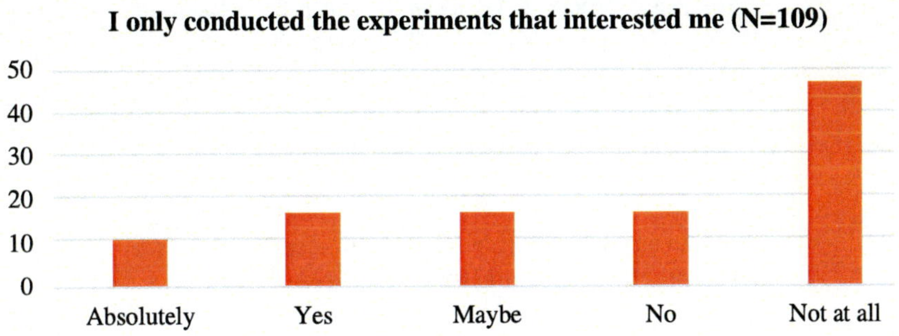


**Table 7.** Did the physics experiments with simple materials help you better understand the laws of physics?

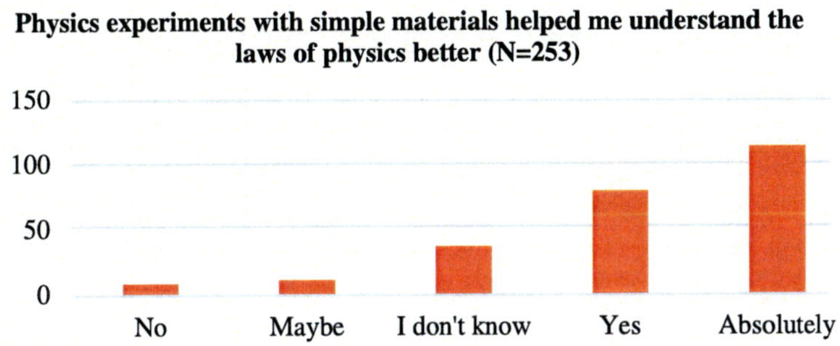


The experiments on physics with simple materials that were taught in the online course were investigated to determine how many of them were implemented in the classroom by the participating science educators.

Out of 102 respondents, 55.7% stated that they tried the light experiments in their schools precisely because the proposed experiments with simple materials can be taught in all ages (Table 8). These findings are of research interest, as it was initially unknown whether the proposed physics experiments would be attempted in classrooms with students, considering that experimentation is not common in Greek schools, as indicated by the literature review. The high acceptance of the proposed teaching method for physics (Tables 8 and 9), even in challenging topics such as light, emphasizes the need for educators to receive training in experiments so that they can acquire experimental skills and methodology to implement physics experiments with simple materials in the classroom.

**Table 8.** Can physics experiments with simple materials be taught at all ages?

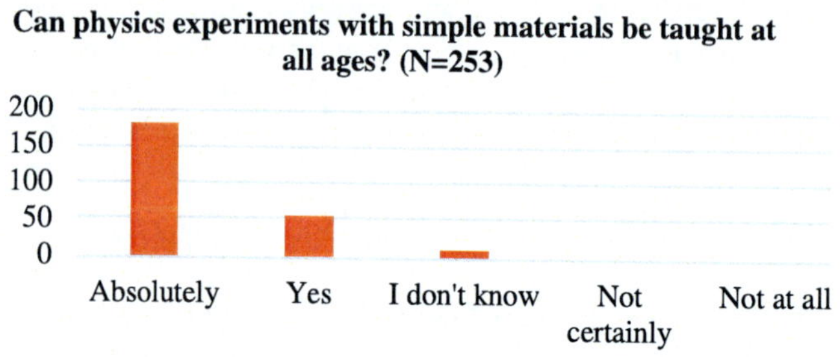


**Table 9.** The course is suitable for your professional development as an educator in the field of physics experiments

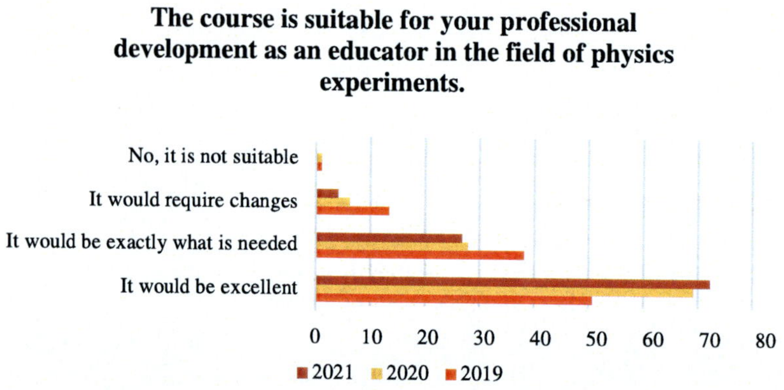


During their interviews, the educators noted that the lab greatly assisted them in their teaching at school and it is suitable for their professional development as educators in the field of physics experiments (Table 9). The interviews and open-ended questions were answered anonymously to ensure absolute freedom in evaluating the educational activities and the outcomes of the lessons. The primary aim of this particular research methodology was to conduct a comprehensive analysis of the quantitative data presented earlier and employ triangulation techniques to validate and strengthen the research findings.

## 6 Discussion and Conclusion

The field of light is inherently intricate, demanding extensive years of study to achieve a comprehensive understanding (McDermott et al. 1990). The proposed Learning-by-Doing teaching method for the fundamentals of light is suitable for students of average learning ability as well as those with learning difficulties and other particular needs (Nantsou et al. 2020). This approach connects the science curriculum with the students' daily lives and personal interests, helping them bridge the gap between theoretical concepts and tangible, everyday experimental data. In traditional teaching methods, students often fail to develop the fundamental skills necessary for learning, remaining passive recipients of knowledge rather than active participants in the educational process. The online laboratory maintained a high level of student engagement and provided results similar to the in-person laboratory. This was possible because students conducted experiments using the LBD method at home under the supervision of their instructor, as detailed in the study. The literature review conducted in this study demonstrates that students acquire and retain the most information and scientific knowledge when they actively participate as researchers in research-oriented projects (Nantsou and Tombras 2022a, 2022b; Elliniadou and Sofianopoulou 2023), such as the proposed light laboratory. International research in the fields of Physics Education Research (PER) and Engineering Education supports the notion that passive attendance in traditional lectures and scientific discussions within the school environment does not yield substantial learning outcomes (Nantsou 2023).

According to this research, the process of designing an online course on the subject of light can offer innovative experimental and educational ideas to science teachers, which can be easily implemented and further explored in classrooms and within the PER and Engineering research community. The specialized approach to distance teaching, both synchronous and asynchronous, is expected to yield enhanced cognitive and research outcomes. It will lead to the emergence of alternative ways to present experiments and educational materials, as there will be variations in teacher-student interaction models. It is important to note that the laboratory proposed in this study should be tested by independent researchers in different learning environments and at various times to review, adapt, and generalize the research findings.

In the future, educational institutions, especially in the post-COVID-19 era, should strive to develop online courses, as they offer significant educational benefits for students and contribute to the professional development of science educators. By creating experimental online courses, educational institutions can meet the modern learning needs of students and contribute to the dissemination of knowledge within the society, thereby enhancing the scientific capital of the country.

**Acknowledgment.** National and Kapodistrian University of Athens (Special Account for Research Grands) is kindly thanked for funding this work. The authors express their gratitude for the support provided by Mathesis (FORTHE) and Perimeter Institute for Theoretical Physics. Special thanks are also due to Greg Dick (Perimeter Institute), E. Nistazakis (NKUA) and S. Trachanas (Mathesis-FORTHE) for their support in this work. The realization of this online course was solely enabled by the invaluable financial support from the Bodosakis Foundation and the Stavros Niarchos Foundation.

## References

American Association of Physics Teachers: Goals of the introductory Physics laboratory. Am. J. Phys. **66**(6), 483–485 (1998)

Coccia, M.: The evolution of scientific disciplines in applied sciences: dynamics and empirical properties of experimental physics. Scientometrics **124**(1), 451–487 (2020)

McDermott Lillian, C., Gardner, M., Greeno, J., Reif, F., Schoenfeld, A.H.: A view from physics. In: Toward a Scientific Practice of Science Education, pp. 3–30 (1990)

Elliniadou, E., Sofianopoulou, C.: A two-year STEM initiative on students' attitudes towards science. In: 2023 IEEE Global Engineering Education Conference (EDUCON), pp. 1–8 (2023). https://doi.org/10.1109/EDUCON54358.2023.10125130

Moeed, A.: Science investigation that best supports student learning: teachers' understanding of science investigation. Int. J. Environ. Sci. Educ. **8**(4), 537–559 (2013). https://doi.org/10.12973/ijese.2013.218a

Nantsou, T., Frache, G., Kapotis, E.C., Nistazakis, H.E., Tombras, G.S.: Learning-by-doing as an educational method of conducting experiments in electronic physics. In: 2020 IEEE Global Engineering Education Conference (EDUCON), Porto, Portugal, pp. 236–241 (2020). https://doi.org/10.1109/EDUCON45650.2020.9125324

Nantsou, T.P., Kapotis, E.C., Tombras, G.S.: A lab of hands-on STEM experiments for primary teachers at CERN. In: 2021 IEEE Global Engineering Education Conference (EDUCON), pp. 582–590 (2021). https://doi.org/10.1109/EDUCON46332.2021.9453915

Nantsou, T.P., Kapotis, E.C., Tombras, G.S.: Greek MOOC of experiments with simple materials for students generates significant findings for teachers and physics education. In: 2021 IEEE Global Engineering Education Conference (EDUCON), pp. 631–636 (2021). https://doi.org/10.1109/EDUCON46332.2021.9453876

Nantsou T.P., Tombras G.S.: Hands-on experiments in electricity for physics teachers and students. In: 2022 IEEE Global Engineering Education Conference (EDUCON), pp. 242–248 (2022). https://doi.org/10.1109/EDUCON52537.2022.9766756

Nantsou, T.P., Tombras, G.S.: STEM lab on climate change with simple hands-on experiments. In: 2022 IEEE Global Engineering Education Conference (EDUCON), 2022, pp. 249–256. https://doi.org/10.1109/EDUCON52537.2022.9766619 (2022)

Nantsou T.P.: Experimentation and educational methods in modern physics. In: NKUA Library of the School of Science Department of Physics Doctoral Dissertation. https://pergamos.lib.uoa.gr/uoa/dl/frontend/en/browse/3330970NKUA (2023)

National Research Council: Research on future skill demands: A workshop summary. In: Margaret, H. (ed.) The National Academies Press, Washington, DC (2008)

# Transferring Analogue Teaching to Digital Delivery: Blended Learning Across an International Network for Socio-cultural Sustainability

Neelakshi Chandrasena Premawardhena[1(✉)], Korakoch Attaviriyanupap[2], Agron Kurtishi[3], Vera Ebot Boulleys[4], and Arnaldo Baltazar Diez[5]

[1] University of Kelaniya, Kelaniya, Sri Lanka
neelakshi@kln.ac.lk
[2] Silpakorn University, Nakhon Pathom, Thailand
attaviriyanupap_K@su.ac.th
[3] Mother Teresa University, Skopje, North Macedonia
agron.kurtishi@unt.edu.mk
[4] Grace Bilingual Educational Complex, Douala, Cameroon
[5] Sprach Institut-Cebu, Cebu City, Philippines
arnaldobaltazardiez@gmail.com

**Abstract.** Sustainability of cultural heritage is pivotal for any society to transfer the wisdom, knowledge, customs, and traditions to the next generation. In most countries oral tradition including folk tales, folk songs and ballads appear to fade away due to the advancement of technology and shift of interest, thus creating a void in the values and one's own heritage to pass down to the future generations. Folklore of a community preserves the oral traditions of yesteryears giving insights into linguistic and cultural aspects. This paper presents the results of a hybrid mode adopted virtually and onsite on a typological study of folktales of more than 20 countries across the world on analogue and digital teaching approaches. The project commenced virtually and concluded in hybrid mode after a period of intensive research from May to September 2022 aimed to focus on socio-cultural sustainability, tracing back to the oral traditions that made a society adhere to different norms and traditions. The significance of folk tales as cultural heritage, their common and unique features across cultures and how awareness could be created for the next generations through a blended approach were highlighted during outcomes of the study.

**Keywords:** Blended learning · Cultural heritage · Cultural sustainability · Folklore · Virtual learning

## 1 Introduction

This paper presents the results of an international project conducted across several countries on analogue and digital teaching under the auspices of the Alumni Academy of University of Siegen, Germany in 2022 titled "Blended Learning across an International


M. E. Auer et al. (Eds.): ICL 2023, LNNS 899, pp. 65–76, 2024.
https://doi.org/10.1007/978-3-031-51979-6_7

Network". The Alumni Office of the University of Siegen conducts frequent academies for alumni of interdisciplinary nature from across the globe with the support of the German Academic Exchange Service (DAAD). During the Alumni Academy related to the present study the team of alumni was divided into four groups according to the proposals submitted for the project which included Engineering Sciences, Language and Communication, Environmental Education and Economy and Society to develop analogue and digital teaching modules. The Alumni Academy focused on three pillars of sustainability - ecology, economy, and social issues which are interconnected. The ecological system and the economy rely on the society and in turn, the society survives on the eco-system and the environment they live in. Thus, the team members of group 2 on Language and Communication with alumni representing Cameroon, Ghana, North Macedonia, Philippines, Sri Lanka, and Thailand designed a teaching project comprising analogue and digital modules aiming to focus on socio-cultural sustainability through discussion and analysis of folklore of different countries which represent oral traditions, unique linguistic and cultural features. While the preparatory work was to be done in the individual countries onsite as well as online with the students the final teaching session was to be delivered onsite at the University of Siegen where all the participating alumni will be present. The significance of cross-cultural understanding and intercultural communication as well as cultural sustainability is recognised as essential to promote global understanding. This has also been highlighted in the Sustainable Development Goals (SDGs) which aim that by 2030, "all learners acquire the knowledge and skills needed to promote sustainable development, including, among others, through education for sustainable development and sustainable lifestyles, human rights, gender equality, promotion of a culture of peace and non-violence, global citizenship and appreciation of cultural diversity and of culture's contribution to sustainable development" [1]. Numerous comparative or typological studies on folktales have been conducted previously [2–5]. Literature was also reviewed in respect of analogue and digital delivery with regard to folklore including digital story telling [6–8]. Furthermore, the merits of both delivery modes were assessed prior to embarking on this project [9]. Nevertheless, a study of this nature involving analogue and digital teaching as well as a typological study of folktales across continents has not been conducted so far to the knowledge of the authors.

The selected theme not only made the team leaders and students from respective countries analyse, relate their folklore, and identify what elements are present in their folktales which are unique, but also find the similarities with other cultures, understand their oral traditions and the significance. Furthermore, it provided a platform to find out what moral lessons folktales of each society convey to make the society more respectful of each other as well as the environment. Thus, the team members and students from different countries connected digitally on a common theme. Due to the limited time frame, only the folktales were the focus of the teaching project.

## 2 Purpose

In present day context it is expected that the young generation turns out to be Global Citizens who are a part of the entire world and not confined to one country. Nevertheless, it is to be understood that each of these global citizens has own roots as the core of his/her existence and outlook towards life. Oral traditions including folklore reflect one's own culture, customs, traditions, norms, myths, beliefs, and superstitions that shape our lives from the childhood. Today less importance is given to these cultural and linguistic treasures due the shift of interest towards more technologically advanced modes of entertainment including computer or online games. The shift from onsite learning to online learning during the Covid-19 pandemic also made a severe negative impact on children and young adults getting addicted to online games, for instance, which made them live in an imaginary world full of characters appearing in the games or showing aggression, anger, aloofness, isolation as well as signs of depression [10]. This is a major issue that the entire world had to deal with following the digital transformation that occurred due to the pandemic. Hence, it was decided that the right time has come to intervene and create awareness in young adults that there are many valuable gems which have been handed down from one generation to the other that needs to be sustained and preserved for the future. Thus, the selected theme of folklore clearly bonded with social issues and socio-cultural sustainability which aptly complemented the aims and objectives of the Module 2 of Alumni Academy of University of Siegen. The project titled "Folklore: The Global Aspects" focused on how a common theme can be analysed and discussed across borders through a blended learning approach and how participants from different cultural backgrounds could come together on a common digital platform to share knowledge and learn from each other. The study also focused on conducting a typological review of folktales to identify unique and similar features since the rest of the members of the alumni academy from more than fifteen additional countries also took part in the final presentation of the project.

## 3 Approach

The project commenced in May 2022 with virtual meetings on how to develop analogue and digital teaching modules for students in the respective countries of the team members, who are attached to universities or language institutions. The analogue teaching modules were introduced to the students of these institutions where both English and German were accepted as languages of communication.

**Table 1.** Worksheet for team leaders and students

| Country | Themes | Charac-ters | Loca-tion/ Setting | Linguis-tic fea-tures | Messag-es/ Moral lessons conveyed | Unique features | Further observa-tions/ remarks |
|---|---|---|---|---|---|---|---|
| | | | | | | | |

Firstly, the team developed a worksheet to discuss different aspects of folktales in their respective countries with the students as illustrated in Table 1. The details required included the themes, settings, characters, message, or moral values conveyed and linguistic features including a discourse analysis.

During the first teaching session, the students were expected to discuss the features of folktales that were required to fill the table in the given worksheet. Thereafter, they were divided into groups to relate folktales they are familiar with. Subsequently the students were requested to act out one selected folktale in each group. Thus, the students were confronted with folktales that they may not have been familiar with during their childhood, which enhanced the knowledge of one's own heritage. It is noteworthy that the team leaders were given the choice of conducting these sessions in English or German in each country since at least four of the members were teaching German at their respective home institutions.

The third stage comprised several sessions where students and team leaders prepared their contribution to the final presentation either as PowerPoint presentations, narration, or video clips of their own dramatized versions of folktales. During each session feedback was obtained from the participants to learn about their experiences and the knowledge gained. Not all team members could get the students involved in all the preparatory stages due to numerous reasons i.e., time constraints, vacation at the universities, connectivity issues if the sessions were scheduled online. The team members met virtually once again to share the results of the work with the students and to plan the final onsite teaching module which was to be completed in September 2022 in Siegen, Germany. The implementation of the presentation of the planned teaching/learning project was limited to four hours for each group. During this time, the alumni as well as a group of students were available onsite to complete the planned activity. In the case of the project selected by the authors it was an interactive teaching session on "Folklore: the global aspects". Since the students and staff of different countries of the team members already took part in the preparatory stage of analysing the folktales and filling out the worksheet given to each member, it was decided to conduct the final four-hour session on teaching and learning in a hybrid mode. While three of the team members representing Philippines, Sri Lanka, and Thailand were present onsite, the other three members joined the session virtually. Furthermore, academic staff and students from the University of Kelaniya, Sri Lanka, representing several disciplines and the students of German Studies from Silpakorn University, Thailand joined the session online. Thus, the analogue teaching session conducted onsite in Siegen could be followed online by all who joined from many different parts of the world.

The team members presented the folktales from home countries following an introduction to the project theme. This included digital story telling of *How the village headman went to heaven* from Sri Lanka, *Legend of the Pineapple* from Philippines, *How Andare the court jestor ate sugar* daramatised online by students from Sri Lanka, and a video recording of the *Story of the Golden Goby* presented by students from Thailand. *The Legend of Rozafa* from North Macedonia was presented as video created for the occasion accompanied by narration and dialogues while *Arah*, a folktale from the Manyu Division of Cameroon representing the Banyang and Ejagham communities was presented as a narration. The folktales were selected by the participants according to their wide popularity in their respective countries. While the Thai presentation was in German, the digital story of North Macedonia as well as the other folktales were presented in English. During the final presentation the teaching modules were further tested with the onsite participants of alumni representing more than 15 countries. The participants were divided into three groups and were requested to accomplish the same tasks as the students from the countries of the team members. Each participant filled the table with information about folktales of the home country, shared the gathered details with each other, related a folktale from the respective countries to the other group members and selected one folktale or more to act out. Thus, there were several onsite dramatisations including Iraqi and Kenyan folktales by alumni members and a narration of *Puss in Boots* from Germany.

The final feedback from onsite and online participants was pivotal to assess the success of the project. The data available for the analysis of this study were of qualitative nature which include the worksheets, recorded feedback obtained during the initial stage of discussing and analysing each country's folktales to fill out the worksheet, recordings of preparatory meetings and the final session delivered in hybrid mode which includes the feedback given by onsite and virtual participants on what they gained from the project.

Furthermore, the evaluation of the project conducted during the final days of the Alumni Academy by the experts invited from different universities in Germany as well as the rest of the three groups of Alumni Academy on Engineering Sciences, Environmental Education and Economy and Society, were also taken into consideration when analysing the final outcomes and merits of this project. Two days were allocated for the evaluation of the projects presented by each group of the Alumni Academy following their presentations of the analogue and digital teaching according to the allocated themes. The invited experts were academics from German universities specialising in analogue and digital teaching. As mentioned above, alumni from the rest of the three groups worked in their respective groups on the evaluation of the project based on a SWAT analysis using Padlet. The expert comments were followed by the presentations of the SWAT analysis done by the three groups of alumni. The Folklore Project received numerous positive comments on its merits of cultural sustainability, intercultural communication and how analogue and digital teaching contributed to accomplishing the assigned task.

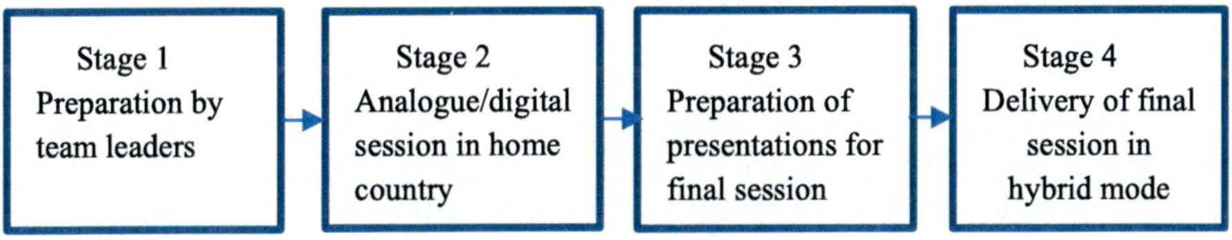


**Fig. 1.** Process of the project

Figure 1 illustrates the process that was adopted from May to September 2022 to make the project a reality by transforming analogue teaching to digital delivery. While stage 1 was completed virtually, stage 2 was conducted onsite or virtually depending on the ode of delivery adapted by the respective institutions of the participating countries. Stage 3 was also completed by individual groups in their home countries onsite or online whereas all the discussions and planning of the final session of four hours were conducted by the team leaders online. Zoom video conferencing tool was used for all the discussions and meetings since the majority of the participants were more familiar with this tool which was also used for the conducting of Alumni Academy Module 2 by the University of Siegen. The final session was conducted onsite as well as online via Zoom with over 75 participants joining virtually including staff and students from University of Kelaniya, Sri Lanka representing the Faculty of Humanities and students of German from Silpakorn University, Thailand. Furthermore, there were more than 35 participants onsite which comprised undergraduates of University of Siegen and alumni from over 15 countries i.e., Albania, Canada, Ghana, India, Iraq, Kenya, Namibia, Pakistan, South Africa, Vietnam, as well as participants from Germany. It is noteworthy that a fifth stage containing the evaluation of the projects that followed all four presentations of the projects by alumni which is not illustrated in Fig. 1.

The Figs. 2, 3, and 4 below bear testimony to the hard work that involved in the preparation and delivery of the final session in September 2022 as an onsite and virtual session close to four hours.

**Fig. 2.** Video on a legend from North Macedonia created for the project

The video of the dramatisation of *The Legend of Rozafa* as depicted in Fig. 2 which also included a narration English was specially created for the final session of the project by the team leader from North Macedonia. This dramatisation and narration of the legendary folktale about a woman buried in the foundation of the castle of Rozafa involved young school children from Skopje, North Macedonia, thus giving the younger generation too an opportunity to learn, understand and appreciate their folklore.

Figure 3 illustrates a dramatisation of the Thai folktale *The Golden Goby* carrying dialogues in German presented by the students of the Department of German, Silpakorn University, Thailand. The dramatisation has been previously performed and recorded for a screening event of the Silpakorn University. The video recording of the dramatisation was presented during the final session in Siegen after an introduction to Thai folktales by the team leader from Thailand.

**Fig. 3.** Dramatisation by students from Thailand

**Fig. 4.** Dramatisation of an Iraqi folktale by the alumni participants

Figure 4 presents the dramatisation of an Iraqi folktale selected during the group activity conducted onsite for the final session by the participants of the Alumni Academy hailing from diverse disciplines and different countries. The teams of alumni present onsite acted out three folktales. Additionally, an English narration of the popular German folktale of *Puss in Boots* was also presented by a participant from University of Siegen.

The active participation of the alumni in the group activity during the final session clearly illustrated their enthusiasm and the motivation during the dramatisation of the selected folktales. The diversity of disciplines, country of origin, language or age did not matter to take part in the assigned group activities and share their knowledge on folklore with each other.

## 4 Actual Outcomes

The project with a blended approach connected young adults with their own cultural heritage as well as with participants from different cultures across the world. The participation of school children in the dramatisation of *The Legend of Rozafa* from North Macedonia and the contribution of undergraduates from Sri Lanka and Thailand were highlights of the project. The involvement of students in the Alumni Academy during analogue and digital teaching sessions is significant because the project was not confined to only the alumni but included contributions from school children and undergraduates of different countries.

The success of the project conducted during the previous module of Alumni Academy of the University of Siegen in 2021 which digitally connected four countries and three continents on a virtual cultural exchange project involving university students from China, Egypt, North Macedonia and Sri Lanka [11] paved way for implementing the present project which is of a much larger scale connecting over 20 countries across the globe. Albeit no quantitative analysis was conducted on the outcomes of the project, the feedback obtained during the sessions provided an insight into the knowledge and exposure they gained and how rewarding an international project of this nature could be. Many participants including the students and the alumni mentioned that they went back to their childhood by revisiting the memories of folktales being related by grandparents or parents and other close relatives. According to their statements working together in their home setting first and later virtually on an international platform presenting their cultural treasures was a unique and rewarding experience. The alumni who were from multi-disciplinary background mentioned that the hybrid session where they worked in groups and related about their folktales evoked positive feelings about their cultural heritage and were thankful for the opportunity given to share their own culture with representatives from over 20 countries. Moreover, there was an opportunity for reflection, to learn numerous aspects of their culture and oral traditions and to identify the common features with the rest of the world. During the presentations of each group, it was evident that many common features were found in the folktales of each country. This gave the participants a sense of belonging and togetherness while identifying their unique features at the same time.

The project that was planned and executed from May to September 2022 with an analogue and digital teaching/learning approach was concluded in hybrid mode with many gains for the participants irrespective of their diverse languages, cultures, disciplines, and age. Several generations were involved in this project aiming at sustaining cultural heritage through analogue and digital efforts including school children, undergraduates, academic staff, and alumni of the University of Siegen representing four continents, thus connecting the participants to share the wisdom of yesteryear with each other in a very effective manner across generations. Many similarities and unique features of different cultures were shared among the participants through this project.

Table 2 illustrates the common features of folktales found in the six countries where the team leaders of the project hailed from. The themes, characters, location or setting,

**Table 2.** Common features of folktales in countries of the team leaders

| Country | Themes | Characters | Location/setting | Linguistic features | Messages/moral lessons conveyed | Unique features | Further observations/remarks |
|---|---|---|---|---|---|---|---|
| Cameroon<br>Ghana<br>North Macedonia<br>Philippines<br>Sri Lanka<br>Thailand | Values, family, ethics. Parenting, love, envy, fantasy, imagination | Village folk, rulers, kings, aristocrats, wise people, animals, super-natural | Village, forest, palace | Narrative, dramatic, spoken language, communication between animals and humans | Gratitude, benefaction, obedience, honesty, humility, respect for others, politeness, heroism, patriotism, respect for nature, environment, greed, social evils | The super-natural, karma, re-birth, transformation from humans to plants, rocks/stones | Lessons for the young and the old |

the style of presentation, the moral lessons conveyed were found to be of similar nature. The unique features found in some countries are listed separately including the belief of rebirth in Thailand and Sri Lanka, the transformation of humans to plants/trees as in the Philippines and humans turning into stone as in the case of North Macedonia.

A comparison with the rest of the participating countries during the final session of the project illustrated that there were many common features among the folktales across borders that were not confined to the six countries the team members represented. These insights made many of the participating students realise that the world is more similar than different, thus promoting cross-cultural understanding and empathy towards other cultures. The final session contained presentations in German and English and the bilingual nature was of much advantage to the students of German who joined online from Sri Lanka and Thailand. It also proved that linguistic barriers could be overcome by adapting a bilingual approach. Furthermore, the blend of analogue and digital teaching was seen as a very positive aspect of the project and many who joined online mentioned that they did not realise they were merely virtual participants of the final presentation at the University of Siegen. It is noteworthy that thanks to technology, the world is no longer too distant and inaccessible, and connecting with each other across borders is no longer a daunting task if appropriate planning and organisation are in place. The study also connected the team members from different parts of the world on a common digital platform at every meeting that did not hamper the collaborative work in any way. Also, during the final presentation of the project only three members of the team were present onsite in Germany while the others made their contributions from different locations. Thus, this project is a very good example of digital collaboration across borders. Furthermore, the participants of the alumni academy who took part in the final session of the project and acted selected folktales from their home countries were from diverse disciplines including Administration, Chemistry, Education, Engineering, Mathematics, and Physics. This diversity did not deter the participants from sharing their cultural heritage with other international participants.

## 5 Conclusions and Recommendations

The analogue and digital teaching modules proved to be a very rewarding experience for the participating students, the team members, and the rest of the alumni to connect with each other at local and international level. Despite the participants of the projects including the alumni and students being of inter-disciplinary nature, the theme of folktales and cultural sustainability appealed to all as seen by the comments during the feedback and evaluation sessions. At first the team members and students worked in their own cultural setting onsite or online, discussing, and analysing their folktales. And sharing their knowledge with each other. Thereafter, the team leaders discussed the findings of their analysis and shared folktales from each country i.e., North Macedonia, Philippines, Sri Lanka, and Thailand during an online meeting, laying the foundation for the transformation of analogue teaching into digital delivery. The final teaching session was conducted on hybrid mode at the University of Siegen with some of the team members and students joining virtually. Thus, not only geographical barriers but also cultural and linguistic barriers could be overcome through this project to promote more effective intercultural communication and cross-cultural understanding. More similarities than contrastive aspects could be found among folktales of 20 countries which provided a common platform to further investigate the aspects of folklore at an international level. When considering the limitations, the project conducted simultaneously at different locations was of short duration as planned by the Alumni Academy of the University of Siegen. Judging by the dividends it brought, more similar projects of longer duration are recommended for the future. Furthermore, an extension of the typological study is envisaged connecting partner universities of the team leaders of this project. Through this study it was evident that it is time and cost-effective than conducting projects onsite, thus facilitating lager groups of participants. The students are in their familiar environment yet have the opportunity to gain international exposure to broaden their horizons. Through this digital collaboration participants from Asia, Europe, Africa, and America including school children, undergraduates and academics connected virtually to enhance their knowledge on cultural heritage, learn from each other, respect other cultures and values and be proud of one's own heritage.

**Acknowledgements.** The authors wish to acknowledge the support of the University of Siegen Germany, Alumni Office of the University of Siegen for organising and the German Academic Exchange Service (DAAD) supporting the Alumni Academy Digitalisation 20+ and Sustainability 2021/2022 Module 2 on Blended Learning Across an International Network through which the present project was realized. The contribution made by Dr. Messan Mawugbe, University of Professional Studies (UPSA) and Centre for Media Analysis (CMA), Ghana, all the alumni of Module 2 and the participating students to the project is much appreciated.

## References

1. Council of Europe. https://www.coe.int/en/web/education/4.7-education-for-sustainable-development-and-global-citizenship. Last accessed 03 June 2023
2. Gulmira, S.: A typological study of the plot lines of Uzbek and Japanese folk tales. JournalNX **6**(07), 174–177 (2020)
3. Hansen, W.: Mythology and folktale typology: chronicle of a failed scholarly revolution. J. Folklore Res. (Indiana University Press) **34**(3), 275–280 (1997)
4. Karsdorp, F., Fonteyn, L.: Cultural entrenchment of folktales is encoded in language. Palgrave Commun. **5**, 25 (2019)
5. Meletinsky, E.: 1 Structural-Typological Study of Folktales, vol. 1, pp. 19–52. In: Maranda, P. (ed.). De Gruyter Mouton, Berlin, Boston (1974)
6. Drozdowicz, J.: Teaching analog skills in a digital world. Docens Ser. Educ. **2**, 66–80 (2022)
7. Ruslan, T.S., et al.: Transformation of folklore texts into interactive multimedia digital forms as blended learning teaching material. In: Proceedings of the Sixth International Conference on Language, Literature, Culture, and Education (ICOLLITE 2022), pp. 575–580. Atlantis Press, Amsterdam (2022)
8. Chatterjee, P., et al.: Digital story-telling: a methodology of web based learning of teaching of folklore studies. In: 2019 21st International Conference on Advanced Communication Technology (ICACT), pp. 573–578 (2019)
9. Lester, P.M., King, C.M.: Analog vs. digital instruction and learning: teaching within first and second life environments. J. Comput. Mediat. Commun. **14**(3), 457–483 (2009)

10. Premawardhena, N.C.: The impact of virtual learning on undergraduate and postgraduate programmes: a Sri Lankan experience. In: Auer, M.E., Pachatz, W., Rüütmann, T. (eds.) Learning in the Age of Digital and Green Transition. ICL 2022. Lecture Notes in Networks and Systems, vol. 633, pp. 312–323. Springer, Cham (2023)
11. Premawardhena, N.C., Saleh, A., Kurtishi, A.: Building a digital bridge across cultures and continents: exploring new vistas in virtual collaboration. In: Auer, M.E., Pachatz, W., Rüütmann, T. (eds) Learning in the Age of Digital and Green Transition. ICL 2022. Lecture Notes in Networks and Systems, vol. 634, pp. 757–768. Springer, Cham (2023)

# Motivations for Becoming a Voluntary Mentor: A Case Study on What Experienced Scholars Gain from Mentoring Their Peers

Ivan Acebo-Choy(✉) and Samira Hosseini

Writing Lab, Institute for the Future of Education, Tecnologico de Monterrey, 64849 Monterrey, NL, Mexico

{iacebochoy,samira.hosseini}@tec.mx

**Abstract.** Writing Lab, at Tecnologico de Monterrey (Mexico), has become a signature center to enhance the academic production of its nationwide faculty. Unlike other writing centers, Writing Lab is aimed at assisting faculty and at-large collaborators. The mentors are volunteer experienced researchers who are helping interested junior faculty members, the mentees, to develop publication and research skills. Much attention has been given to the development of resources and programs for mentees to complement the mentorship process; however, we lack a clear understanding as to the motivations behind mentors' enrolment and what we can do to support their mentoring tasks. The research question is "what are the motivations for peer mentoring at Writing Lab within the framework of the institution?". We created an online survey to collect data from mentors through email (convenient sampling), which were later subjected to text analysis using the SWOT technique approach. The findings report mentors come to Writing Lab seeking recognition, prestige, and new avenues of production without losing sight of their senior status as researchers. Mentorships are structured interactions where the mentor takes a directive role in illustrating and discovering new competences to mentees (traditional model).

**Keywords:** Mentoring · Educational innovation · Peer-mentoring

## 1 Introduction

Since its creation in 2018, Writing Lab has become one of the most productive, and attractive, programs at the Institute for the Future of Education within Tecnologico de Monterrey, Mexico. Wanting to strengthen the culture of research in educational innovation, Writing Lab has devoted resources and expertise to support faculty development within the institution to specifically push for the creation and publication of high-quality materials on educational innovation; the goal of Writing Lab is to assist faculty and collaborators at large in the production of these articles for eventual publication in international indexed conferences and high impact journals. It is important to note, furthermore, that this is not a remedial center. Rather, it is an initiative that seeks to create a network of collaboration and academic support for up-and-coming scholars as they embark in educational research.


M. E. Auer et al. (Eds.): ICL 2023, LNNS 899, pp. 77–83, 2024.
https://doi.org/10.1007/978-3-031-51979-6_8

To this end, Writing Lab has created a signature mentoring program that currently consists of 136 mentors and 408 mentees, involved in 303 live interactions (or mentoring relationships). All registered mentors and mentees are voluntary, and come from diverse academic disciplines; to note, as of May 2023, the majority of the mentors (35%) are associated with the School of Engineering and Sciences, followed by those in administrative roles (20%) and by those from the School of Business (14%). The mentoring program seeks to pair up a mentor with a newly registered mentee to provide accompaniment and guidance as the mentee conceptualizes, executes and reports research findings in educational-innovation related projects.

Current scientific literature on mentoring and its benefits agrees on various tenets. Mentoring is usually described as a relationship [1] defined by the transfer of knowledge and skills from the mentor to the mentees; it is thought that this transaction, the mentoring itself, will provide the required direction and role-models necessary for the mentees to achieve a goal, which in the case of Writing Lab is a publication in indexed platforms. Mentorships are, then, defined by their potential to boost faculty productivity and facilitate collaborative networks [2].

In all these definitions, however, there is a subtle hierarchy of the mentee over the mentor; after all, a mentorship's effectiveness is measured on particular learning outcomes as illustrated by the mentees' performance. But insofar as a mentorship is defined as a relationship, one must also linger on the meaning and benefits of mentorships to mentors [3]. If mentorships are "two-way streets with bidirectional value related to the efforts of both parties" [2], what then is the intrinsic gain or benefit and motivation for mentors in the long run?

At Writing Lab, the nature and availability of its resources answer to the needs of the identified mentees. This is indeed in synch with the goal of the initiative, seeking to attract first timers in indexed publications and providing them with guidance and feedback to steer them into eventually self-sustaining careers in research. But as the center becomes more productive and well-known in the institution attracting larger numbers of mentees, it is important to think about the meaning of mentoring and the caring that mentors need in order to perform successfully [3]. As Hale points out, the lack of conceptual and theoretical foundations in mentoring programs can adversely affect the experience and outcomes of these relationships [4].

This study seeks to clarify the meaning and motivations of mentorship as described by the mentors of Writing Lab. It also purports to understand any needs that solicit attention as we nurture them to become role models for mentees, making them into an integral part of the mission of the program and the institution in its race to spearhead research in educational innovation.

## 2 Methods

This paper focuses on a group of 136 voluntary mentors in order to learn about the perceived strengths and benefits as well as what actions need to be implemented to achieve continued and sustained mentor participation. An online survey was created to collect data from mentors through email (convenient sampling), which were later subjected to text analysis using the SWOT technique approach [5]. We also relied on a case study approach, always recurring to "why" and "how" in interview questions [6]. The survey questions were classified under a SWOT quadrant to yield information on the perceived strength (the current perceived value of belonging to the community of mentors as it exists), weaknesses (the perceived lack of nurturing initiatives towards this community), opportunities (the expressed needs of the mentors) and threats (problems that can be identified and neutralized) of the mentorship system at Writing Lab. The questions are (Fig. 1):

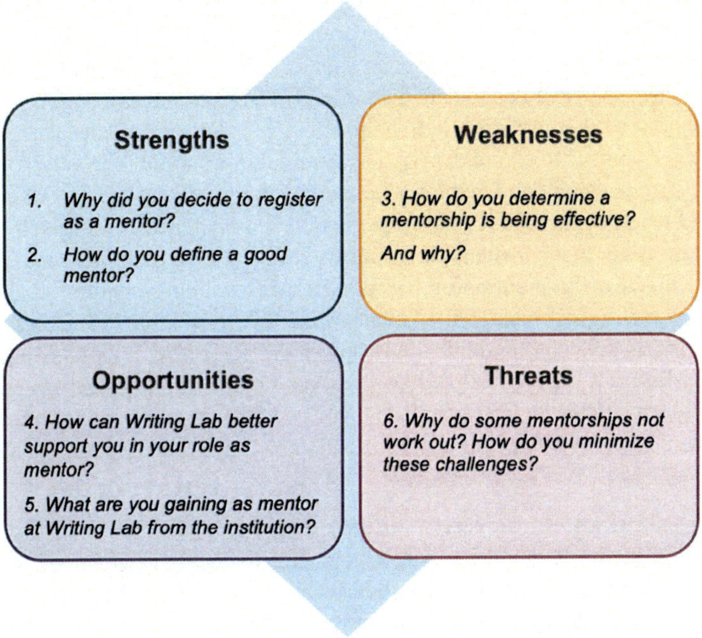


**Fig. 1.** Survey questions classified under SWOT quadrant.

The survey yielded 20 anonymous responses from the group of mentors. Each quadrant was analyzed thematically to hone in on the core of the responses, particularly as they reveal themes and sentiment to the idea of mentorship. The analysis was carried out as follows: First, as the survey was shared through a survey management software (Qualtrics), the author obtained transcripts for each quadrant. Secondly, the author labeled these transcribed responses searching for themes and repetition. Thirdly, the author reviewed and organized the themes in meaningful clusters of information that were later confirmed via visual displays of data such as word cloud. The use of this qualitative approach was adopted to favor detailed views of information sources (mentors) as directly as possible from their social context [7].

## 3 Results and Discussion

All received responses to the survey offered sharp insight into the perceptions of the mentors and their needs. We will discuss the analysis per SWOT quadrant, in order to reflect on essential aspects of the information provided:

**The perceived strength: the current perceived value of belonging to the community of mentors**. In this quadrant, the survey sought to identify the motivations of the current community of mentors, as well as to clarify what a good mentor is, in order to grasp the collective conceptual definition of the existing network which is currently working alongside a diverse group of mentees. When prompted about their motivations to become mentors, the responses yielded three specific themes. The community of mentors decided to engage as such to a) keep up to date with trends in educational innovation and expand their production; b) exercise a desire to help others and share in lessons learned in their research career; and, c) to signal experience acquired throughout the years.

To the participants, a good mentor is defined by their ability to a) provide guidance; b) share and teach skills; and, c) has demonstrable competences in research. We only received one answer that considers the mentor's ability to be a partner, a team player, as important. This is worth mentioning because an overwhelming majority of the replies points out to a traditional conception of mentorship; that is, wherein the mentor is viewed as "wiser", and is given a hierarchical position in the mentoring relationship [2]. Their main responsibility is, thus, to help others discover, guide and show the mentee towards the path of success. This idea, in fact, is in tune with the mentors' reported motivations. They see themselves as holders of expertise and knowledge that they want to share with others [8].

**Perceived weaknesses: the perceived lack of nurturing initiatives towards this community**. In this quadrant, the survey asked a question to clarify what a good mentorship is and how effectiveness is being understood. While Writing Lab considers a published paper as the end, or completion, of a mentorship, there is a lack of systematic measurements of the process of mentoring as priority is given to the final deliverable over the process. In this sense, it is surprising to find out that mentors too prioritize the end product of the mentorship: participants responded that a mentorship is effective when a) mentees are showing knowledge and insight that can eventually make them autonomous researchers; b) mentees are producing papers; and c) mentees show the passion and commitment for their own research and to mentor others.

The idea of mentees as recipients of knowledge and subjects that can be affected by someone's previously acquired expertise is also present in these descriptions. Mentors are concerned by actual results, and by the potential continuation of a chain of knowledge transfer [9].

**Opportunities or the expressed needs of the mentors**. Given the findings in the previous quadrant, it became essential to learn from the group of mentors what the program can improve to support them in their role as mentors both at the program and institutional level. The participants provided clear responses to the question of what the initiative should do to better uphold them in their role: a) training on good mentoring practices and experiences; and, b) specific reward systems for mentors (priority access to Writing Lab services, for example) and minimal compensation for time invested that is not otherwise paid by the institution.

In regards as to what the institution should guarantee to drive their work as mentors, the participants reported: a) access to a collaboration network; b) personal growth; and c) recognition/prestige. It becomes clear that a system of rewards needs to be put in place to create a closer rapport with the mentors in order to make them feel valued for the expertise that they are voluntarily bringing to the program. One of the responses alluded to the notion of "emotional salary", or the non-financial gains that can increase the level of satisfaction and wellbeing of employees within the institution. Recent trends in educational management point to the use of formulating policies envisioned to strengthen job satisfaction and job performance [10]; in this sense, the study signals potential for improvement on the availability of rewards, that do not necessarily need to be financial, but systems to show appreciation and recognition of unpaid work.

**Threats, or problems that can be identified and neutralized in the mentoring process.** Finally, the survey asked participants what the potential threats to their roles as mentors are given the importance they place on mentees' actual production, and the mentors' self-perceived value in the process [4]. The participants reported that the main challenge is a) mentees' expectations, or further clarification is needed as to what each role entails; b) mentees' lack of commitment, or lack of actions to sustain the interest of the mentees to push the work until publication; and, c) lack of incentives for mentors to encourage the work that is otherwise not compensated by the institution.

It is interesting to note that mentors seem to allude to the management of their relationship with the mentees as a potential challenge, particularly as it refers to the ability of the mentees to put in the same effort and commitment that mentors do. While the effectiveness of a mentor is the result of the mentor's own perceived value and willingness to share with others, the mentors seem to indicate that the quality of the learning and the effectiveness of the mentorship is rather defined by the attitude of the mentee [11]. The mentee's input is then crucial towards effectiveness, unburdening the role of the mentors as the sole conductor of the relationship.

## 4 Conclusion

The findings of this study will directly impact future policy affecting the attraction of new mentors and the development of initiatives to nurture and recognize the mentor community. Writing Lab at Tecnologico de Monterrey has increasingly demonstrated that the existing mentoring program is one of its backbones, and has actively contributed in the rising numbers of junior faculty and administrators in the scientific production on educational innovation, making Tecnologico de Monterrey the top reference in indexed platforms such as Scopus [12]. We must not be oblivious, however, to the fact that this is the result of the active transfer of knowledge and skills, and the collaboration networks derived from mentoring sessions.

This study aids in clarifying that the mentors come to Writing Lab seeking recognition, prestige, and new avenues of production without losing sight of their senior status as researchers. Mentorships are structured interactions where the mentor takes a directive role in illustrating and discovering new competences to mentees; to the mentors, the mentorship is effective insofar as the deliverable (e.g., publication) has been attained. In order to sustain mentors' commitment, the program must implement reward systems that uphold the recognition and experience that brought them in the first place, either via financial compensation or through emotional salary.

The findings of this study have also confirmed that the model in place in the mentorship program is a traditional one. The traditional mentorship consists in one part functioning as wiser and leader, holding a hierarchical position, and another part following indications and processing knowledge. Unlike transformational mentorships [1], the mentor and mentee are not considered equal; the interaction aims to complete a goal, a publication, and there is expectation on the mentee to eventually become an autonomous researcher.

### A. Limitations and future work direction

This paper has some limitations. First, we have only explored the perceptions of the mentors; we are planning a survey for the mentees in order to align the insights of both parties. Second, we did not consider how many of the mentors are currently working on a mentee or have recently completed the mentorship. For future work, we believe it may be important to carry out correlations between the definitions reported above and the actual production in the database. Elsewhere, we have reported on a considerable large gap of women assuming mentoring roles and how gender roles affect self-perception and actual achievement as a result of the interactions at Writing Lab [13]. In assuming that gender factors in power relationships, it would be important to study how power, or a leading position as reported by the mentors above, impacts the context, the practices and the achievements of the mentorship.

**Acknowledgments.** The authors would like to acknowledge the financial and the technical support of Writing Lab, Tecnologico de Monterrey, Mexico, in the production of this work. The authors also recognize the guidance and support by Dr. Asad Abbas in the process of writing this paper.

## References

1. Busby, K.R., Draucker, C.B., Reising, D.L.: Mentoring-as-partnership: the meaning of mentoring among novice nurse faculty. J. Nurs. Educ. **62**(2), 83–88 (2023)
2. Cooke, K.J., Patt, D.A., Prabhu, R.S.: The road of mentorship. In: American Society of Clinical Oncology Educational Book. American Society of Clinical Oncology. Annual Meeting, vol. 37, pp. 788–792 (2017)
3. Hudson, P.: Forming the mentor-mentee relationship. Mentor. Tutor.: Partnersh. Learn. **24**(1), 30–43 (2016)
4. Hale, R.: Conceptualizing the mentoring relationship: an appraisal of evidence. Nurs. Forum **53**(3), 333–338 (2018)
5. Helms, M., Nixon, J.: Exploring SWOT analysis – where are we now? A review of academic research from the last decade. J. Strateg. Manag. **3**(3), 215–251 (2010)
6. Yin, R.K.: Case Study Research: Design and Methods. SAGE (2009)
7. Ahmad, S., Wasim, S., Irfam, S., Gogoi, S., Srivastava, A., Farheen, Z.: Qualitative v/s. quantitative research—a summarized review. J. Evid. Based Med. Health **6**(43), 2828–2832 (2019)

8. Allen, T.D., Poteet, M.L.: Developing effective mentoring relationships: strategies from the mentors' viewpoint. Career Dev. Q. **48**(1), 59–73 (1999)
9. Mancuso, C.A., Berman, J.R., Robbins, L., Paget, S.A.: What mentors tell us about acknowledging effort and sustaining academic research mentoring: a qualitative study. J. Contin. Educ. Health Prof. **39**(1), 29–35 (2019)
10. Kumari, J., Kumar, J.: Influence of motivation teacher's job performance. Humanit. Soc. Sci. Commun. **10**(1), 158 (2023)
11. Allen, T.D., Eby, L.T.: Relationship effectiveness for mentors: factors associated with learning and quality. J. Manag. **29**(4), 469–486 (2003)
12. SCOPUS Homepage. https://www.scopus.com/results/results.uri?sort=plf-f&src=s&sid=971715b5f4529a084c25dbba1d18fbfb&sot=b&sdt=b&sl=101&s=%28TITLE-ABS-KEY%28educational+AND+innovation%29+OR+TITLE-ABS-KEY%28education+innovation%29%29+AND+PUBYEAR+%3e+2016&origin=savedSearchNewOnly&txGid=2bbb3365405efab986da4a115e1c2d49. Last accessed 24 May 2023
13. Guemoria, Y., Acebo, I., Rosales-Lopez, M.J., Hosseini, S.: Does gender gap in confidence explain gender gap in academic achievement? In: 24th International Conference on Interactive Collaborative Learning, ICL 202, pp. 177–189, Dresden (2021)

# Integrating Collaborative Annotation into Higher Education Courses for Social Engagement

Mark P. McCormack(✉) and John G. Keating

Maynooth University, Co. Kildare, Ireland
{mark.mccormack,john.keating}@mu.ie

**Abstract.** Collaborative Annotation (CA) is a literacy strategy that engages students in critical reading, critical thinking, writing and collaboration all in one activity [1]. This collaboration amongst students promotes social engagement with course materials and has been shown to be beneficial to higher education by improving learning comprehension [2] and soft skills amongst students [3]. For our study, we will investigate the benefits that CA provides higher education courses by means of social engagement with boundary objects in assessment. We have designed several pedagogical pipelines which illustrate how to integrate Collaborative Annotation into several types of assignments. Our research is concerned with the impact CA has on students' quality of learning. This study aims to design pipelines to integrate Collaborative Annotation into several assessment contexts for social engagement.

**Keywords:** Collaborative Annotation · Online assessment · Social learning

## 1 Introduction

Collaborative Annotation is a pedagogical methodology where students make annotations on digital boundary objects that teachers provide and engage in discourse about their thoughts, interpretations, and the material itself to obtain a thorough understanding of the course material from being exposed to several perspectives. This methodology inherently uses social engagement as a means for interacting with assignments and as such, assessing the potential Collaborative Annotation has for students' learning in a course can lead to a deeper understanding of how social engagement impacts students' interaction with education material and what it can be used for.

Consequently, we have begun the development of "pedagogical pipelines". These pipelines allow us to design a framework for the integration of Collaborative Annotation into several educational contexts. we will use these pipelines to conduct trials of our software in practical pedagogical settings. We will run this experiment for students in various courses, each course will be selected to test one of our given pipelines. We can then gauge from these results the impact each of our pipelines has on both the students' outcome of learning and academic result.


M. E. Auer et al. (Eds.): ICL 2023, LNNS 899, pp. 84–89, 2024.
https://doi.org/10.1007/978-3-031-51979-6_9

## 2 Background/Literature Review

Current research suggests that collaboration not only makes learning more enjoyable for students, but also improves overall learning outcome and promotes social engagement. Dahal [4] suggests that higher education students that use collaborative learning technologies engage more in group activities. Furthermore, we have run UI/UX tests that suggest students would use this software in existing course materials and believed it to be beneficial to their studies.

## 3 Software Implementation

Maynooth University has developed a Collaboration Annotation software which teachers can use to assess the understanding students have of their material. Currently, students may annotate both text and image documents, providing annotation functionality for many disciplines. With this selection of features, we can investigate how these specific tools affect student learning, how Collaboration Annotation is used in various contexts and what changes need to be made to enhance student learning.

There are three categories of users in this software: administrators, teachers, and students. Administrators can control and assign accounts amongst students and teachers. Teachers can assign students material and organize them into groups, also annotating the documents they assign. Students can access these documents to create both personal and collaborative annotations on various materials. Our interface has been verified by extensive UI/UX testing using a small sample of students, where each was assigned to a user story and feedback was assessed using the feedback form, the results for which can be seen in figure one. Teacher testing was also conducted (Fig. 1).

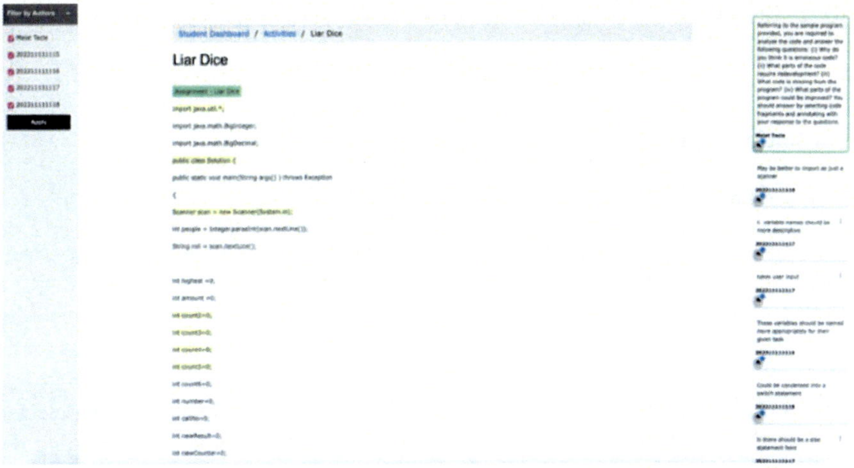


**Fig. 1.** MU Collaborative Annotation software for text annotation

### 3.1 Pedagogical Aspects

The MU Collaborative Annotation software allows for conducting several forms of assessment. The software allows teachers to upload traditional course materials to students for annotation. Once the students have had a dialogue and submitted their annotations, the teacher may then review the assignment. During the review process they can grade and provide feedback to each individual student or to student groups (teams). From the students' perspective, they are provided with the assignment boundary object (currently images and/or text) which are available for annotating as part of a team.

Once complete, the group will meet (online or in person) to discuss their annotations and re-annotate to align with their new viewpoint from their conversations. The group, could for example, collectively decide which annotations are to be submitted for a group assignment, while the rest may be discarded. Once submitted and assessed by the teacher, students can view their grade and feedback to understand where they can improve performance in future assignments (Fig. 2).

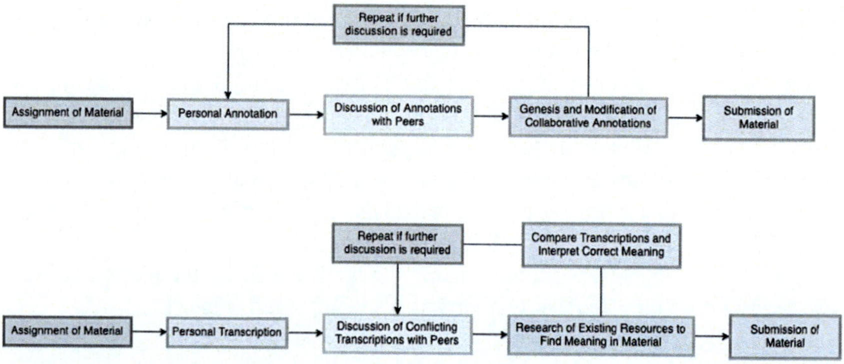


**Fig. 2.** Collaborative Annotation pipelines for text documents and transcription materials

## 4 Research Study

### 4.1 Sampling

For our research study, we would like to run this experiment with multiple groups, each of which containing university students from a single type of degrees. Given different subjects require different levels of social interaction and engagement, we regard it important to assess how each degree makes use of this methodology that promotes social engagement and how it impacts education experience.

English Literature students will make up our first group. This degree will be selected given the subject contains several opportunities for discussions/debates regarding the literature in the classroom, as well as students being more inclined to provide descriptive and substantial dialogue to peers. We believe this will be most evident with Collaborative Annotation as the methodology provides students the space to comment and critique peers' thoughts and opinions.

Our next group will be comprised of undergraduates taking an introductory course in Computer Science. This subject will be selected as learning the basics of programming and more specifically one programming language is often a solitary process, only later exposing students to teamwork once they have built this foundational knowledge. The students will also be rather technology-literate, which helps as they will likely be quicker to understand how to use our software.

Irish Language students will make up our final group for the study. This sample will be selected as while it is rather like the English Literature group above, we also get to explore the aspect translation has in Collaborative Annotation for Social Engagement. Social Engagement around material being discussed in another language may impact how students communicate with each other and may influence engagement with the material, which may provide us with novel insights for our research.

### 4.2 Instrumentation

Given our study assesses the impact of Collaborative Annotation on Social Engagement, we require several instruments to assess both the student's quality of learning using Collaborative Annotation, how this process impacts students' Social Engagement with one another, and what thoughts students have around this methodology.

For assessing students' quality of learning, we will make use of the summative result they receive for their assignment. Given the primary focus of the study is around Social Engagement, we have decided that this metric suffices for assessing the impact of this process on students results. This grade will be used as an indicator to suggest if Social Engagement in the classroom using this technique is constructive towards students' learning.

In assessing Collaborative Annotation's impact on social engagement, we will use Penny's [5] multi-rubric assessment. Penny breaks down these rubrics into four sections in Cognitive, Mechanical, Managerial, and Interactive. Cognitive focuses on the students learning from online discussions with peers, mechanical assesses grammatical mistakes and typos, managerial looks for how often students participate in the online discussions and finally interactive looks at the co-creation of knowledge between students. These rubrics provide us with an excellent framework to assess how well and often students engage with each other in our Collaborative Annotation platform, what impact our methodology has on Social Engagement before and after using our platform and finally which aspects of Social Engagement need to be improved further in the future.

Our final instrument will be the ASPECT questionnaire developed by Wiggins [6], which can be used to measure qualitative results such as the student's opinion on learning experience. We would combine this with our own questionnaire targeting what aspects of the software the students would like to see improved. This is so that we get both an understanding of how the software impacted the students' Social Engagement from the previous instruments, as well as the student's opinion on the process and what they feel could be changed to enhance their experience engaging collaboratively with course materials.

### 4.3 Preparation

Before conducting our experiment, we will meet with teachers for the respective subjects of our study and collaborate to design assignments that students can complete that link into their subject as well as provide an outlet for discussion of the material. This can come in the form of group work; peer assessment or other assessment forms teachers may suggest that are tailored for their subject.

Once these assignments have been created, both students and teachers will be introduced to the software and be provided with a hands-on tutorial which will get them familiar with both the process of Collaborative Annotation and our software facilitating it. We will design a sample assignment which consists of both image and text annotation where students can create and interact with each other's annotations. Once everyone understands how the software and methodology works, the students will be split into groups and assigned the task by their teacher to complete.

This study will take place over the course of one week. Given the asynchronous nature of online discussion forums, we have decided to allow students to respond over a larger timescale to have a more substantial corpus of discussion text for our rubrics to analyze after the study has been complete. Once the students have completed their task, they will be sent the questionnaire to assess their thoughts regarding Social Engagement using Collaborative Annotation.

## 5 Conclusions

We anticipate that students will find our Collaborative Annotation software increases quality of learning and promotes Social Engagement with the course materials. Furthermore, teachers will discover new use cases for our software in their modules from which we can develop more tools. These conclusions will provide us with ample evidence to show that the integration of Collaborative Annotation technology into higher education courses increases students' social engagement with materials and peers, from which we can recommend the best pipelines to implement for a variety of educational contexts. Given there is little research in the existing literature about a strategic approach to the integration of Collaborative Annotation into existing practices, this will enable us to provide and suggest frameworks that educators can use to add this social aspect to teaching. With this, there are now resources for researchers to develop novel CA pipelines for other pedagogical contexts.

## 6 Discussion and Future Work

Regarding the analysis of online discussions between students, there is room to explore how Natural Language Processing (NLP) techniques could be used to automatically assess students' annotations and engagement based on criteria defined by both the teacher, rubrics discussed in Penny's study and guidelines for which the assignment in question may be marked. There is also the opportunity to explore how social engagement impacts several other types of subjects. For our study we only considered three degrees (English Literature, Computer Science, and Irish Language), however there may be a novel finding in subjects we have not explored yet.

## References


1. Schwane, E.: Collaborative Annotation: For Any Text and Any Class. http://www.wcteonline.org/wp-content/uploads/2015/10/Collaborative-Annotation.pdf
2. Razon, S., Turner, J., Johnson, T.E., Arsal, G., Tenenbaum, G.: Effects of a collaborative annotation method on students' learning and learning-related motivation and affect. Comput. Hum. Behav. **28**(2), 350–359 (2012)
3. England, T.K., Nagel, G.L., Salter, S.P.: Using collaborative learning to develop students' soft skills. J. Educ. Bus. **95**(2), 106–114 (2020)
4. Dahal, N.: Understanding and uses of collaborative tools for online courses in higher education. Adv. Mobile Learn. Educ. Res. **2**(2), 435–442 (2022)
5. Penny, L., Murphy, E.: Rubrics for designing and evaluating online asynchronous discussions. Br. J. Edu. Technol. **40**(5), 804–820 (2009)
6. Wiggins, B.L., Eddy, S.L., Wener-Fligner, L., Freisem, K., Grunspan, D.Z., Theobald, E.J., Timbrook, J., Crowe, A.J.: ASPECT: a survey to assess student perspective of engagement in an active-learning classroom. CBE Life Sci. Educ. **16**(2), ar32 (2017)

# Norms for Team Process and Outcome Measures by Race/Ethnicity and Gender

Matthew W. Ohland[1](✉), Emily Redler[2], David J. Woehr[2], and Misty L. Loughry[3]

[1] Purdue University, West Lafayette, IN 47907, USA
ohland@purdue.edu
[2] University of North Carolina Charlotte, Charlotte, NC 28223, USA
{eredler,dwoehr}@charlotte.edu
[3] Rollins College, Winter Park, FL 32789, USA
mloughry@rollins.edu

**Abstract.** This descriptive study reports norms for team process and outcome measures for various populations. We collected the data using the Comprehensive Assessment of Team-Member Effectiveness (CATME) Team Tools system. We observed gender differences for Task Conflict and Psychological Safety with small effect sizes. We observed differences on multiple team process variables with a range of effect sizes based on reported race/ethnicity, with some effect sizes approaching moderate (d > 0.5). We discuss possible explanations for systematic differences, but report no evidence that can support any particular explanation. Instructors should know that such differences exist to consider how those differences might affect their interpretation of peer evaluation data.

**Keywords:** Peer evaluation · Team dynamics · Demographic differences

## 1 Teamwork in STEM Education and the Goal of This Work

### 1.1 The Benefits and Challenges of Teamwork in STEM Education

Undergraduate STEM education, and undergraduate education in general, relies heavily on students working in teams to learn course concepts, build students' team skills, and increase student engagement. Team-based learning approaches such as cooperative learning result in better student learning and higher student engagement than traditional lecture and individualistic learning approaches [1–4].

In spite of the benefits for learning and the necessity of developing team skills for the job market, teamwork is not a universally positive experience for students [5, 6]. Engineering majors especially dislike teamwork, as indicated in a study that found 83% of upper-class students prefer to work alone [7]. The study also found other attitudes that are dysfunctional for teamwork: 68% like to rank themselves against others, 65% try to exclude peers they perceive to have inferior technical skills, and 40% like to have their individual contributions stand out. Free riding can also turn students against teamwork in


M. E. Auer et al. (Eds.): ICL 2023, LNNS 899, pp. 90–101, 2024.
https://doi.org/10.1007/978-3-031-51979-6_10

engineering and other disciplines [8, 9], causing students to have bad team experiences, particularly those perceived as different. For example, people tend to be more tolerant of free-riding by similar (in-group) than dissimilar teammates [10].

### 1.2 The Particular Challenges for Minoritized Populations in Teams

In many STEM majors, including engineering, women are underrepresented and undervalued [11]. Unfortunately, some students and faculty, including some leaders in higher education, believe that women are not proficient in some areas [12, 13]. Women are sometimes treated differently in ways that would likely affect their team experiences; as a result, teamwork can have negative impacts on women [6, 14–18]. For example, teammates sometimes do not respect women's ideas and assign them less-important, non-technical parts of team assignments [19]. In engineering, some male students misinterpret female speech patterns as signs of weakness and lack of ability [20] and lack confidence in female peers' leadership skills [21, 22]. Female students define leadership as facilitating collaboration, being responsible to the team and contributing to the team, whereas males think of leadership as directing team activities, running meetings and overseeing the project [23]. Women sometimes feel they need to prove themselves to be viewed as equal to teammates in skills and abilities [24]. There is much less research on the experience of underrepresented minority students precisely because they are so underrepresented. While the effects of a "chilly climate" on women is discussed more frequently, minoritized students often report feeling marginalized [25] and the "lack of hospitality" undermines their academic performance [26].

It is important to understand the team experiences of all student groups because students who have bad experiences in teams could have worse college outcomes overall [27, 28]. Black students who perceive less racism and discrimination have higher graduation rates and commitment to their institutions [29]. Female students can be marginalized in engineering teams [30]. Minority women find both racism and sexism to be significant barriers to pursuing the sciences, particularly as undergraduates [31–33]. International students at U.S. institutions are less engaged in some areas and more engaged in others and the differences change over the years of undergraduates' studies [34]. Differences in students' team experiences can affect STEM persistence [35, 36]. Meeting often in study groups increases persistence in STEM majors, and students without successful team experiences are less likely to develop those affiliations [37].

### 1.3 The Goal of This Work – Norms for Various Process and Outcome Measures

Through this descriptive study, we identify the extent to which different groups of students report different perceptions of team processes, including task, relationship, and process conflict, psychological safety, cohesion, interdependence, team satisfaction, interpersonal perceptions of warmth and competence, and team viability. Our research extends what is understood about how gender and race/ethnicity affect the team experiences of underrepresented students.

## 2 Team Processes and Outcomes

### 2.1 Conflict

***Conflict*** occurs when members of a team are aware that they have disagreements or incompatible wishes [38]. Team-member characteristics can have both positive and negative effects on team conflict [39–42]. Team-level conflict is typically assessed in three components. Perceptions of ***task conflict*** reflect differences of opinion about how to perform the task. Perceptions of ***relationship conflict*** reflect disagreements related to interpersonal differences. ***Process conflict*** focuses on different preferences for assignment of roles and responsibilities. Data has been collected in the CATME system using the conflict measure by Jehn and Mannix [43] since 2005.

### 2.2 Psychological Safety

***Psychological safety*** involves feeling safe enough to take interpersonal risks and is an important condition for people to learn, contribute, and perform [44, 45]. Meta-analytic results show outcomes including expressing voice, learning and citizenship behaviors, commitment, and engagement [46, 47]. It is especially important in team environments where there is task conflict, which is necessary and common in complex student learning projects [48]. Data has been collected in the CATME system using the psychological safety measure by Edmonson [49] since 2014.

### 2.3 Cohesion

***Cohesion*** reflects the degree to which members identify with and feel part of the team [39, 50, 51]. Members of cohesive teams have emotional and social bonds that link them to one another and to the group. Measures of cohesion typically assess multiple components including: task attraction (e.g., "Team members like the work that the group does"), interpersonal cohesiveness (e.g., "Team members enjoy spending time together") and task commitment (e.g., "Our team is united in trying to reach its goals for performance"). Data has been collected in the CATME system using the cohesion measure by Loughry and Tosi [52] who built on Carless and de Paola [53] since 2005.

### 2.4 Team Satisfaction

***Team satisfaction*** is an emotional state reflecting positive perceptions of events within teams, and includes satisfaction with team outcomes, with other team members, and with being part of the team's project. Data has been collected in the CATME system using the measure by Van der Vegt, Emans, and Van de Vuert [54] since 2005.

### 2.5 Warmth and Liking

***Warmth/Liking*** is one of two fundamental dimensions that form the basis of interpersonal perception [55–59]. The other is competence (described below). These two dimensions are primary drivers of social judgments, which, in turn, drive individuals' emotions and behavior [58, 60]. The first dimension is primarily affective and captures characteristics related to a person's intent (e.g., likability, friendliness, and trustworthiness) [60]. Evidence supports both liking and competence as key antecedents of interpersonal perceptions and outcomes [55–59, 61]. Data have been collected in the CATME system since 2005 using a warmth/liking scale we developed [62].

## 2.6 Competence

***Competence*** reflects an individual's ability (e.g., skill, knowledge, capability, contribution) and the capacity to act on their intentions as reflected in their behavior. We measure competence using CATME's behaviorally anchored rating scale for peer evaluation [62]. This instrument measures five different aspects of team members' contributions to the team: contributing to the team's work, interacting with teammates, keeping the team on track, expecting quality, and having relevant knowledge, skills, and abilities [63]. These data have been collected in the CATME system since 2005.

## 2.7 Task Interdependence

***Task interdependence*** is "the degree to which taskwork is designed so that members depend upon one another for access to critical resources and create workflows that require coordinated action" [64, p.1829]. Task interdependence occurs when members of a team must work together to accomplish their tasks or share materials or information to accomplish the team task [65]. Task interdependence affects the teams' task-related functioning and thereby teams' performance [64]. When teams have higher levels of task interdependence, there is a stronger link between their teamwork processes and team performance [66]. Furthermore, team cohesion matters more for team performance when there is high task interdependence [67]. Recent work suggests that in culturally diverse teams, deep-level diversity is related to team creativity and innovation when tasks are interdependent [68].

## 2.8 Team Viability

***Team viability*** reflects the extent to which team members want to maintain membership in a team. The success of a team is typically conceptualized in terms of two general criteria—task performance and team viability [69, 70]. While task performance is important, it is also imperative that the social processes of the group "...maintain or enhance the capability of members to work together on subsequent team tasks" [70, p.323]. A team that performs well but no longer wants to work together is less successful than one that performs well and maintains its capacity to work together [70–73]. Thus, the extent to which individuals want to continue working with teammates can be used to evaluate teams [70–72]. Data has been collected in the CATME system since 2005 using a team-viability measure developed by our research team [62].

### 2.9 Team Processes and Outcomes in Diverse Teams

Team diversity has complex effects on communication, conflict, cohesion, commitment, creativity, innovation, and performance [74, 75]. Diversity can facilitate learning and lead to more creative and cognitively complex team results by bringing different perspectives, experiences, and knowledge bases [74, 76]. Diverse teams tend to have more divergent thinking, higher productivity, and better-quality end-products [35, 77, 78]. Having more women on teams can improve team processes [79, 80]. Yet, diversity can increase conflict and reduce trust, which can negatively affect team performance [81]. Diverse teams tend to experience more frustration, more sustained conflict, and lower positive feelings [82, 83]. Both gender and racial diversity have small negative effects on team performance in field studies in work organizations [84]. Race, gender, and their intersectionality affect engineering students' self-ratings of their teamwork, communication, and leadership skills [85].

## 3 Methods

### 3.1 Participants

Participants were U.S. college students working in course-related teams between 2007 and 2015. Team members completed peer evaluations of their teammates using the online Comprehensive Assessment of Team Member Effectiveness (CATME) system, which includes a five-item behaviorally anchored rating scale designed to measure different aspects of team members' performance: contributing to the team's work, interacting with teammates, keeping the team on track, expecting quality, and having relevant knowledge, skills, and abilities [62]. The system has been adopted by approximately 25,000 instructors at 2,600 institutions in 91 countries [86, 87].

### 3.2 Data

Each team member rated themselves and each of their teammates on the five CATME dimensions (using a 5-point scale with anchors at levels 1, 3, and 5). It is important to note that these teams were not formed for this study; rather, the course instructors collected the data for class-related purposes but allowed analysis of the data for research purposes with student identifiers removed. Thus, teams were actual teams in which their team's work contributed to their final course grade.

Participants responded to the other team process and outcome measures listed above when those were collected along with peer evaluations. Participants provided demographic information using the Team-Maker system as described earlier. The measures used in this study have been collected from 1,309,005 to 8,610,801 times. The amount of available data available for each varies because (1) CATME users choose which variables to measure and (2) although data released for research purposes is deidentified, CATME users are not required to release their data for research purposes. When we study the intersection of these data with available demographic variables, fewer data are available because we do not collect demographic data with peer evaluations to avoid the potential for stereotype threat [88, 89], so demographic data are only available for students who independently complete a team-formation survey with that question.

# 4 Findings

## 4.1 Findings by Gender

Gender information was collected from more than 100,000 students with three response choices: "Female", "Male", and "Other or prefer not to answer." The last option provided a choice for students who did not identify with a binary gender choice while not communicating the details of their gender identity to their instructor.

We acknowledge that the question formulation is problematic because it both invalidates certain gender identities [90] and "others" individuals who do not choose one of the binary identities [91]. A major revision to the primary system question for collecting gender identity information was designed and tested recently [92]. Males report both higher task conflict (1.58 vs. 1.50) and higher psychological safety (6.07 vs. 5.95).

## 4.2 Findings by Race/Ethnicity

Race/ethnicity was also available for from 194,667 to 348,792 respondents. The response choices were "Hispanic White", "Non-Hispanic White", "Black, African-American", "American Indian or Alaskan Native", "Asian or Pacific Islander", "Other/Mixed-heritage", and "Other/Prefer not to answer". Again, we acknowledge that the question formulation is problematic because it invalidates multiracial respondents and those with a racial/ethnic identity not listed and "others" individuals who choose one of those options. A revision of this question was also designed and tested recently.

Asian students report higher team commitment (2.86 vs. 2.72, $d = 0.27$) and interdependence (3.62 vs. 3.46, $d = 0.25$), but lower psychological safety (5.77 vs. 6.12, $d = 0.48$), compared to White students. Students reporting as Black/African-American report lower interpersonal cohesion (4.07 vs. 4.23, $d = 0.25$) and lower psychological safety (5.97 vs. 6.12, $d = 0.21$). Hispanic/Latino students have lower psychological safety (5.96 vs. 6.12, $d = 0.22$). Native students reported lower psychological safety (5.96 vs. 6.12, $d = 0.23$) and higher process conflict (1.53 vs. 1.35, $d = 0.34$). Students reporting as Other/Mixed-Heritage and Other/Prefer not to Answer also reported lower psychological safety than White students.

# 5 Discussion and Conclusion

## 5.1 Discussion of Findings by Gender

The higher level of task conflict reported by students identifying as male may be explained by gendered differences in communication. Studies of gendered language usage have found that men tend to use language that is more disagreeable [93]. The lower psychological safety reported by women is not surprising given the literature on women's experiences in teams, particularly in STEM majors, as discussed earlier.

### 5.2 Discussion of Findings by Race/Ethnicity

Asian students reported higher team commitment and interdependence, but lower psychological safety, compared to White students. To the extent that the students responding as Asian are more likely to identify with a collectivist culture [94], all three findings would seem to be related. A student with a collectivist mindset would tend to place the needs of the team above their own, and thus have a higher team commitment. They would also work toward a higher level of interdependence. At the same time, as they encounter White students, particularly domestic White students in the United States, the more individualistic cultural norms would seem likely to cause them to feel less psychologically safe in addition to any reduction of psychological safety resulting from anti-Asian racism or stereotyping.

The lower interpersonal cohesion reported by Black/African-American may be due to their severe underrepresentation in higher education. The lower psychological safety reported by Black/African-American, Hispanic/Latino, and Native students are likely symptomatic of underrepresentation, stereotyping, and lack of belongingness. We have no explanation for why Native students report higher process conflict.

### 5.3 Conclusion

This work begins to identify systematic differences in certain team processes. The universality of the higher psychological safety reported by White students is notable and is further evidence of the importance of psychological safety as a critical variable to use in describing team experiences. It should also be noted that while the findings reported have moderate to large effect sizes, many of the means are indicative of positive team experiences. To wit, while all women had lower psychological safety than men and all racial/ethnic groups had lower psychological safety than White students, the psychological safety of all groups was close to 6 on a 7-point scale—even students in groups with lower psychological safety generally felt quite psychologically safe. Similarly, "lower interpersonal cohesion" was still above 4 on a 5-point scale, "higher process conflict" and "higher task conflict" were below 2 on a 5-point scale. Overall, student teams are having positive experiences. As this work continues, other groupings will be explored, such as discipline of study (college major) and class level (1st year, 2nd year, etc.) and considering the intersectional identities of race/ethnicity and gender.

## References

1. Hsiung, C.M.: The effectiveness of cooperative learning. J. Eng. Educ. **101**(1), 119–137 (2012)
2. Johnson, R.T., Johnson, D.W.: Active learning: cooperation in the class-room. The Annual Report of Educational Psychology in Japan, vol. 47, pp. 29–30 (2008)
3. Felder, R.M., Brent, R.: Effective strategies for cooperative learning. J. Coop. Collab. Coll. Teach. **10**(2), 69–75 (2001)
4. Eberlein, T., et al.: Pedagogies of engagement in science. Biochem. Mol. Biol. Educ. **36**(4), 262–273 (2008)
5. Pfaff, E., Huddleston, P.: Does it matter if I hate teamwork? What impacts student attitudes toward teamwork. J. Mark. Educ. **25**(1), 37–45 (2003)

6. Tonso, K.L.: The impact of cultural norms on women. J. Eng. Educ. **85**(3), 217–225 (1996). https://doi.org/10.1002/j.2168-9830.1996.tb00236.x
7. Leonardi, P.M., Jackson, M.H., Diwan, A.: The enactment-externalization dialectic: rationalization and the persistence of counterproductive technology design practices in student engineering. Acad. Manag. J. **52**(2), 400–420 (2009)
8. Tenenberg, J.: Factors affecting free riding on teams: implications for engineering education. Int. J. Eng. Educ. **35**(6A), 1703 (2019)
9. Thom, M.: Are group assignments effective pedagogy or a waste of time? A review of the literature and implications for practice. Teach. Public Admin. **38**(3), 257–269 (2020)
10. Hütter, M., Diehl, M.: Motivation losses in teamwork: the effects of team diversity and equity sensitivity on reactions to free-riding. Group Process. Intergroup Relat. **14**(6), 845–856 (2011)
11. Stoddard, E., Pfeifer, G.: Working towards more equitable team dynamics: mapping student assets to minimize stereotyping and task assignment bias. In: ASEE Conference: 2018 CoNECD-The Collaborative Network for Engineering and Computing Diversity Conference (2018)
12. Mead, P.F., Moore, D., Natishan, M., Schmidt, L., Brown, S.V., Lathan, C., Goswami, I., Mouring, S.: Faculty and student views on engineering student team effectiveness. J. Women Minor. Sci. Eng. **5**(4) (1999)
13. Jaschik, S.: What Larry Summers said. https://www.insidehighered.com/news/2005/02/18/what-larry-summers-said. Last accessed 2023/05/13
14. Tonso, K.L.: Teams that work: campus culture, engineer identity, and social interactions. J. Eng. Educ. **1**(1), 1–13 (2006)
15. Felder, R.M., Felder, G.N., Mauney, M., Hamrin, C.E., Dietz, E.J.: A longitudinal study of engineering student performance and retention. III. Gender differences in student performance and attitudes. J. Eng. Educ. **84**(2), 151–163 (1995). https://doi.org/10.1002/j.2168-9830.1995.tb00162.x
16. Hartman, H., Hartman, M.: Attitudes to group work: gendered differences? In: ASEE Annual Conference and Exposition, Conference Proceedings (2006)
17. Hartman, H., Hartman, M.: Do gender differences in undergraduate engineering orientations persist when major is controlled? Int. J. Gend. Sci. Technol. **1**(1), 62–82 (2009)
18. Rosser, S.V.: Female-Friendly Science: Applying Women's Studies Methods and Theories to Attract Students. Pergamon Press, New York, NY (1990)
19. Wolfe, J., Powell, B.A., Schlisserman, S., Kirshon, A.: Teamwork in engineering undergraduate classes: what problems do students experience? In: ASEE Annual Conference & Exposition (2016)
20. Wolfe, J., Powell, E.: Biases in interpersonal communication: how engineering students perceive gender typical speech acts in teamwork. J. Eng. Educ. **98**(1), 5–16 (2009). https://doi.org/10.1002/j.2168-9830.2009.tb01001.x
21. Rosch, D.M., Collier, D.: Incoming leadership-oriented differences between students in a leadership studies course and a team-based project course. J. Leadership Educ. **12**(2), 103–121 (2014). https://doi.org/10.12806/v12/i2/103
22. Rosch, D., Collier, D., Zehr, S.: Self-vs.-teammate assessment of leadership competence: the effects of gender, leadership self-efficacy, and motivation to lead. J. Leadership Educ. **13**(2), 96–124 (2018). https://doi.org/10.12806/v13/i2/r5
23. Bailey, R.R., Swan, A.K., Coso, A., Creager, M.F., Rowan-Kenyon, H.T.: The role of gender in student perceptions of leadership on interdisciplinary engineering teams. J. Women Minorities Sci. Eng. **18**(2), 97–113 (2012). https://doi.org/10.1615/jwomenminorscieneng.2012003031
24. Natishan, M.E., Schmidt, L.C., Mead, P.: Student focus group results on student team performance issues. J. Eng. Educ. **89**(3), 269–272 (2000). https://doi.org/10.1002/j.2168-9830.2000.tb00524.x

25. Etzkowitz, H., Kemelgor, C., Uzzi, B.: Athena Unbound: The Advancement of Women in Science and Technology. Cambridge University Press, Cambridge, UK (2000)
26. The College Board: Reaching the Top: A Report of the National Task Force on Minority High Achievement. College Board Publications, New York (1999)
27. Rosser, S.V.: Group work in science, engineering, and mathematics: consequences of ignoring gender and race. Coll. Teach. **46**(3), 82–88 (1998)
28. Collins, K.H.: Confronting color-blind STEM talent development: toward a contextual model for Black student STEM identity. J. Adv. Acad. **29**(2), 143–168 (2018)
29. Brown, A.R., Morning, C., Watkins, C.: Influence of African American engineering student perceptions of campus climate on graduation rates. J. Eng. Educ. **94**(2), 263–271 (2005)
30. Marinelli, M., Male, S., Kim, L.V., Sydney, Z.: Development of educators' resources for creating inclusive teamwork in engineering and computer science. In: 30th Annual Conference for the Australasian Association for Engineering Education (AAEE 2019): Educators Becoming Agents of Change: Innovate, Integrate, Motivate, p. 653. Engineers Australia (2019)
31. Clewell, B.C., Anders, B.T.: Women of Color in Mathematics, Science, and Engineering: A Review of the Literature. Center for Policy Studies, Washington, D.C. (1991)
32. Clewell, B.C, Ginorio, A.: Examining women's progress in the sciences from the perspective of diversity. In: C. Davis, A. Ginorio, C. Hollenshead, B. Lazarus, P. Rayman (eds) The Equity Equation. Jossey-Bass, San Francisco (1996)
33. Malcom, S., Hall, P.Q., Brown, J.W.: The Double-Bind: The Price of Being a Minority Woman in Science. AAAS, Washington, D.C. (1976)
34. Zhao, C.M., Kuh, G.D., Carini, R.M.: A comparison of international student and American student engagement in effective educational practices. J. High. Educ. **76**(2), 209–231 (2005)
35. Guimera, R., Uzzi, B., Spiro, J., Amaral, L.A.N.: Team assembly mechanisms determine collaboration network structure and team performance. Science **308**(5722), 697–702 (2005)
36. Ohland, M.W., Sheppard, S.D., Lichtenstein, G., Chachra, D., Eris, O., Layton, R.A.: Persistence, engagement, and migration in engineering. J. Eng. Educ. **97**(3), 259–278 (2008)
37. Green, A., Sanderson, D.: The roots of STEM achievement: an analysis of persistence and attainment in STEM majors. Am. Econ. **63**(1), 79–93 (2018)
38. Boulding, K.E.: Conflict and Defense. Harper & Row (1963). Republished as Conflict and defense: A General Theory, 2018 by Pickle Partners Publishing
39. Webber, S.S., Donahue, L.M.: Impact of highly and less job-related diversity on work group cohesion and performance: a meta-analysis. J. Manag. **27**(2), 141–162 (2001)
40. Stahl, G.K., Maznevski, M.L., Voigt, A., Jonsen, K.: Unraveling the effects of cultural diversity in teams: a meta-analysis of research on multicultural work groups. J. Int. Bus. Stud. **41**(4), 690–709 (2010)
41. Hülsheger, U.R., Anderson, N., Salgado, J.F.: Team-level predictors of innovation at work: a comprehensive meta-analysis spanning three decades of research. J. Appl. Psychol. **94**(5), 1128–1145 (2009)
42. Bowers, C.A., Pharmer, J.A., Salas, E.: When member homogeneity is needed in work teams: a meta-analysis. Small Group Res. **31**(3), 305–327 (2000)
43. Jehn, K.A., Mannix, E.A.: The dynamic nature of conflict: a longitudinal study of intragroup conflict and group performance. Acad. Manag. J. **44**, 238–251 (2001)
44. Edmondson, A.C., Lei, Z.: Psychological safety: the history, renaissance, and future of an interpersonal construct. Annu. Rev. Org. Psychol. Org. Behav. **1**, 23–43 (2014). https://doi.org/10.1146/annurev-orgpsych-031413-091305
45. Newman, A., Donohue, R., Eva, N.: Psychological safety: a systematic review of the literature. Hum. Resour. Manag. Rev. **27**(3), 521–535 (2017)
46. Frazier, M.L., Fainshmidt, S., Klinger, R.L., Pezeshkan, A., Vracheva, V.: Psychological safety: a meta-analytic review and extension. Pers. Psychol. **70**(1), 113–165 (2017)

47. Chamberlin, M., Newton, D.W., Lepine, J.A.: A meta-analysis of voice and its promotive and prohibitive forms: identification of key associations, distinctions, and future research directions. Pers. Psychol. **70**(1), 11–71 (2017)
48. Bradley, B.H., Postlethwaite, B.E., Klotz, A.C., Hamdani, M.R., Brown, K.G.: Reaping the benefits of task conflict in teams: the critical role of team psychological safety climate. J. Appl. Psychol. **97**(1), 151 (2012)
49. Edmondson, A.: Psychological safety and learning behavior in work teams. Adm. Sci. Q. **44**(2), 350–383 (1999)
50. Beal, D.J., Cohen, R.R., Burke, M.J., McLendon, C.L.: Cohesion and performance in groups: a meta-analytic clarification of construct relations. J. Appl. Psychol. **88**(6), 989–1004 (2003)
51. Chiocchio, F., Essiembre, H.: Cohesion and performance: a meta-analytic review of disparities between project teams, production teams, and service teams. Small Group Res. **40**(4), 382–420 (2009)
52. Loughry, M.L., Tosi, H.L.: Performance implications of peer monitoring. Organ. Sci. **19**(6), 876–890 (2008)
53. Carless, S.A., de Paola, C.: The measurement of cohesion in work teams. Small Group Res. **31**, 71–88 (2000)
54. Van der Vegt, G.S., Emans, B.J.M., Van de Vuert, E.: Patterns of interdependence in work teams: a two-level investigation of the relations with job and team satisfaction. Pers. Psychol. **54**, 51–69 (2001)
55. Casciaro, T., Lobo, M.S.: Competent jerks, lovable fools, and the formation of social networks. Harv. Bus. Rev. **83**(6), 92–99 (2005)
56. Casciaro, T., Lobo, M.S.: When competence is irrelevant: the role of interpersonal affect in task-related ties. Adm. Sci. Q. **53**, 655–684 (2008). https://doi.org/10.2189/asqu.53.4.655
57. Cuddy, A.J.C., Fiske, S.T., Glick, P.: Warmth and competence as universal dimensions of social perception: the stereotype content model and the BIAS map. Adv. Exp. Soc. Psychol. **40**, 61–149 (2008). https://doi.org/10.1016/S0065-2601(07)00002-0
58. Cuddy, A.J.C., Glick, P., Beninger, A.: The dynamics of warmth and competence judgments, and their outcomes in organizations. Res. Organ. Behav. **31**, 73–98 (2011). https://doi.org/10.1016/j.riob.2011.10.004
59. Judd, C.M., James-Hawkins, L., Yzerbyt, V., Kashima, Y.: Fundamental dimensions of social judgment: understanding the relations between judgments of competence and warmth. J. Pers. Soc. Psychol. **89**(6), 899–913 (2005). https://doi.org/10.1037/0022-3514.89.6.899
60. Fiske, S.T., Cuddy, A.J.C., Glick, P.: Universal dimensions of social cognition: warmth and competence. Trends Cogn. Sci. **11**(2), 77–83 (2007). https://doi.org/10.1016/j.tics.2006.11.005
61. Asch, S.E.: Forming impressions of personality. J. Abnorm. Soc. Psychol. **41**(3), 258 (1946). https://doi.org/10.1037/h0055756
62. Ohland, M.W., et al.: The comprehensive assessment of team member effectiveness: development of a behaviorally anchored rating scale for self and peer evaluation. Acad. Manag. Learn. Educ. **11**(4), 609–630 (2012)
63. Loughry, M.L., Ohland, M.W., Moore, D.D.: Development of a theory-based assessment of team member effectiveness. Educ. Psychol. Measur. **6**(67), 505–524 (2007)
64. Courtright, S.H., Thurgood, G.R., Stewart, G.L., Pierotti, A.J.: Structural interdependence in teams: an integrative framework and meta-analysis. J. Appl. Psychol. **100**(6), 1825–1846 (2015). https://doi.org/10.1037/apl0000027
65. Van der Vegt, G.S., Emans, B.J.M., Van de Vuert, E.: Patterns of interdependence in work teams: a two-level investigation of the relations with job and team satisfaction. Pers. Psychol. **54**(1), 51–69 (2001)

66. LePine, J.A., Piccolo, R.F., Jackson, C.L., Mathieu, J.E., Saul, J.R.: A meta-analysis of teamwork processes: tests of a multidimensional model and relationships with team effectiveness criteria. Pers. Psychol. **61**(2), 273–307 (2008). https://doi.org/10.1111/j.1744-6570.2008.00114.x
67. Gully, S.M., Devine, D.J., Whitney, D.J.: A meta-analysis of cohesion and performance: effects of level of analysis and task interdependence. Small Group Res. **26**, 497–521 (1995). https://doi.org/10.1177/1046496495264003
68. Wang, J., Cheng, G.H.L., Chen, T., Leung, K.: Team creativity/innovation in culturally diverse teams: a meta-analysis. J. Organ. Behav. **40**(6), 693–708 (2019). https://onlinelibrary.wiley.com/doi/10.1002/job.2362
69. Balkundi, P., Harrison, D.A.: Ties, leaders, and time in teams: strong inference about network structure's effects on team viability and performance. Acad. Manag. J. **49**(1), 49–68 (2006). https://doi.org/10.2307/20159745
70. Hackman, R.: The design of work teams. In: Lorsch, J. (ed.) Handbook of Organizational Behavior, pp. 315–342. Prentice-Hall, Englewood Cliffs, NJ (1987)
71. de Wit, F.R.C., Greer, L.L., Jehn, K.A.: The paradox of intragroup conflict: a meta-analysis. J. Appl. Psychol. **97**(2), 360–390 (2012). https://doi.org/10.1037/a0024844
72. De Cooman, R., Vantilborgh, T., Bal, M., Lub, X.: Creating inclusive teams through perceptions of supplementary and complementary person-team fit: examining the relationship between person-team fit and team effectiveness. Group Org. Manag. **41**(3), 310–342 (2016). https://doi.org/10.1177/1059601115586910
73. Hackman, R.: Leading Teams: Setting the Stage for Great Performances. Harvard Business Publishing, Boston, MA (2002)
74. Jackson, S.E., Joshi, A.: Work team diversity. In: APA Handbook of Industrial and Organizational Psychology, vol. 1: Building and Developing the Organization, pp. 651–686. American Psychological Association (2011)
75. Ely, R.J., Padavic, I., Thomas, D.A.: Racial diversity, racial asymmetries, and team learning environment: effects on performance. Organ. Stud. **33**(3), 341–362 (2012)
76. Curşeu, P.L., Pluut, H.: Student groups as learning entities: the effect of group diversity and teamwork quality on groups' cognitive complexity. Stud. High. Educ. **38**(1), 87–103 (2013)
77. Nemeth, C.J.: Dissent as driving cognition, attitudes, and judgments. Soc. Cogn. **13**(3), 273–291 (1995)
78. McLeod, P.L., Lobel, S.A., Cox, T.H., Jr.: Ethnic diversity and creativity in small groups. Small Group Res. **27**(2), 248–264 (1995)
79. Bear, J.B., Woolley, A.W.: The role of gender in team collaboration and performance. Interdiscip. Sci. Rev. **36**(2), 146–153 (2011)
80. Woolley, A., Malone, T.: What makes a team smarter? More women. Harv. Bus. Rev. **89**(6), 32–33 (2011)
81. Curşeu, P.L., Schruijer, S.G.: Does conflict shatter trust or does trust obliterate conflict? Revisiting the relationships between team diversity, conflict, and trust. Group Dyn. Theory Res. Pract. **14**(1), 66–79 (2010)
82. Kurtzberg, T.R.: Feeling creative, being creative: an empirical study of diversity and creativity in teams. Creat. Res. J. **17**(1), 51–65 (2005)
83. Harrison, D.A., Price, K.H., Bell, M.P.: Beyond relational demography: time and the effects of surface-and deep-level diversity on work group cohesion. Acad. Manag. J. **41**(1), 96–107 (1998)
84. Bell, S.T., Villado, A.J., Lukasik, M.A., Belau, L., Briggs, A.L.: Getting specific about demographic diversity variable and team performance relationships: a meta-analysis. J. Manag. **37**(3), 709–743 (2011)
85. Ro, H.K., Loya, K.I.: The effect of gender and race intersectionality on student learning outcomes in engineering. Rev. High. Educ. **38**(3), 359–396 (2015)

86. Our User Base—CATME. https://info.catme.org/instructor/history-research/our-user-base/. Accessed 2023/05/13
87. History & Research—CATME. https://info.catme.org/instructor/history-research/. Accessed 2023/05/13
88. Spencer, S.J., Steele, C.M., Quinn, D.M.: Stereotype threat and women's math perfor-mance. J. Exp. Soc. Psychol. **35**(1), 4–28 (1999)
89. Sackett, P.R., Hardison, C.M., Cullen, M.J.: On interpreting stereotype threat as account-ing for African American-White differences on cognitive tests. Am. Psychol. **59**(1), 7 (2004)
90. Sue, D.W.: Microaggressions in Everyday Life: Race, Gender, and Sexual Orientation, pp. xxiii, 328. Wiley (2010)
91. Canales, M.K.: Othering: toward an understanding of difference. Adv. Nurs. Sci. **22**(4), 16–31 (2000)
92. Ohland, M.W., Dickerson, D., Masta, S., Pawley, A.L.: Making engineering education more inclusive through the power of defaults. In: IEEE/ASEE Frontiers in Education Conference (2023, October, in review)
93. Park, G., et al.: Women are warmer but no less assertive than men: gender and language on Facebook. PLoS One **11**(5), e0155885 (2016)
94. Hofstede, G.: Culture's consequences: International differences in work related values. Beverly Hill, CA, Sage (1980)

# Learning to Research Through Inquiry-Based Learning - A Field Report from Exploratory Sexual Research in Psychology

Philipp Stang[1] and Yvonne Sedelmaier[1,2(✉)]

[1] SRH Wilhelm Loehe University of Applied Sciences, Merkurstraße 19, 90763 Fuerth, Germany
philipp.stang@srh.de, yvonne.sedelmaier@hs-coburg.de
[2] University of Applied Sciences Coburg, Friedrich-Streib-Straße 2, 96450 Coburg, Germany

**Abstract.** Research is a challenging aspect for students for several reasons: Research is a complex process that is usually addressed in many small-scale steps in a wide variety of courses. However, students often fail to see the big picture. This paper describes a competence-oriented inquiry-based learning approach to improve psychology students' research skills. In a capstone project, students are guided by the instructor through a complete research process in which they define their own research question, decide independently on the research design, and conduct and document the research. They conduct guided interviews on "explorative sexual research". Evaluation shows the high gain of competence for students. At some point, students ask for a little more assistance for example by preparing a research report.

**Keywords:** Inquiry-based learning · Explorative sexual research · Psychology studies · Capstone project

## 1 Introduction

Research is a central competence that is now found in many degree programs and is often reflected in dedicated modules such as "Scientific Work" or "Methods of Empirical Social Research". At the latest in the bachelor's thesis, many students are required to do their own scientific and often research-based work. However, the leap from theoretical subjects such as scientific working, which often involves the correct use of templates or citation rules, or isolated questions that are often only dealt with theoretically, as in the subjects of empirical social research, is often too great for students to really learn to research independently and to apply this holistically and comprehensively in the bachelor's thesis.

This paper presents a didactic concept that represents an intermediate step and transition from theoretical and "manual" research to independent and autonomous research and, as a kind of a guided capstone project, helps students to link individual elements and steps of the research process and to place them in a larger overall context while


M. E. Auer et al. (Eds.): ICL 2023, LNNS 899, pp. 102–110, 2024.
https://doi.org/10.1007/978-3-031-51979-6_11

also working on a concrete research project. Such a project is shown in a course in a psychology degree program.

Section 2 explains the pedagogical basis of our concept before the context of our capstone research project in explorative sexual research is described in Sect. 3. Sect. 4 gives insight into our competence-oriented capstone-project itself including students' results before some conclusions are drawn in sect. 5.

## 2 Competence-Oriented Research and Education and Inquiry-Based Learning

### 2.1 Competence-Oriented Research and Education (CORE)

In the Psychology degree program at our university learning is based on the CORE principle (Competence Oriented Research and Education; [1]) which combines the acquisition of competences and the joy of learning through active, independent learning and the associated experience of self-efficacy [2]. Learning environments are therefore designed in such a way that subject-specific, social, methodological, and personal competences can be trained. Lively learning takes place [3].

### 2.2 Inquiry-Based Learning

In higher education, competence-oriented methods such as inquiry-based learning have now also arrived and are being applied.

Competence-oriented learning has a long tradition, especially in vocational education (see e.g. [4]). During the last years, triggered by the bologna process, competence-oriented learning settings are becoming more popular even in higher education [5]. At some universities, competence-oriented didactic is investigated [6–11] and subject-matter didactic theories are developed, applied, and evaluated [12].

Learning how sciences do work is more than memorizing scientific facts, but rather is really understanding how scientific information is gained and interpreted and how scientific methods and concepts are applied [13]. To achieve learning, applying and understanding science, inquiry-based learning is seen as an opportunity to improve students' research skills [14] and is in line with constructivism [15, 16]. In pedagogy, learning works best when a high practical relation exists and when students recognize a need or identify a problem that has to be solved [17].

Schwab [18] outlined levels of scientific inquiry which were revived by Banchi and Bell [19]. These four levels describe the grade of given structure in the learning setting. At level one a given inquiry setting with known results is given to students. At level two a research setting is given, and students should interpret the results. The third level is a guided inquiry and at fourth level gives students the possibility to decide on the complete research process (see Fig. 1).

In our inquiry-based learning setting students go through a complete research process from the beginning. They develop research questions by themselves, decide about the research design, apply it, collect, and evaluate data. According to Banchi and Bell this corresponds to level four.

For these reasons and in combination with our intended learning outcomes, we applied this approach to our course.

**The four levels of inquiry and the information given to the student in each one.**

| Inquiry Level | Question | Procedure | Solution |
|---|---|---|---|
| 1—Confirmation Inquiry<br>*Students confirm a principle through an activity when the results are known in advance.* | ✔ | ✔ | ✔ |
| 2—Structured Inquiry<br>*Students investigate a teacher-presented question through a prescribed procedure.* | ✔ | ✔ | |
| 3—Guided Inquiry<br>*Students investigate a teacher-presented question using student designed/selected procedures.* | ✔ | | |
| 4—Open Inquiry<br>*Students investigate questions that are student formulated through student designed/selected procedures.* | | | |

**Fig. 1.** Four levels of inquiry and the information given to the student in each one [19].

# 3 Context and Intended Learning Outcomes

## 3.1 Context

It is particularly important for psychology students to be able to plan and conduct own research independently. According to the guidelines of the German Psychological Society and the European Certificate in Psychology, each thesis should contain an active and independent research component [20–22]. In our bachelor's program, an empirical research seminar is therefore offered in the fourth semester. It includes a research project through which students gain impressions of practical work in the professional future in the sense of the CORE principle (see Chap. 2.1) and are prepared for research they need for the bachelor's thesis.

In previous study sections, the basics of scientific work in general as well as research methods such as quantitative and qualitative research were already trained.

## 3.2 Intended Learning Outcomes

In pedagogy, it is indisputable that goals of an educational measure are indispensable [see e.g. 23, 24]. In CORE, and also in this case, we start with defining competences for our course: The students know empirical methods and apply them to a case study they have developed themselves. They analyse the collected data and present the elaborated results in front of an audience. According to Roth's [18] competence model, we address competences at all four competence areas:

- Professional competence: The students know why and with which scientific knowledge strategies they approach application questions, where they can find scientific studies, how they are to be interpreted. They know the latest scientific developments in basic and applied research in psychology and related disciplines. They apply adequate quantitative and qualitative research methods to analyse and interpret data.
- Methodological competence: Students can model and receive questions, simultaneously develop their own questions for an empirical study based on the received research literature and discuss these in the context of consultations. They process or investigate the questions in an independent empirical research approach, collect and

analyse data and document them according to scientific standards. They present the results appropriately.

- Self-competence: Students know about the importance of good scientific practice, the importance of normative-ethical principles of good scientific practice, apply this knowledge to their own research and represent or justify the approach to others.
- Social competence: During the processing of the research question, students work together with other stakeholders and show teamwork skills. When preparing the presentation, they reflect on and discuss different perspectives and interests of other stakeholders and show that they can adequately react to others. They present, discuss, and justify their own research projects and results in front of an expert audience.

# 4 Capstone-Research-Project in Explorative Sexual Research

## 4.1 Teaching and Learning Concept

These are full-day courses in which theoretical knowledge is refreshed or deepened in the morning and implemented on a concrete research project in the afternoon. At the end of each day, the results are compiled in the whole group and subjected to a peer review.

The total of nine students works in small groups of 2 people, which remain constant throughout the entire process. The task is to work independently on a research question including the application of qualitative empirical research methods or a mixed methods design. In doing so, the students should make and implement their own research decisions and gain research experience. The examination performance is a research report. In previous study sections, the basics of scientific work in general as well as research methods such as quantitative and qualitative research were already trained.

Day 1: The guiding theme for the whole project is given: Experiencing and Behaviour of People. After a warm-up, the lecturer gives an overview of the whole research process, including a systematic literature process, before a common overall theme is systematically developed. This process is methodical supported and moderated by the lecturer. Outcome: Sexual satisfaction in relationships with the possible questions: To what extent exist differences in people's sexual satisfaction? And what patterns can be identified?

Day 2: The main topic is fleshed out in the respective small groups, related work is researched, and an interview guideline is developed for each small group. It is possible to set different focal points or perspectives on the jointly chosen main topic in the individual teams. Thus, different interview guidelines are possible for the different small groups. One group also takes next a psychometrical instrument for the data-collection.

This is followed by a phase in which each group conducts and transcribes its interviews.

Day 3 focussed on the evaluation and interpretation of the collected data by means of qualitative research.

The 4th and 5th day are dedicated to the consolidation, processing, and discussion of the results from the small groups as well as the presentation of the research results. The project is concluded with a so-called post-mortem analysis [26].

## 4.2 Four Student Sub-Projects

During the summer term of 2022, students intrinsically motivated to engage with sexuality research from a psychological perspective and developed study designs to conduct learning projects for exploratory, empirical research. The research interest emerged from the intertwining of diverse topics on sexuality, satisfaction, and MBSR and different research approaches during their studies, as well as the students' personal interest and life realities. In contrast to the project of a quantitative longitudinal study of a short-term MBSR intervention in winter 2021, the students in the present project opted for qualitative data collection and predominantly qualitative evaluation methods [27].

In addition, the students have already gained insights into research in sexual science in their previous studies. In various sub-disciplines of psychology, students also dealt with aspects of theoretical and empirical foundations of sexuality and gender, e.g. general psychology, biopsychology, developmental psychology, differential and personality psychology, social psychology, clinical psychology, etc. It is important to create a safe place at our university for sexual and gender diversity. We also address sexual and gender diversity, anti-discrimination, and marginalization in professional days at the university [28, 29].

### 4.2.1 Sexual Satisfaction of Men

Schneider and Doganay [30] addressed the sexual satisfaction [31] of heterosexual men aged 18 and 30 at the time of different relationship stages, as the change in sexual satisfaction of men and women in monogamous romantic relationships is a little studied topic in previous research [32]. Methodologically, an exploratory qualitative interview study was conducted with 5 men at one measurement point who were in at least a one-year relationship. The partially standardized interviews were analysed via qualitative content analysis following Mayring [33]. The results indicate that the sexual satisfaction of the subjects was lower at later relationship stages than at earlier relationship stages. Explanatory approaches could be found on the one hand in a reduction of "the (erotic) tension". On the other hand, subjects reported a process of building up emotional components during the relationship, which in turn had a positive effect on sexual satisfaction and compensated for the lack of tension to a certain extent. For all subjects, frequency of partnered intercourse was positively associated with sexual satisfaction.

### 4.2.2 Sexual Satisfaction of Women

Firnges, Hofer, and Maier [34] addressed the sexual satisfaction [35, 36] of 20-year-old women in monogamous heterosexual relationships, as the specific age range has received little isolated attention in previous research [37]. Therefore, an exploratory study with one point of measurement was conducted via guided interviews ($N = 18$), which were qualitatively analysed via QCAmap (https://www.qcamap.org/ui/de/home). No differences emerged in the sample between the sexual satisfaction of female subjects with short-term and long-term relationships. Female subjects attributed sexual satisfaction to masturbation, partnered sex, and the combination of masturbation and partnered sex. Communication within a relationship was of great importance. According to their own statements, most of the test persons found it easy to express sexual desires and needs.

In addition, the sexual satisfaction of the partner has an influence on their own sexual satisfaction. Quantified, 83% of the test subjects were sexually very satisfied and grateful for their relationships. The remaining 17% stated that sexual satisfaction could be expanded and that more open communication could make this possible.

### 4.2.3 Differences in Sexual Satisfaction Between Single Men and Women

Karl and Schuster [38] addressed differences in sexual satisfaction [31, 39] of single men and women aged 18 to 25 due to a gap in research [36]. Methodologically, an exploratory study with qualitative fully standardized interviews was conducted with 6 subjects (three heterosexual men, three queer women) at one measurement point and analysed discriminatively. Exploratory evidence emerged that most men and women interviewed who are not in a relationship, are sexually satisfied. The men focused on frequency of sexual contact, masturbation, physical closeness, and affection in their assessment of sexual satisfaction. The women operationalized their sexual satisfaction as a state in which they lacked nothing, connectedness, security, consensus, and communication to the intimate partner.

### 4.2.4 Emotional Closeness and Sexual Satisfaction

Reinwald and Tursic [40] addressed emotional closeness and sexual satisfaction, as the current body of research has insufficiently explored this relationship [36, 41]. A mixed-methods design was taken to conduct an exploratory cross-sectional study, a semi-structured guided interview and the Sexual Satisfaction Questionnaire [42]. It was found among the 10 heterosexual subjects (6 females and 4 males, ages 19–21, with current sexual contacts) that individuals who were emotionally close had higher sexual satisfaction scores than participants with a low marked emotional connection. Positive expressions of the subcategories of emotional closeness communication, commitment, and mutual influence appeared to a condition of higher sexual satisfaction scores. The expressions of these subcategories behaved proportionally to the sexual satisfaction scores. It can be assumed that emotional closeness to the sexual partners appears to be related to sexual satisfaction in the present sample. Similarities with other research support the findings [43, 44].

## 5 Conclusion, Summary and Outlook

This paper applies inquiry-based learning to a capstone project in a psychology programme to improve research skills for bachelor students. Qualitative evaluation shows that intended learning outcomes are being achieved very well and inquiry-based learning is a suitable method for improving research skills. In the context of teaching empirical scientific work, it was possible for students to conduct exploratory research projects independently and autonomously. It should be emphasized that the research projects were student projects for application-oriented teaching and learning in order to make first independent experiences with applied research. This also means that the significance of these findings must be interpreted with caution due to methodological limitations.

In the summative qualitative evaluation, the students expressed an increase in subject-specific, social, methodological, and personal competences regarding the teaching responsibility and the pedagogical concept. Among other things, students stated that they had learned how to deal with sexual feelings from a psychological perspective, how to talk seriously about sexuality, and how to adopt a matter-of-fact attitude towards the sexual topic with subjects. They also practiced aspects of qualitative research methodology and mixed methods. In particular, the practical aspects of the pilot projects were positively evaluated as preparation for the bachelor's thesis, which is due in the following term. In addition, the following aspects were positively highlighted: free choice of topic, independence in research, joint preparation of a paper, personal responsibility, teamwork, common overall topic with specific sub-projects, time management and freedom. As a recommendation for the future, the students requested the presentation of a theoretical concept on the research process as well as an outline for the paper in the seminar. With these recommendations the module will be completed and repeated, supplemented by a quantitative evaluation.

Positively, regarding the research results, their hypothesis-generating character can postulate. With the exploratory weight of the results, future research is spaced on these, e.g., replicating studies or hypothesis-testing the highlighted aspects [45, 46].

**Acknowledgement.** We thank all participating students, Cenk Doganay, Anouk Firnges, Hannah Hofer, Natascha Karl, Anne Maier, Tita Reinwald, Daniel Schneider, Isabell Schuster, and Leonie Tursic, as well as the subjects.

## References

1. Rózsa, J., Bauer, S., Edinger, S.: CORE-Gerechte Modulkonzeption. SRH Hochschule Heidelberg, Heidelberg (2021)
2. Bandura, A.: Self-efficacy: toward a unifying theory of behavioral change. Adv. Behav. Res. Ther. **1**, 139–161 (1978). https://doi.org/10.1016/0146-6402(78)90002-4
3. Arnold, R. (ed.): Lebendiges Lernen. Schneider, Baltmannsweiler (1996)
4. Mertens, D.: Schlüsselqualifikationen. Thesen zur Schulung für eine moderne Gesellschaft. Mitteilungen aus der Arbeitsmarkt- und Berufsforschung **7**, 36–43 (1974)
5. Arnold, R.: Bildung nach Bologna!: Die Anregungen der europäischen Hochschulreform. Springer Fachmedien Wiesbaden, Wiesbaden (2015). https://doi.org/10.1007/978-3-658-08978-8
6. Sedelmaier, Y., Landes, D.: Active and inductive learning in software engineering education. In: 37th International Conference on Software Engineering (ICSE), pp. 418–427 (2015)
7. Sedelmaier, Y., Landes, D.: Software engineering body of skills. In: 5th IEEE Global Engineering Education Conference (EDUCON), pp. 395–401. IEEE (2014)
8. Sedelmaier, Y., Landes, D.: Practicing soft skills in software engineering. In: Yu, L. (ed.) Overcoming Challenges in Software Engineering Education, pp. 161–179. IGI Global (2014). https://doi.org/10.4018/978-1-4666-5800-4.ch009
9. Sedelmaier, Y., Landes, D.: A multi-level didactical approach to build up competencies in requirements engineering. In: Penzenstadler, B., Gregory, S., and Landes, D. (eds.) 8th Int. Workshop on Requirements Engineering Education & Training Co-located with 22nd International Conference on Requirements Engineering (RE 2014), pp. 26–34. CEUR Workshop Proceedings, Karlskrona, Schweden (2014)

10. Sedelmaier, Y., Landes, D.: A multi-perspective framework for evaluating software engineering education by assessing students' competencies: SECAT—a software engineering competency assessment Tool. In: 44th IEEE Frontiers in Education Conference (FIE), pp. 2065–2072. IEEE, Madrid (2014)
11. Sedelmaier, Y., Landes, D.: Evaluating didactical approaches based upon students' competences. In: 7th IEEE Global Engineering Education Conference (EDUCON), pp. 527–536. IEEE (2016). https://doi.org/10.1109/EDUCON.2016.7474603
12. Sedelmaier, Y.: Basics of Didactics for Software Engineering. LAP LAMBERT Academic Publishing, London (2019)
13. Bell, T., Urhahne, D., Schanze, S., Ploetzner, R.: Collaborative inquiry learning: models, tools, and challenges. Int. J. Sci. Educ. **32**, 349–377 (2010). https://doi.org/10.1080/09500690802582241
14. Huber, L.: Warum Forschendes Lernen nötig und möglich ist. (2013)
15. Reich, K.: Konstruktivistische Didaktik: das Lehr- und Studienbuch mit Online-Methodenpool. Beltz, Weinheim Basel (2012)
16. Siebert, H.: Pädagogischer Konstruktivismus: Lernzentrierte Pädagogik in Schule und Erwachsenenbildung. Beltz, Weinheim (2005)
17. Holzkamp, K.: Wider den Lehr-Lern-Kurzschluß. In: Arnold, R. (ed.) Lebendiges Lernen, pp. 21–30. Schneider, Baltmannsweiler (1996)
18. Schwab, J.J., Brandwein, P.F. (eds.): The Teaching of Science: The Teaching of Science as Enquiry. Harvard University Press, Cambridge (1962)
19. Banchi, H., Bell, R.: The many levels of inquiry. Sci. Child. **46**, 26–29 (2008)
20. DGP.: Deutsche Gesellschaft für Psychologie e.V. Psychologische Rundschau. **74**, 24–80 (2023). https://doi.org/10.1026/0033-3042/a000619
21. DGPs.: Empfehlungen des DGPs-Vorstands zu Bachelor- und Masterstudiengängen in Psychologie (2021). https://www.dgps.de/fileadmin/user_upload/PDF/Empfehlungen/Empfehlungen_des_Vorstands_Bachelor_und_Master_15_12_14.pdf
22. EFPA.: EuroPsy—the European certificate in psychology. EFPA Regulations on EuroPsy and Appendices (2021). https://www.europsy.eu/_webdata/europsy_regulations_december_2021_virtual_ga.pdf
23. Mager, R.F.: Preparing Instructional Objectives. Kogan Page, London (1992)
24. Arnold, R., Erpenbeck, J.: Wissen ist keine Kompetenz: Dialoge zur Kompetenzreifung. Schneider, Baltmannsweiler (2014)
25. Roth, H.: Entwicklung und Erziehung: Grundlagen einer Entwicklungspädagogik. Schroedel, Hannover (1971)
26. Birk, A., Dingsoyr, T., Stalhane, T.: Postmortem: never leave a project without it. IEEE Softw. **19**, 43–45 (2002). https://doi.org/10.1109/MS.2002.1003452
27. Stang, P., Rico-Dresel, D.: Einfluss eines vierwöchigen achtsamkeitsbasierten Trainings auf die Lebenszufriedenheit. PPmP—Psychotherapie Psychosomatik Medizinische Psychologie (2023). https://doi.org/10.1055/a-2050-3633
28. Stang, P., Köllner, M., Weiss, M.: Fachtag Psyche und Gesundheit im Einklang, Fürth (2023)
29. Stang, P., Wüchner-Fuchs, M.: Sexualität und Geschlecht im Kontext von Menschen mit Behinderung. In: Versorgungsstrukturen von Menschen in vulnerablen Lebenslagen. Kohlhammer, Stuttgart (in press)
30. Schneider, D., Doganay, C.: Sexuelle Zufriedenheit bei jungen Männern in verschiedenen Beziehungsstadien (2022)
31. Derogatis, L.R., Melisaratos, N.: The DSFI: a multidimensional measure of sexual functioning. J. Sex Marital Ther. **5**, 244–281 (1979). https://doi.org/10.1080/00926237908403732

32. Irmer, J.: Die Rolle des Sexuallebens in einer romantischen Partnerschaft für die Beziehungszufriedenheit. Der Fragebogen zum Erleben von Sexualität in engen Partnerschaften (FESP). J. Fam. Res. **20**, 229–246 (2008). https://doi.org/10.20377/jfr-244
33. Mayring, P.: Qualitative Content Analysis. SAGE Publications Ltd, London (2021)
34. Firnges, A., Hofer, H., Maier, A.: Empirisches Forschungsseminar (2022)
35. Bucher, T., Hornung, R., Buddenberg, C.: Sexualität in der zweiten Lebenshälfte: Ergebnisse einer empirischen Untersuchung. Z Sex-Forsch. **16**, 249–270 (2003). https://doi.org/10.1055/s-2003-43537
36. Schönbucher, V.: Sexuelle Zufriedenheit von Frauen: psychosoziale Faktoren. Z Sex-Forsch. **20**, 21–41 (2007). https://doi.org/10.1055/s-2007-960559
37. Prentki, D.: Sexuelle Zufriedenheit von Frauen. Eine Studie zum Vergleich verschiedener Altersgruppen. Diplomica, Hamburg (2010)
38. Karl, N., Schuster, I.: Empirische Forschungsarbeit (2022)
39. Renaud, C., Byers, E.S., Pan, S.: Sexual and relationship satisfaction in Mainland China. J. Sex Res. **34**, 399–410 (1997)
40. Reinwald, T., Tursic, L.: Empirisches Forschungsprojekt zur emotionalen Nähe und sexuellen Zufriedenheit (2022)
41. Øverup, C.S., Smith, C.V.: Considering attachment and partner perceptions in the prediction of physical and emotional sexual satisfaction. J. Sex. Med. **14**, 134–143 (2017). https://doi.org/10.1016/j.jsxm.2016.11.310
42. Soeder, M.: Fragebogen zur sexuellen Zufriedenheit (SZ) (2007). https://www.praxis-fuer-maenner.de/organisation/php/syntax_sz.php
43. Byers, E.S.: Beyond the birds and the bees and was it good for you?: thirty years of research on sexual communication. Can. Psychol./Psychol. Can. **52**, 20–28 (2011). https://doi.org/10.1037/a0022048
44. Schmiedeberg, C., Schröder, J.: Does sexual satisfaction change with relationship duration? Arch. Sex. Behav. **45**, 99–107 (2016). https://doi.org/10.1007/s10508-015-0587-0
45. Büsing, S., Hoppe, C., Liedtke, R.: Sexuelle zufriedenheit von frauen—entwicklung und ergebnisse eines fragebogens. Psychother. Psychosom. Med. Psychol. **51**, 68–75 (2001). https://doi.org/10.1055/s-2001-10757
46. Lawrance, K.-A., Byers, E.S.: Sexual satisfaction in long-term heterosexual relationships: the interpersonal exchange model of sexual satisfaction. Pers. Relat. **2**, 267–285 (1995). https://doi.org/10.1111/j.1475-6811.1995.tb00092.x

# Collaborating Towards Humanizing Pedagogies in Teaching and Learning: Case of Universities of Technology in South Africa

Nereshnee Govender[1](✉) and Elisha Didam Markus[2]

[1] Durban University of Technology, Durban, South Africa
nereshneeg@dut.ac.za

[2] Central University of Technology, Free State, South Africa
emarkus@cut.ac.za

**Abstract.** The complex history of the South African higher education sector has led to a growing need for universities to engage in humanistic approaches to teaching and learning. This paper discusses ways in which key agents including academics can challenge the status quo in higher education and promote collaborative and transformative practice. Universities in modern society aim to educate and develop students to cope with the complexities of the economy and social change, and to produce new knowledge, while strengthening a commitment to society and democracy. Owing to the current low pass rate of students in the South African higher education sector, there has been a growing call to rethink ways of learning and teaching to ensure greater success for the diverse student body. Students enter university with rich histories, experiences, knowledge, and ways of viewing reality and require enabling environments to develop their identity and succeed at university. The authors acknowledge that students need these supportive spaces to navigate their way through university. Using a collaborative auto-ethnography inquiry approach, we reflect on how our experiences shape our practice in higher education. We value a social constructivist approach where learning is constructed through interaction, and not solely transmitted through instruction. The principles of Communities of Practice (CoP), Paulo Freire's idea of a Humanizing Pedagogy (1970) and ideas on Critical Reflexivity in teaching and learning are adopted as a framework underpinning our reflections. The discussion is located within these tenets to explore the significance of collaborative approaches to teaching and learning in higher education in South Africa, and more specifically in a University of Technology. This paper argues that it is important to authentically engage with practices from the past, reflect on them and look towards future practices to provide quality educational experiences for students. Our reflective discussion on teaching and learning has suggested that teaching-learning are intertwined, and that carefully planned collaboration and interaction would bring us a step closer to offering students a holistic university environment that supports humanisation and the development of the 'whole' student.

**Keywords:** Collaboration · Communities of practice · Humanizing pedagogy


M. E. Auer et al. (Eds.): ICL 2023, LNNS 899, pp. 111–123, 2024.
https://doi.org/10.1007/978-3-031-51979-6_12

## 1 Introduction

Students have rich experiences, ideas, beliefs and knowledge, and need spaces to be able to learn and grow socially, emotionally, and academically. This study acknowledges that students need supportive, flexible and inclusive spaces to navigate their way through university. The aim of this paper is to understand how to make student learning more effective and to meet the needs of the diverse student population. The reflective narratives, through the telling of our experiences on teaching and learning within two Universities of Technology (UoT) in South Africa (SA), suggests that teaching and learning are intertwined and that engagements and interactions that are carefully considered and context-based bring us towards a university environment that promotes and values humanisation in higher education. The authors are academics at two UoTs in SA. Author one is a writing centre practitioner and author two is an associate professor in electronic engineering. Our reflections in this paper points to ways in which key agents, including academics, can challenge the status quo in higher education and promote collaborative and transformative higher education practice and quality educational experiences for students. In telling our stories, we engage with knowledge and voices that draw not only on the work of scholars from the global North, but also from the global South, which sets the tone to understanding what it means to be involved in an academic project in an African institution. In a university context, there is a need to have a shared understanding among academics, academic leaders and managers, students, and academic developers, and to create enabling environments that focus on students' learning and shape responsible, global citizens that can enter the workplace and influence society. Reflecting on our practice is important to be able to move towards teaching approaches that encourage and embrace active learning and the production of students' own form of learning. In this paper, the principles of Communities of Practice (CoP), Paulo Freire's idea of a Humanizing Pedagogy (1970) and ideas on Critical Reflexivity in teaching and learning are adopted as a framework underpinning our reflective discussion to explore the significance of collaborative approaches to teaching and learning in a UoT context. In our attempt to create more inclusive and humanizing educational experiences for students, our collaborative autoethnography (CAE) focuses on three main objectives: Analysing how academic agents can promote and sustain humanizing pedagogies in higher education; Analysing Communities of Practice (CoP) as a tool to promote learning in UoTs; Examining the role of critical reflection in teaching and learning in SA.

The authors are both from historically disadvantaged higher education institutions having been affected by the country's education transformation process. The democratic dispensation in SA led to the South African government having to make complex changes to the higher education sector. The sector saw amalgamations of higher education in SA and universities were reduced from 36 to 24 (McKenna 2004). The restructuring was as a result of mandatory and not voluntary processes. Durban University of Technology (DUT) and Central University of Technology (CUT) were formed through restructuring and mergers as part of the government's transformation plan. DUT and CUT now form part of the seven UoTs in the country. DUT has approximately 33 000 students and CUT has more than 21 000 students, most of whom are first generation entrants to university. These students are from disadvantaged communities and depend on government funding and support for their studies.

## 2 Overview of Higher Education in South Africa

The democratisation of SA in 1994 allowed for more black students to participate in the tertiary sector and enabled epistemological access (Morrow 1993). It also brought significant changes to the issues of qualifications and staffing and resources. The national plan for higher education was intend to "promote equity of access and fair chances for success, to all who are seeking to realize their potential through higher education, simultaneously eradicating all forms of unfair discrimination and advancing redress for past inequalities" (Department for Higher Education and Training 2013, 27). However, planned changes to curricula were not implemented and historically "little attention has been paid to the transformation of highly problematic institutional cultures" (Quinn and Vorster 2017, 36). This is affirmed by the Council on Higher Education (CHE) report that calls for the evaluation and review of institutional cultures in SA (CHE 2016). Higher education has an important role in furthering social democracy, and "a transformed higher education system would play a critical role in an emerging, non-racial, progressive democracy, in producing critical, independent citizens as well as skilled and socially committed graduates who would be capable of contributing to social and economic development" (CHE 2016, 22). Badat proposes collective approaches to "research, scholarship, learning and teaching, curriculum, pedagogy" (Badat 2015, 2) in ensuring justice and human rights in the South African higher educator sector. One of the key roles of universities in modern society is to train and educate students, produce new knowledge and equip them with the capabilities to navigate through social change while strengthening social justice, equity, and democracy (Badsha and Cloete 2011). Higher education has a significant role in the development of a modern economy (CHE 2016, 17). As expressed by Boughey (2019), we are experiencing a "new conjuncture" in higher education in our country and affirms the need for knowledge workers. Her analysis urges those in the university project to understand students as social beings. "Learning and teaching is not neutral - it is culturally, socially, and politically motivated" (Govender 2021, 11). It is against this backdrop that we locate our reflective discussion on promoting collaborative and transformative higher education practice that values humanisation, social justice, and social and academic transformation. Developing critical thinkers needs to be at the centre of teaching and learning in the modern age as students need to be able to adapt to changing demands in university and the broader society.

At this juncture in higher education majority of students are not succeeding in their studies. Quinn and Vorster (2019, 7) explain that throughput rates remain racially skewed "with a majority of black students dropping out or failing". The turbulent student protests of 2015 and 2016 in SA called for the decolonisation of institutional structures, cultures, and curriculum. The protests foregrounded students' feelings of isolation and alienation from institutional structures and cultures and the curriculum. The South African higher education curriculum has been a contentious issue with calls for changes in delivery modes of the curriculum, content and purpose. Researchers have called for "a South African education system that challenges colonial influences on education and a transformation of curricula that enables the holistic development of students" (Leibowitz and Bozalek 2015 and Govender 2021, 11). Perhaps one way of getting there is through collaborative approaches to teaching and learning. Sadly, while there have been engagements on how to transform discourse there has been a gap in terms of "not engaging in significant structural and cultural changes beyond changing staff and student demographics" (Quinn and Vorster 2017, 37). This paper focuses on how academics can use humanizing pedagogies in working with students to authentically engage with past and current practices to provide quality educational experiences for students and to meet the needs of the diverse student body.

## 3 Methodology

Using collaborative autoethnography (CAE) we critically reflect on how our experiences in academia shape our practice in UoTs in SA. We recount our experiences and critically reflect on ways in which we can enhance our practice and the teaching and learning agenda at our universities. Autoethnography allowed us to reflect on and write about our past and cultural experiences within our academic workspaces (Adams et al. 2017). This approach enabled us to assert our agency as researchers and academics to narrate and analyze our own lived experiences in our own voice. Ellis et al. (2011) assert that the autoethnography approach plays a vital role in enabling authors to reflect on their personal lives and share their academic experiences. Moreover, autoethnography allowed for reflexivity, enabling us to embed our perspectives and reflect on how these perspectives are influenced by the wider behaviours, habits and processes as insiders within our universities (Adams et al. 2017). CAE allowed us to reflect on our lived experiences in UoTs in SA and how we negotiated and continue to negotiate our roles within our universities to contribute towards driving a sustainable and transformative teaching and learning agenda. CAE has particular strengths including "self-reflexivity associated with autobiography, cultural interpretation associated with ethnography, and multi-subjectivity associated with collaboration" (Chang et al. 2013, 17). CAE enables researchers to "break through the dominant representations of professional practice, creating new knowledges" (Denshire 2014, 838) and explore the complexity of teaching and learning in higher education (Barkhuizen 2017). In conducting this CAE inquiry (May 2022 to May 2023) we were guided by Chang et al.'s (2013) five steps in using CAE for research which included: 1) forming our team, 2) deciding on the research focus area, 3) selecting a collaborative model, 4) defining our roles and boundaries, and 5) discussing ethical principles in the research. In terms of *forming our team* or group we decided to collaborate on this research project two years ago through informal conversations on teaching and learning at our respective UoTs. Our discussions centred on our individual practices; and the contestation on whether we needed to be producing students who pass exams or students who are able to contribute to societies in which they live and will one day work. These conversations began during our engagements in a Higher Education Studies post graduate course that we had undertaken (2019-2021). We subsequently decided that we would endeavour to write up our 'stories' of lived experiences of working with students in UoTs in SA. In 2022 we decided to set up regular virtual meetings due to our geographical proximity. This enabled us to formally continue the dialogue that was initiated during our studies together. The on-line meetings were scheduled bi-weekly, and we maintained in regular contact, exchanging information, readings and ideas via emails and telephone calls. We used an MS Teams folder to collate all meetings, discussions and resources.

In *deciding on the research focus area,* we wanted to focus on how our learnings during the post graduate course enabled us to reshape our thinking on teaching, learning, research and assessment in higher education. When we began our initial discussions, we kept going back to the idea of the value of reflection and how it is important to look back in order to move forward and transform our practice. As a result, we began reading on autoethnography and subsequently CAE as a potential methodology with its affordances to explore the nexus between the personal and the cultural (Chang et al. 2013; Roegman et al. 2020). The regular meetings resulted in common understanding of CAE inquiry, and we discussed how we will use it to explore our lived experiences in teaching and learning by writing up and analyzing our 'lived' stories. Through our discussions on our trajectories in higher education, we identified three concepts that resonated with both our work and identities as academics. We then shared literature on these conceptual frameworks, namely CoPs; humanizing pedagogies; and critical reflexivity. We engaged in discussion on the texts we shared and made notes; discussed how they related to our individual work as writing centre practitioner and engineering lecturer and then wrote up our 'draft' narratives and shared them using our MS Teams folder. We read each other's narratives, asked questions, and challenged each other's thinking using the comments feature on MS word. This enabled us to reread our stories, refine and engage in more critical reflection of our practice. We also took into consideration Spry's (2001) argument for a good autoethnography and how it "is a provocative weave of story and theory" (p. 713). In considering this view, we acceded to focus on the patterns in our accounts and we decided to develop three key questions that would underpin our study. We crafted the questions that aligned seamlessly with three theoretical frameworks. The iterative writing process allowed us to revise and strengthen our narratives while developing the flow, organisation and tone of the writing. We also agreed that our analysis would be integrated so as to not appear as an isolated analysis of experiences and recounts of our stories, but rather as a collective analytical process that involved collaborative exchanges and engagements through regular meetings, discussions, questioning and analysis. The analysis presented in section 4 is the outcome of a combined process and reflect our learning from the collaborative critical dialogues through synchronous and asynchronous exchanges. We were able to engage in discussion and question each other's ways of thinking about the needs of students in higher education.

The third step was *selecting a collaborative model.* As the CAE involved two researchers, we agreed that we would both be involved in all stages of the project. This meant that we both were integral to the data gathering, data analysis, interpretation of the narratives and the final written lived experiences. The online meetings, which were stored on MS Teams, were revisited as they served as data that informed the analysis and discussion. The researchers also set out to *define research roles and boundaries.* Given that we both shared a common interest in this research area, we discussed particular aspects in which either of us would assume the lead role. We discussed preferences, timeframes, communication styles and set targets in terms of meeting our research timelines. The fifth step was a *discussion on ethical principles.* We engaged in discussion on ethical concerns and maintained to not reveal the identities of any individual/s implicated in our stories. The next section provides the authors' reflections which are interwoven with the theoretical tenets of CoP, Humanizing Pedagogy (Freire 1970) and ideas on Critical Reflexivity in teaching and learning in higher education.

## 4 Our Stories Through Academic Lenses

As academics, we practice in a complex world where teaching and learning are seamlessly interconnected. Furthermore, students require enabling, flexible, collaborative environments and partnerships to flourish and succeed in their academic journeys. Through our reflective narratives we tell our stories of how we introspect, reflect and self-define to shape our practice in order to be able to provide quality learning experiences for students.

DUT's Writing Centre was launched in 2013 to provide reading and writing support to students and staff. This was an important initiative for DUT as academic discourse and reading and writing remains a huge concern and often a fearful task among students, majority of whom are first generation students at the institution. Students from disadvantaged backgrounds may experience such difficulties to a greater degree than other students. Writing centre tutors are from diverse disciplines who take on the role of an outside reader to facilitate inclusive, collaborative learning. In working towards developing critical thinkers, consultations focus on developing students' ideas, assisting them with formulating arguments and synthesising information. Tutors use a variety of techniques such as brainstorming and free-writing to enable students to unlock their writing potential. "Those that engage with the writing centre, grow in an informal way by dialogical engagement and interactions with tutors, students, writing practitioners, staff, and other members of the university community" (Govender 2021, 13). Badat proposes collaborative and deep discussions relating to "research, scholarship, learning and teaching, curriculum, pedagogy" (2015, 2) and ensuring justice and human rights. *"In my work with students I try to engage with them in ways that will build their confidence in writing in a safe academic environment. I often encourage students to talk about their ideas in an informal, relaxed way before moving onto their academic writing tasks. A simple thing like sitting beside a student and affording them the time to talk through their ideas can open ways for more engagement and deep learning"* (A1). Dialogic conversations that are well-planned and carefully considered can facilitate powerful learning experiences for students. Writing centre work further focuses on how students negotiate the content and requirements of their disciplines and develop their voices in response to the difficulties they experience at university. Writing Centre work is steeped in the principles of a "humanizing pedagogy" (Freire 1970) that values attentive listening to students. This contextualised approach brings us a step closer to a university sphere that promotes and values humanisation, social justice, and social and academic transformation. *"Too often I see students that are fearful of communicating and when consulting with them I always explain how our centre is for all students, not just those struggling with writing – I try to get students to understand that everyone needs collegial writing support and that writing is learned through doing"* (A1). *"At CUT, at the faculty of Engineering, students are orientated and introduced to their lecturers, facilities and the requirements of their course. When I engage with first year engineering students, I make sure that I get students to understand what their career paths require and what they need to succeed. I have found that students tend to be quite shy initially and I encourage students to ask questions, engage and express themselves in class while making sure that I listen attentively. In this way, I also get to understand who my students are"* (A2). Working in this way is aligned to Boughey and McKenna's (2015) view of understanding students as "social beings".

*"During writing centre consultations, we engage in deep listening and open up spaces for students to share their lived experiences and realities and to generate learning"* (A1). The creation of safe spaces is essential, and the responsibility should be "on academics 'hegemonic ear' to listen and respond to black students' evaluations and moral reasoning about the regulative discourse of the curriculum and how it affects their sense of agency and becoming" (Luckett et al. 2019, 38). Such understandings, when made explicit, provide valuable opportunities for teaching and learning and to examine how such literacy practices relate to epistemological concerns (Lea and Street 1998). Writing centre work is rooted in a humanizing approach that cultivates a pedagogy that respects the human, inter-personal side of teaching (Salazar 2013). At the writing centre we acknowledge that students can learn through "doing" and we are aware of the valuable process of "becoming" (Huerta and Brittain 2010 in Salazar). *"Students that work with writing centre tutors learn together while engaging in writing tasks such as drafting their writing, brainstorming ideas, or discussing discipline concepts. These delivery methods encourage learning from each other and producing their own forms of knowledge. It is through this respectful, responsive, and supportive engagement that students come to understand not only their writing strengths and challenges, but also develop their identities and confidence in their student roles"* (A1).

A humanizing pedagogy underpins writing centre practice through enabling resiliency (Salazar 2013; Franquiz and Salazar 2004) and promoting social justice and democracy which is aligned to universities vision to develop globally portable graduates with the acumen to initiate and/or respond to change. CUT's mission statement highlights the institution's aim to "promote access with success in attracting potentially successful students and supporting them to become employable graduates" (CUT Vision 2030: 2023). CUT's core values support innovative thinking by creating enabling environments and nourishing a commitment to provide "supportive relationships that are flexible and responsive to the institutional and student needs and to the interests of society at large" (CUT Vision 2030: 2023). Author 2 further affirms that this vision promotes a humanizing higher education environment by fostering a collaborative approach to learning. *"When I interact with my students, either in class or outside, I approach them as innovative thinkers and co-creators of knowledge. I encourage them to come up with ideas beyond their learning outcomes and I have seen them transform from just seeking marks to thinking beyond"* (A2).

*"Working at the Writing Centre has enabled me to recognise the value of the long overdue educational shift to teaching and learning and I have further identified how dialogic, innovative, and contextualised approaches to education will contribute to developing globally portable, critically conscious graduates that will use their values and knowledge to contribute to the societies in which they live and work. Through my interactions I have found that students come to university with their own experiences, knowledge and ideas and need spaces to develop these capabilities"* (A1). Those

involved in the academic project in SA need to recognise the value of understanding students' knowledge, literacies, and identities and create enabling environments for them to collaborate, learn and grow. *"I have come to value that it's important to spend time getting engineering students to understand and develop engineering competencies. In my course, students are often tasked with providing solutions to complex world problems given to them. I have found that there is huge value in creating spaces for students to collaborate with their module cohorts and others outside their discipline to arrive at workable solutions. These types of engagements foster collaboration, creativity and problem-solving skills which are much needed in industry"* (A2). When students enter university, they are expected to learn how to engage with knowledge and they need to be supported with a humanizing approach. *"Students require supportive environments to work with the new ways of thinking, learning, and doing in higher education and need to be scaffolded to familiarise themselves with the academic culture. This requires care, trust and a humanistic approach"* (A2). Working with students in this way creates a "pedagogical atmosphere of care, trust, support, dialogue, respect, fairness, tolerance, inquiry, freedom, commitment, responsibility, multiculturalism and reciprocity" (Aloni 2014, 5).

*"My work is guided by the notion that the writing centre is a communities of practice and the ways of teaching and learning inherent in our work enables the holistic development of our diverse students. The tutors at the centre afford students the opportunity to talk about their writing needs and we often find that students feel at home at the centre"* (A1). Furthermore, the White Paper on post school education and training (South Africa, 2013, 27) advocates for "the strengthening of social justice, equity, and democracy in South African universities". The CoP inherent in writing centre work cultivates a humanizing pedagogy which can meaningfully influence the lives of students (Salazar 2013). *"Writing centres are well positioned to support this call and through working with students we aim to provide opportunities for social mobility. Writing centre praxis is guided by the idea as suggested by Lewanika and Archer (2011) that writing centres function as communities of practice. We understand that learning does not take place as an isolated individualised activity but also as a socially mediated process. Writing centre work is guided by the view that writing is a social interaction."* (A1). Writing centres aim to develop students' understanding of "writing as a social act that allows readers and writers to negotiate meaning and their social identities" (Gee 2008, 13). CoP as defined by Wenger (1998, 2000), Wenger-Trayner and Wenger-Trayner (2015) exist when learning is collective and takes place in a shared domain by a group of people who interact regularly. Author 2 maintains *"This is more eminent in engineering modules where students often engage in constructivist approach to knowledge. Mostly in laboratory settings, students are required to work in groups, design their own circuits based on certain assumption they make and then report on the workability or otherwise of their designs. Working in these groups or CoPs enables students to feel confident and contributes to life-long learning"* (A2).

UoTs have had to rethink curriculum in a way where students are supported and prepared with life-long learning within their local context and a holistic education that goes beyond learning in their respective discipline. A humanizing pedagogy is based on "critical approaches to teaching, valuing the individual and taking into account the student voice" (Govender and Alcock 2018, 4). Older curricula were "designed to be applied to any classroom context regardless of the historical, cultural, and socioeconomic differences that characterise various schools and students" (Giroux 2009, 442). Other authors believe that "our higher education system needs to be aligned to social justice, and students regardless of economic background who should have access to transformative teaching and learning that affords them epistemological access" (Morrow 1993, 2007; Govender 2021). Quinn and Vorster (2017, 39) explain "For black students, curricula and pedagogic processes are often not aligned with who they are as people and it is not possible to divorce themselves – their being – from what is taught and how it is taught. What they are arguing for is greater recognition that teaching and learning is not only an epistemological project, but, in essence, also an ontological one." *"At the centre we acknowledge the economic and social difficulties faced by our students and see the value of Communities of Practice and a humanistic approach to our work which will assist students in coping with their challenges through the support provided"* (A1). *"As an academic and researcher, I often seek 'new' ways of embedding particular values and thinking in terms of the engineering discipline. I believe that learning can be taken beyond the classroom and one way of doing this is via service learning. This is at the centre of community service and development of active citizens. Service learning has a close nexus between learning and community service, and I believe that service learning correctly applied to the teaching and learning especially at UoTs would sensitise both academics, corporate agents and students to solving societal needs. This transformative strategy to teaching and learning would motivate agencies to strive towards meeting these needs via education"* (A1). The incorporating of service learning has the potential to develop critically conscious citizens and afford them opportunities "to address complex needs in their communities through the application of knowledge, and to form creative partnerships between the university and the community" (Bringle 1996, 236). *"As a writing centre practitioner, I try to engage with lecturers and encourage them to see and understand the link between epistemology and ontology and how through our work we can open-up opportunities for resonance with our students. It is important to understand our diverse student body and one way is to link knowledge to the histories of our students and engage with them on how it impacts their lives. I have come to understand that when I critically reflect on my work and engagement with students, I am slowly able to scaffold students into new ways of being that will prepare them to become well-rounded students"* (A1).

The heightened university protests in SA in 2015 and 2016 brought a sharp focus on knowledge and knowledge production. It raised issues on inclusion and exclusion and on how knowledge and education is valued. The protests throughout South African universities sparked the decolonisation debate (Heleta 2016). Significant concerns were raised on the promotion of certain epistemologies from the global north and narrow scholarship on Africa. Student protests also led to academics reviewing and reflecting on their current practice. During the protests the traditional curriculum was interrogated for its role in alienating and marginalising black students in particular. The concept of multilingualism has surfaced as an important aspect in the decolonisation discussion in SA, and this is particularly imperative for DUT and CUT students, 80% of whom speak English as a second language and come from disadvantaged communities. The concept of multilingualism or multilingual teaching spaces have come to the fore at both institutions. English is the official medium of instruction, and this is challenging for students who are expected to engage in academic writing, synthesising of information and building arguments in their writing when at university. Acknowledging this challenge, as academics it is imperative that we heed the call for the re-thinking of curriculum and embracing of multilingualism at the university to facilitate what Bigg and Tang refer to as deep learning (Biggs & Tang 2007). "*I encourage students to bring in their assignment briefs and rubrics as a basis from which to begin our conversations with students. I have seen that students struggle to explain what they understand about their written tasks. It empowers students when we support the use of the mother tongue as this allows a student the comfort of a language they know and are confident in. This unlocks their creativity, critical thinking and articulating themselves without fear of being alienated or censored. In this way, using two languages is a vehicle for meaning making to facilitate learning*" (A1). This type of learning, by moving between languages is also supported by Henricks (2016, 20) who believes that "when students work in pairs, multilingualism enables them to engage in dialogical conversations freely". At the writing centre, this approach to learning is valued and encouraged, as it enables explorative, flexible, collaborative deliberations. "*In our work with students we must reflect on our delivery methods and move away from monological teaching to teaching that encourages and embraces active learning and the production of students' own form of learning*" (A1). "Dialogic conversations are a necessary disruption of colonial curriculum practices but need to be well planned, theorised, considered and facilitated in ways that enable, rather than shut down what can be a moving experience of disrupting our normality to make us more human" (Behari-Leak 2019, 84). 'Disruption' allows for the questioning and challenging of the traditional ways of 'being' in higher education. Furthermore, Freire (1970) in his work, encourages those working with students to listen to their students and encourage engagement in contextualised, dynamic, reflective, and personalised educational practices that is humanistic in nature and furthers the goals of social justice and social transformation.

"*When I joined CUT, my engineering classes were structured in the traditional way of theory classes and laboratory sessions where students get to put into practice their theoretical knowledge of the subject. Over the years through my reflections and learning, I have been able to shift my approach to encourage more dialogue and engagements via service learning. This approach has created more discussion and questioning of status quo thereby contributing to critical citizens and students that can contribute to society*" (A2). As supported by Hayes (1996), service learning is a learning theory that enables students to see their culture in different ways, respect for others' cultures and allows them to be more critically conscious of societal issues and power relations and can foster in them a vision for a more democratic society.

Higher education institutions play an essential role in contributing to the nurturing and building of participatory, knowledgeable citizens that can make an impact locally and globally. We live in an increasingly globalised and connected world and graduates should be able to strengthen our still nascent democracy. "*Through my lectureship I aim to provide students with the kind of learning opportunities that enables their growth as rounded, productive, responsible human beings in a globalised world. I connect them to international engineering audience by discussing my international collaborative research projects."* (A2). This means that there will be disruption to the curriculum with care and consideration of students' understanding of academic knowledge. The higher education sector in South Africa is in need of a focus on decolonial possibilities and practices and discourse that are inclusive (Markus and Govender 2023). It is imperative that academics reflect on their practice in higher education environments and purposively respond to structures and cultures in order to create transformative educational spaces. "Human beings have the powers of critical reflection upon their social context and of creatively redesigning their social environment, its institutional or ideational configurations, or both….and collaborating with others in bringing about transformation" (Archer 2000, 308). Archer (2000) affirms that people often engage in reflexivity or "the internal conversation" and we support this view and advocate for critically reflexivity in our higher education spaces to question our own attitudes and ways of being and to understand our roles more fully in relation to student needs. *"In the CUT context, there are many academics who are committed to addressing the "concerns" and "projects" in their specific teaching and learning environment. The reality is that there are also many academics who resist changing their practice or embracing new ways of doing things in higher education. What is required is providing key role-players the space and means to recognise the importance of understanding who our students are, where they come from and embracing a pedagogy of being and doing. In my work I often emphasise the importance of collaboration and critical reflection of our practice in our socio-cultural context to understand how our personal identities, agency and criticality are developed and can be transformed to contribute to enhancing the learning that takes place in higher education. We need to reflect on our role in higher education and develop personal agency, after all, we have an ethical obligation to students to develop quality education in our country"* (A2). With major changes in curricula being implemented, it is important to work towards developing "teaching and assessment methods that take into account the legitimate learning needs of all students" (Quinn and Vorster 2017, 32). Many authors affirm the importance of those in the academic project challenging the status quo in higher education and to push for transformative practices (Quinn and Vorster 2017, 34; Markus and Govender 2023). Critical reflection can enable us to engage in transformative approaches to slowly scaffold students and prepare them to become critically engaged globally portable students. Reflection enables one to negotiate new dimensions of our roles in higher education in SA. The CAE has afforded us the lens to analyse the intricate relationship between "self" and "cultures/discourses" when recounting our storied experiences in higher education.

## 5 Conclusion

This research suggests that for students to thrive academically and navigate this complex world as human beings, they need to feel valued and included and be able to express and articulate themselves. The central argument presented in this paper is that collaborative approaches are needed to disrupt restrictive practices and to prepare critical thinkers who can professionally and competently contribute to the societies in which they live and work. We value a social constructivist approach to our work in higher education where learning is constructed through interaction and collaboration and not just transmitted through instruction. A further argument is that learning at university is not only about gaining knowledge, it is about developing the whole individual - one that can meaningfully contribute to not only the community one comes from, but to the global society. Those in the academic project need to find meaningful ways through critical reflection to effect changes, to build an inclusive and socially just higher education system that is well placed to contribute to social justice, equality, transformation, epistemological repositioning, innovation, and redress. We need to create a culture of valuing; being accountable for quality teaching and learning in our institutions and transformation for a more equitable dispensation in higher education.

## References

Adams, T.E.: Critical autoethnography, education, and a call for forgiveness. Int. J. Multicult. Educ. **19**(1), 79–88 (2017)

Archer, M.S.: Being Human. The Problem of Agency. Cambridge University Press, Cambridge

Aloni, N.: Humanistic education. In: Peters, M., Ghiraldelli, P., Žarnić, B., Gibbons, A., Besley, T. (eds.) Encyclopedia of Educational Philosophy and Theory, pp. 1–5 (2014)

Badat, S.: Putting an End to Causes of Pain. Rhodes University, Graduation Speech (2015)

Badsha, N., Cloete, N.: Higher education: contribution for the NPC's national development plan (unpublished paper) (2011)

Barkhuizen, G. (Ed.).: Reflection on Language Teacher Identity Research. Routledge (2017)

Behari-Leak, K.: Disrupting Single Stories Through Participatory Learning and Action. In: Quin, L. (ed.) Re-Imagining Curriculum—Spaces for Disruption. African Sun Media (2019)

Biggs, J., Tang, C.: Teaching for Quality Learning at University, 3rd edn. Open University Press, Berkshire, England (2007)

Boughey, C., McKenna, S.: Analysing an Audit Cycle: A Critical Realist Account. Studies in Higher Education (2015)

Boughey, C.: Overview of the History of Academic Development in South Africa. Presentation to PG Dip HE for Academic Developers. Rhodes University (2019)

Bringle, R.G., Hatcher, J.A.: Implementing service learning in higher education. J. High. Educ. **67**(2), 221–239 (1996)

Central University of Technology (CUT) Vision 2030 (2023). https://www.cut.ac.za/vision-2030

Council on Higher Education (CHE).: South African Higher Education Reviewed: Two Decades of Democracy. CHE, Pretoria (2016)

Chang, H., Ngunjiri, F.W., Hernandez, K.C.: Collaborative Autoethnography. Left Coast Press (2013)

Department of Education (DOE).: Education White Paper 3: A Programme for the Transformation of Higher Education. Pretoria: Department of Education (1997)

Ellis, C., Adams, T.E., Bochner, A.P.: Autoethnography: an overview. Qualitative Sozialforschung/Forum: Qual. Soc. Res. **12**(1), 273–290 (2011)

Fránquiz, M., Salazar, M.: The transformative potential of humanizing pedagogy: addressing the diverse needs of Chicano/Mexicano students. High School J. **87**(4), 36–53 (2004)

Freire, P.: Pedagogy of the Oppressed. Continuum, New York (1970)

Leibowitz, B., Bozalek, V.: Foundation provision—a social justice perspective. South African J. High. Educ. **29**(1), 8–25 (2015)

Gee, J.P.: Social Linguistics and Literacies. Ideologies in Discourses, 3rd edn. Routledge, London (2008)

Giroux, H.A.: Youth in a Suspect Society: Democracy or Disposability? Palgrave-Macmillan, New York (2009)

Govender, N., Alcock, A.: Humanising writing centre practice: peer tutor reflections at a UoT. Africa Educ. Rev. (2018). https://doi.org/10.1080/18146627.2018.1467735

Govender, N.: Collaborative approaches to teaching and learning in a South African University of Technology. Postgraduate Diploma in Higher Education: For Academic Developers. Rhodes University (2021)

Heleta, S.: Decolonisation of higher education: dismantling epistemic violence and Eurocentrism in South Africa. Transform. High. Educ. **1**(1), 1–8 (2016)

Henricks, M.: Using languages for transformative teaching: small steps that can lead to big moves. In: Vorster, J.A. (ed). Curriculum in the context of transformation: reframing traditional understanding and practices. Rhodes University, Grahamstown, South Africa (2016)

Hernandez, K.C., Ngunjiri, F.W., Chang, H.: Exploiting the margins in higher education: a collaborative autoethnography of three foreign-born female faculty of color. Int. J. Qual. Stud. Educ. **28**(5), 533–551 (2015)

Lea, M.R., Street, B.: Student writing in higher education: an academic literacies approach. Stud. High. Educ. **23**(2), 157–172 (1998)

Lewanika, T., Archer, A.: Communities of practice: reflection on writing, research and academic practices in a writing centre. In: Archer, A., Richards, R. (eds.) Changing Spaces: Writing Centres and Access to Higher Education, pp. 147–158. SUN Press, Stellenbosch (2011)

Hayes, E., Cuban, S.: Border pedagogy: a critical framework for service learning (1996)

Markus, E.D., Govender, N.: Teaching and learning transferable skills in engineering education via service learning: case study of a university of technology in South Africa. In: Auer, M.E., Pachatz, W., Rüütmann, T. (eds.) Learning in the Age of Digital and Green Transition. ICL 2022. Lecture Notes in Networks and Systems, vol. 634. Springer, Cham (2023). https://doi.org/10.1007/978-3-031-26190-9_35

Markus, E.D., Govender, N.: Can universities of technology in South Africa achieve transformation by promoting a culture of social responsibility among academic and student agents? Public Organ. Rev. (2023). https://doi.org/10.1007/s11115-023-00701-9

McKenna, S.: The intersection between academic literacies and student identities. South African J. High. Educ. **18**(3), 269–280 (2004)

Morrow, W.: Epistemological access in the university. AD Dialogues **1**, 3–4 (1993)

Quinn, L., Vorster, J.: The "decolonial turn": what does it mean for academic staff development? Educ. Change **21**(1), 31–49 (2017)

Quinn, L., Vorster, J.: Why the focus on "curriculum"? why now?—the role of academic development. In: Quinn, L. (ed.) Re-imagining curriculum—spaces for disruption. African Sun Media (2019)

Roegman, R., Reagan, E., Goodwin, A.L., Lee, C.C., Vernikoff, L.: Reimagining social justice-oriented teacher preparation in current sociopolitical contexts. Int. J. Quali. Stud. Edu. 1–23 (2020)

Spry, T.: Performing autoethnography: an embodied methodological praxis. Qual. Inq. **7**, 706–732 (2001)

Wenger, E.: Communities of practice and social learning systems. Communities Pract. **7**(2), 225–246 (2000)

Wenger-Trayner, E., Wenger-Trayner, B.: Communities of practice: a brief introduction (2015)

# Gamification Based Collaborative Learning: The Impact of Rewards on Student Motivation

Sonia Sahli[1](✉) and Thierry Spriet[2]

[1] Higher Institute of Technological Studies, City Erriadh – B.P.135, Sousse, Tunisia
sonia.sahli@kairouan.r-iset.tn

[2] Avignon University, 74 rue Louis Pasteur, 84 029 Avignon cedex 1, France
thierry.spriet@univ-avignon.fr, lncs@springer.com

**Abstract.** Motivation is still required in higher education. Therefore, collaborative learning is highly recommended as an instructional method. To provide a more engaging learning experience for higher students education, a gamification can be introduced. Rewards present a foundational element of gamified learning. It included points, badges, animated feedbacks and even gifts in order to increase the enjoyment of a game. The aim of the current study is to investigate the effect of reward on the attitudes of students towards gamified group learning environments. The proposed method aims to rise the benefit of gamified rewards such as collaboration between team members, motivation and evidently challenge applying gamification based collaborative learning. In this context, a comparative study between 2 gamified groups of students without rewards and with rewards, was conducted. The results indicate that the gamified group with rewards remains the most relevant having the highest interest rate. Results confirm that the reward has enhanced Student Motivation (SM = 98.75%), group unity and collaboration between team members (CTM = 97.11%) and increasing student challenge (SC = 95.56%). The propose method highlights high-quality rewards performance in the collaborative gamification.

**Keywords:** Gamification · Motivation · Rewards

## 1 Introduction

Collaborative learning is defined as a method that students work together to achieve a common goal. It consists of the creation of small groups of students that interact with each other and share knowledge and data [1]. With the development of soft skills et its importance, the collaboration practical advantages for student in the learning are multiplying all the time. Ramos et al. considered it as a social interaction that develops common ground, communication and shared knowledge [3]. The communication within the group was considered valuable to share ideas and work on development of the collective spirit. As a result, it increase understanding than before [4].

To provide a more engaging learning experience for higher students education, a gamification, that is the use of game elements and techniques in non-gaming environments, can be introduced. Fahid et al. indicate that collaborative game-based learning


M. E. Auer et al. (Eds.): ICL 2023, LNNS 899, pp. 124–130, 2024.
https://doi.org/10.1007/978-3-031-51979-6_13

environments increase engaging group learning experiences [2]. Azita Iliya et al. confirm the positive impact of game-based learning at both cognitive and emotional levels and its influence on engagement and learning. Game elements include levels, points, winners and challenges. On the other hand, rewards are considered as part of game dynamics [7], knowing that rewards included points, badges, feedback messages, animated feedbacks and even gifts. But they are not essentially critical conditions of games. The use of rewards in an educational game can create a competitive environment and boost the learners motivations [8]. Right use of competition, challenge, level and scoring can motivate students to win, help learners to develop a sense of achievement, satisfaction and motivation [9].

This article offers the influence of rewards in collaborative educational games. A comparative study, between the gamified group with rewards and gamified group without rewards, has been used to investigate the effect of rewards on the attitudes of students towards gamified group learning environments. The objective is to report a comprehensive view of the student attitude, such as collaboration between team members, motivation and evidently challenge, face to collaborative learning game, that includes rewards.

He paper is organized as follows: Sect. 2 describes the proposed method starting with the description of the work environment then the course of the collaborative base learning activity, Sect. 3 deals with results of the methodology, the discussion is the subject of Sects. 4 and 5 is dedicated to the conclusion section.

## 2 Methodology

### 2.1 Participants

In this work, an overall of 180 students of second year of a Bachelor's degree in computer science, web development specialty, were selected to participate in this pilot study. The study's participants were students attending the Higher Institute of Technological Study in Tunisia. The gamified group without rewards consisted of 98 students, whilst the gamified group with rewards consisted of 82 students. The UI/UX design course was the context of the case study that will be presented. The mean age of the participants was 22 years old, ranging from 20 to 23 years. Just over half of the participants (53%) are females. Data collection took place at the second semester of the professional year, between February and May.

### 2.2 Collaborative Game-Based Learning Activity

The aim of the study was to investigate the effect of rewards in the gamification on the attitudes of students towards working as small groups. The methodology is described through the steps presented in the Figure bellow. In the formative assessment framework, at the end of the course, the class is split into two groups of 8 students. They formed their own groups, choosing the group members and their group's pseudonym. The professor informed students by preparing three track questions about the lesson for the other group in order to assess the achievement of course objectives (peer assessment). In the gamified group with rewards, one point will be awarded for each correct answer. The group's pseudonym and points number are noted on the board. At the end, the winning team will get a relevant gift.

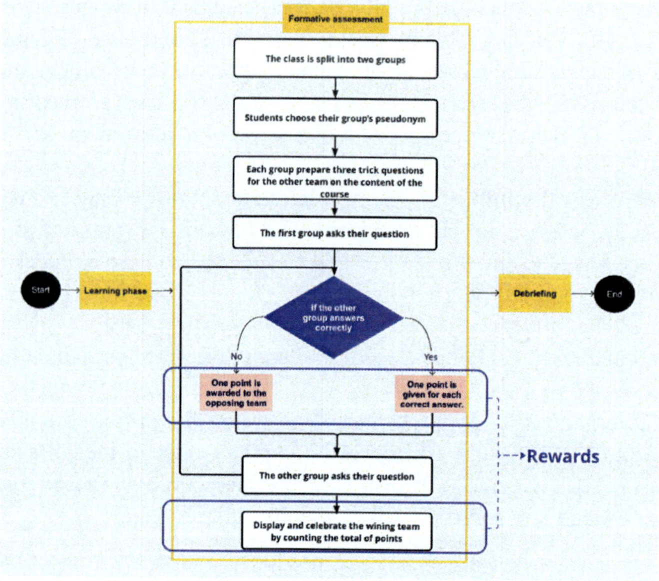


**Fig. 1.** Flowchart of the proposed method.

The proposed method aims to rise the benefit of gamified rewards such as collaboration between team members, motivation and evidently challenge applying gamification based collaborative learning. Therefore we tested the same formative evaluation with and without rewards. In the gamified group without rewards, the same game is happening, formative peer evaluation without points and without gift for the winning team.

This paper purposes to answer the following research questions:

- Q1. Is there a difference between the gamified group with rewards and gamified group without rewards in term of student motivation?
- Q2. Is there a difference between the gamified group with rewards and gamified group without rewards in term of increasing student challenge during the game?
- Q3. Is there a difference between the gamified group with rewards and gamified group without rewards in term of collaboration between team members?

## 3 Results

In this experimental design study, gamified group without rewards consisted of 91 students and gamified group with rewards consisted of 89 students were compared to improve the performance of rewards in within gamification based collaborative learning.

In order to evaluate the effects of gamified rewards using points and gift, the proposed evaluation is focused on given questionnaire to assess Student Motivation (SM), collaboration between team members (CTM) and increasing student challenge (CH). The same questionnaire is proposed to collect data of the gamified group with and without rewards. Answers are on a scale range one to ten per student. The ten represent the most positive, the one the most negative value and five a neutral evaluation.

The mean total score of Student Motivation (SM), collaboration between team members (CTM) and increasing student challenge (CH) in two groups are presented in Tables 1 and 2.

**Table 1.** Training process in terms of SM, CH, CTM focused on group without rewards.

| Students | SM (%) | CH(%) | CTM (%) |
|---|---|---|---|
| Student 1 | 68,56 | 83,76 | 77,03 |
| Student 2 | 86,81 | 86,10 | 76,20 |
| Student 3 | 70,60 | 80,59 | 78,57 |
| Student 4 | 76,50 | 90,10 | 75,51 |
| Student 5 | 77,88 | 86,14 | 78,10 |
| Mean (%) | 77,75 | 85,33 | 77,11 |

**Table 2.** Test process in terms of SM, CH, CTM focused on group with rewards.

| Students | SM (%) | CH(%) | CTM (%) |
|---|---|---|---|
| Student 1 | 98,56 | 93,76 | 96,11 |
| Student 2 | 96,81 | 96,10 | 96,20 |
| Student 3 | 97,88 | 96,84 | 98,10 |
| Student 4 | 97,96 | 94,53 | 98,11 |
| Student 5 | 98,97 | 96,57 | 97,03 |
| Mean (%) | 98.75 | 95,56 | 97,11 |

As shown in Fig. 2, question1 compute the difference between the gamified group with rewards and gamified group without rewards in term of (SM). The score of SM in gamified group with rewards is higher than SM in gamified group without rewards. It is clear from Fig. 3 that, unlike gamified group without rewards, increasing (CH) during the game have been remarkably high compared to gamified group with rewards. We can notice that the proposed study can aids students to increase CTM; as reveled in Fig. 4.

## 4 Discussion

The gamified group with rewards remains the most relevant having the highest interest rate. Results confirm that the reward has an enhanced effect results of Student Motivation (SM = 98.75%), group unity and collaboration between team members (CT = 97.11%) and increasing student challenge (CH = 95.56%).

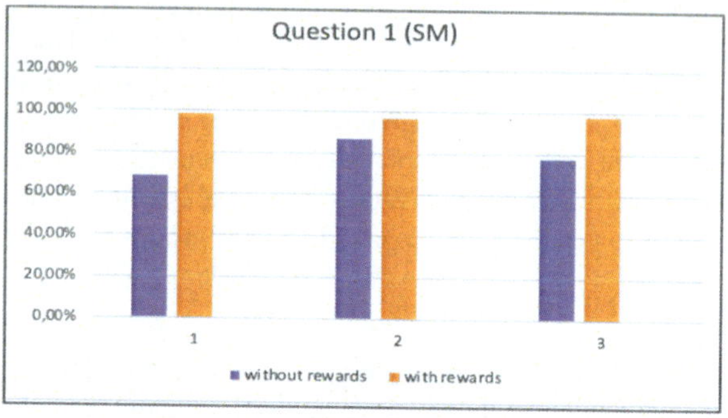


**Fig. 2.** Response to question 1 in term of SM

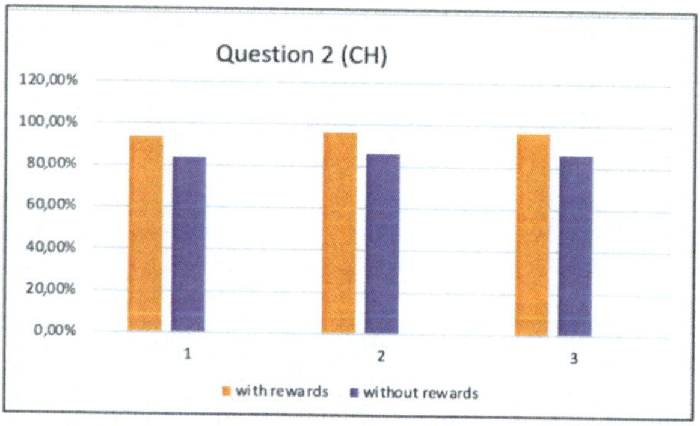


**Fig. 3.** Response to question 2 in term of CH

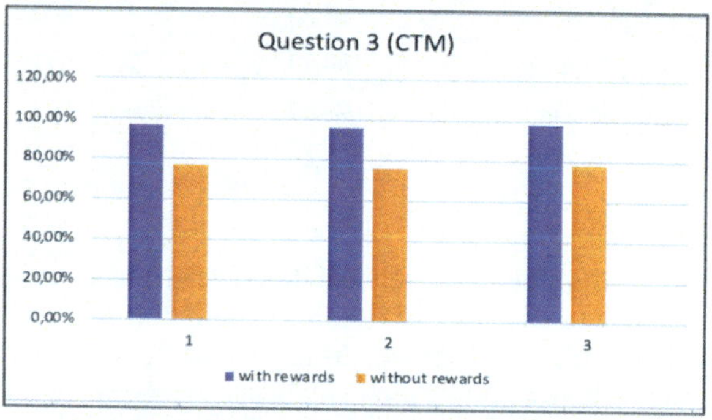


**Fig. 4.** Response to question 3 in term of CTM

### 4.1 Student Motivation Outcomes

As exposed in Fig. 2, it is proved that rewards have given high impact on students motivation. The total score of SM is 97,75% compared to collaborative game based learning without rewards 77,75%. We found that the students have more interaction and engagement by having points during the game and wining a gift at the end. In addition, discussing their answers with their peers makes them being competitive and motivated. Students precise that rewards make the game more excited and enjoyable.

### 4.2 Collaboration Between Team Members Outcomes

The number of students answers produced has seen a high score (97,11%)in term of relation between team members during the game; as presented in Fig. 3. Rewards help students to increase their engagement during the game, such as discussing and answering questions with peers allowing communication to win. When they have a common goal, they collaborate, communicate and increase their group unity to achieve ang win.

### 4.3 Increasing Students Challenge

We notice that the results obtained by gamified group with rewards are more accurate and reliable (98.95%) when compared to gamified group without rewards in term of growth challenge. Students have more behavioral and emotional connection when they have point for each correct answer. The challenge is increased in order to win.

## 5 Conclusion

The proposed method highlights high-quality rewards performance in terms of increasing motivation and challenge during the game, and developing collaboration between team members. The collaborative gamification provides a higher performance using rewards. It is recommended to use rewards in a collaborative game as well as simple game, principally to develop students soft kills. In future work the proposed rewards method can be enhanced by adding additional terms and evaluation metric.

## References

1. Prince, M.: Does active learning work? A review of the research. J. Eng. Educ. **93**(3), 223–231 (2004)
2. Fahid, F.M., et al.: Effects of Modalities in Detecting Behavioral Engagement in Collaborative Game-Based Learning (2023)
3. Ramos, J.L., Cattaneo, A.A., de Jong, F.P., Espadeiro, R.G.: Pedagogical models for the facilitation of teacher professional development via video-supported collaborative learning. A review of the state of the art. J. Res. Technol. Educ. **54**(5), 695–718 (2022)
4. De Jong, F.P.C.M.: Knowledge In-(ter)-action. Responsive learning as knowledge building. Aeres Applied University Wageningen and Open University Heerlen (2020a). https://doi.org/10.46884/2020.2
5. Cai, Z., Mao, P., Wang, D., He, J., Chen, X., Fan, X.: Effects of scaffolding in digital game-based learning on student's achievement: a three-level meta-analysis. Educ. Psychol. Rev. **34**(2), 537–574 (2022)
6. Abdul Jabbar, A.I., Felicia, P.: Gameplay engagement and learning in game-based learning: a systematic review. Rev. Educ. Res. **85**(4), 740–779 (2015)
7. Bunchball.: Gamification 101: an introduction to game dynamics [White paper] (2010). Retrieved from http://jndglobal.com/wp-content/uploads/2011/05/gamification1011.pdf. Accessed 2 Jan. 2018

8. Kalogiannakis, M., Papadakis, S., Zourmpakis, A.I.: Gamification in science education. A systematic review of the literature. Educ. Sci. **11**(1), 22 (2021)
9. Jagušt, T., Botički, I., So, H.J.: Examining competitive, collaborative and adaptive gamification in young learners' math learning. Comput. Educ. **125**, 444–457 (2018)
10. Eltahir, M.E., Alsalhi, N.R., Al-Qatawneh, S., AlQudah, H.A., Jaradat, M.: The impact of game-based learning (GBL) on students' motivation, engagement and academic performance on an Arabic language grammar course in higher education. Educ. Inf. Technol. **26**, 3251–3278 (2021)
11. Wang, M., Zheng, X.: Using game-based learning to support learning science: a study with middle school students. Asia-Pac. Educ. Res. **30**, 167–176 (2021)
12. Beiranvand, S., Foladvandi, M., Mokhayeri, Y., Khodaei, S., Hasanvand, S., Hoseinabadi, R.: The effect of simulation education based on flipped learning on academic engagement, motivation, and performance of first-year nursing students (2023)

# Job Lab Collaborative Approach: An Innovative Model for Enhancing Graduates' English Language Skills

Olga Kissová(✉) and Jiří Tengler

University of Žilina, Univerzitná 8215/1 Žilina, Slovakia
{olga.kissova,tengler}@uniza.sk

**Abstract.** The study examines the effectiveness of the innovative Job Lab collaborative approach in teaching graduates English. The Job Lab model, introduced by the University of Zilina, Slovakia, prepares students for the global market through student-centred learning. We focus on improving learners' language proficiency, teamwork, and communication skills. Fifty graduates were randomly taken and divided into two groups (T-test). The study compared the experimental group (N = 25), which received Job Lab collaborative training, to the control group (N = 25), which received traditional teaching. Pre/post-tests and questionnaires were used to collect data, which showed that the experimental group had better testing results towards professional topic-based English language learning. Data was examined statistically using SPSS software (M, SD). Findings revealed that the experimental group, mainly due to students' active collaborative participation and technology tool implementation, achieved better testing results and showed higher cognitive skills in language learning after the intervention. The study recommends a Job Lab approach in teaching English to reduce speaking anxiety and enhance students' communicative higher-level cognitive skills.

**Keywords:** Collaborative approach · JobLab · Speaking anxiety · FLCAS questionnaire · Higher-order thinking

## 1 Introduction

The Job Lab's collaborative learning (CL) presents a modern trend in learning English as the global language for business and communication to express ideas clearly, persuade others, and engage in meaningful discussions. The study examines reducing speaking anxiety and developing higher-order skills by comparing collaborative and traditional teaching approaches. The study employs the Foreign Language Classroom Anxiety Scale questionnaire (FLCAS) before and after the intervention, and pre/post-tests based on higher-order thinking (revised Bloom's taxonomy), verbal creativity tests, and critical thinking tests to collect data. After the intervention, the data analysis was conducted to compare the results of the two groups and identify statistically significant differences [1].


M. E. Auer et al. (Eds.): ICL 2023, LNNS 899, pp. 131–142, 2024.
https://doi.org/10.1007/978-3-031-51979-6_14

## 2 Collaborative Approach

Collaborative learning (CL) considers learning a naturally social act in which participants work together based on well-defined learning outcomes, requirements, principles, and activity structures. Its goal is to achieve higher levels of cognitive learning and social interaction among students working in small groups (3–5 students) where each member's effort is necessary and required for the group to achieve a common goal (solve a problem, complete a task, or create a product through the mutual engagement of participants) in a coordinated effort. The demand for professional English language preparation requires implementing modern educational strategies that enhance critical thinking, social and interpersonal skill development, effective communication, and problem-solving skills.

### 2.1 Five Principles of Collaborative Learning (CL)

Jonson and Johnson (2009) and Oxford (1997) presented an overview of five critical CL principles integrated into the instructional design process:

1. **Individual accountability**: Assigning specific roles and responsibilities to each student promotes a sense of ownership and responsibility for the team's success. It leads to increased motivation and commitment.
2. **Positive interdependence**: Focusing on achieving shared goals rather than individual efforts fosters a cooperative, not competitive, environment. This promotes collaboration and mutual support among students, and improves problem-solving abilities and collective achievement, where participants celebrate positive milestones.
3. **Promotive interaction**: Enhancing active participation and dialogue among students, where they actively share ideas, provide constructive feedback, and collectively problem-solve. This interaction enhances students' language proficiency and nurtures their critical thinking and communication skills.
4. **Group processing**: Conducting small group evaluations and reflecting on the teamwork process helps students improve their communication skills and subject-specific language. It allows them to assess their collective progress, identify strengths and weaknesses, and make necessary adjustments. This reflective practice enhances self-awareness, teamwork abilities, and overall performance.
5. **Social skills**: Interacting effectively with peers is vital for students' development. Collaborative learning allows students to practice and refine their social skills, such as communication, empathy, and cooperation. These skills are beneficial academically and in future personal and professional settings [2, 3]

Implementing a collaborative approach to foreign language learning can enhance language skills, promote cultural understanding, increase motivation, and provide valuable peer support. It creates an interactive and dynamic learning environment that simulates real-world language use and fosters a deeper appreciation for language and culture.

It involves real-world tasks and projects, such as role-plays, debates, presentations, and problem-solving activities. This connects language learning to practical and meaningful real-life and professional contexts.

# 3 JobLab – Innovative English Learning Model

The Institute of Lifelong Learning, Department of Foreign Languages at the University of Zilina, Slovakia, has recently introduced an innovative English language teaching model for engineering students called JobLab by implementing a collaborative approach for practical preparation for the business world. It focuses on one semester-long professional topic-based English communication skills training for graduates.

Mårtensson et al. claim that university teaching appears "peculiarly resilient to all sorts of reform efforts made by managers, educators, and politicians". Therefore, the driving force for change should be the university teachers who enhance effective changes and know how to meet their student's needs. Moreover, the collaborative JobLab approach created by university teachers adopts different learning styles, ensuring that all students are engaged and can contribute to the learning process. Through the collaborative approach, students are trained to develop negotiation strategies, persuasive and speaking skills essential for their future careers [4].

## 3.1 Five Pillars of Job Lab's Collaborative Approach

Pleschová et al. introduce the following five essential conditions leading to the adequate transformation of teachers' conceptions and practices that were implemented in JobLab: "cross-disciplinary participation, trustful relationships, conducive spaces, caring attitudes, and co-construction collaborative practices" adopted in JobLab training [5].

Job Lab's collaborative approach adopted for the experimental group in intervention employs a five-pillar system in its tofi-based semester English for graduates with clear objectives, supportive teaching methods, and assessments aligned with the expected learning outcomes.

**Create a positive learning environment**. Implementing strategies ensures that all students can participate and share their ideas in a non-threatening environment. They overcome barriers and achieve personal and professional success through positive reinforcement, constructive feedback, and group collaboration. Teachers provide clear explanations through supportive approaches (scaffolding, layering, modelling by guided practice before students' independent practice.

Effective error treatment creates a positive speaking environment and enhances students' confidence. By implementing strategies like self-correction, teacher assistance, and pointing out errors through gestures and facial expressions, teachers can help students overcome their fear of speaking and improve their declarative knowledge and enjoyment of English learning based on fluency, students 'added value, creativity, and student-based cognitive and collaborative learning.

Job Lab used that positive emotional stimuli persist longer in memory and are recalled with greater accuracy and immediacy, which increases declarative word knowledge and enjoyment of using a foreign language [6].

**A student-centred approach**. Involving this approach helps to meet each learner's needs and learning style (visual, auditory, verbal, and kinesthetic), bringing learner autonomy and active participation to enhance learning. Teachers activate learning and form diverse groups to promote active and flexible participation based on creativity and interactivity to alleviate stress and help students maintain a healthy work-life balance.

**Digital technology (educational technology, Edtech).** Promoting environmental awareness through the JobLab online platform creates flexible, competitive, and student-centred learning, allows students to collaborate on a single document/project in real-time and provides a self-paced learning environment. Such autonomy will enable students to take control of their learning, promoting a deeper understanding. Digital technologies like e-learning and m-learning provide new opportunities for affordable, accessible, open, and flexible tools that increase motivation, adaptability, and student learning process regulation. The current generation of "digital natives" comprises intuitive technology users who enjoy integrating digital technology into life and education.

**Higher-order thinking skills development.** Focusing on developing higher-order thinking skills (based on Anderson et al.'s revised Bloom's taxonomy) involves the three-step interview, brainstorming, mind mapping, peer assessment, and think-pair-share design. These are Job Lab strategies for promoting collaboration and developing topic-based verbal creativity and critical thinking skills (reasoning, decision-making, problem-solving, and negotiation in real-life business settings) [7].

**Students' portfolio.** Recording materials help quickly assess each student's progress, identify areas where they may need additional support or guidance, and provide flexibility in their learning process.

### 3.2 Speaking Anxiety

Foreign Language Anxiety (FLA) negatively impacts foreign language learning as it creates a mental block against learning a foreign language, triggers negative emotions, blocks memory, and brings a subjective feeling of tension, nervousness, and worry associated with the arousal of the autonomic nervous system. Students with weak anxiety levels in a foreign language face great difficulty demonstrating proficiency.

The FLCAS (Foreign Language Classroom Anxiety Scale) questionnaire measures students' speaking anxiety. Job Lab training is based on overcoming it (breathing exercises, visualisation), encouraging risk-taking, self-reflection, and coping strategies.

According to Horwitz, "the FLCAS aims to assess the degree of anxiety evidenced by negative performance expectations and social comparisons, psychological symptoms, and avoidant behaviour" [8].

The discomfort, stress, and anxiety of speaking English are usually triggered by a lack of linguistic ability and a fear of making mistakes, which one perceives as threatening to one's identity [8].

### 3.3 JobLab Collaborative Approach

The key strategies in JobLab training provide support through designed learning activities to ensure that even lower-achieving students are successfully involved and challenged. The experimental group is expected to achieve the expected learning outcomes based on higher levels of cognitive learning within interactions (discussions, role-play, group work), essential learning needs (active listening, mind mapping, brainstorming, peer assessment based on criteria), and advanced higher-order thinking strategies (Table 1).

**Table 1.** The higher-order thinking strategies implemented in the Job Lab approach.

| Task | Strategy description/high-order thinking benefits |
|---|---|
| Reciprocal peer assessment and round table commission | Encourages assessment using reciprocal peer assessment and committees. Students estimate each other's work in pairs or groups of five, and peer committees evaluate based on criteria. Benefits include promoting collective assessment criteria and inter-subjective control, stimulating active learning and a critical thinking approach in a supportive environment |
| Negotiation strategies | Encourages critical discussion to exchange ideas through the negotiation matrix strategy. Group members prepare arguments and support or reject ideas to reach a consensus. Benefits include improving problem-solving skills, clear expression of thoughts, and training the ability to communicate persuasively via logical and compelling viewpoints |
| 3-step interview strategy (icebreakers, team building tasks) | Encourages problem-solving in groups for complex issues without clear solutions to develop higher-order cognitive abilities that align with learning outcomes and social skills, such as analytical, synthetic, and evaluative thinking<br>Benefits include productive learning experiences that help students apply their acquired knowledge and skills in a positive environment |
| Think-pair-share strategy | Encourages students to process their thoughts during instruction, enabling them to think individually to answer questions and share ideas, including analysis, evaluation, and synthesis. Benefits include developing the ability to clarify information, draw conclusions, and consider another point of view. It empowers students to participate actively in their learning process and cultivates life skills such as open-mindedness |

*Source* Processed by authors

## 4 Methodology

The qualitative and quantitative methods were chosen to obtain the most accurate and objective information regarding the set goals and verify the research hypotheses.

### 4.1 Experiment

The research was conducted at the University of Zilina, Slovakia, based on different teaching procedures. The experimental group (N = 25) received Job Lab collaborative training, and the control group (N = 25) received traditional teaching. For both groups, the intervening variables are the same (teaching goal, teacher, five topics, 13-week long training, and the same number of students with approximately the same English level). After the end of this experiment, the results are compared for both studied groups.

The JobLab intervention was implemented in a 13-week long training (1x100 minutes per week) course during the winter term of 2022/2023. The JobLab aimed to enhance cognitive learning through interactions on five professional topic-based subjects related to instructional technologies. Students were divided into two groups: traditional (n = 25) and collaborative (n = 25). Data collection ensured participant privacy. The intervention involved group work, role-play, and discussions in pairs and small groups. Teaching materials included a Benchmark glossary, interactive videos, and resources on Branding, Marketing mix, Negotiation strategies, Ansoff matrix, and Business plan. Collaboration among teachers, students, and communities is influential for future education.

### 4.2 Participants

The research sample was based on 50 graduates of Economics and management teaching English as a second language with the same level of English (B2 based on CEFR, 2020) who were randomly taken from the engineer's degree of study and divided into two groups with the same teacher, study materials and the same length of training.

Before the investigation, the 50 graduates of the economics study were randomly divided into two homogeneous groups: experimental (n = 25) and control (n = 25). Their English language level is B2 by the Common European Framework of Reference. In evaluating homogeneity, two criteria were implemented. Firstly, we considered each student's final English grade in the control and experimental groups. Secondly, the percentage of their present results was added and determined by the student's English Placement Test (CEFR, B2 Level) [9].

In university semester education, instructors award grades based on a fixed percentage scale that lists the grade and percentages: A (100–93%), B (92–85%), C (84–77%), D (76–69%) and E (68–61%) and F (less than 60%). A passing grade is above 61%; a failing grade is below 60% (FX).

Table 2 shows students' scores from A++ to E grade (excluding FX). The grade represents a specific percentage range assessing the students' performance. By employing this three-sub-level percentage interval, we ensure a more specific allocation of percentages, facilitating a comprehensive data evaluation; the difference between intervals is 2.5% (e.g., A++ = from 100 to 97.50%, A+ = from 97.49 to 95%, A from = 94.99 – 92.50%),

**Table 2.** Students' scores' descriptions in the pre-intervention phase.

| Ev. mark | A++ | A+ | A | B++ | B+ | B | C++ | C+ | C | D++ | D+ | D | E++ | E+ | E |
|---|---|---|---|---|---|---|---|---|---|---|---|---|---|---|---|
| CG | 1 | 4 | 2 | 2 | 4 | 2 | 3 | 3 | 3 | 1 | 0 | 0 | 0 | 0 | 0 |
| EG | 2 | 3 | 2 | 3 | 5 | 3 | 2 | 4 | 0 | 1 | 1 | 0 | 0 | 0 | 0 |

*Note* Ev. Mark - Evaluation mark, CG - control group, EG - experimental group
*Source* Processed by authors

with the number of students who achieved each grade and the frequency distribution of evaluation marks for the control (CG) and the experimental group (EG).

The Shapiro-Wilk normality test was separately performed using SPSS software for both groups, indicating that both groups exhibited a normal distribution. This is an essential prerequisite for using parametric tests in subsequent research.

The next step involved assessing whether the two independent samples were approximately equal to the previous evaluation. Since the data from both groups showed a normal distribution, it was possible to use a parametric t-test to compare two independent samples. This part of the research aimed to determine if the two groups had statistically significant differences before the intervention. That is, whether the groups were suitable for this research.

Table 3 shows the results of Levine's F-test, a T-test for two independent samples computed using SPSS software.

**Table 3.** Independent samples test comparison before intervention.

| Independent Samples Test before intervention | | | Comparation | |
|---|---|---|---|---|
| | | | Equal variances assumed | Equal variances not assumed |
| Levene's Test for Equality of Variances | F | | ,260 | |
| | Sig. | | ,613 | |
| t-test for Equality of Means | t | | ,512 | ,512 |
| | df | | 48 | 47,576 |
| | Sig. (2-tailed) | | ,611 | ,611 |
| | Mean Difference | | ,88000 | ,88000 |
| | Std. Error Difference | | 1,71833 | 1,71833 |
| | 95% Confidence Interval of the Difference | Lower | -2,57494 | -2,57574 |
| | | Upper | 4,33494 | 4,33574 |

*Source* Processed by authors

The t-test itself is conducted in two steps. The first step examines whether the two probability random samples have equal variances. Levine's F-test is used for this purpose. The result of this test provides information on which variant of the T-test to use for comparison. In Table 3, we can see the results of Levine's test, where the significance of

F is 0.613, higher than the chosen alpha significance level of 0.05. This indicates that the variances in both groups are approximately equal. We use the column "Equal variances assumed" (grey background) to select the results based on this information.

The means agreement is already tested using the t-test in the second step. In Table 3 in the t-test section, the significance of the t-statistic is 0.611, which is again higher than the chosen significance level of 0.05. Based on these results, we cannot confirm a significant difference between the groups, suggesting that the groups are likely to be the same.

## 4.3 Study Design

In the study, qualitative and quantitative approaches were involved. We used the pre/post-questionnaire before and after the intervention (FLCAS) to collect information from the students. Quantitative data were collected using a questionnaire that measured students' communication anxiety, higher-order thinking skills based on creativity, critical thinking, higher-level reasoning, and the communication development decision-making process. Qualitative data were collected through open-ended questions in the questionnaire, which were coded and analysed using Google Analytics. The data were analysed using SPSS software to identify differences between the groups.

## 4.4 The Aim and Objectives

The research aimed to determine the impact of using a collaborative approach in teaching/learning English at the University Zilina, Slovakia, in the JobLab model of topic-based professional English language training for graduates.

Objectives were intended to:

O1: Find out if students with the JobLab collaborative approach in the experimental group deal better with public speaking anxiety than those with traditional teaching/learning in the control group.

O2: Find out if students with JobLab's collaborative approach, impact the higher-level cognitive development of presentation skills (PowerPoint) than those with a traditional approach based on better topic-based speaking.

## 4.5 Hypotheses of the Research

In connection with objective O1, we hypothesised that experimental learners, by engaging in collaborative learning activities, could reduce students' anxiety better than the traditional approach.

Verifying hypothesis H1 - overcoming speaking anxiety:

- $H1_0$: Learners who use a collaborative approach in the experimental group do not overcome speaking anxiety better than the traditional control group approach.
- $H1_A$: Learners who use a collaborative approach in the experimental group overcome speaking anxiety better than the traditional control group approach.

**The Foreign Language Classroom Anxiety Scale Questionnaire (FLCAS).** The research employed a widely used anxiety and self-assessment instrument – a standardised questionnaire design (FLCAS) to investigate the effectiveness of a collaborative learning intervention in measuring participants' speaking anxiety. The questionnaire represents reliable and valid scaling anxiety and self-assessment instrument that measures FLA (Foreign Language Anxiety). It consists of 33 items. Participants rated each item on a five-point Likert scale ranging from 1 (strongly disagree) to 5 (strongly agree) as a reliable and valid scaling anxiety and self-assessment instrument. Data was analysed statistically using SPSS software.

**Verification of Hypothesis H1.** Descriptive statistics were used to calculate the mean (M) and standard deviation (SD) of the FLCAS scores for the experimental and control groups before and after the intervention. The mean values of the level of anxiety reported by the respondents were calculated. The results indicated a significant decrease in FLCAS scores after the intervention, in favour of the experimental group, with a p-value of less than 0.01. Specifically, the SD decreased from 0.88 to 0.61 in the experimental group and from 0.85 to 0.41 in the control group. The intervention's effect size was also more significant for the experimental group, with Cohen's d value of 1.10, compared to 1.06 for the control group. Furthermore, the paired t-test revealed a significant decrease in FLCAS scores for the experimental group ($t = 8.86$, $p < 0.01$), indicating a positive effect of the collaborative learning intervention in reducing speaking anxiety. However, the control group showed a slight decrease in FLCAS scores, but the difference was not statistically significant ($t = 1.08$, $p = 0.266$). The independent t-test showed a significant difference between the mean FLCAS scores of the two groups after the intervention ($t = 8.72$, $p < 0.01$), with a large effect size (Cohen's $d = 1.10$). The Cronbach's alpha coefficient value ($\alpha = 0.79$) indicated good internal consistency of the FLCAS questionnaire.

The JobLab experimental group with the collaborative approach reached better outcomes in overcoming speaking anxiety than the control group with a traditional approach based on comparing pre/post-questionnaire group results. It supports hypothesis $H1_A$ that collaborative learning significantly reduces speaking anxiety. Hypothesis $H1_A$ is accepted, and the null hypothesis $H1_0$ is rejected.

## 4.6 Hypotheses of the Research

Verifying hypothesis H2 – developing higher-level cognitive skills:

- $H2_0$: Learners who use a collaborative approach do not reach higher-level cognitive skills development than a traditional approach.
- $H2_A$: Learners who use a collaborative approach develop higher cognitive skills than traditional ones.

**Pre/post-intervention tests.** It involves measuring both groups' performance before and after receiving the intervention in an online testing platform to measure higher-order thinking skills regarding verbal creativity and critical thinking activities.

**Verification of Hypothesis H2.**

$H2_A$: Learners who use a collaborative approach reach higher-level cognitive skills influencing presentation skills development than a traditional approach. The knowledge gained by the course participant was calculated according to the formula below.

$$E = \frac{V_{post} - V_{pre}}{V_{max} - V_{pre}}.100 \quad (1)$$

$V_{pre}$ the participant's knowledge before the intervention.
$V_{post}$ the participant's knowledge after the intervention.
$V_{max}$ the maximum possible knowledge the participant could acquire.

The comparison of pre-test and post-test results based on verbal creativity and critical thinking tasks between the control group (CG) with traditional teaching and the experimental group (EG) with collaborative teaching is presented in Table 4.

The average pre-test score for the control group was 10.54, with a success rate of 70.27%. After the intervention, the average post-test score improved to 13.18, with a success rate of 86.86%. The participants in this group exhibited an average point gain of 2.64, resulting in an improvement index of 1.24.

In contrast, the experimental group, which received collaborative teaching, had an average pre-test score of 10.87, with a success rate of 72.78%. Following the intervention, the average post-test score increased to 13.96, with a success rate of 89.97%. The collaborative teaching group participants demonstrated an average point gain of 3.09, resulting in a higher improvement index of 3.01.

**Table 4.** Pre-and post-test results comparison

| Group | Pre-test score | Average success rate (%) | Pos-test score | Average success rate (%) | Gain | Average point gain | Improvement index |
|---|---|---|---|---|---|---|---|
| CG | 10.54 | 70.27 | 13.18 | 86.86 | 2.64 | 1.24 | 1.24 |
| EG | 10.87 | 72.78 | 13.96 | 89.97 | 3.09 | 3.01 | 3.01 |

*Source* Processed by authors

Table 4 provides a clear comparison between the two groups, showing the positive effects of both teaching approaches on knowledge and performance. However, the collaborative teaching group displayed a more significant improvement, as indicated by the higher improvement index and average point gain values. This suggests that learners with the collaborative approach significantly developed higher-level cognitive skills, particularly in presentation skills, compared to the traditional approach.

Furthermore, statistical analysis using a non-parametric Wilcoxon Test revealed significant differences in scores between the pre-test and post-test in both the experimental and control groups ($p < 0.01$, $\alpha = 0.01$). However, the differences were more pronounced in the experimental group, emphasising the more significant impact of the collaborative approach in enhancing higher-order skills than the traditional approach.

Based on these findings, we accept hypothesis $H2_A$, which indicates that the collaborative teaching approach leads to a more significant improvement in success rates and cognitive skills. Conversely, the rejected null hypothesis $H2_0$ confirms that the traditional teaching approach is less effective in promoting such advancements. Hypothesis $H2_A$ is accepted, and the null hypothesis $H2_0$ is rejected.

## 5 Discussions

A study limitation is the small sample size and the need to create more extensive research and reach more enormous sampling data. Moreover, the necessity for future research to delve deeper into this subject matter and examine other groups of students cannot be overstated. Expanding the sample size and conducting more extensive studies will provide a more comprehensive understanding of the phenomenon under investigation. This will enable researchers and educators to make informed decisions and develop tailored strategies to address the needs of diverse student populations.

## 6 Conclusion

The works focus on finding and adopting new collaborative trends to enhance the demands of the global business world.

Our CA model incorporates five pillars: cross-disciplinary participation, trustful relationships, conducive spaces, caring attitudes, and co-construction collaborative practices. It aims to create a positive learning environment where all students can actively participate and share ideas, overcome barriers, and achieve personal and professional success through positive reinforcement, constructive feedback, and group collaboration. The model also adopts effective error treatment to enhance students' confidence in speaking and improve their declarative knowledge and enjoyment of English learning. It promotes digital technology through an online platform for flexible and student-centred learning, focuses on developing higher-order thinking skills, and uses students' portfolios for progress assessment and direct support. Implementing CL in the JobLab English learning group requires sharing authority and responsibility among members for the group's actions to promote compelling student learning experiences; overcoming speaking fear in a stress-free learning environment stands a better chance of developing their language proficiency, developing higher-order thinking skills in terms of verbal creativity and critical thinking development in the English language topic-based learning for university students.

In conclusion, the research demonstrated the positive impact of the collaborative approach on reducing speaking anxiety and developing higher-order thinking skills among English learners. JobLab empowers students to navigate the complexities of the professional world, adapt to changing environments, and excel in their careers, ultimately contributing to their personal growth and the success of their companies.

It aims to remove barriers to language anxiety for students by setting positive and stimulating incentives to enhance active involvement. Secondly, technology-based interactive teaching fosters learning for self-paced, more independent study. Finally, it cultivates essential social skills such as communication, teamwork, negotiation, and critical thinking to develop valuable soft skills, including social, academic, psychological, and enhanced cultural understanding as the 21st life skills.

**Acknowledgement.** This research and paper publishing was supported by the financial support of the Slovak National Grant Agency KEGA under Project: KEGA 007ŽU-4/2023 Job Labs Foreign language training for global market needs at UNIZA.

## References

1. Torrance, E.P.: Tests of Creative Thinking. Personnel Press, Lexington, MA (1974)
2. Oxford, R.L.: Cooperative learning, collaborative learning, and interaction: three communicative strands in the language classroom. Mod. Lang. J. **81**, 443–456 (1997)
3. Johnson, D.W., Johnson, R.T., Smith, K.A.: Cooperative learning: improving university instruction by basing practice on validated theory. J. Excell. Univ. Teach. **25**, 1–26 (2014)
4. Mårtensson, K., Roxå, T., Olsson, T.: Developing a quality culture through the scholarship of teaching and learning. High. Educ. Res. Dev. **30**(1), 51–62 (2011)
5. Pleschová, P., Roxå, T., Thomson, K.E., Felten, P.: Conversations that make meaningful changes in teaching, teachers, and academic development. Int. J. Acad. Dev. **26**(3), 201–209 (2021)
6. Králová, Z., Kamenická, J.: Foreign Language Anxiety: Post-communist Country Context. Verbum, Praha (2019)
7. Anderson, L.W., Krathwohl, D.R., Airasian, P.W.: A Taxonomy for Learning, Teaching, and Assessing: A Revision of Bloom's Taxonomy of Educational Objectives, Complete ed. Longman, New York (2021)
8. Horwitz, E.K., Horwitz, M.B., Cope, J.: Foreign language classroom anxiety scale. Foreign language classroom anxiety. Mod. Lang. J. **70**(2), 125–132 (1986)
9. Council of Europe: Common European Framework of Reference for Languages: Learning, Teaching, Assessment—Companion Volume. Council of Europe Publishing, Strasbourg (2020)

# Integrating DGBL and Collaborative Learning in Enterprise Resource Planning Courses for Students with Engineering Background

Ming-Der May(✉)

Department of Industrial Management, Lunghwa University of Science and Technology, Taoyuan, Taiwan, Republic of China
mdmay@mail.lhu.edu.tw

**Abstract.** Although the current ERP (Enterprise Resource Planning) course has the latest information system software as in the industry that can be operated in class, and with the refinement of PBL and teaching methods, students can learn from the perspective of system implementation, rather than the traditional system operation level. However, for most students, they are still more inclined to work data input, and cannot connect to process-oriented knowledge, or even processes analysis, causing students to fall into repetitive and monotonous system operations. Therefore, this study intends to apply the digital online game competition in the ERP courses of the technical University students. To evaluate the "game-based teaching plan and course design strategy", a statistical comparative analysis is carried out on the impact of the ERP course plan. It is hoped that through such an instructional design integrating PBL and game-based learning, it can encourage students' spontaneous and interesting learning motivation and improve the effect of learning participation.

**Keywords:** Digital game-based-learning · Online business game · Enterprise resource planning

## 1 Introduction

The theme of this program is to apply Digital Game-based Learning (DGBL) method, cooperate with the existing problem-based learning (PBL) teaching materials and teaching methods, combine physical and virtual methods, simulate the information and operation processes between real enterprises, so as to enhance the fun of course learning and teamwork, and let students understand the specific impact of ERP on enterprise profit and cost control through a highly real-time online operation platform. The team competition guides students to think about how to improve profits and reduce costs from the execution and strategy aspects to improve operational performance and achieve better competition results.

Combining classroom ERP learning with game competitions requires redesigning the teaching plan and overcoming the interface conflict between the two. For example, the time frame of real-time interactive games is running day by day, and a day is usually


M. E. Auer et al. (Eds.): ICL 2023, LNNS 899, pp. 143–154, 2024.
https://doi.org/10.1007/978-3-031-51979-6_15

set to 30 s respectively, so a competition game of 60 days will be over a short span of 30 min. But the ERP system often needs to arrange a number of data in a single process, it is common to take at least 2–3 min to establish a simple purchase order (PO) by the operator, then it will be 4–6 days going past in the online game. So that the two systems need some modification to be fully synchronized with each other before further implementation in the class. In case of extending the time frame for a day, for example: 120 s a day, the overall integration of the enterprise simulation game will seriously affect the immediacy and interactive effect of the game design. The game may become no fun at all in such a slow-motion style. Therefore, from the perspective of learning, the way of system integration and the teachers and students interaction need to be rearranged carefully.

## 2 Literature Reviews

### 2.1 DGBL Pedagogy and Related Application

Game-based learning is a type of serious game that has its function in learning effectiveness, and most scholars have a positive attitude towards the learning effectiveness of game-based learning [4]. The digital game teaching platform or system related to this study, such as the ERP enterprise simulation game (https://erpsim.hec.ca) developed by HEC Montréal Institute in Canada, this simulation game is a team of three to five participants, through the same simulation economic environment as the real world, integrated with the famous German ERP system manufacturer SAP, users must use SAP to create a system form to purchase raw materials, production and sales, participants simultaneously go through a complete manufacturing information system cycle or sales system cycle. In order to perform these tasks, participants must be able to use the ERP system - SAP to support decision-making. On the other hand, at present, the more famous simulation learning game system in Taiwan is the Business Operation Simulation System (BOSS), the BOSS platform was jointly developed by Guanghua Management Strategy Foundation and the Institute of Business Administration of National ChengChi University in 1984, and has been commercialized with product content including 14 different business simulation games such as Beer games, Strategic Retail Management, Restaurant Management Guru, and so on. So far, there are more than 100 colleges and Universities, and about 31 domestic enterprises in Taiwan; and about 70 schools and 25 enterprises in the mainland China, have used this system. Many schools also use the BOSS system to hold business simulation competitions to attract students to compete with each other and exchange business experience.

Ke and Grabowski [5] used a standard experimental design to explore the impact of games, first team using a "group game competition method", one team using the "individual competition method", and the last team not using games, and taught fifth-grade elementary mathematics. The results showed that the learning outcomes of students under the Group Game Competition Method and the Individual Competition Method were better than those who did not use games, and the learning attitude of the students under the Group Game Competition Method was best of the all teams.

In this paper, the advantages of games for learning proposed by Hogle [3] are used for the characteristics of effective gaming-learning systems: Stimulating Motivation and

Interest, Improving Retention, Effects of Practice and Feedback, and finally Improving Higher Order Skills. Prensky [6] points out the characteristics of digital game learning, as shown below, and this study applies them to design appropriate assessment methods.

1. Fun: The game presents a fun atmosphere, allowing learners to feel fun and enjoyable during the game.
2. Play: Provides a form of play. Deepen the motivation and high level of fun for learners.
3. Rules: The content of the game is structured, so that learners can easily understand and organize the content of the game, play the game through the field, and interact with the game.
4. Goals: Specific goal tasks in the game that can clearly guide learners to play.
5. Human-computer interactivity: The game design interface allows learners to play games through computer operation and interaction.
6. Outcome and Feedback: The game provides feedback mechanisms for learners to have the opportunity to learn and review.
7. Adaptive: The design of the game can be adapted to different appropriate tasks according to the learner's ability.
8. Sense of Victory (Win states): During the game, learners gain success experiences and provide learners with a sense of self-satisfaction.
9. Conflict/competition/challenge/opposition: Competition and challenge make learners feel excited during the game (adrenaline).
10. Problem solving: Set up problems in game situations and stimulate learners' thinking to solve problems.
11. Social Interaction: Let learners form a game community to produce interactivity.
12. Representation and story: Through pictures and storylines, learners can gain emotions and memories.

From the above literature, it can be seen that digital game learning system is the way of combining computer software with education and then matching competition. A good competitive digital game learning system allows users to complete tasks through "cooperation" and "competition", thereby enhancing knowledge, building confidence and deepening their sense of honor.

## 3 Research Methods

### 3.1 Digital Situation Operation

Through the competition-based experience game lesson plan designed by the teacher, students can run a business in an online team, and can more realistically experience the specific steps of various sales, procurement, production and logistics in the operation process of each day, with on-site synchronous leadership, so that students can use the digital situation of the computer, combined with real-time operation simulation and the operation of the WF ERP system (https://www.digiwin.com/tw/), to experience how the ERP system support the daily routine operations and exchange cross-departmental information. Let the game be more than just a game, and make students feel like they are in the actual company's system running in an office, factory, or warehousing center. The

Input-Process-Outcome Game Model of Garris, Ahlers, and Driskell [1] are adapted here to study the impact of institutional games on learning. The leftmost part of this model is the input part, which includes instructional content and game characteristics, which are the specialized knowledge or specialized training content that the educator wants the learner to learn. The characteristics of the game are explained in the following sections. The middle is the process of the game cycle, including user judgments or reactions such as enjoyment or interest, user behaviors such as greater persistence or time on task, and further system feedback. If the teacher are successful in matching instructional content with appropriate game features, this game cycle results in recurring and self-motivated game play. Finally, this engagement in game play leads to the achievement of training objectives and specific learning outcomes. The main purpose of this model is twofold. The first is to design a teaching plan that combines the characteristics of the game. Second, these characteristics trigger the user's judgment or response as much as fun or interest, as well as the user's ongoing behavior and other feedback on the task. Through this cycle, it can be judged whether the combination of game and education is successful, resulting in a cycle of regular and spontaneous active play, as shown in the Fig. 1.

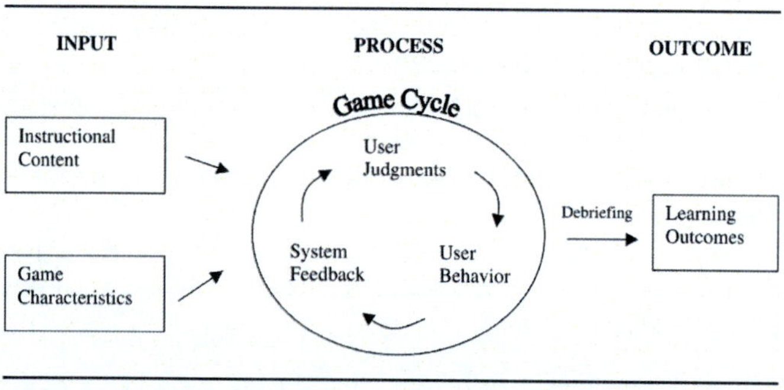


**Fig. 1.** Input-process-outcome game model [1]

## 3.2 Digital Enterprise Resource Management Operation Simulation Platform

MonsoonSim, located in Australia, follows experiential learning and business process teaching to develop an online real-time business simulation web-based platform (URL: http://www.monsoonsim.com/) to experience real-time operations online, including corporate processes and decision-making in 13 functional modules.

1. Finance, purchasing, retail: financial statements, cash flow, market demand analysis, prices
2. Marketing, warehouse, logistics: market analysis, market investment return analysis, logistics analysis, capacity planning, data analysis
3. Wholesale, production: cost accounting, profit analysis, sales analysis, operation management, bill of materials, asset procurement, production capacity, production planning, raw material management, tendering

4. Material Requirements Planning (MRP) Forecasting: Forecasting, Material Requirements Planning (MRP), Production Analysis
5. Asset Management, Human Resources: Employee Recruitment, Consulting, Dismissal, Competency. Wealth maintenance (periodical, forecasting). Impact of maintenance
6. Service management: service arrangement, customer service, overall operation management

Figure 2 illustrates the student screen while the system is running, and Fig. 3 displays the comprehensive ranking and KPI values after each game. Learners initially operate the game before learning ERP, engaging in real-time simulation. This approach aligns with the experiential learning principle of "first move and then know". Therefore, this research combines the platform with ERP software, enabling students to gain a detailed understanding of the enterprise's operational process. They can experience how the different departments of functional organization affect the overall profit and cost performance of the enterprise.

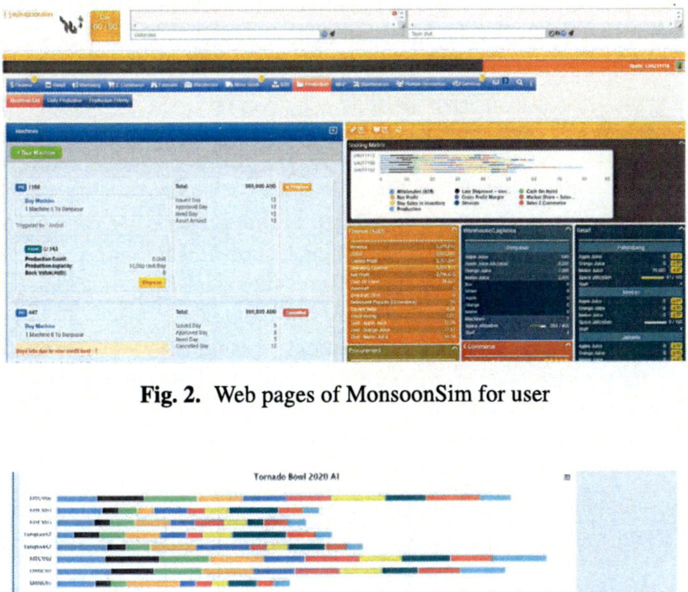

**Fig. 2.** Web pages of MonsoonSim for user

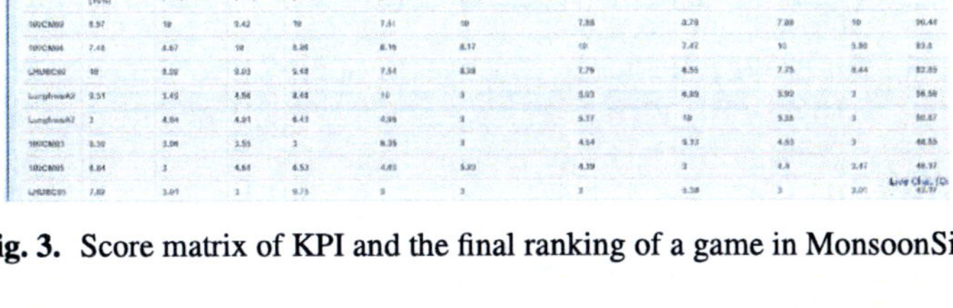

**Fig. 3.** Score matrix of KPI and the final ranking of a game in MonsoonSim

Since its establishment in 2012, the company has expanded its presence to Australia, Singapore, Hong Kong, Malaysia, Thailand, the Philippines, and Indonesia. In 2019, they made the first entry into Taiwan and organized a nationwide competition in November for college and high school students. The following year, our University purchased and integrated this gaming web platform into the curriculum of relevant courses.

# 4 Instructional Content and Game Characteristics Design for ERP Learning

## 4.1 Instructional Content of ERP

ERP, with Material Requirement Planning (MRP) and Manufacturing Resource Planner (MRPII), integrates key processes and information, including finance, distribution, production, human resources, and supply chain. It has been globally accepted by commercial enterprises as the core system for operations management and as a tool for e-enterprise and process reengineering. Students from various engineering and management backgrounds have enrolled in this class to learn how to use ERP systems. In this study, a trading module is used as an example to illustrate the interaction between ERP and the data generated during the online business simulation game.

## 4.2 Game Characteristics of ERP with MonsoonSIM

The general sequence to produce finished goods to fulfill customer demand starts with renting space for a warehouse and factory. Once the space is available, students need to familiarize themselves with the product's Bill of Materials (BOM) and purchase raw materials accordingly, based on the desired quantity of finished goods their company intends to produce. Subsequently, students must acquire at least one machine, which will automatically produce goods as long as there are sufficient raw materials. Following these steps, the produced finished goods are automatically placed into the warehouse.

The game characteristics of MonsoonSIM include the stress of real-time operation, where each day lasts for 30 or 40 s, and the processing time for each step depends on the students' skill level. Another game characteristic is the ranking of team performance. The Scoring Matrix is set up by the teacher according to different scenarios and is continuously updated to display the rankings as the game progresses. Students become engaged in the scoring system and experience excitement as their rankings rise or fall.

## 4.3 Develop New Teaching Materials and Plans

The new lesson plan, which utilizes PBL and online game competition learning methods, is expected to be published and promoted in a revised version after practical application and evaluation of this project. It aims to enhance the ability to solve practical problems by incorporating existing PBL-ERP teaching materials. Additionally, the plan aims to engage students by making the learning process enjoyable, thus encouraging them to dedicate more time to studying the course and achieve effective learning outcomes.

**Game time setting and lesson running.** Such a design ensures that the game characteristics and the lesson teaching in the class would be operated coherently.

1. Teaching mode: Run one day at a time, the game will stop every day, so the teacher can demonstrate the ERP data entry according to the newly events in the real-time business game. The game is designed to run on a daily basis, with each day representing a discrete unit of time. At the end of each day, the game will pause, allowing the teacher to demonstrate the process of ERP data entry based on the events that occurred in the real-time business game. This approach enables students to observe and understand how the principles and practices of ERP systems are applied in response to the day's events. By providing this real-time demonstration, students can gain practical insights into the integration of ERP systems and business operations.
2. Competition mode one: In order to facilitate the simultaneous execution of ERP and MonsoosSIM, the daily time within the game can be extended to be two or three times longer. This means that each day in the game will have a duration of up to 120 s, allowing for more comprehensive and in-depth exploration of both systems. By extending the daily time, students will have more opportunities to engage with the ERP system and MonsoosSIM, enabling them to gain a deeper understanding of their functionalities and interplay within a business context.
3. Competition mode two: In an event-driven manner, students are assigned to complete at least one event, such as purchasing goods or participating in a bidding activity in a B2B market. The teacher has the flexibility to manage the lesson by either pausing the activity until the event is completed or scheduling a break and automatically resuming after a specified duration of 10 or 20 min.

This allows for a dynamic and interactive learning experience, ensuring that students actively engage in real-world scenarios and promoting their problem-solving skills.

**Business model**. The game with 13 functional modules could be divided into four kinds of Business Scenario Model: *The retail B2C* business model, *Wholesale B2B* business model, *Manufacturing & operation* model, and *Combination or Full Enterprise* mode. The snapshots of B2B business models in Fig. 4 are listed as an example.

By integrating MonsoonSIM as a virtual company with the ERP system, students can apply their theoretical knowledge of resource management in a practical business setting. They will actively search for and utilize the necessary data from the game, thereby deepening their understanding of how ERP systems streamline and optimize resource management processes.

### 4.4 Integrate Online Business Simulation Competition, ERP, and Data Analysis Tools into Teaching Plan

To develop new teaching methods, we changed the existing ERP courses that focused on system operation and data input into ones that increase the value of data application, understand the importance of information collection and database establishment, and learn more advanced skills step by step.

**Integrate online business simulation game and ERP**. The students have been learning the ERP system in the first six weeks of this 18-week course, followed by another six weeks dedicated to MonsoonSIM. Finally, the remaining weeks involve a mixture of the two systems. The interactions between the online business game and ERP are compared

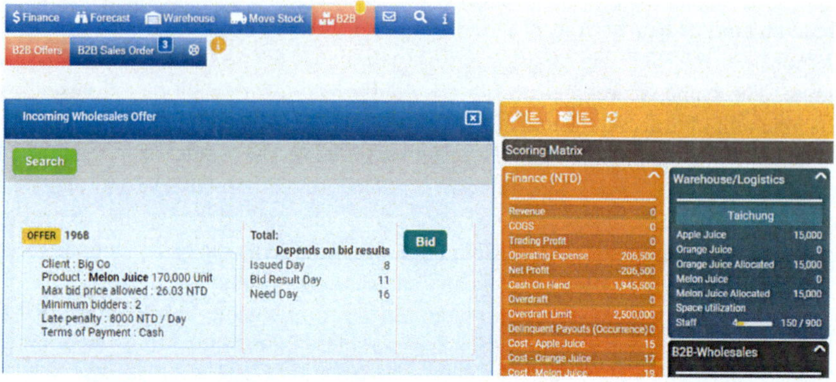


**Fig. 4.** Web pages in MonsoonSim (*Wholesale B2B business model*)

in Table 1. Five students in a group work together to run a company, aiming to generate profits and compete with nine other companies in real-world countries and cities. The real-time data items generated in the online game serve as realistic problems for the students. After every transaction is completed in the online game, the corresponding process in the ERP system must also be finalized. These immediate tasks require students to apply ERP based on the transactions from MonsoonSIM, simulating the daily operations of a real company. However, before engaging in the daily operations, several prerequisites must be completed in advance. For example, the items data and bill of materials for the commodities (Melon Juice, Apple Juice, and Orange Juice) sold in MonsoonSIM must be created in the ERP. These interactive steps mirror the process a company undergoes when initially implementing ERP and managing its daily business operations. Figure 5 displays snapshots of the finished goods purchase job in MonsoonSIM and the related purchase order in ERP. Once such exercises are completed, a report containing all these pictures from both MonsoonSIM and ERP is submitted for evaluation.

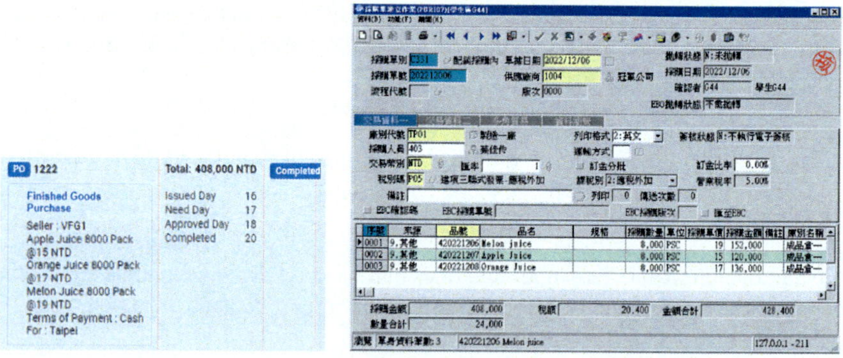


**Fig. 5.** Purchasing goods in MonsoonSIM and add PO in ERP

**Table 1.** Interactions between online business game and ERP

| Way of working | MonsoonSIM B2B transaction game | ERP implementation of trading system |
|---|---|---|
| Prerequisites | A. Find and rent a warehouse location | A. Create items data<br>B. Create the bill of materials (BOM) |
| Instant jobs | B. Purchasing goods<br>C. Delivery to the warehouse<br>D. Bid offer<br>E. Win the bid<br>F. Shipping from warehouse | C. Add purchase orders (PO)<br>D. Add receipts<br>E. Add quotations<br>F. Add sales orders<br>G. Add new shipments |

**Data Analytics Live Access.** Although the current ERP course has the latest information system in the industry that can be operated in practice, and with the refinement of PBL and teaching methods, students can learn from the perspective of strategical system implementation, rather than the traditional system operation level. For a relatively large number of students, it is still easy to flow into homework data input, become a mentality to cope with teaching progress, and unable to achieve decision-making knowledge management, and even strategic data analysis, resulting in students are easy to fall into repetitive and monotonous system operations. It is difficult to continue the enthusiasm and fun of learning.

All gaming data from MonsoonSIM can be exported to an Excel file or accessed live through MS Power BI or Looker Studio (Google) for data visualization and dashboard presentations. To facilitate this process, a set of slides, similar to the one shown in Fig. 6, is provided to the students. These slides outline the process of connecting these tools to the MonsoonSIM online database, allowing students to independently analyze the data.

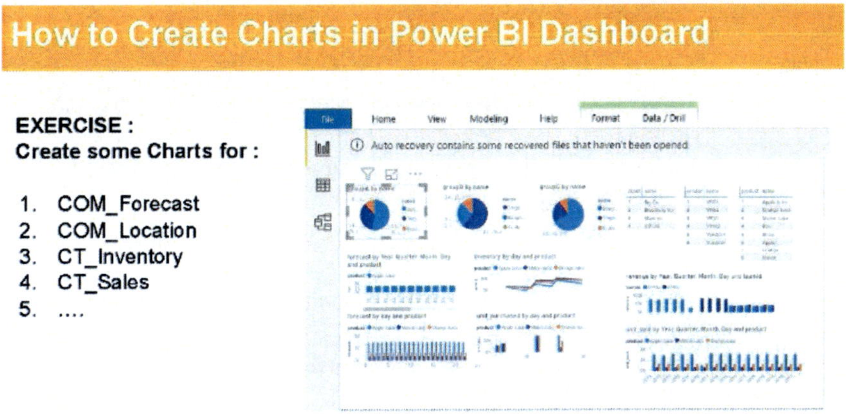


**Fig. 6.** Slide for student to apply data analytics with online business game results

## 5 Results and Analysis

For third and fourth-year university students enrolled in the 'Application of ERP System' course, MonsoonSIM is used in the classroom. To assess the effectiveness of this approach, a questionnaire was developed based on the one designed by Huang et al. [2] for evaluating the teaching and practice of simulators (sim man) among medical interns. This questionnaire was modified for the purpose of this study and comprises three components: self-confidence, knowledge, and user experience satisfaction assessment (including two sets of reverse questions). The Likert 5-point scale was used, and the average scores were calculated from the 11 valid questionnaires completed by the 19 students in the class. The results are presented in Table 2.

**Table 2.** DGBL for ERP learning satisfaction

| | |
|---|---|
| ***a) Self-confidence assessment*** | |
| 1) **After the MonsoonSIM + ERP simulation exercise, I was able to learn the courses related to business operations with more confidence** | 3.8 |
| 2) **After the MonsoonSIM + ERP simulation exercise, I was able to form a team with other team members more confidently and work together** | 4.1 |
| 3) **After the MonsoonSIM + ERP simulation exercise, I can more accurately determine where the main problems of enterprise operation and management are** | 3.9 |
| 4) After the MonsoonSIM + ERP simulation exercise, I can design business operation processes and steps with more confidence | 3.4 |
| 5) After the MonsoonSIM + ERP simulation exercise, I can solve problems related to improving business operations with more confidence | 3.5 |
| 6) After the MonsoonSIM + ERP simulation exercise, I can participate in business operations with more confidence | 3.5 |
| ***b) Knowledge learning assessment*** | |
| 7) I think this exercise can teach basic enterprise process problem solving techniques | 3.7 |
| 8) I feel that through this exercise, I can learn about business processes and help me make decisions at work | 3.6 |
| 9) **I think I can learn the characteristics of enterprise processes (ex. Procurement) and strategy (ex. Low cost) in practice** | 4.0 |
| 10) **I think I can learn the characteristics and conditions of Blanket Order in practice** | 3.8 |
| 11) **I feel that I can learn how to use ERP systems and understand the importance of ERP in the process** | 4.0 |
| 12) **I think I can better understand the correlation between ERP and enterprise processes in the process** | 4.0 |

(*continued*)

**Table 2.** *(continued)*

| | |
|---|---|
| 13) **I think I can learn how to interpret the significance of indicators such as bottlenecks, equipment utilization, demand incidence, and cost-effectiveness of enterprise processes, and understand their importance** | 3.8 |
| 14) I feel like I can learn how to apply ERP systems to do the work that the business needs | 3.5 |
| 15) I feel that I can grasp the specific methods of business strategy related to actual profitability | 3.6 |
| ***c) User Experience Satisfaction Assessment*** | |
| 16) I think the design of the interface of this software is helpful for the progress of the game | 3.5 |
| 17) **I think the virtual business situation simulated by this software is more interesting than I thought** | 4.0 |
| 18) I feel that using this software to simulate the business model of the enterprise, I cannot learn the functions and processes of the enterprise | 3.1 |
| 19) I think this software is old and ineffective, and should be updated or replaced | 3.2 |
| 20) **I find it more attractive to use this virtual situation to learn ERP than to learn from a book or teacher** | 3.9 |

## 6 Conclusion

This study try to apply DGBL on the production processes of a ERP system, incorporate the two approaches into one integrated pedagogy. An online web-based business simulation site, MonsoonSIM is served as the game part to provide the major characteristics of fun and play. A legacy ERP system that has been adapted to provide instructional content are running concurrently with the online business game. The two inputs are applied in the game cycle process and a class with 24 students are evaluated with the questionnaire after three to four weeks course by the above integrated learning approach. The learning satisfaction evaluated outcomes show that the knowledge assessment is quite positive and the feedback from students also expressed the fun from a competitive gaming, and so much knowledge of a company is helpful for the students to learn the higher skill of ERP.

In this study, the online gaming is used to make students get familiar with basic business processes and lead them to learn why ERP is designed and how ERP is collecting data of a real company. Likewise, the further study is recommended for more and deeper integration with the gaming approach and ERP modules. For example, if the students is well experienced for the basic functions, then an even tighter time settings or some specific and harder scenario would bring more challenge and edutainment into the class. A tighter time setting needs well trained skill both on the MonsoonSIM function and the ERP system to concurrently execute all commands in time. Furthermore, the specific scenario, such as economics crisis or the pandemics impact from COVID-19, would require strategical thinking and responsive action on the proper time of the game, and the ERP would serve as the tactical or short-term operational tools to implement the strategical decision. The goal of DGBL should be providing a practical and realistic platform for the students and teachers to interact with each other. So in further study, ERP data would be very timely and dynamically retrieved from the online running business game, and the correct response of students in ERP system would be just like the role of ERP in every real world company.

## References

1. Garris, R., Ahlers, R., Driskell, J.E.: Games, Motivation, and Learning: A Research and Practice Model. Simul. Gaming **33**, 441–467 (2002). https://doi.org/10.1177/1046878102238607
2. Huang, G.-S., et al.: Effects of education on emergent and critical case-oriented simulation practice (in Chinese). J. Taiwan Crit. Care Med. **9**, 228–241(2008)
3. Hogle, J.G.: Considering Games as Cognitive Tools: In Search of Effective "Edutainment." Working paper, Department of Instructional Technology, University of Georgia (1996)
4. Kirriemur, J., McFarlane, A.: Literature Review in Games and Learning. NESTA Futurelab Series. NESTA Futurelab, Bristol (2004)
5. Ke, F., Grabowski, B.: Game playing for maths learning: cooperative or not? Br. J. Edu. Technol. **38**(2), 249–259 (2007)
6. Prensky, M.: Digital Game-Based Learning. McGraw-Hill, New York (2001)

# Collegial Video-Based Reflection on Teaching in Teacher Education - Reflection Processes and Levels of Reflection Quality

Kerstin Göbel[1] (✉), Lisanne Rothe[1], and Marie Christin Schwark[2]

[1] University of Duisburg-Essen, Universitätsstraße2, 45141 Essen, Germany
kerstin.goebel@uni-due.de

[2] Heinrich-Böll Gesamtschule, Agnesstraße 33, 44791 Bochum, Germany

**Abstract.** The ability and willingness to reflect on one's own teaching is seen as an important characteristic of teachers' professional competence [1]. In the contemporary discourse about teacher professionalization in Germany, great importance is attached to the training of reflective competence [2]. In order to bring up further perspectives on the event to be reflected, collaborative reflection seems to be a promising way [3] and further, video-based reflection formats can help student teachers to diversify their perspectives on teaching [4]. Results from a video-based collaborative reflection intervention show that reflection-related attitudes of student teachers can be improved by video-based collaborative reflection settings, especially when reflecting on their own videos [5]. However, the quality of collegial video-based reflection process has rarely been studied. This research gap is addressed in the present paper, which deals with the quality of reflection processes in the context of a collegial video-based reflection on teaching of pre-service teachers in a special format called *Reflecting Teams*. Results show that reflection and feedback quality are quite elaborated, as students are already capable of in-depth engagement with the teaching video [6, 7], students seem to benefit from both, taking and giving feedback [8]. However, students have to become familiar with the structure of *Reflecting Teams*.

**Keywords:** Video-based reflection on teaching · Reflection process · Collegial reflection · Teacher education

## 1 Theoretical and Empirical Background

As teachers' everyday professional lives can neither be standardized nor conclusively planned or predicted, the conscious mediation between experiences in the school context, the professional knowledge, and the teachers' own expectations and beliefs play a crucial role [9]. Reflection can be a central connecting element between experience, knowledge, expectations and beliefs, Combe and Kolbe [10] see reflexivity as a key competence of the teaching profession. Reflection can be described as a main facet of teachers' pedagogical professionalism and an "emergent condition of pedagogical skills" along with knowledge, experience, and personality [11]. Korthagen [12] defines reflection broadly


M. E. Auer et al. (Eds.): ICL 2023, LNNS 899, pp. 155–165, 2024.
https://doi.org/10.1007/978-3-031-51979-6_16

as "[…] the mental process of trying to structure or restructure an experience, problem, or existing knowledge or insight." [12, p. 63]. Reflection processes seek alternative patterns of thought and action to change the perspective of the event being reflected upon [13]. Reflection-on-action, as retrospective reflection on a meta level (like an external observer) [14], is a common practice in teacher education as it can be done from a distance and is thus free from an immediate pressure to act [14, 15]. Reflection should be systematic, well-structured and should help to expand subjective perspectives to reach multiple perspectives on relevant situations in school [4].

According to Eysel [16], the quality of reflection can be divided into two dimensions: the breadth of reflection, which reflects the variety of references and contexts taken into account and the depth of reflection, which describes the comprehensiveness of the examination of the situation reflected on. A high reflection quality can be assumed if an individuum finds more than one explanation for an observed fact, recognizes different perspectives of a situation, elaborates structures in events, links observations from different levels, formulates further questions, develops more than one possible solution for an issue, and uses theoretical and empirically based arguments to analyze a situation [17].

As reflection on teaching is meant to be a systematic, in-depth examination based on theoretical knowledge [18, 19], video-supported reflection on teaching can play a relevant role in bridging the gap between theory and practice [20]. In teacher education, videos on teaching experiences have been increasingly used as a learning and reflection medium [21]. The benefits of teaching videos to enhance reflection and to foster professional vision of teachers (or student teachers) have already been empirically confirmed [22–24]. As teaching videos represent real situations in practice, the complexity of teaching and learning interaction can be experienced more intensively than is possible with any other media [25]. Therefore, videos are very useful for deep reflection on teaching, as most relevant aspects of a classroom situations may be documented [26].

As different perspectives and different explanations for an observed situation can enrich the quality of reflection, cooperative reflection is seen as a means for enhancing reflection and considered especially beneficial for reflection in teacher education [3]. Meschede and Colleagues [8] point out the potential of peer feedback in the context of video-based classroom analysis, as students gain insight into the process of peer analysis both when giving and receiving feedback, when comparing with their own analysis for further differentiation. Moreover, they argue that differences in prior knowledge and diverse attitudes of peers may lead to different perspectives on what is happening in the classroom, which can broaden the range of perceived relevant features of teaching and enable a change in perspective [4, 8]. Besides the potential of collaborative reflection for the quality of reflection, student teachers tend to hold positive attitudes towards collegial reflection and are hence motivated to engage in this kind of reflection setting [27, 28]. The quality of student teachers' reflection might further be improved by offering highly structured, guided reflection methods [29–31]. Sherin and Van Es [34] propose the division of the reflection process into 1) describing, 2) interpreting, and 3) developing alternative actions. Further, learning opportunities to reflect on practical experiences which link with theoretical concepts seem to be particularly promising for professional development in the context of teacher education [32, 33]. In the context of video-based collegial reflection settings, common reflection on a specific teaching video can also be understood as a feedback to the person who presents their own teaching video. According to the interactive-two-feedback-loops model of Narciss [7], reflection depends as well on internal feedback (like E.G. Internal reflection processes) as on external feedback (E.G. From peers), both feedback perspectives have to integrated by the individual to foster the learning process. Although there are several arguments for structured and collegial reflection on teaching in teacher education, previous studies show that pre-service teachers tend to reflect less systematically on their own teaching as the analysis on teaching is predominantly done by mentors [35]. Collaborative reflection on teaching videos are studied in teacher training [36] and in pre-service teacher education [37] to some extent. Still, the potential of collaborative video-based reflection settings in teacher education has yet not been addressed in a detailed way [38, 39].

The intervention study (FLECTT - collegial video-based reflection on teaching in ReFLECTing teams; a pre-post control group study) addresses the research field of collegial video-based reflection in teacher education by offering a structured, theory-based peer-reflection format. The intervention is based on the concept of *Reflecting Teams* (see Fig. 1), adapted from systemic counselling [40]. This cooperative form of reflection pursues the goal of enabling multiple perspectives on the teaching situation to be reflected on in a respectful atmosphere and with the help of structured guidelines [41]. first results show positive effects of reflection on student teachers' attitudes [5]. All participants assign a high relevance to feedback and reflection on teaching is highly appreciated in all groups. students in the intervention groups show a high appreciation of and low concerns about video-based reflection on teaching. A positive change in attitudes in the sense of decreasing concerns about video-based reflection on teaching and an increase in positive evaluation of peer feedback after the intervention could be identified particularly for student teachers who reflected on their own teaching video [5]. The reflection of one's own videos might be accompanied by greater emotional involvement, an increasing motivation and deeper reflection [42]. To further investigate the quality of reflection of collegial video-based reflection in *Reflecting Teams* the present paper addresses the analysis of reflection processes in the context of the before mentioned FLECTT- study.

## 2 Research Subject and Research Questions

The paper addresses the following research questions:

1. What quality of reflection do teacher students achieve in the collegial video-based reflection setting *Reflecting Teams*?
2. How do teacher students perceive reflection and feedback in the collegial video-based reflection setting *Reflecting Teams*?

## 3 Method

### 3.1 Design

In the context of the project FLECTT (Collegial video-based reflection on teaching in reFLECTing Teams), as part of the project ProViel funded by the German Federal Ministry of Education and Research (BMBF, funding code: KZ 01JA1910), pre-service

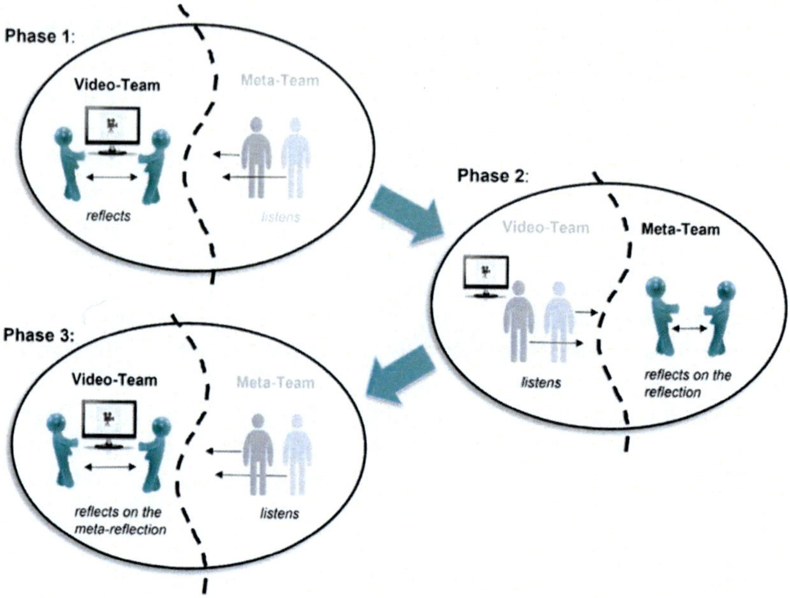


**Fig. 1.** The reflection process of the collegial video-based reflection setting – *Reflecting Teams* [5, p. 5]

teacher students in their practical term were introduced to a reflection setting for the collegial reflection on teaching videos. The reflection setting is based on the idea of a systemic counselling approach, providing a clear structure for the reflection process and enabling a targeted discussion [5, 40]. Short scenes of teaching videos of the pre-service teachers formed the content of reflection, while a specific dimension of instructional quality was addressed as theoretical focus [18, 43, 44]. Two equally important and inter-dependent teams were formed (see Fig. 1): The video-team consists of a person seeking advice and introducing scenes from an own teaching video and an interviewer, who is supposed to stimulate in-depth reflection based on reflection supportive questions. The meta-team consists of peers forming a supporting system to intensify the examination of the content of reflection. The discussion about the video scenes is divided into different phases, the teams alternate in discussing. The discussion is realized within the teams only. The *Reflecting Teams* took place in presence of all participants of the two teams. Both teams were provided with reflection supporting questions adapted to the respective content of reflection [34].

It is important to mention that the students selected for the study came into contact with the method described above for the first time in the project related seminar. To support the reflection process and to reflect and to foster the compliance of the communication rules students received a short training and a set of reflection supporting questions (reflection prompts). Before the reflection process started, rules of communication were introduced to clarify the structure of *Reflecting Teams* and to foster a positive and resource-oriented feedback culture within the discussion about the video scenes. Further, reflection supporting questions as general impulse questions were provided to support structure, which were specified according to selected theoretical topics of reflection. The reflection prompts describing, interpreting and developing alternative actions, as proposed by Sherin and van Es [34], have been tailored to the role within the *Reflecting Team* in advance.

### 3.2 Sample

The sample of the present exploratory analysis comprises a group of five students of the Master of Education study program at the University of Duisburg-Essen (2 male, 3 female). The students had own teaching experiences in the course of their practical term. They were supported by the FLECTT-project team to record their own teaching with videos during their practical term. The videography of teaching was realized with the consent of the parents of students in class.

### 3.3 Qualitative Analysis

The reflection process in *Reflecting Teams* was video and audio recorded with the consent of the teacher students involved. The recordings were literally transcribed [45] and participants were anonymized in the transcript. A total of approximately 90 min of video material comprising three rounds of *Reflecting Teams* was analyzed using qualitative content analysis according to Mayring [46]. The analysis of the transcript of the videotaped reflection session was realized with the program for qualitative data analysis MAXQDA [47] Focusing on the reflection process, quality of reflection, and perception of the setting. A category-guided qualitative content analysis was chosen to summarize the content of the material, to enable an adequate description and in-depth interpretation of the data to arrive at a meaningful answer to the research questions [46].

The coding frame for the qualitative content analysis was based on different theoretical perspectives on reflection, the *ERTO*-model by Krieg & Kreis [6] and the reflection model by Sherin and Van Es [34]. The reflection levels of the *ERTO*-model (**e**vent, **r**eflection, **t**ransformation, and **o**ption for a New action) were used as indicators for reflection quality. While the *descriptive level* involves situational descriptions and at least one evaluation of the situation or a recognition of a problem, the *explicative level* includes further interpretations. At the *introspective level*, various justifications as well as descriptions of own experiences and assumptions are made. Finally, at an *integrative level*, reflection includes scientific theories that may be useful in processing the event. As categories like "description of the event" as well as "reflection-deepening questions" and "transformative reflection" have been fundamental for the reflection process and had formed part of the reflection prompts [34], They have also been integrated into the coding frame. Further, as the perception of participants of the reflection setting was of interest for the analysis, categories of evaluation have been integrated into the coding frame. additionally, categories addressing the surface structure of reflection like reflection content, the reflection phase and rule compliance in reflecting teams were added. The generated coding frame contains the following categories:

1. Surface structure of <u>reflection process</u>: Reflection contents (1); Reflection phases (2); Rule compliance of the method (3)

2. Deep structure of reflection process (quality): descriptive description (0); descriptive reflection (1); explicative reflection (2); introspective reflection (3); integrative reflection (4); transformative reflection (5); reflection deepening questions/ suggestions (6)
3. Perception of the reflection setting: Positive added value (1); perceived difficulties in the reflection process (2); changes in the perception of the setting in the course of the reflection phases (3)

## 4 Results

The analyzed reflection process in *Reflecting Teams* comprises three rounds of reflection, addressing the topics classroom-climate, motivation of students and classroom-management [43]. The analysis of reflection quality, reveals that students reach different levels of reflection: descriptive description, descriptive reflection, explicative reflection and introspective reflection. The results hint at a deep engagement with the teaching video, as students reach the level of introspective reflection several times [6]. In the analyzed reflection process, students named different explanations for the events they reflected upon and took different perspectives into consideration. Furthermore, students connect observations from different levels and generate new reflection-deepening questions. In the further course of the reflection process, students did not only describe the teaching situation but showed an increased engagement, generating more intensifying questions and several solutions for future teaching situations. The feedback of the interviewing person and the meta-team have been evaluated as being extremely positive and helpful as considered by the person who brought the video into the reflection session.

However, the analysis also shows that there is a lack of inclusion of theoretically and empirically based arguments throughout the reflection process, as the integrative level involving scientific theories in the reflection could not be consistently achieved by the students. Nevertheless, in the present study, it could be shown that reciprocal feedback and stimulation within the collegial reflection could enhance the quality of reflection. It can be stated that although the students had difficulties in finding their way methodically at the beginning, the participating students subsequently showed a high level of reflective discourse. The participating students succeeded to reach all reflection stages at least twice, except for the integrative stage. Thus, with regard to the development of reflection processes, an increase in reflection quality can be observed in all course participants. Looking at the different actors in the *Reflecting Teams*, the meta team expressed statements of higher reflection quality than the interviewer in the video team and hence showed deep engagement with the teaching video.

With regard to students' perception of the reflection process, the analysis shows that participants successfully maintained an appreciative and positive attitude towards the person seeking advice (person with teaching video). The participants in both teams stated that interactions have been supportive and helpful for amplifying perspectives.

However, especially at the beginning of the reflection process, participants express difficulties in adhering to the conversational rules of the setting *Reflecting Teams*. It was unfamiliar to the students to reflect on the event only in their own system (video-team or meta-team) in a predefined order. However, in the course of the process the students succeeded to clarify the structure. The initially existing ambiguities and difficulties with the methodology of the structured reflection processes indicate a high complexity of the reflection concept. Nevertheless, the analysis shows that the participants have managed the ambiguities and difficulties in the further course of the reflection process. Positive statements about the cooperative reflection process in *Reflecting Teams*, such as "high quality event", "very interesting" and "fun".

## 5 Discussion

As previous studies on video reflection with *Reflecting Teams* focused on changes in reflection-related attitudes [5, 41], the present paper aims to investigate the processes and quality of reflection in the collegial cooperative video-based reflection setting with *Reflecting Teams*.

Concerning reflection and feedback quality, results hint at students' capacity to a deeper engagement with the object of reflection (introspective reflection and external feedback) [7, 48]. The qualitative analysis supports the assumption that the setting *Reflecting Teams* can support a deepened and positively perceived examination of the teaching video. Students in the intervention were able to reach complex levels of reflection, which is consistent with previous studies on quality of reflection, showing that teacher students reach more complex levels of reflection when assisted by highly structured reflection methods [30, 49]. As Participants in the presented sample develop the prerequisites for a high reflection quality in the course of reflection, the cooperative reflection settings seem useful in promoting reflection skills [17]. The use of collegial reflection structures seems particularly beneficial for teaching contexts [5, 13], since collegial reflection supports the integration of diverse observations, interpretations, suggestions, theoretical knowledge, technical knowledge, and hence leads to a strengthening of reflection quality [50].

The use of videos in teacher training is meant to promote the link between theoretical knowledge and practice [51–54], to focus on relevant aspects of teaching and thus improve the professional perception of teaching [25, 55]. The results of our study hint at further potential for the development of reflection competences. Still, as well-developed reflection competence is meant to integrate theoretically and empirically based arguments [18, 26], the lack of integration of theoretically and empirically based arguments in the reflection of the teaching video in the analyzed reflection process hints at the necessity to further prepare students to establish this link. The complete absence of the inclusion of scientific models in the reflection suggests that the students' competences regarding the link between theory and practice may still be developed. It might be helpful to further address this connection more strongly in the reflection supporting prompts.

The results of the exploratory video study also demonstrate that with increasing experience in the method, participants' reflection quality increases over the course of the three phases. The use of collaborative video reflection seems to motivate deeper reflection, but should be practiced intensively and regularly for participants to optimally benefit from it. The increasingly positive perception of the reflection setting, the influence of the collaborative reflection on the persons with own videos, and the increase in reflection quality indicate a connection between the positive attitude toward the reflection setting and the intensity of engagement with the object of reflection [5, 40, 56].

A limitation of the present study is its exploratory character, given the very small sample size. Due to this fact, results might be very selective and conclusions should be drawn with great caution. The analysis is limited to the subjective reflection skills of the selected students and their attitude towards the setting. The present study could not consider previous pedagogical experience of the participating students, which might have had an influence on their reflection process and the quality of their statements.

Concluding from former studies and from the presented results, for sustainable promotion of reflection skills of prospective teachers, cooperative reflection on teaching videos, seem promising. Still, it seems advisable to apply structured reflection methods regularly, starting during the studies and continuing in the legal profession, since it is necessary to become accustomed to the structures and the sequence of reflection settings like e.g. Reflecting Teams before the concentration can fall completely to the intensive examination of the subject of reflection. Teaching sequences from expert teachers could also be helpful to reflect upon good teaching standards and to train the reflection structure. Since in-depth reflection is meant to be based on theoretical knowledge [18], this link should be supported and encouraged more intensively. For this reason, further research on a larger sample with a theoretically further refined reflection setting is needed to understand better how to improve the quality cooperative video-based reflection on teaching.

## References


1. Darling-Hammond, L., Hyler, M.E., Gardner, M.: Effective Teacher Professional Development. Learning Policy Institute, Palo Alto, CA (2017)
2. Te Poel, K., Heinrich, H.: Professionalisierung durch (Praxis-)Reflexion in der Lehrer_innenbildung. HLZ **3**(2), 1–13 (2020)
3. Brown, C., Poortman, C., Gray, H., Groß-Opphoff, J., Wharf, M.: Facilitating collaborative reflective inquiry amongst teachers: what do we currently know? J. Int. Educ. Res. **105**(3), 10165 (2021)
4. Körkkö, M.: Towards meaningful reflection and a holistic approach: creating a reflection framework in teacher education. Scand. J. Educ. Res. **65**(2), 258–275 (2019)
5. Göbel, K., Bönte, J., Gösch, A., Neuber, K.: The relevance of collegial video-based reflection on teaching for the development of reflection-related attitudes. Teach. Teach. Educ. **120**, 103878 (2022)
6. Krieg, M., Kreis, A.: Reflexion in Mentoringgesprächen - ein Mythos? Zeitschrift für Hochschulentwicklung **9**(1), 103–117 (2014)
7. Narciss, S.: Conditions and effects of feedback viewed through the lens of the interactive tutoring feedback model. In: Carless, D., Bridges, S.M., Chan, C.K.Y., Glofcheski, R. (eds.) Scaling up Assessment for Learning in Higher Education, pp. 173–189. Springer, Singapore (2017)
8. Meschede, N., Adl-Amini, K., Hardy, I.: "Ich sehe was, was du nicht siehst"—Profitieren Studierende bei der videobasierten Unterrichtsanalyse von Peer Feedback? In: Donie, C., et al. (eds.) Grundschulpädagogik Zwischen Wissenschaft und Transfer, pp. 108–113. Springer VS, Wiesbaden (2019)
9. Leonhard, T., Rihm, T.: Erhöhung der Reflexionskompetenz durch Begleitveranstaltungen zum Schulpraktikum?—Konzeption und Ergebnisse eines Pilotprojekts mit Lehramtsstudierenden. Lehrerinnen- und Lehrerbildung auf dem Prüfstand **4**(2), 240–270 (2011)

10. Combe, A., Kolbe, F.-U.: Lehrerprofessionalität: Wissen, Können, Handeln. In: Helsper, W., Böhme, J. (eds.) Handbuch der Schulforschung, pp. 833–851. Springer Fachmedien, Wiesbaden (2004)
11. Neuweg, G.H.: Emergenzbedingungen pädagogischer Könnerschaft. In: Heid, H., Harti, C. (eds.) Verwertbarkeit. Ein Qualitätskriterium (erziehungs-) Wissenschaftlichen Wissens?, pp. 215–228. Springer VS, Wiesbaden (2005)
12. Korthagen, F.: Eine Reflexion über Reflexion. In: Korthagen, F., Kessels, A.J., Koster, J., Lagerwerf, B., Wubbels, T. (eds.) Schulwirklichkeit und Lehrerinnen- und Lehrerbildung. Reflexion der Lehrertätigkeit, pp. 55–73. EB Verlag, Berlin (2002)
13. Wyss, C.: Unterricht und Reflexion. Eine mehrperspektivische Untersuchung der Unterrichts- und Reflexionskompetenz von Lehrkräften. Waxmann, Münster (2013)
14. Schön, D.A.: Educating the Reflective Practitioner, 3rd edn. Jossey-Bass, Hoboken, NJ (1988)
15. Fraefel, U.: Wo ist das problem? Kernideen des angloamerikanischen Reflexionsdiskurses bei Dewey und Schön. In: Berndt, C., Häcker, T., Leonhard, T. (eds.) Reflexive Lehrerinnen- und Lehrerbildung Revisited, pp. 56–73. Klinkhardt, Leipzig (2017)
16. Eysel, C.: Interdisziplinäres Lehren und Lernen in der Lehrerinnen- und Lehrerbildung—Eine Empirische Studie zum Kompetenzerwerb in Einer Komplexen Lernumgebung. Logos, Berlin (2006)
17. Imhof, M., Schlag, M.: Herausforderungen im Lehramtsstudium: Beobachten und Reflektieren im vertrauten Terrain? Das Beispiel der Praktikumsdokumentation Uni cum. In: Pilypaitytė, L., Siller, H.-S. (eds.) Schulpraktische Lehrerprofessionalisierung als Ort der Zusammenarbeit, pp. 45–68. Springer VS, Wiesbaden (2018)
18. Hatton, N., Smith, D.: Reflection in teacher education: towards definition and implementation. Teach. Teach. Educ. **11**(1), 33–49 (1995)
19. Wegner, C., Remmert, K.: Unterrichtsreflexion 2.0—Ein Wiki zur Professionalisierung der Selbstreflexion von Lehramtsstudierenden. Zeitschrift für Medienpädagogik **58**(2), 74–81 (2014)
20. Blomberg, G., Stürmer, K., Seidel, T.: How pre-service teachers observe teaching on video: effects of viewers' teaching subjects and the subject of the video. Teach. Teach. Educ. **27**, 1131–1140 (2011)
21. Weber, K.E., Gold, B., Prilop, C.N., Kleinknecht, M.: Promoting pre-service teachers' professional vision of classroom management during practical school training: effects of a structured online- and video-based self-reflection and feedback intervention. Teach. Teach. Educ. **76**, 39–49 (2018)
22. Gold, B., Pfirrmann, C., Holodynski, M.: Promoting professional vision of classroom management though different analytic perspectives in video-based learning environments. J. Teach. Educ. **72**(2), 431–447 (2021)
23. Brouwer, N.: Was lernen Lehrpersonen durch die Arbeit mit Videos? Ergebnisse eines Dezenniums empirischer Forschung. Beiträge zur Lehrerinnen- und Lehrerbildung **32**(2), 176–195 (2014)
24. Miller, K., Zhou, X.: Learning from classroom video: what makes it compelling and what makes it hard. In: Goldmann, R., Pea, R., Barron, B., Derry, S.J. (eds.) Video Research in the Learning Sciences, pp. 321–334. Lawrence Erlbaum, Mahwah, NJ (2007)
25. Seidel, T., Thiel, F.: Standards und trends der videobasierten Lehr-Lernforschung. Z. Erzieh. **20**(1), 1–21 (2017)
26. Thissen, A., Scheid, V., Albert, A.: Unterrichtsreflexion und Videofallarbeit. Eine Studie zur Reflexionsfähigkeit von Sportstudierenden im Schulpraktikum. Sportunterricht. Monatsschrift zur Wissenschaft und Praxis des Sports mit Lehrhilfen **68**(5), 207–213 (2019)
27. Göbel, K., Wyss, C., Neuber, K., Raaflaub, M.: Student feedback in practical phases of teacher education as a source for reflection. In: Rollett, W., Bijlsma, H., Röhl, S. (eds.)

Student Feedback on Teaching in Schools. Using Student Perceptions for the Development of Teaching and Teachers, pp. 173–190. Springer Nature, Heidelberg (2021)
28. Göbel, K., Neuber, K.: Verändern sich reflexionsbezogene Einstellungen von Studierenden nach der Nutzung von Schülerrückmeldungen im Praxissemester? Befunde einer Interventionsstudie. Zeitschrift für Erziehungswissenschaft **25**, 721–744 (2022)
29. Cojorn, K., Sonsupap, K.: Investigating student teachers' reflections on early field experiences. J. Educ. Issues **8**(1), 279–291 (2022)
30. Moore-Russo, D.A., Wilsey, J.N.: Delving into the meaning of productive reflection: a study of future teachers' reflections on representations of teaching. Teach. Teach. Educ. **37**(1), 76–90 (2014)
31. Husu, J., Toom, A., Patrikainen, S.: Guided reflection as a means to demonstrate and develop student teachers' reflective competencies. Reflective Pract.: Int. Multidiscip. Perspect. **9**(1), 37–51 (2008)
32. Brouwer, N., Korthagen, F.: Can teacher education make a difference? Am. Educ. Res. J. **42**(1), 153–224 (2005)
33. Porsch, R., Gollub, P.: Potentiale von Langzeitpraktika im Lehramtsstudium: Ein systematisches review. Lehrerbildung auf dem Prüfstand **14**(2), 241–269 (2022)
34. Sherin, M., Van Es, E.: Effects of video club participation on teachers' professional vision. J. Teach. Educ. **60**(1), 20–37 (2009)
35. Nowak, A., Ackermann, P., Borowski, A.: Rahmenthema Reflexion im Praxissemester Physik—Potsdamer Beiträge zur Lehrerinnen- und Lehrerbildung und Bildungsforschung. Potsdamer Universitätsverlag, Potsdam (2018)
36. Van Es, E.: Examining the development of a teacher learning community: the case of a video club. Teach. Teach. Educ. **28**(2), 182–192 (2012)
37. Gröschner, A., Seidel, T., Kiemer, K.: Facilitating collaborative teacher learning: the role of "mindfulness" in video-based teacher professional development programs. Gr. Organ. **45**, 273–290 (2014)
38. Beisiegel, M., Mitchell, R., Hill, H.C.: The design of video-based professional development: an exploratory experiment intended to identify effective features. J. Teach. Educ. **69**(1), 69–89 (2018)
39. Svojanowsky, P.: Supporting student teachers' reflection as a paradigm shift process. Teach. Teach. Educ. **66**, 338–348 (2017)
40. Andersen, T.: Das Reflektierende Team. Dialoge und Dialoge über Dialoge. Verlag Modernes Lernen, Dortmund (1990)
41. Göbel, K., Gösch, A.: Die Nutzung kollegialer reflexion von Unterrichtsvideos im Praxissemester. In: Degeling, M., Franken, N., Freund, S., Greiten, S., Neuhaus, D., Schellenbach-Zell J. (eds.) Herausforderung Kohärenz: Praxisphasen in der Universitären Lehrerbildung. Bildungswissenschaftliche und Fachdidaktische Perspektiven, pp. 277–288. Verlag Julius Klinkhardt, Bad Heilbrunn (2019)
42. Blomberg, G., Renkl, A., Gamoran Sherin, M., Borko, H., Seidel, T.: Five research-based heuristics for using video in pre-service teacher education. J. Educ. Res. Online **5**(1), 90–114 (2013)
43. Helmke, A.: Unterrichtsqualität und Lehrerprofessionalität—Diagnose, Evaluation und Verbesserung. Erhard Friedrich Verlag, Hannover (2015)
44. Pianta, R.C., Hamre, B.K.: Conceptualization, measurement, and improvement of classroom processes: standardized observation can leverage capacity. Educ. Res. **38**, 109–119 (2009)
45. Kuckartz, U.: Mixed methods. Methodologie, Forschungsdesigns und Analyseverfahren. Springer VS, Wiesbaden (2014)
46. Mayring, P.: Qualitative inhaltsanalyse. In: Mey, G., Mruck, K. (eds.) Handbuch Qualitative Forschung in der Psychologie, pp. 601–613. Springer VS, Wiesbaden (2015)

47. MAXQDA: Software für Qualitative Datenanalyse 1989–2021, VERBI Software Consulting Sozialforschung GmbH, Berlin (2019)
48. Narciss, S.: Informatives Tutorielles Feedback. Waxmann, Münster (2006)
49. Poom-Valickis, K., Mathews, S.: Reflecting others and own practice: an analysis of novice teachers' reflection skills. Reflective Pract.: Int. Multidiscip. Perspect. **14**(3), 420–434 (2013)
50. Leijen, Ä., Allas, R., Toom, A., Jukka, H., Marcos, J.-K.M., Meijer, P., Knezic, D., Pedaste, M., Krull, E.: Guided reflection for supporting the development of student teachers' practical knowledge. Procedia—Soc. Behav. Sci. **112**(7), 314–322 (2014)
51. Gaudin, C., Chaliès, S.: Video viewing in teacher education and professional development: a literature review. Educ. Res. Rev. **16**(1), 41–67 (2015)
52. Groß-Mlynek, L., Graf, T., Harring, M., Feldhoff, T.: Unterrichtsvideos als Element der Theorie-Praxis-Verzahnung im Lehramtsstudium. Journal für LehrerInnenbildung **18**(3), 56–61 (2018)
53. Kleinknecht, M., Schneider, J.: What do teachers feel when analyzing videos of themselves and other teachers teaching. Teach. Teach. Educ. **33**, 13–23 (2013)
54. Scheidig, F.: Unterrichtsvideos. Neue Szenarien digitaler Praxisbezüge. Journal für LehrerInnenbildung **20**(1), 28–41 (2020)
55. Stürmer, K., Seidel, T., Holzberger, D.: Intra-individual differences in developing professional vision: preservice teachers' changes in the course of an innovative teacher education program. Instr. Sci. **44**(3), 293–309 (2016)
56. Darling-Hammond, L.: How teacher education matters. J. Teach. Educ. **51**(3), 166–173 (2000)

# Digital Transition in Education

# Digitization in the Field of Engineering Teacher Training

Andreas Probst[1,3](✉), Ralph Dreher[2], and Klaudia Lettmayr[3]

[1] HTL Wels, Fischergasse 30, 4600 Wels, Austria
Andreas.Probst@bildung.gv.at
[2] University of Siegen, Breite Str. 11, 57076 Siegen, Germany
[3] University of Education Upper Austria, Kaplanhofstraße 40, 4020 Linz, Austria

**Abstract.** In order to teach in German vocational colleges/vocational schools or Austrian vocational schools so called HTL both secondary level II, engineers (usually with a BA degree) need an additional study of vocational pedagogy and technology didactics in addition to the subject study and a professional practice, which should prepare the students for the requirements of everyday school life in the best possible way. In the area of teacher training at the University of Siegen and the Upper Austrian University of Teacher Education, specific concepts and content for the digitization of such courses have been developed and are currently being used in teaching, continuously evaluated and further developed. This paper examines the extent to which the concepts, which were initially developed independently, can be transferred between the universities and what content can be used for teaching.

**Keywords:** Digitization · Teacher training · New learning formats · COMET

## 1 Introduction

As many teachers in both countries will retire in the next few years, there is an enormous need for new teachers in the field of industrial-technical education. In order to make the training courses attractive and to bring current topics such as Industry 4.0, digitalization, the mobility revolution and the zero-carbon goal into the training courses, there is a lack of resources and expertise at the universities for technical teacher training. One approach is to coordinate existing resources across university boundaries and to use them specifically in courses at other universities. This should not happen unreflectively, but should be scientifically prepared, investigated and accompanied. An important aspect is also the survey and opinion of the students, which will be investigated. A first attempt will be made with the University of Siegen and the Upper Austrian University of Teacher Education to gather first experiences and impressions. This first phase is designed for two academic years. For this purpose, it is first necessary to examine the different situations and boundary conditions at the universities, to find commonalities and to point out differences.


M. E. Auer et al. (Eds.): ICL 2023, LNNS 899, pp. 169–177, 2024.
https://doi.org/10.1007/978-3-031-51979-6_17

# 2 Problem Statement and Research Question

## 2.1 Problem Statement

For future teachers of technical schools, the content of training is undergoing significant change. In addition to training in the classic engineering disciplines such as mechanical engineering, electrical engineering, chemistry- and civil engineering, etc., as well as pedagogical content and aspects, new content from the field of digitalization, which affects all engineering disciplines, is becoming increasingly important. Likewise, there is an increasing convergence of once separate disciplines (for example, mechanical and electrical engineering to mechatronics), which requires a more holistic technical teaching. For example, future engineers in companies, regardless of discipline, should also have knowledge of Augmented Reality (AR) [1] or Internet of Things (IoT) and at least have worked with tools related to this [2, 3]. Figure 1 gives an AR example of merging the real world with the digital information on the car windshield.

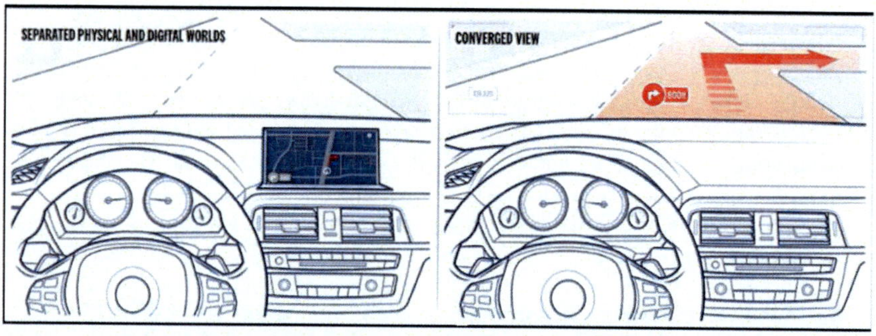


**Fig. 1.** Converging physical and digital using Augmented Reality [1]

This situation is exacerbated by the availability of artificial intelligence (AI), which makes the memorization and recitation of pure factual knowledge, which was already not very useful, obsolete. The discussion of whether ChatGPT or similar programs will be used is not intended here. With the rise of these tools, they will be used, rather, a future paper will consider how they can be usefully integrated into training. The integration of AI into various tools such as AR and IoT cannot yet be assessed at the current state of the art, but this topic will be considered in one of the future papers.

## 2.2 Research Question

The research question is whether students and teachers consider the use of digitization content from a transfer of digitized teaching to be useful despite the different course structures and conditions of the respective other universities.

Based on this, it is planned to regularly exchange the digitization concepts and content across the institutions in the case of positive feedback and thereby create a broader offering in the sense of OER (Open Education Resources) for the students.

For this purpose, this first contribution to a planned paper series will present the status quo in Siegen and Linz and identify useful overlaps and additions. In addition, a functioning network between learners and teachers is to be formed in this academic year, which will serve as a basis for future developments.

To this end, the research design will first be worked out and prepared for the coming academic years, and the performance of the courses will be measured using the existing and proven COMET system. COMET is a competence measurement procedure aimed at the assessment of holistic design competence, which uses the criteria of functionality, clarity, utility orientation, business process orientation, economic efficiency, social compatibility, environmental compatibility and creativity. The evaluation is highly reliable and thus very objective and is carried out by at least two independent raters. COMET was developed in particular as an evaluation and feedback instrument for project learning by a Bremen research team led by Felix Rauner [4].

## 3 Technical Vocational Teacher Education at University of Siegen and Upper Austrian University of Teacher Education

The following Table 1 displays the differences and similarities of Technical Vocational Teacher Education at University of Siegen and Upper Austrian University of Teacher Education.

**Table 1.** Comparison of technical vocational teacher education Siegen and Upper Austria

| University | Upper Austria | Siegen |
|---|---|---|
| State | Upper Austria (1.5 million inhabitants) | North Rhine-Westphalia (17.9 million inhabitants) |
| Demand (students) | Low; high competitive pressure from engineering programs | Low; high competitive pressure from engineering programs and need for 18 months of preparatory service |
| Demand (vocational schools) | High nationwide; retirement wave of baby boomers | Extremely high nationwide; demand gap of 3,500 technical teachers/a |
| Student numbers 2022/23 | 609 in Upper Austria/Linz | 25 in Siegen |
| Teaching forms | Lectures<br>Exercises<br>Hybrid seminars<br>Online teaching and e-learning components etc. | Project seminars (AL);<br>Flanking hybrid seminars<br>Practical semester<br>Online tutorial B4U-Basics for you<br>Online tutorial T3A - Things to think about<br>Mentoring for workshop-experience |

(*continued*)

**Table 1.** *(continued)*

| University | Upper Austria | Siegen |
|---|---|---|
| Qualification lecturers | University professors and professors from higher education institutions (HTL, HLW, HAK, SOB and BAfEP etc.), ideally doctoral degree in the relevant subject area | One university professorship each for vocational/business education, instructional development (inclusion), and didactics of technology; teaching assignments for practicum semesters (recruited faculty) |
| Practical phase | 2 or 3 years part-time | 1 Semester (3rd semester of the master's program) |
| Number of courses in engineering | Bachelor of Education (BEd); Optional Master of Education (MEd)<br>Dual education and technology and trades (DATG)<br>Specialized Studies<br>Supplementary Studies (FSES) | 6 Courses, 3 in main Mechanical engineering, 3 in main electrical engineering<br>(Model A: undergraduate vocational specialty/subject, Model B: undergraduate major/small vocational specialty; C: advanced model to BA Eng.) |
| Fields of study | Mechanical Engineering, Mechatronics, Electrical Engineering, Civil Engineering, Chemistry, Computer Science Food Technology, etc. | Major vocational specialties:<br>Mechanical Engineering<br>Electrical Engineering |
| | Education - Developmental Support (BAfEP and BASOP)<br>Social work (SOB<br>Information and Communication (Applied Digitalization) | Small vocational specialties:<br>Manufacturing technology,<br>Automotive technology;<br>Computer engineering |
| Lateral entry as teacher of vocational education | Usual way, persons with at least two or three years of professional experience and corresponding qualification<br>At least 8 h of teaching | Possible; on request as dual model with 14 h teaching obligation |
| Preparation service | Introductory courses before starting school<br>With all content required:<br>LBV, SchUG, SchOG, classroom management, lesson planning;<br>Introduction to MenTut program | Following MA studies, 18 months |

*(continued)*

Table 1 shows that the two countries have a lot in common: Both countries have a high demand for teachers, so they primarily work with a postgraduate course with the option of a "Master of Education" degree, and both courses already have a high practical component and focus on a direct link between theory and practice.

**Table 1.** *(continued)*

| University | Upper Austria | Siegen |
|---|---|---|
| Formal degree | Foreman for vocational schools (3 years)<br>Engineer for technical schools (4 years)<br>Graduate engineer for HTL (5years) | Master of education BK (vocational college) |
| Extent of facultas | In the respective subjects for all types of schools<br>Vocational schools (3 y)<br>Technical schools (3/4 y)<br>BMHS (5 y) | In the respective subjects for all school types at the vocational college:<br>Adolescents without apprenticeship<br>Basic vocational training year (lower secondary school leaving certificate)<br>Vocational school (intermediate school leaving certificate)<br>Technical classes (skilled worker's certificate incl. Middle school leaving certificate)<br>Berufsoberschule (University of Applied Science entrance qualification)<br>Berufsgymnasium (General university entrance qualification)<br>Technician school (after apprenticeship, incl. University entrance qualification) |
| Qualification-work | Bakk thesis and Bakk presentation - without examination character at the end of education<br>Master thesis and master examination at the end of the optional master program | Master-thesis |
| Portfolio-system | Evaluation meetings with the head of the study program, direct feedback to lecturers by study program directors | Feedback on COMET design tasks for instructional development in the dimensions of functionality, presentation, utility orientation, business process orientation, economic efficiency, social compatibility, environmental compatibility, creativity |

## 4 The Siegen Model of Teacher Education: The Principle of Academic Learning (AL)

The Siegen Model of Teacher Education see Fig. 2 is based on the thesis that teachers prefer to reproduce those lessons that they have found particularly conducive to learning in their own learning biography. (compare Grounded Theory and its effect on teacher action according to Trautwein/Merkt [5, p.187]).

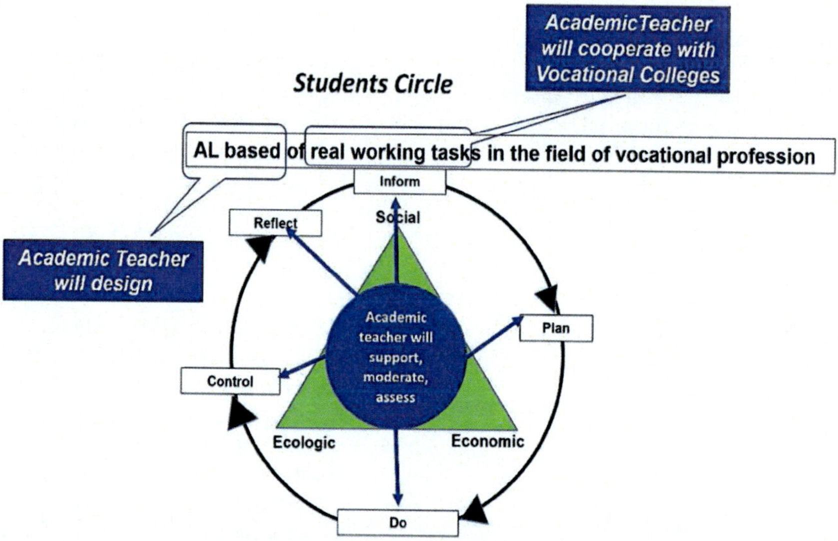


**Fig. 2.** Siegen Model of Teacher Education

If the existing learning biography of student teachers is now to be expanded to include the aspect of design-based learning with its possibility of parallelizing theory, empiricism, quality of action, and reality of action on the one hand [6, p.335] and the occasions for reflection present therein as moments for promoting design competence on the other, the following (obvious) conclusion is possible: The implementation of a learning field in concrete learning situations as well as the corresponding instructional planning represent in their entirety a learning situation on the part of the teacher (hereafter dubbed "academic learning situation-AL"), which opens up corresponding decision-making spaces and thus satisfies the salience criterion for indeterminacy described by Neuweg as essential [7, p.32], which is a prerequisite for sufficient occasions for reflection.

Table 2 below shows the concrete phases and subtasks as well as reflection inserts for such a technology didactic AL seminar.

**Table 2.** Structure of a technology didactic project seminar according to the Siegen Model

| Project phase<br>Student task | Subtasks | Lecturers |
|---|---|---|
| In advance:<br>Definition of a portfolio-appropriate professional task with corresponding curricular reference (curricular designated learning field)<br>(not a student task) Task definition from professional practice; Real-life testing in the technical-didactic laboratory | | Identification of tasks from professional practice;<br>Real testing in the tech-didactic laboratory |
| Project phase I:<br>Finding the master-solution | In the technical didactic laboratory:<br>Analysis of the technical problem;<br>Design and practical testing of the work process;<br>Checking of the work process for sustainability;<br>Documentation of the work process as process step documentation or progress plan | Support in work process analysis (together with workshop foreman);<br>Establishing contacts with skilled workers/experts to compare practical solution and master solution;<br>Assumption of the role of "critical friend" in the assessment of sustainability |
| *First reflection phase*:<br>Sustainability check of the master solution | | |
| Project phase II:<br>Development of internally differentiating learning situations finding salience factors to make the task more binding or open | Finding salience fac-tors to make the task more binding or open | Presentation of the model of the "learning situation matrix" as a finding tool;<br>Assisting in defining typical student groups;<br>Assuming the role of "critical friend" in reviewing the learning situation matrix |
| *Second reflection phase*:<br>What criteria for internal differentiation were used and why?<br>Which groups of student clientele are targeted by each differentiation measure? | | |

(*continued*)

## 5 Further Procedure Regarding Collaboration and Content Transfer

In both universities, teaching materials were selected to be used in the other university starting in the academic year 2023/24. This year can be described as a pilot year, as it will be tested to see whether and how the teaching materials are received by teachers and students. Scientific support is provided by the proven COMET system of the University of Siegen.

**Table 2.** (*continued*)

| Project phase<br>Student task | Subtasks | Lecturers |
|---|---|---|
| Project phase III:<br>Development of a les-son plan Development of a phase-differentiating | Lesson plan for a learning situation (in-form, plan, decide, implement, control, reflect) with the follow-ing information in each case:<br>– Time required for the phase,<br>– social form;<br>– aids;<br>– means of documentation;<br>– Teacher role,<br>– Learner role;<br>– Benchmark for the phase (what needed to be accomplished);<br>– Criteria for teaching performance evaluation Supplementary notes on the (non)justification of the choice of methods; | Critical inquiry about planning the implementation phase, as often not all student groups can be hands-on at the same time;<br>Moderation of the presentation of the process phase |
| *Third reflection phase*:<br>Open discussion on time allocation, choice of methods, role of the teacher, use of media and design of the implementation phase | | |
| Project closing | Project conclusion Individual reflection student - lecturer: Feedback analysis for the lecturer | |

On the part of the Technical Vocational Teacher Education at University of Siegen videos on different topics are provided on a platform. These are available for the following topics:

- Genesis of vocational didactics,
- choice of methods,
- School performance measurement,
- Methods of vocational science,
- Methods of competence measurement,
- Simulation work with cyber-physical systems (via AR).

The Upper Austrian University of Teacher Education provides content on the topic of augmented reality that is specifically intended for the introduction of AR to students, as this technology is relatively new in the companies from which the students come but is to be taught in the technical schools. In July 2024, an evaluation of the COMET evaluations is planned and the definition of further steps in the cooperation between the two universities.

## 6 Conclusion and Outlook

An initial comparison of the two study program concepts reveals, on the one hand, a great deal of agreement with regard to the objectives of the study programs: application-oriented, but at the same time science-based teaching that aims to solve problems in vocational education and training in a theoretical-practical context. On the other hand, it is clear that both universities complement each other in the courses they offer by setting different focal points, which results in a meaningful, mutually complementary range of courses. Based on the results of this comparison, a mutual exchange of online courses is therefore intended. The portfolio system installed at both universities should then also take on the function of a quality check.

## References


1. Porter, M.E., Heppelmann, J.E.: Why every organization needs an augmented reality strategy. Harvard Business Review, no. November-December (2017)
2. Elsaadany, A., Soliman, M.: Experimental evaluation of internet of things in the educational environment. Int. J. Eng. Ped. **7**(3), 50–60 (2017)
3. Heidling, E. et al.: Ingenieurinnen und Ingenieure für Industrie 4.0. https://impuls-stiftung.de/wp-content/uploads/2022/05/Ingenieurinnen-und-Ingenieure-fuer-Industrie-4.0.pdf (2019). Accessed 2023/01/24
4. Rauner, F.: Methodenhandbuch. Messen und Entwickeln beruflicher Kompetenzen. p.23f. wbv, Bielefeld (2018)
5. Trautwein, C., Merkt, M.: Struktur und Entwicklung von Lehrkompetenz im Spannungsfeld von Überzeugungen, Konzepten, und Praxis von Lehren und Lernen. In: Heiner, M., Wildt, J. (eds.) Professionalisierung der Lehre. Perspektiven formeller und informeller Entwicklung von Lehrkompetenz im Kon-text der Hochschulbildung, pp.179–210. Bertelsmann-Verlag, Bielefeld (2013)
6. Oser, F.: Standards: Kompetenzen von Lehrpersonen. In: Oser, F., Oelkers, J. (eds.) Die Wirksamkeit von Lehrerausbildungssystemen. Von der Allrounderbildung zur Ausbildung professioneller Standards, pp 225–243. Rüegger, Chur (2021)
7. Neuweg, G.H.: Könnerschaft und implizites Wissen. Zur lehr-lerntheoretischen Bedeutung der Erkenntnis- und Wissenstheorie Michael Polyanyis. Waxmann, Münster (1999)

# Students' Views on the Internet of Things in Engineering Education

Andreas Probst[3], Reinhard Bernsteiner[1,2](✉), Wolfgang Pachatz[4], Christian Ploder[2], and Thomas Dilger[2]

[1] HTL Jenbach, Schalsertrasse 43, 6200 Jenbach, Austria
reinhard.bernsteiner@mci.edu
[2] MCI | The Entrepreneurial School, 6020 Innsbruck, Austria
[3] HTL Wels, Fischergasse 30, 4600 Wels, Austria
[4] Federal Ministry of Education, Science and Research, 1010 Vienna, Austria

**Abstract.** Entrepreneurship and innovation are thriving in the Internet of Things (IoT) era. IoT can enable businesses to discover new opportunities and create IoT-based solutions or services. The IoT is a major technological transformation impacting various domains, such as healthcare, transportation, manufacturing, and agriculture. The IoT is a fast-growing field with many job prospects. Therefore, it is essential to integrate IoT technologies and their industrial applications into the Higher Vocational Education curriculum, as the graduates will be the future workforce. In Austria, some departments of Higher Vocational Colleges have introduced IoT technologies. To ensure practical implementation, seminars were developed to provide lecturers with comprehensive knowledge and training materials. The main goal of this study is to examine the students' perception of IoT education. To collect empirical data a questionnaire was developed based on hypotheses to gather empirical data, which will be analyzed using statistical methods. According to the findings, IoT is generally well-received by students as a component of their vocational education. The analysis highlighted the positive attitude of students from various departments regarding IoT education. Students also expressed some interest in pursuing their diploma thesis in this domain. But it is necessary to conduct a more extensive examination of the variations between departments. The insights gained from this research can then be utilized to tailor and enhance the work of lecturers, curriculum, and instructional resources.

**Keywords:** Higher vocational education · Internet of things · Product development · Engineering education · Digital skills · Empirical survey

## 1 Introduction

The continuous integration of the global economy has resulted in more intricate value networks. The need for customized products and services further amplifies this complexity. Additionally, the current global trends have led to a surge in several product variations as customers increasingly seek personalized options [1, 2].


M. E. Auer et al. (Eds.): ICL 2023, LNNS 899, pp. 178–188, 2024.
https://doi.org/10.1007/978-3-031-51979-6_18

The Internet of Things (IoT) may help to cope with these challenges because IoT technologies provide innovative solutions to complex problems. Scientific research has focused on developing novel applications and systems using IoT. For instance, a study presents an overview of IoT applications in different domains, including healthcare, transportation, and agriculture, highlighting the potential for problem-solving [3].

In light of these circumstances, educational institutions at all levels must revise their curricula to align with these developments. The integration of IoT education has been widely implemented in Higher Vocational colleges (HVCs) throughout Austria. HVCs are structured into departments that align with their vocational focus areas. Consequently, IoT has been introduced in select departments. This paper presents the initial empirical findings on the perception of IoT education in HVCs.

This paper is part of a comprehensive research project which evaluates education in Augmented Reality (AR) [4, 5] and IoT from the viewpoint of students and teachers. To allow for further analysis, the structure of the questionnaires and the hypotheses are the same for AR and IoT.

The paper follows the following structure: Sect. 2 introduces the related literature, serving as a basis for the empirical analysis. Additionally, this section presents an overview of the research's related work. Section 3 outlines the problem statement and research questions. Section 4 describes the methodology and design of the empirical survey. Section 5 presents the results obtained. Finally, Sect. 6 concludes the paper, highlighting its limitations and providing an outlook for future research.

## 2 Internet of Things

Science has developed different definitions of the IoT, capturing different viewpoints. A rather general definition comes from Lee and Lee, "[IoT] is a new technology paradigm envisioned as a global network of machines and devices capable of interacting with each other" [6].

IoT is defined as "an open and comprehensive network of intelligent objects that can auto-organize, share information, data, and resources, reacting and acting in the face of situations and changes in the environment" [7]. Through IoT, enterprises can collect and use data from different objects or industrial machines, thus being enabled to automate and adjust machine performance [8].

Dorsemain et al. define IoT as "a group of infrastructures, interconnecting connected objects and allowing their management, data mining and the access to data they generate" where connected objects are "sensor(s) and/or actuator(s) carrying out a specific function that are able to communicate with other equipment" [9].

All definitions include the connection of devices, which collect environmental data and transfer them to other devices or central IT systems. This environmental data is usually collected via sensors and includes for example humidity, temperature, pressure, CO2 emission, and proximity. Other widespread sensors include gyroscopes, accelerometers, and location (e.g., by GPS coordinates). The data can further be processed, and decisions can be derived and then executed by actuators.

### 2.1 Industrial and Consumer Internet of Things

IoT can further be divided into several streams based on the application areas with their specific characteristics. One separation between the industrial IoT (IIoT) and consumer (CIoT) can be drawn. The key difference between these two areas is the context in which devices and applications are used.

The IIoT is the further development of the term "Industrial Internet", which was first used by General Electric. From General Electric's point of view, the Industrial Internet is IoT primarily about connecting sensors and actuators of industrial machines and further integrating with other industrial networks. The IIoT highlights the use of IoT in industry, especially in industrial production [10]. CIoT is related to consumer products, ranging from elementary and cheap applications to high-end smart home applications [11].

IoT devices usually face restrictions, at least in comparison to non-IoT devices, which need special attention when IoT-based systems are developed and implemented. The most common of these restrictions are [12–15]:

- Low computational power
- Limited power supply
- Small memory
- Limited data rate

### 2.2 IoT Architectures

As mentioned above, IoT is not only about smart things (physical objects with sensors and actuators). The term IoT includes the data transfer from things to applications or services that use these data. Since IoT offers a wide range of applications and services, IoT solutions architectures support system design and development. Several IoT architectures and related security concepts have been developed in the last few years [16–19]. They all have four major building blocks or layers in common:

- Things, which comprise of sensors and actuators
- Connectivity, network gateways, and internet connections, sometimes with integrated data pre-processing capabilities
- Middleware, like IoT platforms or data processing at the edge of the network
- Applications provided in on-premise data centers or the cloud

At the IoT World Forum 2014, Cisco, IBM, and Intel presented an IoT solutions reference model. This reference model, which is used in the context of this project, consists of seven layers, from the "things layer" to the "business layer" [20]:

- The layer "Physical Devices and Controllers" represents the things equipped with sensors and actuators, also referred to as "the Edge".
- "Connectivity" offers data and application transfer from physical devices to the edge and between all other layers.
- "Edge Computing" refers to data management activities that are close to the physical devices. Due to short distances from the things to the edges of the network, decisions with low latency can be performed.
- The "Data Accumulation layer" serves as an intermediary between the edge and the Data Abstraction or Application layer. Data are pre-processed for further usage.

- In the "Data Abstraction layer" data are merged from different things or edges. The Data Abstraction layer also receives data from the Application layer, which are then prepared and transferred to actuators.
- The "Application layer" is the platform in which services and applications are managed and executed.
- "Collaboration and Processes" offers applications and services to users. At this layer, several applications provided in the Application layer can be combined to support entire processes or user-oriented activities.

## 3 Problem Statement and Research Overview

The teaching and learning IoT-related subjects in Higher Vocational Education is relatively recent. A seminar series has been established to enhance teachers' knowledge in IoT and provide instructional materials for their courses. IoT topics are incorporated into various departments and typically commence in the third year of study. Therefore, it is crucial to determine the most suitable year for IoT education. Students from different departments may have distinct perspectives on IoT education. If variations exist, adjustments must be made to the content and structure of the seminar series. The obtained results will enable the formulation of conclusions aimed at enhancing the educational environment for students and refining the seminar series for teachers.

Consequently, the following research question arises: How do students in Higher Vocational Colleges perceive IoT education? To gain more comprehensive insights, differences among students in terms of a) academic year and b) departments are further examined and analyzed.

## 4 Methodology and Design of the Empirical Survey

This section outlines the methodological approach to gathering empirical data for addressing the research questions. The initial part focuses on developing questionnaires designed for both students and teachers. Subsequently, hypotheses are formulated to investigate potential variations in the perceptions between students and teachers.

### 4.1 Design of the Questionnaire

The central aim of this survey is to identify the perception of IoT education in Higher Vocational Colleges from the student's point of view. All items had to be rated on a Likert scale from 1 ("I totally agree") to 5 ("I totally disagree"). Finally, all collected data were analyzed with the statistical software platform SPSS. Table 1 presents questions for students.

### 4.2 Hypotheses

The following hypotheses are formulated to address the second research question, which examines the perception of IoT among students across different academic years and departments:

**Table 1.** Questions for students

| Variable | Question |
|---|---|
| Department | I attend the following department |
| Year | I currently attend the following year |
| IoT_Thesis | I can imagine writing my diploma thesis in the field of IOT |
| IoT_enjoy | I enjoy learning IOT |
| IoT_interesting | I am personally interested in IoT |
| IoT_important | I think IoT will be important in the future |
| IoT_understanding | I think IoT helps understand technical concepts |
| IoT_attractive | I think IoT makes technical education more attractive |

H1 Students from different years equally enjoy IoT.
H2 Students from different years equally interested in IoT.
H3 Students from different years equally think IoT will be important in the future.
H4 Students from different years equally think IoT helps understand technical topics.
H5 Students from different years equally think IoT makes technical education more attractive.
H6 Students from different departments equally enjoy IoT.
H7 Students from different departments equally interested in IoT.
H8 Students from different departments equally think IoT will be important in the future.
H9 Students from different departments equally think IoT helps understand technical topics.
H10 Students from different departments equally think IoT makes technical education more attractive.

### 4.3 Selection of the Participants

The research targets students who have participated in IoT classes, and a systematic sampling method was employed to gather data from this group. The lecturers who attended the seminar series were provided with links to the questionnaire and were requested to distribute it to their students accordingly.

## 5 Results

In this section, the findings of the empirical survey are presented. Initially, the participants are described, followed by offering insights gained from this preliminary exploratory study, considering both perspectives. Additionally, the results of hypothesis validation are presented as a foundation for addressing the research question.

**Table 2.** Cronbach's Alpha – questionnaire

| Cronbach's Alpha | N of items |
|---|---|
| 0.860 | 5 |

## 5.1 Reliability of the Questionnaire

Cronbach's Alpha was calculated to assess the internal consistency of the questionnaire variables as presented in Table 2.

The results for students with Cronbach's alpha of .860 show a satisfying internal consistency.

## 5.2 Description of the Participants

As previously stated, Higher Technical Vocational Colleges have a five-year duration, with students typically ranging in age from 15 to 19 years. IoT education has been primarily concentrated in the last two years, with initial steps taken in the third year. A total of 172 students from various departments completed the questionnaire, and a detailed breakdown can be found in Table 3.

**Table 3.** Number of students – department by year

| | | Year in higher vocational college | | | Total |
|---|---|---|---|---|---|
| | | 3 | 4 | 5 | |
| Department | Industrial engineering | 3 | 21 | 13 | 37 |
| | Informatics | 7 | 0 | 0 | 7 |
| | Mechanical engineering | 0 | 58 | 70 | 128 |
| Total | | 10 | 79 | 83 | 172 |

The department of Mechanical Engineering accounts for a significant majority of the students. This distribution can be attributed to two main reasons. Firstly, Mechanical Engineering has the highest number of students in Austria's Higher Vocational Education system. Secondly, the training initially commenced with teachers from this specific department.

## 5.3 Student Specific Results

The perception of IoT education across different academic years and departments is presented in Table 4.

In general, a positive attitude towards IoT can be inferred from the fact that all means are higher than 2.5 on a scale of 1 to 5, where 1 indicates "I totally agree" and 5 indicates "I totally disagree."

**Table 4.** Means across department and year

| | Department | | | Year | | |
|---|---|---|---|---|---|---|
| | Industrial engineering | Informatics | Mechanical engineering | 3 | 4 | 5 |
| | Mean | Mean | Mean | Mean | Mean | Mean |
| IoT_enjoy | 2.14 | 2.14 | 2.40 | 1.78 | 2.38 | 2.37 |
| IoT_interesting | 2.19 | 1.71 | 2.58 | 1.67 | 2.53 | 2.49 |
| IoT_important | 1.41 | 1.71 | 1.88 | 1.33 | 1.78 | 1.82 |
| IoT_understanding | 2.08 | 2.43 | 2.33 | 2.00 | 2.32 | 2.28 |
| IoT_attractive | 2.00 | 1.71 | 2.32 | 1.56 | 2.31 | 2.24 |

To address the first part of the second research question, which examines differences between students across years, hypotheses H1 to H5 need to be validated. Initially, it is necessary to assess the data distribution to determine the appropriate statistical analysis method. The results of a Shapiro-Wilk test indicate that none of the items exhibit a normal distribution ($p < 0.05$). Consequently, a non-parametric test is required [21]. Subsequently, the Kruskal–Wallis test should be employed based on the available data.

**Table 5.** Kruskal–Wallis test with grouping variable year

| | IoT_enjoy | IoT_interesting | IoT_important | IoT_understanding | IoT_attractive |
|---|---|---|---|---|---|
| Asymp. Sig | 0.160 | 0.085 | 0.175 | 0.700 | 0.073 |

As shown in Table 5, there are no significant differences (significance level 0.05) across the different years of the students. This means that all null hypotheses have to be retained. It can be concluded that no specific year in the educational journey can be identified as the ideal stage for IoT education.

To address the second part of research question two (departmental differences among students), the hypotheses H6 to H12 need to be tested. A Kruskal–Walis test is used to analyze the data. Table 6 shows the results of the test.

**Table 6.** Kruskal–Wallis test with grouping variable department

| | IoT_enjoy | IoT_interesting | IoT_important | IoT_understanding | IoT_attractive |
|---|---|---|---|---|---|
| Asymp. Sig | 0.378 | 0.040 | 0.039 | 0.529 | 0.084 |

The results indicate significant differences in the variables of IoT_interesting and IoT_important. Further analysis is necessary to determine the relevant departments in this regard. A pairwise comparison of these variables across departments needs to be conducted (Table 7).

**Table 7.** Pairwise comparisons of IoT_interesting across department

| Sample 1-Sample 2 | Test Statistics | Std. Error | Std. Test Statistics | Sig | Adj. Sig.[a] |
|---|---|---|---|---|---|
| Industrial Engineering Informatics | 20.218 | 19.554 | 1.034 | 0.301 | 0.903 |
| Industrial Engineering Mechanical Engineering | −36.274 | 18.423 | −1.969 | 0.049 | 0.147 |
| Informatics-Mechanical Engineering | −16.056 | 8.871 | −1.810 | 0.070 | 0.211 |

Each row tests the null hypothesis that the Sample 1 and Sample 2 distributions are the same. Asymptotic significances (2-sided tests) are displayed. The significance level is 0.05.
a. Significance values have been adjusted by the Bonferroni correction for multiple tests.

The pairwise comparison results (adjusted significance) show no differences between groups. Students of the three departments equally find IoT education interesting. The same analysis has to be conducted for the variable IoT_interesting.

**Table 8.** Pairwise comparisons of IoT_important across department

| Sample 1-Sample 2 | Test Statistics | Std. Error | Std. Test Statistics | Sig | Adj. Sig.[a] |
|---|---|---|---|---|---|
| Industrial Engineering Informatics | −12.664 | 18.552 | −0.683 | 0.495 | 1.000 |
| Industrial Engineering Mechanical Engineering | −21.355 | 8.409 | −2.540 | 0.011 | 0.033 |
| Informatics-Mechanical Engineering | −8.691 | 17.475 | −0.497 | 0.619 | 1.000 |

Each row tests the null hypothesis that the Sample 1 and Sample 2 distributions are the same. Asymptotic significances (2-sided tests) are displayed.
The significance level is 0.05.
a. Significance values have been adjusted by the Bonferroni correction for multiple tests.

Again, the analysis (Table 8) shows a significant difference (adjusted significance of 0.033 with a significance level of 0.05) for the variable IoT_imortant between the department of Industrial Engineering (mean 1.41) and Mechanical Engineering (mean 1.88).

One factor that can indicate students' interest in IoT is whether they are inclined to write their diploma thesis in that field. However, this question does not apply to students in their final year, as they have already completed their thesis. 83 students from the third and fourth year of their program answered the question about their interest in writing their diploma thesis in IoT (Table 9).

**Table 9.** Results of the question "I can imagine writing my diploma thesis in the field of IoT"

| | Frequency | Percent |
|---|---|---|
| Yes | 18 | 21.7 |
| No | 34 | 41.0 |
| Maybe | 31 | 37.3 |
| Total | 83 | 100.0 |

The results show that 21.7% have the intention, and another 37.3% are thinking about writing their thesis in the field of IoT. A share of 41.0% have different favorite topics for their diploma thesis.

### 5.4 Answer to the Research Question

This empirical research aims to answer the following question: "What are the perceptions of Higher Vocational Colleges students on IoT education?" To explore this question further, the study also examines the differences among students from different a) years and b) departments.

The results indicate that students have a positive attitude towards IoT as part of their curriculum, as shown by the means and standard deviations. Moreover, these variables have no significant differences across years and departments. This implies that IoT education is perceived similarly by students from all the departments and years in HVCs.

## 6 Conclusions, Limitations, and Recommendations

Drawing upon the findings, it can be inferred that IoT education generates student interest in this emerging technology. As IoT plays a pivotal role in Industry 4.0, this heightened interest may attract attention and curiosity toward this field. Additionally, students gain a comprehensive understanding, equipping them to enter their professional careers promptly and effectively upon graduation. These endeavors contribute to addressing the imminent shortage of skilled workforce, a concern substantiated by numerous scholarly and practical publications.

The selected quantitative approach was appropriate for addressing the research question in this survey. However, it is crucial to acknowledge the limitations of this research when interpreting its findings. Firstly, the survey involved 172 participants from three departments. Higher Vocational Colleges in Austria offer diverse vocational directions, resulting in an unbalanced distribution of students across departments and years of education.

Scientific literature also reports on positive learning outcomes in IoT education. A study conducted by [22] shows that an adequate learning and teaching environment helps to improve learning effectiveness. Practical guidelines on effectively introducing undergraduate students to the emerging technologies of IoT are presented by [23].

To improve the generalization of this research, additional research is required. Gathering and analyzing more empirical data from students representing a broader range of departments in Higher Vocational Colleges is necessary. This will provide more comprehensive insights. Furthermore, examining the design of IoT education across different years of study can potentially uncover further valuable information. Implementing a year-specific approach to IoT education may enhance its effectiveness.

Another avenue of research involves seeking input from lecturers regarding their impressions and perceptions. As educators who interact with students daily, they possess valuable knowledge about the optimal delivery of IoT education. Consequently, the design of the seminar series should be adjusted based on their feedback and insights.

## References


1. Büchi, G., Cugno, M., Castagnoli, R.: Smart factory performance and industry 4.0. Technol. Forecast. Soc. Chang. (2020). https://doi.org/10.1016/j.techfore.2019.119790
2. Osterrieder, P., Budde, L., Friedli, T.: The smart factory as a key construct of industry 4.0: a systematic literature review. Int. J. Prod. Econ. (2020). https://doi.org/10.1016/j.ijpe.2019.08.011
3. Gil, D., Ferrández, A., Mora-Mora, H., Peral, J.: Internet of Things: a review of surveys based on context aware intelligent services. Sensors (Basel, Switzerland) (2016). https://doi.org/10.3390/s16071069
4. Bernsteiner, R., Probst, A., Pachatz, W., Ploder, C., Dilger, T.: Augmented reality in engineering education in austrian higher vocational education from the students' perspective. In: Auer, M.E., Hortsch, H., Michler, O., Köhler, T. (eds.) Mobility for Smart Cities and Regional Development—Challenges for Higher Education, vol. 389. Lecture Notes in Networks and Systems, pp. 535–545. Springer International Publishing, Cham (2022)
5. Bernsteiner, R., Probst, A., Pachatz, W., Ploder, C., Dilger, T.: Augmented reality in engineering education—a comparison of students' and teachers' perceptions. In: Auer, M.E., Pachatz, W., Rüütmann, T. (eds.) Learning in the Age of Digital and Green Transition, vol. 633. Lecture Notes in Networks and Systems, pp. 207–219. Springer International Publishing, Cham (2023)
6. Lee, I., Lee, K.: The Internet of Things (IoT). Applications, investments, and challenges for enterprises. Bus. Horiz. (2015). https://doi.org/10.1016/j.bushor.2015.03.008
7. Madakam, S., Ramaswamy, R., Tripathi, S.: Internet of Things (IoT): a literature review. JCC (2015). https://doi.org/10.4236/jcc.2015.35021
8. Leminen, S., Rajahonka, M., Westerlund, M., Wendelin, R.: The future of the Internet of Things: toward heterarchical ecosystems and service business models. JBIM (2018). https://doi.org/10.1108/JBIM-10-2015-0206
9. Dorsemaine, B., Gaulier, J.-P., Wary, J.-P., Kheir, N., Urien, P.: Internet of Things: a definition & taxonomy. In: 2015 9th International Conference on Next Generation Mobile Applications, Services and Technologies. 2015 9th International Conference on Next Generation Mobile Applications, Services and Technologies (NGMAST), Cambridge, United Kingdom, 09.09.2015–11.09.2015, pp. 72–77. IEEE (2015). https://doi.org/10.1109/NGMAST.2015.71

10. Sadeghi, A.-R., Wachsmann, C., Waidner, M.: Security and privacy challenges in industrial internet of things. In: Proceedings of the 52nd Annual Design Automation Conference. DAC '15: The 52nd Annual Design Automation Conference 2015, San Francisco California, 07 06 2015 11 06 2015, pp. 1–6. ACM, New York, NY, USA (2015). https://doi.org/10.1145/2744769.2747942
11. European Commission: Preliminary Report—Sector inquiry into Consumer Internet of Things. SWD(2021) 144 final (2021). https://ec.europa.eu/competition-policy/system/files/2021-06/internet_of_things_preliminary_report.pdf. Accessed 21 Oct. 2021
12. Shafique, K., Khawaja, B.A., Sabir, F., Qazi, S., Mustaqim, M.: Internet of Things (IoT) for next-generation smart systems: a review of current challenges, future trends and prospects for emerging 5G-IoT scenarios. IEEE Access (2020). https://doi.org/10.1109/ACCESS.2020.2970118
13. Čolaković, A., Hadžialić, M.: Internet of Things (IoT): a review of enabling technologies, challenges, and open research issues. Comput. Netw. (2018). https://doi.org/10.1016/j.comnet.2018.07.017
14. Mohamad Noor, M.B., Hassan, W.H.: Current research on Internet of Things (IoT) security: a survey. Comput. Netw. (2019). https://doi.org/10.1016/j.comnet.2018.11.025
15. Alladi, T., Chamola, V., Sikdar, B., Choo, K.-K.R.: Consumer IoT: security vulnerability case studies and solutions. IEEE Consumer Electron. Mag. (2020). https://doi.org/10.1109/MCE.2019.2953740
16. Singh, S.K., Rathore, S., Park, J.H.: Block IoT intelligence: a blockchain-enabled intelligent IoT architecture with artificial intelligence. Fut. Gener. Comput. Syst. (2020). https://doi.org/10.1016/j.future.2019.09.002
17. Pape, S., Rannenberg, K.: Applying privacy patterns to the Internet of Things' (IoT) architecture. Mobile Netw Appl (2019). https://doi.org/10.1007/s11036-018-1148-2
18. Ren, J., Guo, H., Xu, C., Zhang, Y.: Serving at the edge: a scalable IoT architecture based on transparent computing. IEEE Netw. (2017). https://doi.org/10.1109/MNET.2017.1700030
19. Shahinzadeh, H., Moradi, J., Gharehpetian, G.B., Nafisi, H., Abedi, M.: IoT Architecture for Smart Grids. In: 2019 International Conference on Protection and Automation of Power System (IPAPS). 2019 International Conference on Protection and Automation of Power System (IPAPS), Iran, 08.01.2019–09.01.2019, pp. 22–30. IEEE (2019). https://doi.org/10.1109/IPAPS.2019.8641944
20. El Hakim, A.: Internet of Things (IoT) System Architecture and Technologies, White Paper (2018)
21. van Hecke, T.: Power study of anova versus Kruskal–Wallis test. J. Stat. Manag. Syst. (2012). https://doi.org/10.1080/09720510.2012.10701623
22. Rahayu, M., Hariyanto, T., Fadhlan, M.Y.: IoT trainer kit training for vocational school teachers as preparation towards the 4.0 industry era. REKA ELKOMIKA (2020). https://doi.org/10.26760/rekaelkomika.v1i2.98-110
23. Khanafer, M., El-Abd, M.: Guidelines for teaching an introductory course on the internet of things. In: 2019 IEEE Global Engineering Education Conference (EDUCON), Dubai, United Arab Emirates, 08.04.2019–11.04.2019, pp. 1488–1492 (2019). https://doi.org/10.1109/EDUCON.2019.8725186

# A Collaborative Learning Model in Engineering Science Based on a Cyber-Physical Production System Line

Yosr Ghozzi[1,2](✉) and Asma Karoui[1]

[1] ESPRIT School of Engineering, Tunis, Tunisia
yosr.ghozzi@esprit.tn
[2] Paul Valéry University Montpellier 3, Montpellier, France

**Abstract.** The terminology of Cyber-Physical Production System (CPPS) has been well received by the industrial community and specifically appropriated in educational circles. In this context, we built a collaborative learning model through an international market study that has led us to place ourselves at the heart of this technology. To align with these findings, a competencies-based learning approach (CBL) study was conducted. Thus, this article deals with the development of the CPPS pedagogical device according to a generated curriculum and specific pedagogical activities while respecting the competency referential adopted via compliant and adapted laboratories and the international CDIO framework for a better practical and pedagogical implementation. In this context, the modular cyber-physics technology platform must be adapted to engineer work environment and fed by real industrial projects for better professional integration. A documentary analysis of written and cognitive traces is in progress to list data types and anticipate recursive evolutions of the model.

**Keywords:** CPPS implementation · CBL approach · CDIO framework

## 1 Theoretical Implementation

Faced with students' preoccupation with teaching practice, we might well wonder about the professional motivations that drive them to pursue graduate studies. Are they really looking for training in scientific research when they can also be trained in work situations in their academic environment. In this context, we approach the notion of professionalization not only in the field of training, but also in the field of work, so that teachers and learners confront work situations and articulate work and training more closely within the framework of multiple specializations and in a dynamic context. To analyze this dialogue between the different situations encountered by the teacher, it is necessary to trace the epistemological logic underlying the training system and the constituent elements of its "professionalizing" structure, studying the different situations encountered and experienced by the learner and drawing up a list of specific professional skills. In this sense, we note the aspirations of electromechanical engineering students in relation


M. E. Auer et al. (Eds.): ICL 2023, LNNS 899, pp. 189–197, 2024.
https://doi.org/10.1007/978-3-031-51979-6_19

to the skills required by the job market. In this way, a professionalization action enables the implementation of the skills in the referential through involvement in the action. To this end, we conducted a study based on a competency-based approach.

## 1.1 Competencies-Based Approach

Thanks to an international market study, we have compared the competencies of the reference job repository of an electromechanical engineer with the learning outcomes to try and cover the training areas. We also took care to establish the link between skills and learning outcomes on the one hand and learning outcomes and activities on the other. Particular attention has been paid to the articulation between disciplinary fields and professional profiles. In this respect, particular attention has been paid to integrated projects as a specific follow-up to competencies. Learning is therefore presented in the form of a logic model inspired by "Business Skills", enabling the project's inputs, outputs, and multi-disciplinary impact to be visualized, leading to an assessment of the knowledge, attitudes and skills acquired according to the project and the learner's level throughout the course.

The training profile of our electromechanical engineer department is mainly focused on the conception/design/implementation/operation of solutions for the mechanical, electrical and information sciences sectors, particularly to deal with coupling issues. To acquire these skills, the engineering student must study the fundamental subjects and the engineering sciences with in-depth studies proposed for different optional specific courses starting from the 4th year. From this year, the student accrues more specialized methodological and technological skills and have several professional activities. For these activities, the necessary skills that flow from the survey analysis process is shown in Table 1.

**Table 1.** Competencies list

| N° | Competency |
|---|---|
| 1 | Mastering a solid body of knowledge in the basic sciences |
| 2 | Mastery of a solid body of knowledge in engineering sciences and techniques |
| 3 | Deployment of a wide range of knowledge to imagine, design, build and operate adapted, robust, and innovative systems |
| 4 | Planning, management, contribution, in team, to the realization of a multidisciplinary project |
| 5 | Considering environmental and societal constraints |
| 6 | Effective teamwork and structured and contextualized communication |

These skills are interdependent and describe a basic academic professional profile, geared towards the efficiency of the electromechanical engineer, and can be applied to other engineering courses. Competency 1 deals mainly with the requirements of a transversal initial training cycle focusing on the basic sciences. Central skills 2 and 3 characterize electromechanical training and form the core of the engineering profession. Skill 3 adds the CDIO aspect (Conceive, Design, Implement and Operate). Basically, the engineer analyzes complex situations (competency 2), designs, implements, evaluates, and performs a financial analysis of typical and innovative robust solutions (competency 3). Therefore, he must master the basic sciences (competency 1), master a solid body of knowledge in engineering sciences and techniques (competency 2), plan and manage activities, projects, and people (competency 4). He must consider economic, societal, and environmental issues while acting as a responsible professional (competency 5) and a communicator and confirmed writer (competency 6).

These competencies will be transformed during the developing process of the CPPS pedagogical platform [2, 5]. According to Table 2, the evolution of the generated competencies is quite clear when moving from one learning level to another.

**Table 2.** Extract of a scenario development according to a specific skills' level

| Course for CPPS/competency | 1 | 2 | 3 | 4 | 5 | 6 |
|---|---|---|---|---|---|---|
| Robotics | x | x | | x | x | x |
| Additive manufacturing | | x | x | x | x | x |
| Power electronics | x | x | x | x | | x |
| Quality control | | x | x | x | x | x |
| Electrical energy | x | x | x | | x | x |

### 1.2 Approach Implementation

For each course, a scenario for implementing the learning path is generated to match the targeted educational objectives through the technological aspect defined via the CPPS platform in an iterative process. Table 3 below shows a correspondence approach between competencies, level of depth of learning for a reliable implementation. To respect the distribution of technical skills and their chronological and practical aspects, we set up an integrated professional learning path with its own pedagogical integration.

This work is carried out for each learning outcome integrated into each course integrated into the CPPS learning path to align the training objective, learning situation and teaching method.

## 2 Pedagogical Implementation

At this level, the aim is no longer to start from a specialist issue, but rather to map out a learning path and the acquisition of well-defined skills.

**Table 3.** Courses chain for CPPS implementation/competencies matrix

| Example of a technical competency | Skill level | Learning situation |
|---|---|---|
| Ability to understand and model physical phenomena related to electrical energy production, transportation, distribution, conversion, storage, and management | – Know (the standards of electrical networks) | – Interactive course |
| | – Understand (Analyze an electrical diagram) | – Problem based Learning |
| | – Evaluate (compliance of an electric installation standards) | – Practical work |
| | – Create (design an electric installation based on specifications) | – Project |

## 2.1 Pedagogical Professionalization

The method implemented is an organizational model designed to target and evaluate business skills while optimizing human and material potential. This method enables us to redefine our needs and to make the student aware of the need to respect the project's phases by integrating validation phases during the realization stages. The project is therefore presented in the form of a logic model inspired by "Business skills", enabling us to visualize the project's inputs, outputs, and multi-disciplinary impact, leading to an assessment of the knowledge, attitudes and skills acquired according to the project and the learner's level throughout the project. It should be noted that generated case studies are based on CDIO standards providing active learning tactics and illustrating how the ideas were put into effect in actual projects [1]. We demonstrate the basic concepts of the reflexive paradigm of reflection in action, on action, on the system of action [3] and proceed to relate them to reflections on the analysis of practices and work on habitus based on the theories of situated action and the practical unconscious of work [6]. It elaborates a conceptual retrospective in an analysis of the rhetoric of student professionalization. Based on these theories, an action plan was used to initially study the feasibility of the project, then refine it and use it to steer its implementation. We have defined the expected qualitative and quantitative results, specifying, where appropriate, the timeframe within which they are to be achieved. Finally, the activities to be implemented have been planned over time. Building and equipment requirements were identified. Next, control and problem-solving activities must be considered, to deal with a series of malfunctions, for example, and to launch activities linked to the reorganization of work and the computerization of certain functions. The place given to full-scale, accompanied experiential learning regarding future professional activity; skills embodied in action and tacit skills. Reflecting on action enables the development of more intellectualized skills that can be transferred to new situations, promoting the professionalization of professional actions [7].

### 2.2 Collaborative Model

Although the work described in this manuscript focuses on system design, it is essential to consider all phases and requirements of the product life cycle. The different types of product integration presented below are also a source of organizational complexity. They require a high level of collaboration, which not only leads to organizational complexity, but also to the diversity of the domains involved. Thus, a multi-disciplinary collaborative integrated system for mechatronics training was designed and implemented in an evolving professional environment described below in the Fig. 1.

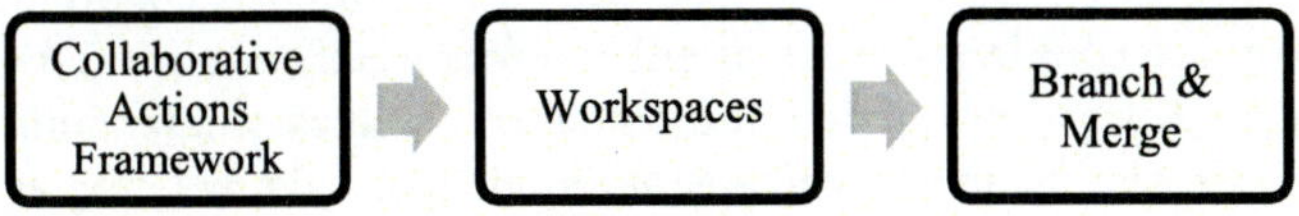


**Fig. 1.** Agile method for multidisciplinary collaborative design

The model improve collaboration between the various disciplines by reducing the existing compartmentalization between the players involved. It is a direct response to the problem of the correlation between lack of collaboration and the resulting low level of product integration. As a result, laboratory equipment is based on an integrated approach, i.e., sequential step-by-step development and a multi-level approach to differentiating educational levels. We have taken care to balance the criteria of duality, versatility, and adaptability of the chain through modular cells that can be arranged according to the objective and level of depth of the activity. To perform this task, the teaching-learning approach [4] is adopted as a flexible and dynamic system open to the changing demands of the world of work and the specialized and/or multi-disciplinary trends all within the framework of the CBL approach performed by the problem-based learning model [5].

### 2.3 Pedagogical Process

The orientation phase of this approach is based on the creation of groups of students supervised by multi-disciplinary teachers who assign them different actions to achieve different learning goals. Students and expert teachers' study, discuss and, if necessary, modify the action requests to draw up precise specifications. The implementation phase is based on group work and weekly meetings to present the progress of the work and assess the learning outcomes according to their degree of depth. This step calls on the second notion of this agile method, which is the workspace to reference every modification made and every data updated. The close-out phase is defined by the completion of tests on the machines and the start-up and communication of the processes in consideration. Validation of collaborative actions considers all assessments made during implementation of the requested actions, which are weighted according to the learning outcome and its level of depth. Indeed, during the implementation stage, if validation is not achieved, the action is transferred back to the student group, with recommendations. Completion of this action is marked by passing the tests.

## 3 Technical Implementation

The approach is aligned with the corresponding industrial-scale processes and systems, and the respective skills required for development and operation. The approach described has been validated by application in a team project at university as part of an electromechanical engineering module. For this purpose, a realistic model-scale process was selected from among others and enhanced to suit the learning objectives previously. To implement the practical aspects of the CPPS [8], hardware and software products have already been implemented. We aligned our CPPS implementation strategies [9] on the teaching-learning approach based on the CBL approach and pedagogical alignment already developed in the Sect. 2. The laboratory workstation is controlled by a personal computer via a USB interface, using either off-the-shelf software tools or software created through advanced integration in the field of graphics programming. Different learning strategies were applied to teach-learn, simulation, and robot programming. The CPPS line was built around the operative part as illustrated in Fig. 2 and based on four processes as shown in Table 4. To ensure pedagogical and technological consistency, the two main parts of the cyber-physical platform have been developed. Indeed, the "Cyber" part is implemented through "The Industrial Performance Starter Pack 4.0" (Fig. 2.a). The "Physical" part of the CPPS platform is the robotic cell "MITSUBISHI RV-2FR-D-S25 6-axis system" (Fig. 2.b). It is the most complex and flexible handling device in an industrial environment upgraded with RFID, Energy Meter, and SERVO DRIVE Sinamics S210.

The illustrated learning device handles an application involving the assembly of 3D-printed parts. By another hand, collaboration between technological aspects has been integrated through digital twins. This technological framework has enabled a multidisciplinary approach to the design of machines, robots, and production lines, removing the barriers between different technical courses. We continue to improve the digital twin design process, helping you to design faster and with better quality, and targeting predictive maintenance aspects. The practice scenarios in Table 4 are based on guidelines. Each practice session is divided into several sections. Each section is made up of several practical workshops that include theory, homework, practice, and instructions to cover the learning outcomes. Preliminary design work precedes each workshop, helping the student to better understand the operation of the elements studied.

The practical work contains diagrams that enable the student to delve deeper into a relevant subject. Thus, for their understanding, it is important that the student has the appropriate background knowledge of the main disciplines, a good knowledge of the basic principles of technical analysis of mechatronic and robotic systems based on the learning paths and their degrees of deepening described Sect. 2.1. The result of this work is the last version of the guidelines, specially developed for the CPPS platform.

## 4 Pre-seminar Survey

To analyze feedback on post-installation pedagogical activities, a qualitative method was used. This involved direct or participative observations (professional or formative situations), explanatory or comprehensive interviews with learners or their tutors, and

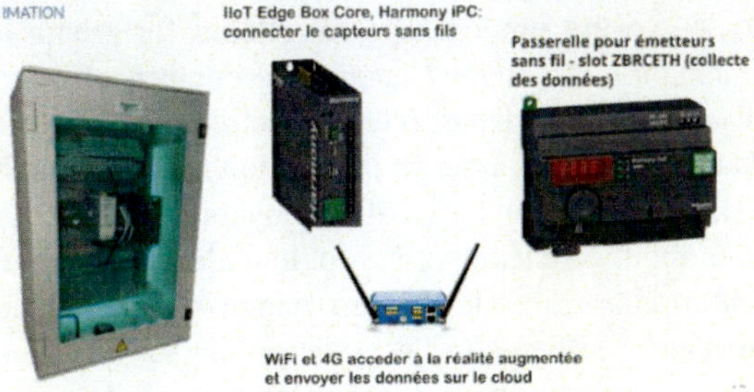


a. SCHNEIDER Industrial Performance Starter Pack 4.0

(Real photo for the CPPS part implementation)

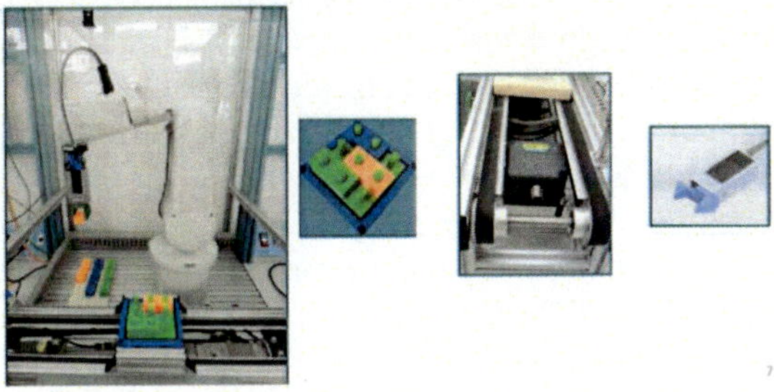

b. MITSUBISHI RV-2FR-D-S25 6-axis robot

**Fig. 2.** The operating CPPS line

**Table 4.** Rating of processes regarding CPPS/competencies fulfilment

| Process regarding CPPS components | Physical | Data Acquisition | Cyber | Feedback/control |
|---|---|---|---|---|
| 3D printed additive manufacturing | xxx | x | xx | x |
| 3D printed assembly line | xx | x | xxx | x |
| Quality control of parts via a manufacturing execution system | x | xx | xx | xxx |
| Procurement and inventory management | x | xx | xx | xxx |

documentary analysis of written traces (learners' productions). We then cross-reference discourses on professional or formative activity, observations of behavior in professional or pedagogical situations, and written thematic files produced by learners. This step enables us to list the types of data and anticipate recursive evolutions of the model. A pilot class of 30 students was chosen to evaluate the cognitive integration of the platform, throughout the pedagogical process around specific workshops. The predefined learning elements were supported by milestones for groups of students. Through the CPPS implementation the student acquires a comprehensive understanding not only of the core disciplines, but also of related disciplines. The student's interest increases as clear interdisciplinary links emerge. In addition to analyzing the student's motivation, a study of the skills acquired in terms of employability was carried out. The main employability skills acquired were teamwork, problem identification and solving, research, time management, organization, leadership, analysis, entrepreneurship, innovation, and project management. Skills such as negotiation, persuasion, flexibility, and proactivity did not score well, which is common among students adapting to the university environment. As a result, we were able to observe that decision-making in a design project is facilitated by more frequent and regular feedback of precise operational information. This information may concern tasks in progress, requirements considered, or difficulties encountered by the various design teams. We also noted the traceability between the changes made to the system definition data and the decisions taken throughout the development project. The aim of this traceability is not only to improve design project management but also to be able to draw lessons from design projects to better anticipate future projects. The next step will be to set up an appropriate scientific method for student assessment, to generate results from the training process and program content, in line with norms 11 and 12 of the CDIO standards.

## 5 Conclusion

The modular cyber-physics technology platform was adapted to work situations and fed by real industrial projects as part of a skills-based approach. The pedagogical model implemented enabled students to develop skills in the field of electromechanical systems. Introducing this practice into the educational process gave students the opportunity to apply theoretical knowledge in a practical environment, namely the development of CPPS. By another hand, application of CDIO standards in the teaching of CPPS was a solid example of the practical implementation of international engineering training standards within electromechanical department workshops. The scientific analysis of educational situations as part of the professionalization of trainee teachers cannot be limited to the study of results at the level of immediate knowledge alone; it must be carried out on an ongoing basis at the various levels and phases of the study, in accordance with standard 10 of the CDIO framework. It is essential to consider the various cognitive, cultural, and economic aspects involved in the search for relevant interpretations of the observations made.

## References

1. Bankel, J., et al.: Benchmarking engineering curricula with the CDIO syllabus. Int. J. Eng. Educ. **21**(1), 121–133 (2005)
2. Kannengiesser, U., Frysak, J., Stary, C., Krenn, F., Müller, H.: Developing an engineering tool for Cyber-Physical Production Systems. e & i Elektrotechnik und Informationstechnik **138**, 1–11 (2021)

3. Le Bellu, S.: Apprendre les secrets d'une profession au travers de l'expérience temps-réel des experts: Capturer et transférer aux novices les savoirs professionnels tacites d'expérience. Perspectives interdisciplinaires sur le travail et la santé, 18-1 (2016)
4. Mäkiö-Marusik, E., Colombo, A.W., Mäkiö, J., Pechmann, A.: Concept and case study for teaching and learning industrial digitalization. Procedia Manuf. **31**, 97–102 (2019)
5. Meng, N., Dong, Y., Roehrs, D., & Luan, L.: Tackle implementation challenges in project-based learning: a survey study of PBL e-learning platforms. Educ. Technol. Res. Dev. (2023)
6. Perrenoud, P.: De la pratique réflexive au travail sur l'habitus. Recherche & formation **36**(1), 131–162 (2001)
7. Poumay, M., Tardif, J., Georges, F.: Organiser la formation à partir des compétences: Un pari gagnant pour l'apprentissage dans le supérieur. De Boeck Supérieur (2017)
8. Thramboulidis, K.: A cyber-physical system-based approach for industrial automation systems. Comput. Ind. **72**, 92–102 (2015)
9. Wu, X., Goepp, V., Siadat, A.: Cyber physical production systems: a review of design and implementation approaches. In: 2019 IEEE International Conference on Industrial Engineering and Engineering Management (IEEM), pp. 1588–1592 (2019)

# Digital Learning Environments to Support Autonomous Learning Processes of Mathematically Creative and Gifted Students

Matthias Brandl[1](✉) and Attila Szabo[2]

[1] Faculty of Computer Science and Mathematics, Didactics of Mathematics, University of Passau, Passau, Germany
matthias.brandl@uni-passau.de

[2] Faculty of Humanities, Department of Teaching and Learning, Stockholm University, Stockholm, Sweden
attila.szabo@edu.stockholm.se

**Abstract.** Analyses performed by the German Federal Ministry of Labour and Social Affairs indicate that the educational system must address lifelong education in a context of digitalization, which is a major challenge of the transforming world of work and societal belonging. Mathematically gifted pupils or students are often characterized as independent, non-conformist and autonomous thinkers who are also displaying considerable levels of creativity. When focusing the mathematical content, gifted pupils enjoy working with challenging problems that offer diversified solutions and are explorative about a specific mathematical area. Accordingly, it is indicated that traditional classroom settings do have limited potential to address the needs of the gifted. Hence, the present paper discusses the possibilities to meet the developmental needs of mathematically gifted pupils via digital online learning environments, example here in the form of Digital Interactive Mathematical Maps (DIMM), a freely accessible digital tool developed at the University of Passau, Germany. The DIMM is based on a visualization of the historical development of different mathematical areas, such as geometry, algebra and calculus. Last and importantly, the DIMM is deepened by tasks from mathematical competitions and fostering settings that are connected to mentioned timelines and that pupils can work with in respective contexts.

**Keywords:** Digital transition in education · Digital interactive mathematical maps · Mathematical creativity and giftedness

## 1 Introduction

Creative and gifted pupils or students seem to prefer learning environments that allow for a sufficient degree of autonomy [1]. In the context of Digital Transition in Education, innovative digital online tools can offer optimal scaffolding with big potential for successful learning and fostering processes. In the following, we will address this issue by presenting specific examples for the case of mathematics.


M. E. Auer et al. (Eds.): ICL 2023, LNNS 899, pp. 198–205, 2024.
https://doi.org/10.1007/978-3-031-51979-6_20

# 2 Context

## 2.1 Digital Interactive Mathematical Maps (DIMM)

Motivated by different aspects related amongst others to transition problems in mathematics education (conf. The "Double Discontinuity" [2–4]) and defragmentation aspects in mathematics teacher education [5], the so called Digital Interactive Mathematical Maps (DIMM) have been developed at the Professorship for Didactics of Mathematics at the University of Passau, Germany, based on the original idea in [6] and in the context of a project fostered by the German Federal Ministry for Education and Research [7].

The DIMM represent a digital two- and three-dimensional visualization of the historical development of mathematics in an illustrated timeline and net structure, respectively. The 3d-tool has some didactical useful functionalities like horizontal and vertical cuts through the visualized knowledge net, leading to representations of knowledge relatedness or development, respectively, with reduced complexity to work with in teaching or learning (see Fig. 1 for an example screenshot). The DIMM are freely accessible [8]. For further details concerning the iterative design process and more specifics of the DIMM see [9] and [10].

**Fig. 1.** Screenshot of the Geometry DIMM with example content of a selected node.

## 2.2 Autonomy of the Mathematically Creative and Gifted

Creative individuals are described by A. Runco as essentially "independent, nonconformist, rebellious, unconventional, norm-doubting, and contrarian" (cited in [1], p. 53), and are further discussed in the context of mathematical giftedness [1]. Mathematically gifted pupils are mainly characterized as "autonomous" in problem-solving.

Nevertheless, there seems to be a substantial impediment of the main elements of creativity and mathematical giftedness in traditional classroom settings – connected to a strict curriculum and to tasks with limited possibilities of creative solutions – which are encouraging a conformist behavior and imitative working methods for all pupils (e.g., [11, 12]).

A very first idea for the possibility of a potential use of the DIMM to satisfy the needs of the creative and gifted ones was sketched in [13] and will be elaborated here in the following.

### 2.3 Work and Education 4.0

According to an analysis performed by the German Federal Ministry of Labour and Social Affairs [14] two of the major trends in the current transformation of the world of work – among globalization, demographic change, and migration – are education and digitalization. Further, in the context of education and digitalization, the right to autonomous decision making and to lifelong education are seen as key attributes for the individual's commitment to the society. However, it seems that integrating socio-culturally underprivileged groups into the education system and the labor market will be one of the main challenges in the upcoming years – a task that should be addressed through increased opportunities offered by digitalization [14].

## 3 Purpose

### 3.1 Traditional Fostering

When it comes to the fostering of mathematically gifted pupils, main approaches recommended by the research field are acceleration or "curriculum compacting" (by studying mathematical content for older pupils), enrichment programs in ability grouping (by deepening the teaching curricula) or combinations of both mentioned approaches [15–17]. Regardless of organizational aspects of mentioned approaches, it is suggested that gifted pupils should encounter rich learning tasks and meaningful, multi-layered mathematical problems that can be solved at different levels and, importantly, they should also be given the possibility to generalize mathematical information, which is considered a pivotal ability of the gifted [11, 16, 18, 19].

### 3.2 Digital Transition

One of the main challenges of the research field on gifted education is that identification programs cannot reliably identify gifted students from socio-culturally underprivileged groups; consequently, a considerably share of gifted students will remain unidentified [20–22]. Taking this in account, for instance, an online mentorship program has been offered for mathematically interested pupils living in urban areas of Sweden with the consideration of also meeting the needs of the gifted. Due to the limited number of mentors, 109 pupils were accepted in 2020 and 144 in 2021. During the program, pupils participate in mathematical discussions and problem-solving activities – designed in order to appeal to the gifted – through a digital environment, i.e., they do not meet physically. The program is still ongoing (May 2023) and the evaluations during 2020–2022 [23] indicate that among pupils who answered the questionnaire:

- 17% were very satisfied, 81% extremely satisfied with online mentoring,
- 46% experienced much motivation, 52% very much motivation towards mathematics,
- 61% experienced more well-being, 38% much more well-being, and
- 90% want to continue with the program.

Even though there was no formal testing of mathematical giftedness before entering the program – the acceptance was based on pupils' interest for mathematics – taking in account the design and complexity of tasks that pupils are working with and the level of their mathematical argumentation, it is reasonable to assume that many of the pupils enrolled in the program are actually mathematically gifted.

Consequently, it seems that pupils interested in mathematics working with online challenges that are designed for the gifted, are enjoying the possibilities offered by the digital environment. Thus, it is not unreasonable to assume that gifted pupils will find it meaningful to solve well-designed mathematical problems in an online environment.

### 3.3 Technology Acceptance

With nowadays pupils belonging to Generation Z, i.e. the "Digital Natives" [24], who "'live and breathe' technology" [25, p. 190], new technological tools intending to support learning processes have to meet high standards of usability and ease of use. The degree of technology acceptance can be measured by the widely used Technological Acceptance Model (TAM) of Davis [26, 27].

The TAM was used in a study concerning the evaluation of the technological quality of the DIMM in a previous version [9, 28] and we present a selection of results in the outcomes Sect. 5.1.

## 4 Approach

### 4.1 Qualitative Research Methodology by Design Based Research

The methodological approach in the development of digital resources like the DIMM is Design Based Research belonging to the area of qualitative research methods [29, 30]. Within this paradigma the technological design of a new tool is always followed up by an evaluation in a practical field and a corresponding analysis of the data. Other methods from qualitative research like qualitative content analysis might be used within this analysing process, too. After the interpretation and discussion of the analysed evaluation results, significant elements are addressed and changed in another evolutionary re-design process, which is tested again and so on. An overview of the iterative design based research cycles so far for the DIMM is illustrated in [9].

### 4.2 Enrichment of the Geometry DIMM with Competition Tasks

By enriching the DIMM with tasks from mathematical competitions and fostering settings, we provide a suitable and historically connected way through different levels of mathematics for independent and self-organized learning processes. Instead of presenting isolated tasks grouped by topics, the selected tasks are linked to the nodes of the DIMM which in turn are connected to each other in a historical and thematical manner. Future research questions include both implementation aspects with regard to the design evolvement of the DIMM and application aspects amongst others concerning the perceived usefulness by the pupils and students.

### 4.3 Autonomous and Collaborative Learning

Even though mathematically gifted pupils enjoy working autonomously during problem-solving and are skeptical towards collaborative learning in heterogeneous groupings [31–33], it is indicated that they appreciate collaborative working if engaged in challenging problems which offer more depth and less breadth in a specific mathematical area [17].

## 5 Outcomes

### 5.1 Evaluation Results on Technology Acceptance

The evaluation of technology acceptance (of the non-enriched DIMM without the tasks from competitions added in the meantime) was part of a larger study conducted in a geometry course at Karlstad University, Sweden, and used the TAM to measure the perceived ease of use and the perceived usefulness of specific design features and components of the DIMM [9, 28]. Results showed a positive picture, as essential elements of the DIMM were granted a high level of usefulness, like timeline (for more than 80% useful), searching and filtering (for 89% useful), preview (for almost 80% useful), vertical cut (for 95% useful), horizontal cut (for 93% useful), see [9, 28] for more details. Negative remarks mainly concerned the partly complex usability of the 3d-Map and one limited functionality on touch devices [9, 28]. We expect similar outcomes for the enriched DIMM containing the additional mathematical competition tasks.

### 5.2 Example Way Through the Geometry DIMM

As mentioned before, users of the DIMM are free to choose an individual way through the nodes and their corresponding information and tasks. For reasons of illustration we suggest one possible way in Table 1, starting with competition tasks related to contents already to be found on the Moscow Papyrus 1850 BC and ending in the interim at tasks dealing with Theorems of Menelaus, Thales and Appolonius.

Although these example nodes are not directly, but at least partly, connected in historical development, all are part of a vertical cut for the Theorem of Menelaus (Fig. 2).

**Table 1.** Chronologically ordered way through the Geometry DIMM containing selected competition tasks (from German mathematics competitions).

| Year | Mathematical content | Mathematicians |
|---|---|---|
| 90 AD | Theorems of Menelaus, Thales, Appolonius | Menelaus |
| 520 BC | Congruence, Perimeter Angle Theorem | Pythagoras |
| 580 BC | Intercept Theorem | Thales |
| 320 BC | Angles in equilateral triangles | Euclid |
| 1850 BC | Area of a triangle | Moscow Papyrus |

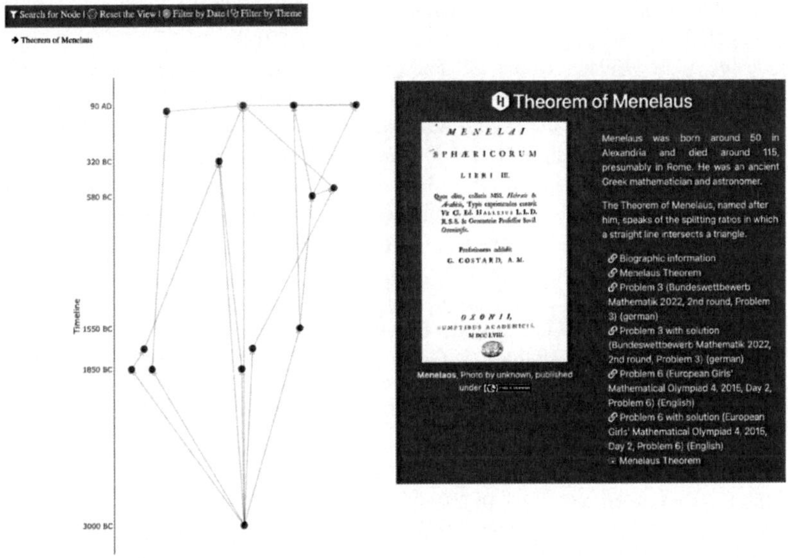


**Fig. 2.** Screenshot of the vertical cut of the 3d-map for the theorem of Menelaus.

## 6 Summary

Digital online tools like the DIMM can provide a useful scaffolding for interested learners to establish individually matching autonomous learning experiences, possibly in a collaborative context. They can especially address the needs of mathematically creative and gifted students and show that the characteristics of an Work and Education 4.0 learning environment are favourable for creative and gifted individuals.

**Acknowledgement.** This project is part of the German "Qualitätsoffensive Lehrerbildung", a joint initiative of the German Federal Government and the Länder which aims to improve the quality of teacher training. The programme is funded by the German Federal Ministry of Education and Research (funding reference 01JA1624).

## References

1. Brandl, M., Szabo, A.: Overexcitability, iconoclasm and mathematical creativity & giftedness. In: Including the Highly Gifted and Creative Students—Current Ideas and Future Directions: Proceedings of the 11th International Conference on Mathematical Creativity and Giftedness. Conference Proceedings in Mathematics Education, 5, pp. 53–58. WTM (2019)
2. Klein, F.: Elementary mathematics from higher standpoint. Volume I: Arithmetic, Algebra, Analysis (G. Schubring, Trans.). Springer, [Original work published 1924] (2016)
3. Winsløw, C., Grønbæk, N.: Klein's double discontinuity revisited: contemporary challenges for universities preparing teachers to teach calculus. Recherches en Didactique des Mathématiques **34**(1), 59–86 (2014)
4. Hefendehl-Hebeker, L.: Doppelte Diskontinuität oder die Chance der Brückenschläge. In: Ableitinger, C., Kramer, J., Prediger, S. (eds.) Zur doppelten Diskontinuität in der Gymnasiallehrerbildung, pp. 1–15. Springer Spektrum (2013)
5. Datzmann, A., Przybilla, J., Brandl, M., Kaiser, T.: New teaching techniques aiming to connect school and university mathematics in geometry. In: Donevska-Todorova, A., et al. (eds.) Mathematics Education in the Digital Age (MEDA), pp. 37–44. HAL, Villeurbanne (2020)
6. Brandl, M.: The vibrating string—an initial problem for modern mathematics; historical and didactical aspects. In: Witzke, I. (ed.) Mathematical Practice and Development throughout History: Proceedings of the 18th Novembertagung on the History, Philosophy and Didactics of Mathematics, pp. 95–114. Logos Verlag, Berlin (2009)
7. SKILL.de Homepage, https://www.skill.uni-passau.de. Accessed 09 May 2023
8. DIMM Hompage, https://math-map.fim.uni-passau.de. Accessed 09 May 2023
9. Przybilla, J., Brandl, M., Vinerean, M., Liljekvist, Y.: Digital mathematical maps—results from iterative research. In: Hodgen, J., Geraniou, E., Bolondi, G., Ferretti, F. (eds.) Proceedings of the Twelfth Congress of the European Society for Research in Mathematics Education (CERME12), pp. 4793–4800. Free University of Bozen-Bolzano and ERME, Bolzano (2022)
10. Przybilla, J., Brandl, M., Vinerean, M., Liljekvist, Y.: Interactive mathematical maps—a contextualized way of meaningful learning. In: Nortvedt, G.A. et al. (eds.) Bringing Nordic Mathematics Education into the Future, Proceedings of NORMA 20. The Ninth Nordic Conference on Mathematics Education, pp. 209–216. Oslo, SMDF, Göteborg (2021)
11. Krutetskii, V.A.: The Psychology of Mathematical Abilities in Schoolchildren. University of Chicago Press (1976)
12. Runco, M.A.: History of Creativity Research. Talk at the Conference "Empowering Creativity in Education". University library, Utrecht, Netherlands (2016)
13. Brandl, M., Kaiser, T.: Digital mathematical maps. In: Chamberlin, S.A. (ed.) On the Road to Mathematical Expertise and Innovation, Proceedings of the 12th International Conference on Mathematical Creativity and Giftedness, Conference Proceedings in Mathematics Education, vol. 8, pp. 297–299. WTM, Münster (2022)
14. Federal Ministry of Labour and Social Affairs.: Re-imagining work—White Paper Work 4.0. Homepage, https://www.bmas.de/EN/Services/Publications/a883-white-paper.html. Accessed 09 May 2023
15. Renzulli, J.S., Reis, S M., Smith, L.H.: The Revolving Door Identification Model. Creative Learning Press (1981)
16. Singer, F.M., Sheffield, L.J., Freiman, V., Brandl, M.: Research on and activities for mathematically gifted students. ICME-13 Topical Surveys. Springer Int. (2013)
17. Szabo, A.: Matematikundervisning för begåvade elever—en forskningsöversikt [Mathematics education for gifted pupils—a survey of research]. Nord. Stud. Math. Educ. **22**(1), 21–44 (2017)

18. Nolte, M., Pamperien, K.: Challenging problems in a regular classroom setting and in a special foster programme. ZDM Math. Educ. **49**(1), 121–136 (2017)
19. Sheffield, L.J.: Extending the Challenge in Mathematics Developing Mathematical Promise in K-Students. Corwin Press, Thousand Oaks (2003)
20. Almeida, L.S., Araujo, A.M., Sainz-Gomez, M., Prieto. M.D.: Challenges in the identification of giftedness: issues related to psychological assessment. Anales de Psicologia **32**(3), 621–627 (2016)
21. Borland, J.H.: Gifted education without gifted children—the case for no conception of giftedness. In: Sternberg, R.J., Davidson, J.E. (eds.), Conceptions of Giftedness, pp. 1–19. Cambridge University Press (2005)
22. Subotnik, R.F., Olszewski-Kubilius, P., Worrell, F.C.: Rethinking giftedness and gifted education: A proposed direction forward based on psychological science. Psychol. Sci. Public Interest **12**(1), 3–54 (2011)
23. Intize.: Utvärdering och statistik för Intize mentorskap i matematik på distans för särskilt begåvade barn i glesbygd. [Evaluation and statistics for Intize online mentoring in mathematics for gifted children in rural areas]. Gothenburg (2021, 2022)
24. Dauksevicuite, I.: Unlocking the Full Potential of Digital Native Learners. Henley Business School, Mc Graw Hill Education Handouts (2016)
25. Cilliers, E.: The challenge of teaching generation Z. PEOPLE: Int. J. Soc. Sci. 3, 188–198 (2017)
26. Davis, F.D.: A Technology Acceptance Model for Empirically Testing New End-User Information Systems: Theory and Results (PhD). Massachusetts Institute of Technology, Cambridge, MA (1985)
27. Scherer, R., Siddiq, F., & Tondeur, J.: The technology acceptance model (TAM): a meta-analytic structural equation modeling approach to explaining teachers' adoption of digital technology in education. Comput. Educ. 128 (2019)
28. Vinerean, M., Liljekvist, Y., Brandl, M., Przybilla, J.: Didactical usefulness of interactive mathematical maps—designing activities supporting student teachers' learning. In: Misfeldt, M., et al. (eds.) Nordic Studies in Mathematics Education—Nordisk Matematikkdidaktikk, NOMAD—Thematic Issue Digital Resources in Mathematics Education, vol. 28(1), (in press)
29. Bakker, A.: Design Research in Education. UK, Routledge, London (2018)
30. Bikner-Ahsbahs, A., Knipping, C., Presmeg, N. (eds.): Approaches to Qualitative Research in Mathematics Education—Examples of Methodology and Methods. Advances in Mathematics Education. Springer (2015)
31. Hunt, B.: The effect on mathematics achievement and attitude of homogeneous and heterogeneous grouping of gifted sixth-grade students. J. Adv. Acad. **8**(2), 65–73 (1996)
32. Li, A.K.F., Adamson, G.: Gifted secondary students' preferred learning style: cooperative, competitive, or individualistic? J. Educ. Gift. **16**(1), 46–54 (1992)
33. Robinson, A.: Cooperation or exploitation? The argument against cooperative learning for talented students. J. Educ. Gift. **14**(1), 9–27 (1990)

# Improving the Efficiency of Students' Independent Work During Blended Learning in Technical Universities

Maryna Miastkovska[1(✉)], Sofiia Dembitska[2], Vitalina Puhach[3], Iryna Kobylianska[2], and Oleksandr Kobylianskyi[2]

[1] Kamianets-Podilskyi Ivan Ohiienko National University, 61 Ohiienko St.Khmelnytskyi Region, Kamianets-Podilskyi 32300, Ukraine
marinenka1@gmail.com
[2] Vinnytsia National Technical University, 95 Khmelnitskoe Shose St, Vinnytsia 21027, Ukraine
[3] Vinnytsia Educational and Scientific Institute of Economics Western Ukrainian National University, 37 Honty St, Vinnytsia 21100, Ukraine

**Abstract.** Peculiarities of students' independent work arrangement during mixed classes are considered. The relevance of the research is not in doubt, because in the conditions of the pandemic, higher educational institutions were forced to switch to a blended learning environment, which led to the need to find effective methods of ensuring the quality training of future specialists. At the same time, blended learning has proven to be effective in maximizing the individual educational needs of students. Taking into account modern perspectives, blended learning requires the search for effective means and methods that enable the formation of professional competences in future professionals. *The purpose of the research* is to develop ways to improve the independent work of students of technical universities in the conditions of two-way education. *The object* of the research is the study of the educational process in higher educational institutions of a mixed format. *The subject* of the research is scientific substantiation of the model of organizing students' independent work in conditions of mixed learning and increasing its effectiveness. In the study, the authors used the following methods: theoretical (theoretical analysis of scientific sources and regulatory documentation) and empirical (survey, testing, pedagogical experiment). The study of a number of scientific and methodical sources on the topic of the research and the conducted questionnaire made it possible to identify the main shortcomings of the organization of independent work in institutions of higher education in mixed classes (lack of proper methodological skills of teachers; lack of developed algorithms), ineffective organization of independent work of students in modern conditions, insufficient motivation of some students to study new material). According to the results of the conducted pedagogical experiment, it was found that the results of success of students from the control groups in the disciplines of professional direction are significantly lower than in the control groups.

**Keywords:** Competence of specialists in technical specialties · Blended learning · Improvement of the educational process


M. E. Auer et al. (Eds.): ICL 2023, LNNS 899, pp. 206–214, 2024.
https://doi.org/10.1007/978-3-031-51979-6_21

## 1 Problem Statement

Blended learning in the process of training specialists cannot be successful without high-quality independent study of students. This is caused by the requirements of the modern globalized labor market for the training of competent specialists who are able to improve their qualification independently throughout their lives and involves determining the specifics of professional activity in conditions of the uncertainty inherent in the market system of economically developed countries with the aim of improving independent work in the process of studying at universities.

The study showed that the organization of independent work of students in a mixed learning format does not provide adequate conditions for the full development of independence and contributes to the formation of their ability to self-education. The main reason for the lack of methodological skills of teachers in the organization and control of the independent work of students remotely is that the majority of technical higher education teachers are specialists in technical fields, and the organization of distance learning requires certain pedagogical skills and knowledge, the ability to adapt developed material for the format of such training. The article attempts to analyze methodical and scientific sources, the available practical experience of organizing students' independent work in the conditions of blended learning and the determination of ways for the opotimization of this process.

## 2 Methodology of Scientific Research

The research consisted of two stages: analytical-diagnostic and ascertaining. As the part of the analytical and diagnostic stage, a study was conducted to identify and analyze the organization of students' independent work in a blended learning format, and the current state of independent work of students, the peculiarities and shortcomings of its organization at Vinnytsia National Technical University (VNTU) were characterized.

Studying the difficulties students and teachers faced in the process of blended learning in quarantine and war a survey of teachers and students, visiting online classes, analyzing the general results of student learning and their performance of course and qualification works was carried out at VNTU. In order to identify the efficient ways students' independent work arrangement in the ormat of blended learning, the available experience in this field in Ukraine and abroad and the possibilities of its practical implementation in higher education institutions were analyzed.

Within the framework of the ascertaining stage of the study, an analysis of the effectiveness of independent work was conducted with VNTU students under the conditions of blended learning format applying mathematical and graphic methods of mathematical statistics and computer data processing methods for the analysis of research results was summarized.

Based on own experience and the conducted survey, the properties that are required for the efficient organization of independent work of higher education students under the conditions of blended education were identified: knowledge of the ways and characteristic features of planning the independent work of students in the process of distance learning; ability to organize the interaction of the educational process participants during distance learning; ability to carry out psychological and pedagogical diagnostics of the state of students during remote communication; ability to control the educational achievements of students during distance and blended learning, using various methods; the ability to identify materials that need to be studied during mixed and distance learning, as well as to define tasks for independent study by students.

## 3 Analysis of Recent Research and Publications

The Covid-19 pandemic has significantly intensified pedagogical research in the field of finding the most effective measures and tools for distance learning. The problems of organization, methodological support, and quality assurance of blended learning are reflected in a number of scientific studies. Researchers N. Megahed and E. Ghoneim believe that the system of higher education has undergone radical changes, entered a new stage of its development. In the publication [1], the possible directions of the development of blended learning, their features and prospects were investigated.

The general characteristics of blended learning and the analysis of the educational institution's work planning are presented in the publication [2]. We agree with the statement that the introduction of blended learning has significant prospects in the sphere higher education, but it requires a radical restructuring of the philosophy of education and its total digitalization, especially in developing countries [3]. Separate practical studies are observed in the publications [4, 5] and others. The overview of means for the implementation of blended learning is given in the publications [6] and others.

A thorough analysis of the philosophical and educational-methodical literature on the topic of our research and the peculiarities of the organization of the blended format of education at VNTU proves that there exists the following obstacles for the effective independent work of students: significant part of teachers does not have the skills to control their remote work; problems dealing with performingthe independent work remotely, as there are currently no algorithms for the its organization in the conditions of a blended learning format; there is no proper motivation of a greater part of students to acquire new knowledge and develop professional skills in the process of distance learning.

## 4 Theoretical Substantiation of the Principles of Organization of Students' Independent Work in Blended Learning

According to scientific research and practice of leading countries, blended learning has significant opportunities for effective student training. However, at the same time, the readiness of both teachers and students for this type of work should be formed, as well as the corresponding legislative and methodical base should be developed. But in the conditions of the pandemic, educational institutions found themselves in a situation of forced transition to a mixed format of education, without having a sufficient amount of developments and experience in this sphere.

Accordingly, to reduce the shortcomings, connected with the educational process organization, it is necessary to arrange methodical support for teachers. At VNTU, such support is carried out by the methodical seminars, held by the departments on a regular basis, which consider problematic issues related to conducting classes, measures to ensure the educational process with students in a blended format, the "JetIQ Electronic Management System" is used, which is the university's global information base for: management of the educational process, analysis of students' knowledge, collection of information regarding students' learning activities, knowledge testing, etc. In addition, the university systematically conducts seminars on the organization of distance learning on the JetIQ platform.

The efficiency of learning in a mixed format in higher education establishments directly depends on its organization. It is clear that in the process blended learning, the volume of work of teachers increases significantly. Therefore, it is necessary to review and adapt educational disciplines to the conditions of such training, to model actions during their study, to develop algorithms needed for the organization of students' independent work and its evaluation, etc. At the same time, control over the learning process, interaction between all its participants, as well as automation of the teacher's work must be ensured, since the amount of control measures automatically increases during the study of any educational course.

However, this approach is not successful, since the arrangement of independent work in a mixed format requires a clear structuring of educational content. Therefore, in order to improve the arrangement of the students independent work in blended learning format, the authors suggest using the approach of reverse design in learning, which was proposed by the Center for Innovative Teaching and Learning of Bloomington University of Indiana [7]. For its implementation, it is necessary to structure the content of the educational discipline with an orientation towards assessment, the results of which should demonstrate students‘ achievements to the outlined course goal.

In accordance with this approach, the authors offer the following scheme for designing students' independent work: define the goal and tasks, the implementation of which will allow the achievement of program learning outcomes. The presence of the feedback is mandatory: it will allow to adjust the content of tasks to achieve the best learning results. Main problem of the organization of student's independent work in a mixed form of education is the lack of motivation to learn new material among greater part of students. The problem is mainly related to the inability of students to work independently, the complexity of the new material and lack of understanding its importance for professional growth, which requires additional teacher‘s explanations, etc.

Therefore, to form motivation during the realization of any independent task, it is necessary to formulate in its preamble the purpose, abilities and skills, the students are to master and which will be useful in their further professional activities. The objective of performing the independent work should correlate with the content of professional competence to be formed. The next step should be the analysis of the relevance of both the content of each academic discipline as a whole and each task offered to the student. The process of updating humanity's knowledge base and related technological changes has become extremely dynamic, therefore, in a few years, the knowledge acquired by a student at a higher education institution becomes obsolete. Accordingly, the content of educational disciplines needs constant adjustment in accordance with new national and industry normative legal acts, developed innovative technologies, teaching methods, etc., which ensure student motivation [8].

In order to organize the work efficiently, students should be informed of the entire structure of the course, the list of all tasks for the independent work to be completed, evaluation criteria and deadlines for each topic or type of the work. The convenience of the mixed learning format is that students have round-the-clock access to all professionally oriented materials of the academic discipline. The final point that must be taken into account, organizing the students‘ independent work in a mixed format is the presence of the feedback, which should be carried out both in the form of tests and reports on the completed work, but also in the format of online communication. In general, the proposed approach to the organization of students' independent work can be summarized in the form of a structural-functional scheme (Fig. 1).

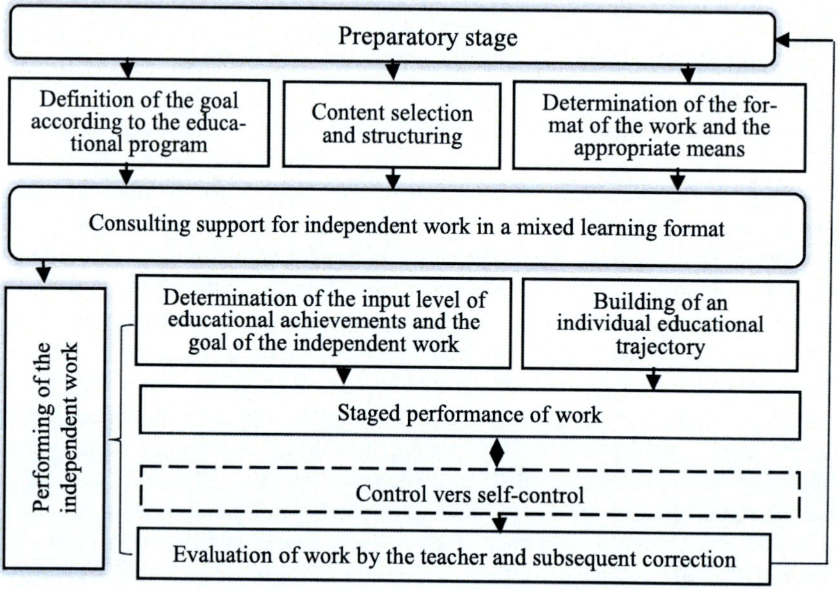


**Fig. 1.** Structural-functional model of students' independent work organization in conditions of blended learning

The implementation of the proposed scheme involved several key stages. At the preparatory stage, in accordance with the purpose of the educational program, the content of the educational material was selected and structured, and the types of work to be implemented remotely and those to be implemented in a face-to-face format were determined. Appropriate tools and methods were selected for each type of work. Consulting support for independent work in a blended learning format is also mandatory. This means that students' independent work at each stage was guided by the teacher. In particular, the teacher analyzed the input level of students' academic achievements and the goals of independent work, and adjusted their individual educational trajectory. Deadlines and evaluation criteria were set for each type of work. This motivated students to complete their work on time. The results were monitored and corrected during the course of the work. These measures helped to improve the quality of students' independent work. Students realized the importance of independent work for their professional development, and also understood that the teacher was always ready to help them. This helped to increase students' motivation to study and the quality of their training.

## 5 Analysis of the Results of the Experiment

To test the hypothesis about the significance of the results, the Fisher's angular transformation was used. The expediency of using Fisher's angular transformation $\varphi$ at this stage is that it allows to establish the statistical significance of the obtained data and the presence of the positive changes in the studied phenomenon.

The initial hypotheses are formulated as follows: $H_0$—is the portion of the students-specialists where a positive result is observed regarding the realization of independent work after the implementation of the created model and methodological conditions in the experimental group (EG) is not greater than in the control group (CG); $H_1$ is greater than in the CG. The formulated hypotheses are checked in the following way:

(1) Four-step scale for measuring students' educational achievements is converted into a two-step scale ("there is an effect", "there is no effect"). The category "there is an effect" comprises students who have demonstrated a sufficient and creative levels of readiness to use the projective technologies in their professional activities, "no effect" category includes the students with elementary and intermediate levels, respectively (Table 1).

**Table 1.** Results at the end of the experiment on a two-step scale

| Groups | Levels of formation | | | | Number of people in the group |
|---|---|---|---|---|---|
| | "No effect" | | "There is an effect" | | |
| | k-t | % | k-t | % | |
| CG-1 | 13 | 72,22 | 5 | 27,78 | 18 |
| EG-1 | 10 | 50,00 | 10 | 50,00 | 20 |
| CG-2 | 14 | 70,00 | 5 | 30,00 | 20 |
| EG-2 | 10 | 52,63 | 9 | 47,37 | 19 |
| CG (total) | 27 | 71,05 | 11 | 28,95 | 38 |
| EG (total) | 20 | 51,28 | 19 | 48,72 | 39 |

(2) Percentage shares according to the points "there is an effect" and "there is no effect" are converted into radians according to the Fisher angular transformation formula:

$$\varphi_1 = 2\arcsin\sqrt{P_1},\ \varphi_2 = 2\arcsin\sqrt{P_2},$$

where $P_1$ and $P_2$ are the shares that are compared (Table 2);

**Table 2.** Fisher's φ-test calculations

| Groups | Levels of formation preparedness of the studied preparedness | | | | Number of people in the group |
|---|---|---|---|---|---|
| | "No effect" | | "There is an effect" | | |
| | Fraction | $2\arcsin\sqrt{P}$ | Fraction | $2\arcsin\sqrt{P}$ | |
| CG-1 | 0,72 | 4,058 | 0,28 | 2,230 | 18 |
| EG-1 | 0,50 | 3,146 | 0,50 | 3,146 | 20 |
| CG-2 | 0,70 | 3,964 | 0,30 | 2,318 | 20 |
| EG-2 | 0,53 | 3,262 | 0,47 | 3,022 | 19 |
| CG (total) | 0,71 | 4,008 | 0,29 | 2,274 | 38 |
| EG (total) | 0,51 | 3,182 | 0,49 | 3,102 | 39 |

(3) Calculate the value of Fisher's $\varphi$-test for the CG and EG using the following formula:

$$\varphi = (\varphi_1 - \varphi_2)\sqrt{\frac{n_1 n_2}{n_1 + n_2}},$$

where $n_1$ and $n_2$ are the volumes of samples (control and experimental groups) under investigation.

The results of calculating Fisher's $\varphi$-test are displayed in Table 3.

**Table 3.** Fisher's φ-test value at the end of the chamber experiment

| Groups | Calculated value of Fisher's φ-test | | By significance levels | |
|---|---|---|---|---|
| | "No effect" | "There is an effect" | 0,01 | 0,05 |
| CG-1 and EG-1 | 2,807 | 2,819 | 2,31 | 1,64 |
| CG-2 and EG-2 | 2,161 | 2,167 | | |
| CG and EG (total) | 3,624 | 3,633 | | |

Based on the data in Table 3, the hypothesis $H_1$ was confirmed, according to which the proportion of the students who have higher level of educational achievements after the implementation of the proposed model of independent work organization is greater than in the control group and these data are statistically significant.

## 6 Conclusions

Therefore, according to the results of the analysis of theoretical and educational and methodological sources and the authors' own pedagogical research, the purpose of which was to increase the efficiency of independent work of students of higher education, taking into account the peculiarities of blended learning at technical universities gives reasons to state the following:

(1) Training of specialists in technical specialties at the current stage is carried out in blended and distance forms. This has both advantages and disadvantages. The advantages include the possibility of using ICT tools for training specialists during classes, convenient mode of work, the automation of the part of the tasks. At the same time, among the significant shortcomings is the lack of developed methodical means for carrying out such work. The authors proposed the generalized results of research on improving the efficiency of independent work of higher education students in the conditions of a mixed format of conducting classes, analyzed possible mistakes and identified ways to improve this process;
(2) In order to increase the effectiveness of the independent work of students of higher education in a mixed format, it is necessary to evaluate the levels of their educational achievements at the beginning of studying of each academic discipline. The purpose of this assessment is to determine the initial level of students' knowledge, as well as to clarify their expectations from studying this course: outlining the purpose of studying the educational course on the whole and students independent work in particular, taking into account aspects of the future professional activity of students; familiarization of students with the structure of the entire course and the defined terms: conducting online classes, general and individual consultations; implementation of educational modules, course projects and works, calculation-graphic and other types of tasks; familiarization of students with forms of reporting, assessment and feedback, etc. Assessment should be understandable for the students;
(3) Based on the approach of reverse design in education, the technology of structuring the independent work of students in conditions of blended learning and the structural-functional model of students independent work organization was proposed and implemented in the educational process of training applicants of higher technical education establichment Vinnytsia National Technical University to ensure its quality and compliance with labor market requirements;
(4) The results of the conducted comparative experiment, organized in order to determine the efficiency of the proposed structural and functional model of organizing students' independent work organization, proved that the level of educational achievements of students of the control groups is lower than the corresponding level of the experimental groups. The statistical significance of the obtained results was established using Fisher's $\varphi$-test. Since the value of the criterion $\varphi$ is greater than the critical one ($3.633 > 1.64$), we consider the proposed model to be effecient.

## References

1. Megahed, N., Ghoneim, E.: Blended learning: the new normal for Post-COVID-19 Pedagogy. Int. J. Mob. Blended Learn. **14**(1), 1–15 (2022). https://doi.org/10.4018/IJMBL.291980
2. Basilaia, G., Dgebuadze, M., Kantaria, M. & Chokhonelidze, G. Replacing the classic learning form at universities as an immediate response to the COVID-19 virus infection in Georgia. Int. J. Res. Appl. Sci. Eng. Technol. **8**(III), 101–108 (2020). https://doi.org/10.22214/ijraset.2020.3021
3. Mayisela, T.: A Practice-Based approach to developing First-Year higher education students Digital Literacy: A case study in a developing country. Int. J. Mob. Blended Learn. **14**(3), 1–14 (2022). https://doi.org/10.4018/IJMBL.314582
4. Asarta, C., Schmidt J.: The effects of online and blended experience on outcomes in a blended learning environment. The Internet and Higher Education 44, (2020). https://doi.org/10.0708.10.1016/j.iheduc.2019.100708
5. Holik, I., Kersánszki, T., Molnár, G., Sanda, I.D.: Teachers' digital skills and methodological characteristics of online education. Int. J. Eng. Pedagog. (IJEP) **13**(4), 50–65 (2023). https://doi.org/10.3991/ijep.v13i4.37077
6. Chiu T. K. F.: Digital support for student engagement in blended learning based on self-determination theory. Comput. Hum. Behav. 106909. Advance online publication (2021). https://doi.org/10.1016/j.chb.2021.106909
7. Backwards Course Design. URL: https://citl.indiana.edu/teaching-resources/course-design/backward-course-design/index.html
8. Kuzmenko, O., Dembitska, S., Miastkovska, M., Savchenko, I., Demianenko, V.: Onto-oriented information systems for teaching physics and technical disciplines by STEM-Environment. Int. J. Eng. Pedagog. **13**(2), 139–146 (2023). https://doi.org/10.3991/ijep.v13i2.36245

# The Learning Gate: A Case Study of Virtual Continuous Education Based on Andragogic Principles

Mayela García-Rodríguez, Patricia Vázquez-Villegas, Jorge Membrillo-Hernández, Jorge Limón-Robles, and Noe Miranda-Becerra(✉)

Tecnologico de Monterrey, 64849 Monterrey, NL, Mexico
{paty.vazquez,noemiranda}@tec.mx

**Abstract.** In 2020, Tecnológico de Monterrey launched LIVE programs in Continuing Education to serve all those professionals looking for courses and diplomas in the virtual version. The Learning Gate (TLG), a personalized platform, was developed under andragogic principles, where the participant can choose what, when, and how much to study. This work describes the TLG case study, including its design, covered competencies, student perceptions surveys and two participants' case studies, and the discussion of the results. The paper also describes the personalized virtual approach, discussing web-based platforms' potential to promote continuous education. The customized platform consists of two parts: the content area, comprising a diagnostic evaluation and simple self-directed learning, and the practical part, where a challenge-based learning approach is employed. The program consists of 4 big areas (leadership, data science, finance, and marketing and sales), accounting for 123 sub-competencies. Until November 2022, there were 602 students enrolled in these programs. Participants who have already taken the courses and challenges and the development of sub-competencies ensure that they are satisfied with the training they have received since the tools newly acquired are effectively applied in the context of the reality of their organization. Continuing education is a tool that higher education institutions should promote in their collaborations with organizations and an essential lifelong learning element in the development of the current profession.

**Keywords:** Andragogy · Educational innovation · Knowledge management · Higher education

## 1 Introduction

### 1.1 Role of Universities in Continuous Education

The evolving concept of universities' "third mission" includes providing knowledge, resources, and renewal/upskilling organizations' capabilities [1]. This contribution to society engages non-academic stakeholders, such as practitioners, with knowledge transfer activities [2], an important source of innovation for society [3]. Nurturing specialized


M. E. Auer et al. (Eds.): ICL 2023, LNNS 899, pp. 215–223, 2024.
https://doi.org/10.1007/978-3-031-51979-6_22

human capital through Continuous Education programs is one of those activities [4]. In this regard, universities can design courses tailored to the needs of professionals and address specific skills by accessing experts and scholars providing valuable insights, high-quality instruction, and mentorship to professionals in different fields [5]. Moreover, universities can offer certification or credentials to demonstrate such skills and provide technology and infrastructure that may not be available to professionals in their workplace (i.e., access to online learning platforms) [5]. However, the current world challenges (i.e., digitalization) require universities to rethink how their third mission can most effectively be fulfilled [1, 6].

### 1.2 Andragogy and Lifelong Learning in Continuous Educational Settings

Andragogy refers to the study of how adults learn [7]. It emphasizes the importance of the learner's autonomy and self-directedness in the learning process [8], assuming internal factors, such as personal growth and relevance to their lives and work [9]. On the other hand, lifelong learning refers to the ongoing learning and personal development process throughout one's life [10, 11]. It includes formal education or training, informal learning experiences, and self-directed learning [12], adapting to the changing demands of work and society [13]. Actualization courses can be considered a lifelong learning strategy.

Taking actualization courses can be a great way to enhance personal and professional development, fostering lifelong learning. Facilitators must implement effective learning strategies that provide valuable experiences and ensure participants acquire knowledge in a relevant and practical context. This is particularly significant given the current social and educational changes resulting from the pandemic's impact on organizations.

### 1.3 Potential of Active Learning Digital Platforms in Continuous Education

To generate differentiated and high-impact teachings in organizations, it is necessary to enable the use of digital tools and work with andragogic instruction. Participants in a continuous education program do not have the priority of taking the course, but what they learn is applicable. If this does not happen in the program, the course generates dissatisfaction, despite the teaching commitment. Continuous Education suppliers must create conditions so that learning is student-centered. That is, develop a virtuous cycle centered on the student's impact on their professional work. In this regard, active learning self-directed methodologies can play a significant role.

Prioritizing the needs of students in a constructivist environment is the essence of active learning [14]. Active, self-directed learning in digital platforms is a growing phenomenon [15]. It facilitates the expansion and flexibility of training offerings [16, 17]. However, the selection of these technologies should focus, in addition to the value of their pedagogical contributions, on analyzing the results and product of the advantages and disadvantages of their application [17]. This work aims to deepen the contribution of "The Learning Gate" (TLG), a continuous education initiative, to consolidate transformative educational processes in organizational environments.

### 1.4 Digital Continuous Education at Tecnologico De Monterrey

In 2020, Tecnológico de Monterrey, a Mexican private university, launched LIVE programs in Continuing Education to serve all those participants looking for courses and diplomas in the virtual version. TLG was developed as an online platform where participants can choose what, when, and how much to study. It is based on a personalized virtual approach. The customized platform consists of a content area comprising a diagnostic evaluation, simple self-directed learning, and a practical part where a challenge-based learning approach is employed. The program consists of 4 big areas (leadership, data science, finance, and marketing and sales), accounting for 123 sub-competencies. TLG offers organizations the upskill and reskill of their team, with accreditation endorsed by Tecnológico de Monterrey. Digital badges, supported by Blockchain Technology, can be shared on social networks to have the endorsement and support of certification of competencies or professional profiles.

This study focuses on designing TLG virtual courses for working adults and analyzing the competencies developed by participants. We also examined the student population and interviewed three participants and a company expert for insights on their experience and satisfaction with TLG courses.

## 2 The Learning Gate

Following the four-level Kirkpatrick model, TLG provides professional development courses and certifications for individuals and companies to improve their skills and knowledge. This model is known to be adaptable to formal online courses [18]. For instance, the four levels are i) reaction (satisfaction), ii) learning (acquired knowledge), iii) behavior (acquired competencies), and iv) results (impact) [19]. Covering all four levels of the model is crucial for continuous education in today's workforce since the objective is that the participant, individually or as part of a business group, shifts their skills to impact the indicators of the organization.

On the other hand, TLG is also based on the Marzano taxonomy of learning [20]. This taxonomy is employed to write learning objectives, incorporating principles from cognitive psychology [21]. This allows the instructor to infer a cognitive process from an observable behavior from the participant or develop new tasks to comply with the expected competencies. Following this approach, TLG course designs follow the INSPIRA (Inspire, Nourish, Significance, Practice, Integrate, Real challenges, and Advice) model of Tecnologico de Monterrey, as shown in Fig. 1.

The INSPIRA model is employed in four-course designs of the Vice Rectory for Continuous Education: face-to-face courses for organizations and individuals, LIVE (100% online), and virtual classroom (50% online). For this, the facilitators of continuous education are trained in this andragogic model. The INSPIRA's effect on facilitators' and students' perceptions will be discussed in future works.

Experts and architects create TLG's courses to ensure high quality. There's a team for content design, production, and quality control. Each course has two parts: knowledge and practical application. Experts test and challenge participants to ensure their newly acquired knowledge is implemented (Fig. 2).

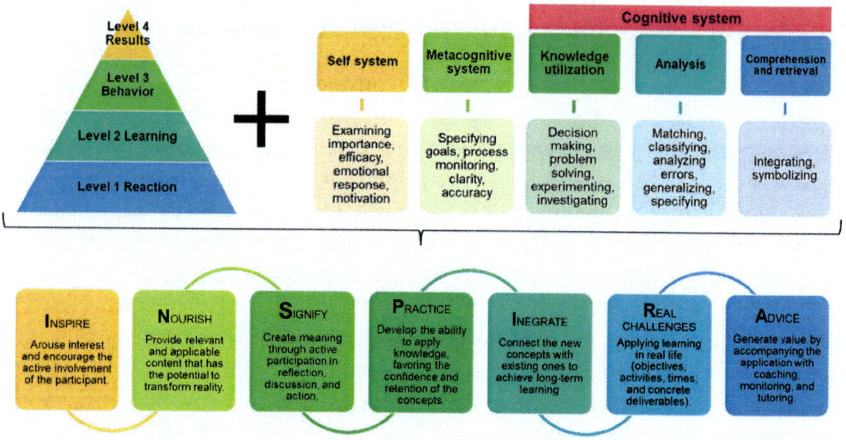


**Fig. 1.** The INSPIRA model of the Vice rectory for continuous education of Tecnologico de Monterrey is based on Marzano's taxonomy [22] and Kirkpatrick's model [18].

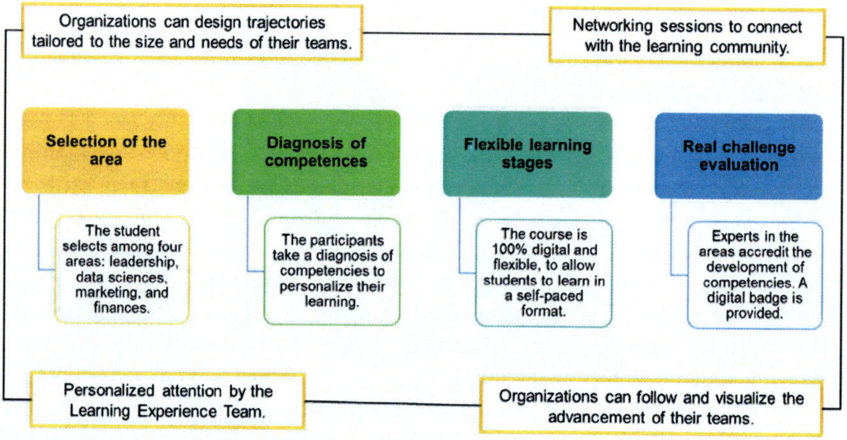


**Fig. 2.** The TLG experience is designed for individuals that want to earn competence in their field or companies that want to upskill their teams.

The actual curricular offer of TLG consists of 4 areas and 12 courses where participants can develop different competencies. The thematic areas of TLG were chosen by analyzing LIVE demands by January 2022. For instance, Leadership and Data Science thematic areas consist of three profiles and 17 and 23 trajectories, respectively. In addition, also in 2023, three other new routes will be developed: Digital Transformation, Operational Excellence, and Projects. Some competencies sought to be developed in some areas are presented in Table 1.

## 2.1 Demography of Participants from TLG

Until November 2022, there were 602 students enrolled in these programs. According to information collected from TLG registrants until October 2022, the average age of the participant is 36 years. Among the generations, we have centennials (20%), millennials (56%), Generation X (23%), and Baby boomers (2%), where 62% are men, and 38% are women; 56% of them are married and 35% single. 60% of participants have at least a bachelor's degree, with 34% having a master's degree. Almost 40% work in areas of engineering and manufacturing, and 95% of the participants currently work in different sectors such as commerce (15%), Professional services (13%), financial services (11%), food, beverages, and tobacco (10%); health services, transportation, engineering and hospitality, and tourism, house one 8%; where 79% of companies have more than 250 employees, 35% of them hold a managerial position, ranging from interns (2%) to directors (5%). They perform in different jobs such as IT (16%), HR and projects (each 10%), and sales and finance (8% each). There is also a smaller percentage of administration, production, business, and research. While most enter these courses looking to develop skills (22%), some seek some certification (17%), specialization (13%), or enroll, looking to update themselves with the support of a course from Tec de Monterrey (9%).

According to the survey, the typical participant is a 36-year-old man with a bachelor's degree in engineering, manufacturing, or construction. The most sought-after positions are "High Impact Manager" and "Data Scientist," with a strong interest in online courses. However, there is a need to promote gender equality in continuing education, as only one-third of the participants are women. Additionally, newer career paths such as "Team Leader" aim to develop a range of skills for everyone beyond just those with access to professional careers. In terms of career growth, trajectory profiles play a crucial role. Typically, younger individuals start with junior roles, while experienced ones aim for senior positions. Respondents value keeping themselves informed on digitalization to remain competitive, particularly those who are not digital natives.

## 2.2 Results of the Interviews

From the surveys conducted, we took the case of five participants and two experts to learn more about their experience.

The five TLG participants were interviewed about their experience before, during, and after taking the Data Science course. Three had bachelor's degrees, one had a master's, and one had a doctorate. They enrolled in TLG for its prestige, course content, and language. They wanted to develop skills in data processing, management, virtual interfaces, machine learning, applied statistics, and programming. A questionnaire with a Likert scale was used to gather opinions from respondents on 20 elements of TLG. The highest value was for valuable learning, followed by the definition of competencies, achievement of objectives, clarity of rubric, and usefulness of information; time destinated to lessons had the lowest average value. The Cronbach alpha was 0.9495, indicating high consistency between responses. TLG courses have helped guide participants on what Data science means while catering to each organization's needs. Students appreciate obtaining badges, tutor reviews, flexibility, clear evaluation rubrics, and enjoyable,

**Table 1.** Selected TLG areas of knowledge, participant profiles, and competencies.

| Area | Profile | Competencies |
|---|---|---|
| Leadership | High impact management | Talent development, accountability skills, effective communication, strategic focus, innovation management, and team well-being |
| | Potential leader | Talent management, engagement, positive leadership, strategic thinking, innovator's DNA, personal wellbeing |
| | Senior leader | Organizations focused on talent, organizational engagement, organizational collaboration, organizational strategy, innovation culture, and organizational well-being |
| Data science | Senior data science | Data science fundaments, Phyton data processing, manipulation, and visualization, machine learning, machine learning for big data |
| | Data scientist | Data science fundamental concepts, extraction, processing, and manipulation of data, Web apps for data science, deep learning |
| | Citizen data scientist | Fundamentals of statistics for data science, data processing and handling with Python, interactive data visual interfaces, and linear data modeling |
| Marketing and sales | Digital marketing strategist | Digital consumer, digital marketing strategy |
| | High impact sales | Future commercial vision, commercial intelligence, account management, leadership, and commercial evaluation |
| | Marketing starter | Marketing as a function in the organization, strategy, operation, measurement, and evaluation |
| Finances | Financial Professional | Economic environment, financial, accountable, corporate, strategic financial vision, risks and sustainability, financial ethic |
| | Financial strategist | Sensitivity in quantitative models, evaluation of financial states, inversion, and financing, finance communication, and conduct |

practical, and summarized courses. Suggestions include including videos, downloadable files, more real cases, improving navigation, simplifying instructions, and adjusting dedicated hours. Some comments suggest improving color usage, faster response times, and increased accompaniment. Participants can take the classes independently since they're 100% online. On the other hand, an area of opportunity is the definition of competencies and emphasizing that TLG is about virtual and flexible learning, in which synchronous sessions are not included, or, vice versa, include synchronous sessions in TLG courses. The use of explanatory videos can replace synchronous classes and make courses easier.

As for the experts who have already reviewed the challenges of the sub-competencies, they assure that they are satisfied with the documents they have received since the tools recently acquired by the participant are effectively applied in the context of the reality of their organization. One of them said that from the 28 challenges he has reviewed of two sub-competencies of Leadership, they have carried out excellent projects, where he saw that they take great care. They promote reflection in the participants, complying with what is requested in the evaluation and instruction rubrics. The other expert interviewed has reviewed 30 Data Scientist challenges (Data visualization with Python), and she tells us that she is also satisfied with the results obtained. Of the 30 challenges, only three of them had a second opportunity. In the same way, she considers the deliverables to be of good quality, denoting a clear understanding and assimilation of the concepts reviewed. Clear and specific instructions help the participants because of the confidence generated by producing a replica of the graphics shown.

## 3 Final Remarks and Future Work

The virtual platform offered by TLG provides essential support for professionals to pursue lifelong learning. Our emphasis on competencies, including strategic focus, innovation management, engagement, leadership, strategic thinking, innovation culture, organizational well-being, data science, and financial ethics, ensures that our users have the tools to succeed. Indirectly, we also emphasize developing self-management, learning to learn, initiative, information acquisition, decision-making, effective communication skills, cultural awareness, and sustainability, ensuring our users are well-rounded and prepared for any challenge.

Although the results still need to be conclusive due to the small number of participants who have completed the courses, we have received positive feedback from both participants and experts. Moving forward, we aim to evaluate the impact of these courses on student learning and company value while promoting equity and sustainable competencies among participants. Continuing education is crucial in developing one's profession and should be supported by higher education institutions. We hope this work will generate greater interest in universities in developing new updating strategies for organizations and, in general, for lifelong learning of adults interested in increasing their capacities for professional development or other purposes.

**Acknowledgments.** The authors would like to acknowledge the financial support of Writing Lab, Institute for the Future of Education, Tecnologico de Monterrey, Mexico, in the production of this work.

## References


1. Markkula, F., Kune, H.: Making Smart Regions Smarter: Smart Specialization and the Role of Universities in Regional Innovation Ecosystems. Technol. Innov. Manag. Rev. **5**(10), 7–15 (2015)
2. Compagnucci, L., Spigarelli, F.: The Third Mission of the university: A systematic literature review on potentials and constraints. Technol. Forecast. Soc. Chang. **161**, 120284 (2020)
3. Perkmann, M., Walsh, K.: University–industry relationships and open innovation: Towards a research agenda. Int. J. Manag. Rev. **9**(4), 259–280 (2007)
4. Markuerkiaga, L., Caiazza, R., Igartua, J.I., Errasti, N.: Factors fostering students' spin-off firm formation: An empirical comparative study of universities from North and South Europ. J. Manag. Dev. **35**(6), 814–846 (2016)
5. UNESCO, https://unesdoc.unesco.org/ark:/48223/pf0000374112, Last Accessed 2023/05/05
6. Gregory, J., Salmon, G.: Professional development for online university teaching. Distance Educ. **34**(3), 256–270 (2013)
7. Knowles, M.S.H., III, E.F., Swanson, R.A., Robinson, P.A.: The Adult Learner. Routledge, England (2020)
8. Kruszelnicki, W.: Self-Directedness and the question of autonomy: from counterfeit education to critical and transformative adult learning. Stud. Philos. Educ. **39**(2), 187–203 (2019). https://doi.org/10.1007/s11217-019-09697-6
9. Collins, J.: Education techniques for lifelong learning. Radiogr. **24**(5), 1483–1489 (2004)
10. Billett, S.: Lifelong learning and self: work, subjectivity and learning. Stud. Contin. Educ. **32**(1), 1–16 (2010)
11. Carlson, E.R.: Lifelong learning: a higher order of consciousness and a construct for faculty development. J. Oral Maxillofac. Surg. **77**(10), 1967.e1-1967.e8 (2019)
12. Sterns, H.L., Walker, R.V.: Lifelong Learning and Work. In: Pachana, N.A. (ed.) Encyclopedia of Geropsychology, pp. 1417–1424. Springer, Singapore (2017)
13. Bejaković, P., Mrnjavac, Z.: The importance of digital literacy on the labour market. Empl. Relat.: Int. J. **42**(4), 921–932 (2020)
14. Fussell, R.: Gather us in: building meaningful relationships in catholic schools amid a COVID-19 Context. J. Cathol. Educ. **23**(1), 149–161 (2020)
15. Curran, V., et al.: Adult learners' perceptions of self-directed learning and digital technology usage in continuing professional education: An update for the digital age. J. Adult Contin. Educ. **25**(1), 74–93 (2019)
16. Wang, A.T., Sandhu, N.P., Wittich, C.M., Mandrekar, J.N., Beckman, T.J.: Using social media to improve continuing medical education: a survey of course participants. Mayo Clin. Proc. **87**(12), 1162–1170 (2012)
17. García, R.V.: La educación continua en México: hacia la transición a la captación a distancia. Edutec. Rev. Electrónica Tecnol. Educ. **20**(20), (2006)
18. Hamtini, T.M.: Evaluating E-learning programs: an adaptation of Kirkpatrick's model to accommodate E-learning Environments. J. Comput. Sci. **4**(8), 693–698 (2008)
19. Acuña, M.: https://www.evirtualplus.com/modelo-de-evaluacion-kirkpatrick/, Last Accessed 2023/05/05
20. Marzano, R.J., Kendall, J.S.: The new taxonomy of educational objectives, 2nd edn. Corwin Press, Thousand Oaks, CA (2007)
21. Valenzuela, J.: Quality standards in distance education course design: Lessons learned from the work of a Task Force from the Tecnologico de Monterrey's Virtual University. In Proceedings of Distance Learning and the Internet 2006, annual conference for the Association for Pacific Rim Universities, pp. 328–337. University of Tokyo, Japan (2016)

22. Irvine, J.: Marzano's new taxonomy as a framework for investigating student affect. J. Instr. Pedagog. **24**, 1–31 (2020)
23. Qureshi, M.I., Khan, N., Raza, H., Imran, A., Ismail, F.: Digital technologies in education 4.0. Does It Enhanc. Eff. Learning? Int. J. Interact. Mob. Technol. **15**(4), 31–47 (2021)
24. Kaplan, A.: Lifelong learning: conclusions from a literature review. Int. Online J. Prim. Educ. (IOJPE) **5**(2), 43–50 (2016)

# Challenges and Opportunities for Open Educational Resources in Higher Mathematics Education

Anne Uukkivi[1](✉), Oksana Labanova[1], Elena Safiulina[1], Anna Šeletski[2], and Tatjana Tamberg[2]

[1] TTK University of Applied Sciences, Tallinn, Estonia
{anne.uukkivi,oksana,elena}@tktk.ee
[2] Tallinn University, Tallinn, Estonia
{annat,tatjana}@tlu.ee

**Abstract.** Open Educational Resources (OERs) can overcome barriers in higher mathematics education through cost reduction, increased accessibility, and adaptable materials for teachers. The purpose of this paper is to provide a deeper understanding of the need for OERs in higher mathematics education, as well as the potential for collaboration and readiness for their use in Estonia. OERs can benefit teaching and learning in various ways. The adoption and utilization of OERs in Estonia is limited. The authors conducted an online survey with mathematics teachers from different universities and institutions to investigate the reasons for the absence of OERs, the necessity of a digital library of OERs, and the willingness to cooperate in its creation and use. The survey results showed that most respondents considered the digital library necessary and were ready to collaborate in various forms, such as creating, testing, reviewing, and translating materials. However, some respondents were reluctant to share their materials for free or did not find the digital library useful for their specific courses. In conclusion, collaboration in the development of OERs would help overcome the barriers and challenges associated with OERs in higher mathematics education in Estonia.

**Keywords:** OER digital library in mathematics · Cooperation in higher education in mathematics

## 1 Introduction

Open Educational Resources (OERs) have the potential to alleviate barriers to learning in higher mathematics education by reducing costs, increasing accessibility, and enabling the adaptation of educational materials to the needs of teachers [1].

OERs are learning, teaching and research materials in any format and medium that reside in the public domain or are under copyright that have been released under an open license, that permit no-cost access, re-use, re-purpose, adaptation and redistribution by others [2].


M. E. Auer et al. (Eds.): ICL 2023, LNNS 899, pp. 224–230, 2024.
https://doi.org/10.1007/978-3-031-51979-6_23

Estonia is a small country located on the eastern coast of the Baltic Sea, the fourth from the end populous country in the European Union. Despite OERs widespread use worldwide for over 20 years, the adoption and utilization of OERs in higher mathematics education in Estonia is still limited and Estonian higher education currently lacks governmental initiatives and interest in supporting the creation and implementation of OERs [3].

To address this issue, many countries and international initiatives have developed strategies and policies to promote the use of OERs in education [4]. Therefore, it is important for educators and policymakers in Estonia to recognize the benefits of OERs and take steps toward their wider adoption in higher mathematics education.

However, the adoption and utilization of OERs in higher mathematics education in Estonia is still limited. Therefore, it is important to examine the current state of readiness and opportunities for collaboration in the development of OERs in mathematics education in Estonia.

The purpose of this paper is to provide a deeper understanding of the need for OERs in higher mathematics education, as well as the potential for collaboration and readiness for their use in Estonia. Also, article aims to investigate the reasons for the absence of OER mathematical materials in higher education in Estonia, according to the perspectives of current teachers. By examining the opportunities and challenges associated with OERs, educators and policymakers can develop effective strategies to promote their use in higher mathematics education, ultimately improving access to education and leading to better student outcomes. In light of the above described, an international Erasmus + project Gate2Math, the aim of this project is to create a smart multilingual library of OERs in the field of mathematics for high-level education, has been established to facilitate the creation and development of OERs, thereby aiming to enhance the current situation and ensure sustainability in the field of OER, especially in Estonia.

## 2 Background

Digital educational resources are a priority of Estonian educational policy, as coordinated by the Ministry of Education and Research in accordance with the Lifelong Learning Strategy for 2014–2020 [5] and the Estonian Education Strategy 2021–2035 [6]. The three strategic goals of the latter are: (1) diverse and accessible learning opportunities, and an education system that ensures a smooth transition between levels and types of education; (2) competent and motivated teachers and directors, a diverse learning environment, and a student-centered approach to learning and teaching; (3) the relevance of training options to the needs of the development of society and the labor market. It should be noted that openness is not the main reason for focusing on digital educational resources.

The OER infrastructure in Estonia consists of educational object repositories, educational resource development tools, evaluation platforms, virtual learning environments (VLE), and auxiliary systems such as metadata application profile and single sign-on (SSO) [3]. The concept of learning objects arose in Estonia in connection with e-learning support programs such as BeSt program (for universities, lasting from 2008–2013), VANKeR (for vocational education institutions and universities of applied sciences, lasting from 2008–2013), e-Jump, e-VÕTI and Primus (lasting from 2008–2015, financed from European structural funds and implemented by the Archimedes Foundation). The goal of this last program was to support the quality of higher education and increase the competitiveness of graduates, and to cepti the goals, it worked closely with 23 partners, including 19 higher education institutions.

A learning object is defined as a complete digital resource that can be reused in different learning contexts and support learning, but changing them may not be possible due to copyright or design considerations. The most common types of e-learning objects in Estonia are animation, audio lectures, presentations, exercises (tasks), training videos, simulators, content packages (short courses), dictionaries, tests, and video lecture/multimedia synopses. Thus, the idea of learning objects (LO) is similar in nature to the idea of open educational resources (OER). Both LO and OER are types of digital educational resources that have associated metadata, including digital rights and learning design information such as learning goals and context.

Founded in 2016 under the auspices of the Ministry of Education and Research, the e-Koolikott (e-Schoolbag) platform presently serves as a repository for over 18,700 educational resources. These resources encompass a wide range of materials, including open source resources, unlicensed resources, and select commercial content. Educators are afforded the opportunity to curate content collections utilizing existing resources while also incorporating their own material into the repository. A significant proportion of the resources available on the platform were created using interactive H5P templates as part of the national Open Educational Resources (OER) initiative known as the "Digital Learning Property" project (DigiÕppeVaramu, 2017–2018). This project entailed the collaborative efforts of Tallinn University and resulted in the production of over 10,000 interactive educational objects. Rigorous quality control measures were implemented, and the materials underwent extensive testing by teachers from 30 schools. The materials on the e-Koolikott platform are freely accessible and editable by all teachers and students [3]. Learning objects can be annotated, recommended, and combined into shared collections, with optional login functionality facilitated through Estonian national authentication methods. Ministry of Education and Research, in conjunction with the European Social Fund, has allocated funding for initiatives aimed at developing digital OER specifically tailored to students with special needs. These projects prioritize the creation of simplified teaching aids, workbooks, worksheets, and other instructional materials. Furthermore, various projects and communities have independently established their own open source repositories. Notable examples include KAE Kool, which hosts over 150 educational videos licensed under the Creative Commons Attribution-NonCommercial-ShareAlike (CC BY-NC-SA) license. Additionally, a physics-themed repository has been collaboratively developed by Project Tiger Leap, the University of Tartu, the Institute of Physics of the University of Tartu, and Ministry of Education and Research. Teachers also employ standard office software and Web 2.0 tools in conjunction with platforms such as Weebly, Blogger, WordPress, and Google Sites to create open educational resources. These resources often integrate external materials sourced from platforms such as YouTube, SlideShare, LearningApps, and Geogebra.

In higher and professional education, the repository of educational objects operated by the Foundation for Information Technology for Education (HITSA) went live in 2009 [HITSA]. Less than 5% of the resources on the platform are unassociated with any funded content development project. The repository contains over 4,600 learning resources, all licensed under a Creative Commons license as required by the projects. The main purpose of this repository was to store resources that were developed in various content development projects funded by the European Social Fund. Põldoja and Laanpere [3] note that after the end of financial support for content development, the growth of the repository has decreased significantly. As a result of the reorganization of tasks between foundations operating in the educational field, the repository of the HITSA Innovation Center has not been available to users since July 1, 2020. All materials in the eLearning repository will be archived until the end of 2025. Now, in the field of higher education, there are currently no initiatives at the state level to support the development and implementation of OER. Materials created for higher education are now stored in institutional repositories. The digital archive Dspace is mostly used as a repository, containing educational material, learning objects, and educational videos prepared by the teaching staff of higher and professional educational institutions. Access to the content of the digital archive is guaranteed to all interested parties through the user interface and API (Application Programming Interface) standards. Dspace content is publicly available for both download and reuse, except in exceptional cases where the work is closed, restricted, or embargoed. All file formats are accepted for the preservation of publications and databases. The amount of data that can be stored in Dspace is unlimited. The main virtual learning environment used in vocational and higher education in Estonia is Moodle, which also stores OERs created during various projects; for example, materials created within the international project "EngiMath—Mathematics on-line learning model in Engineering education" (2018–2021, 2018-1-EE01-KA203-047098).

Analyzing the situation of OER in higher education, Põldoja and Laanpere [3] conclude that although there are lecturers who use open licenses and have created web pages for their OER, for the most part, higher education teachers are ceptical about the adoption of OER and are not ready to share their resources in repositories. The authors believe that this is due to the general lack of awareness of OER and open licenses.

## 3 Methods and Results

The collection of data was carried out using an online survey consisting of 11 questions. The survey comprised close-ended questions, but there was always an option to provide additional information. The survey consisted of questions about the need for a digital library, reasons why it is missing, willingness to create it in a collaborative way, and by whom it could be used.

The survey was compiled using a Google Form and conducted in spring 2023. The survey was distributed to mathematics teachers in universities and universities of applied sciences through personal contacts or the universities' academic affairs units.

Despite the fact that Estonia has a population of only 1.33 million, it has over 15 different higher education institutions, which can be divided into 3 main groups: public comprehensive universities, public specialized universities and private universities and institutes. Two universities train mathematicians and teachers of mathematics, so they represent more different branches of mathematics. Basically, specialized higher educational institutions teach the basics of mathematical analysis and algebra; statistics and economic mathematics are also very popular. However, some schools do not offer mathematical courses at all. It is very difficult to count the exact number of mathematics teachers in higher education institutions, since their number may fluctuate depending on the period of the year and since some teachers work simultaneously in two or three different educational institutions.

33 responses were received from mathematics teachers from different universities and institutions: 8 from both TalTech and Tallinn University, 7 from Tartu University, 4 from Estonian University of Life Sciences and TTK University of Applied Sciences, and 1 each from Estonian Entrepreneurship University of Applied Sciences and Tallinn Health Care College. These respondents represent 5 major areas of mathematics: algebra and geometry, discrete mathematics, mathematical analysis, probability theory and mathematical statistics, and applied mathematics.

First, the survey investigated the reasons for the absence of a digital library in higher education in the field of mathematics (see Table 1).

**Table 1.** The reasons for the absence of a digital library.

| Reasons | Number of responses |
|---|---|
| Lack of government support | 18 |
| There are many specific subjects taught by one or two teachers | 14 |
| Reluctance to share educational materials for free | 12 |
| There is a shortage of mathematicians | 10 |
| Do not want to share my own property or that of the university | 8 |
| Everyone has sufficient study materials of their own | 8 |
| Foreign OER digital collections are used | 2 |
| In mathematics, it is not possible to properly digitize study materials | 2 |
| I use my university's collection of study materials | 2 |
| A general digital collection does not make sense | 2 |
| There is a lack of people for developing such a digital collection | 1 |
| There is a preference for developing of university's LMS | 1 |

The most commonly cited reason was the lack of government support (including platform, funding, and administrator). This response option was chosen by 18 respondents. The next most important response options were the abundance of subject-specific courses taught by only one or two teachers (14 responses), reluctance to share teaching materials for free (12 responses), and a shortage of mathematicians (10 responses). 8 respondents believed that the reasons were reluctance to share their own or university property and having sufficient teaching materials.

24 respondents acknowledged the necessity of a digital library, while 9 respondents stated that such a thing is not needed. Out of those nine, five named the main reason as everyone having sufficient teaching materials of their own. One of the reasons for this

attitude again lies in the size of the country and the number of educational institutions. It is rare to find the same specialty in different educational institutions, and if it is, most likely there are different learning objectives, different requirements for admission and graduation, and different levels of student skill. Accordingly, mathematics is likely taught differently in these different institutions. Therefore, there are few universal educational materials in mathematics, and those that there are often altered by teachers for their students and for the specifics of the subject.

The respondents were informed that we are developing such a digital repository and asked them which volunteer activities they would be willing to contribute to. 24 respondents agreed to cooperate in the development of a digital library. The responses were divided as follows: providing error feedback (12 responses), testing materials (11 responses), creating materials (9 responses), reviewing materials (9 responses), involving learners to collaborate (6 responses), and translating materials (4 responses). Out of the nine individuals who did not consider the digital library necessary, four of them are nonetheless willing to contribute to its creation for free. Two of them are willing to contribute by translating materials, and two by reporting errors.

Those who were not willing to contribute to the creation of a free digital library were asked to justify their response. The main reasons given were lack of time (6 respondents), their specific approach to teaching the subject, which may not be suitable for another teachers (4 respondents), willingness to contribute only for a fee (3 respondents), and contribution to the development of their university's digital repository (2 respondents).

Next, the survey asked about the readiness to use such a digital library. 27 people were ready and 6 were not ready to use it. 30 respondents would recommend using this digital library for learners, whereas 3 would not. Four respondents justified their refusal by stating that the learning materials made by others do not align with their teaching logic, and one respondent mentioned already having sufficient materials.

## 4 Discussion and Conclusion

In conclusion, the findings of this study provide valuable insights into the topic of OER in the field of mathematics in Estonia. The survey reveals that most respondents suppose the absence of an OER in higher mathematics in Estonia is due to a lack of state funding, specific courses only being taught by a few lecturers, and a reluctance to share materials for free. Põldoja and Laanpere [3] also pointed to a link between the lack of financial support and the lack of open educational materials.

The study showed that teachers consider the existence of such a library necessary.

Additionally, the results of the survey indicate a significant willingness of the majority of respondents to cooperate. The study showed that educators are open to various forms of collaboration, including creating materials and testing content developed by other colleagues. Notably, even teachers who do not consider the library necessary expressed their willingness to contribute to its establishment. The majority of teachers expressed a desire to use the library when it is available, and even more recommend it to their students. The authors attribute the limited willingness to share existing materials with the skepticism of teachers towards open educational materials; the same fact is emphasized by Põldoja and Laanpere [3].

Reasons for teacher reluctance to cooperate include lack of time and teachers' specific approaches to teaching a subject that may not be suitable for others.

However, the authors believed that collaboration in the creation of OERs would help mitigate the aforementioned barriers in the field, including the issue of faculty overload, which is a significant concern in higher education.

## References


1. Stines, A.: Faculty perceptions of open educational resources in cyber curriculum: a pilot study [Doctoral Dissertation, Dakota State University], (2020)
2. UNESCO. Recommendation on Open Educational Resources (OER). https://unesdoc.unesco.org/ark:/48223/pf0000373755/PDF/373755eng.pdf.multi.page=3, Last Accessed 2023/05/26
3. Põldoja, H., Laanpere, M.: Open educational resources in estonia. in: current state of open educational resources in the "Belt and Road" Countries, Lecture Notes in Educational Technology pp. 35–47. Springer Nature Singapore Pte Ltd, (2020)
4. Hylén, J.: Open educational resources: opportunities and challenges. Tufts OCW Q. Newsl., **1**(2), (2006)
5. Ministry of Education and Research. The Estonian lifelong learning strategy 2020, (2014), https://www.educationestonia.org/wp-content/uploads/2022/12/estonian_lifelong_strategy.pdf, Last Accessed 2023/05/26
6. Ministry of education and research. Estonian education strategy 2021–2035, (2021), https://www.educationestonia.org/wp-content/uploads/2022/12/haridusvaldkonna_arengukava_2035_kinnittaud_vv_eng_0-1.pdf, Last Accessed 2023/05/26

# A Remote Lab for School Students that Explores the Function of the Human Eye

Thomas Klinger[1(✉)], Christian Kreiter[1], Judith Klinger[2], Thomas B. Steinmetz[3], and Ingrid Krumphals[3]

[1] Carinthia University of Applied Sciences, Villach, Austria
t.klinger@fh-kaernten.at
[2] Established Ophthalmologist, Hermagor, Austria
[3] University College of Teacher Education Styria, Graz, Austria

**Abstract.** The function of the human eye is firmly established in biology and health sciences curricula. Functional models are often used for visualization and education and are usually available in limited numbers for many students. We have developed and implemented a remote lab that allows any student to study and understand the essential functions of the human eye. The basis of the remote lab is an exercise setup for simulating the functioning of the human eye as used in schools. The parameters are adjustable over a web interface, and phenomena such as far-sightedness and near-sightedness, accommodation to objects at different distances, as well as refractive error correction using lenses of different powers can be studied. A basic exercise for the students is to adjust the commonly sighted eye so that the object, which is realized by an object on an OLED display in the remote lab, is sharply displayed on the simulated retina.

**Keywords:** Virtual labs · Remote labs · Biology · Human eye

## 1 Introduction

Remote labs enable students to perform practical exercises at a time and place of their choosing, as they are accessible via a web interface 24 h a day, seven days a week [1]. In addition, a very large number of students can work on the same exercise set-up, even simultaneously under certain conditions; this is especially important when the exercise set-up is costly or has limited physical access [2]. Another aspect is the repeatability of remote labs, which can be used to better understand or verify theoretical content.

Our research project aims to develop remote labs for science education and training in schools since eye training and eye health for school children are important aspects of ophthalmology [3]. Accompanying science education research ensures that the learning environments using the remote labs are designed according to the learners' perspectives.


M. E. Auer et al. (Eds.): ICL 2023, LNNS 899, pp. 231–238, 2024.
https://doi.org/10.1007/978-3-031-51979-6_24

## 2 Remote Lab Design and Development

The basis for the remote lab developed is an experimental set-up for upper secondary school, which is intended to explain the function of the human eye and enables various experiments (Fig. 1), Cornelsen Experimenta [4]. In order to implement a remote lab, the functions of the exercise arrangement must first be made remote controllable.

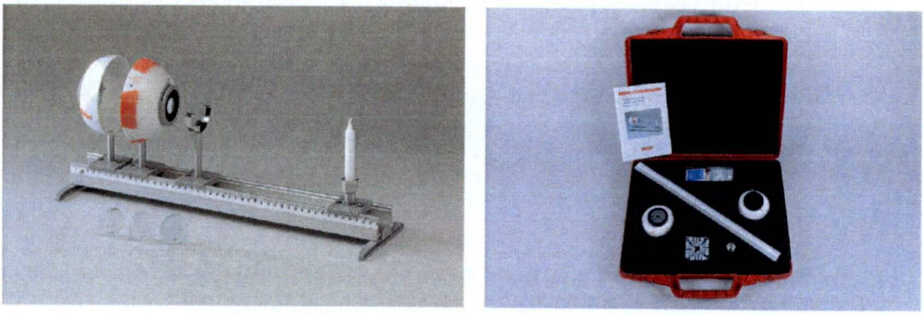

(a) (b)

**Fig. 1.** Functional model of the eye for face-to-face teaching in upper secondary education; it forms the starting point for the remote lab; (a) model built up, (b) carrying case [4].

### 2.1 Experiment Modification

The functions that we defined to be remotely controllable are the following:

- Distance eye lens (or foreground) – eye fundus (or background). This adjustment is necessary for the simulation of a short- or far-sighted eye and is realized in the remote lab with a linear motor (Fig. 2).
- Distance object – eye. This setting is used to simulate the accommodation of the eye and is also implemented with a linear motor.
- Setting the iris diameter with the diaphragm. This setting can be set manually in the original model; a stepper motor is used in the remote lab (Fig. 3).
- Use of corrective lenses as an eyeglass simulation. This function is realized with a turret wheel and a stepper motor.
- Refractive factor of the eye lens. The model includes a silicone lens whose refraction can be changed with a syringe and a variable amount of water. In the remote lab, the syringe is also controlled by a stepper motor (Fig. 4).
- Alternative object display instead of the candle in Fig. 1.

The software to control these functions runs locally on a microcomputer Raspberry Pi 4 Mod. B, the lab controller; it controls also the object display, a Raspberry Pi Pico with an OLED display. The web interface, required to control the model as a remote lab, is described later in this article.

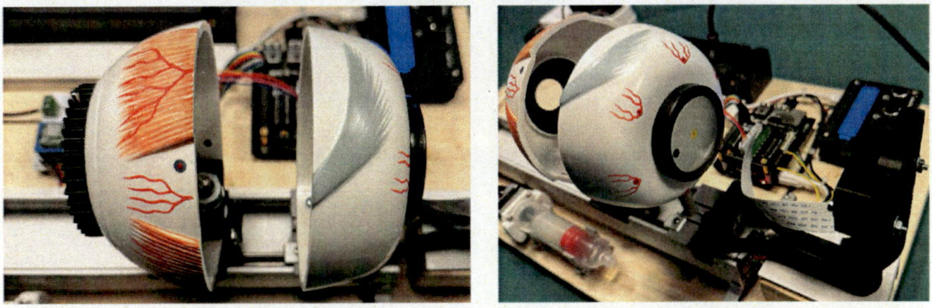

(a) (b)

**Fig. 2.** Foreground of eye (silicone lens removed) and background; (a) side view, (b) rear view. The eye background and thus the distance to the eye foreground can be adjusted by means of a linear motor.

(a) (b)

**Fig. 3.** Iris adjustment mechanism controlled by a stepper motor; (a) iris diaphragm open, the frosted glass retina in the back of the eye is fully visible; (b) iris diaphragm closed, limit switch triggered.

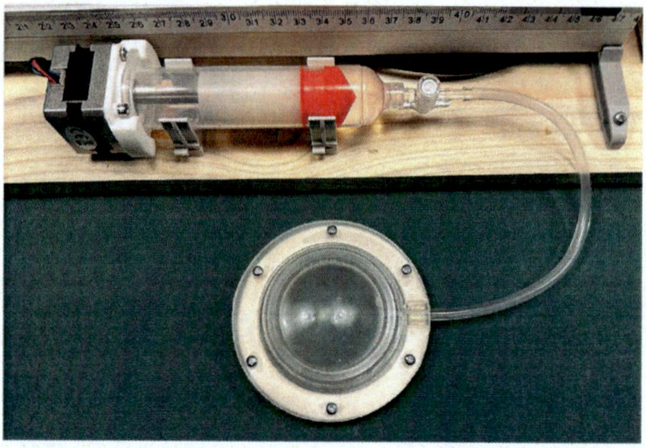

**Fig. 4.** Silicone lens (detached) with stepper motor-controlled syringe.

## 2.2 Object Display

A special point regarding the implementation in a remote lab is the display of the objects, which should then be made visible on the retina of the eye. In the original model, a candle is used for this purpose (see also Fig. 1), which is not an option for a remote lab for obvious reasons. In order to be able to do more advanced exercises in the future, ophthalmologically relevant symbols are also used, among others. Figure 5 shows the objects that are displayed on a Raspberry Pi Pico that can be controlled via the lab controller:

- The Landolt ring is a symbol for determining visual acuity (Fig. 5 a). If the subject can recognize the position of a recess corresponding to one minute of arc, the visual acuity is 1.0 [5].
- The Snellen hook or Snellen E is also used to determine visual acuity (Fig. 5 b). Here, the test person must reliably recognize the orientation of the apertures.

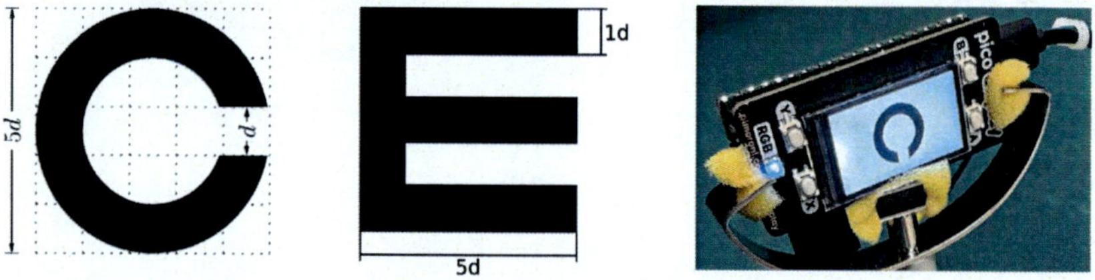


(a) (b) (c)

**Fig. 5.** (a) Landolt ring (by NielsKarschin—own work, CC BY-SA 4.0, https://commons.wikimedia.org/w/index.php?curid=47176443), (b) Snellen hook (public domain), (c) Landolt ring display on Raspberry Pi Pico.

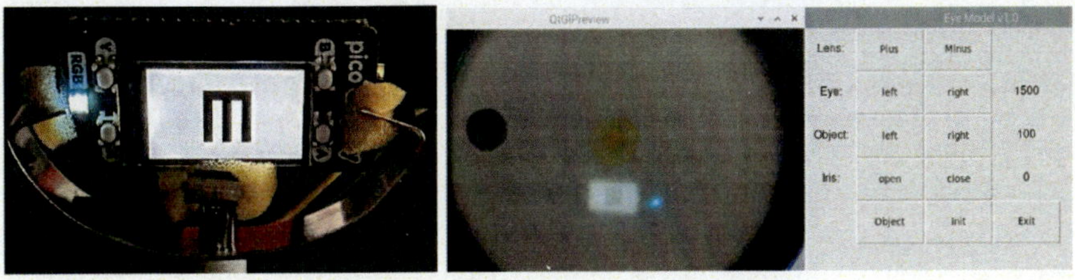


(a) (b)

**Fig. 6.** (a) Object (Snellen hook); (b) object projected (upside down) on the retina, preliminary control interface on the lab controller.

Other objects are, for example, a simple cross, or simple objects of everyday life, such as a house or a tree, as used for the medical examination of children. We also consider using objects which are not symmetrical to be able to observe the change of orientation of the object and the image considering all directions. Figure 6 shows a Snellen hook with open ends at the bottom as well as the image projected on the retina. Both these windows in Fig. 6b are displayed on the lab controller. We will be able to present a fully functional web interface at the conference.

## 3 Web Client

Since the lab is an interactive lab with real-time control, only one user at a time can have access to it. The allocation takes place according to the first come - first served principle. Data transmission between the lab controller and web client is established via the WebSocket protocol [6], and the video stream uses the JSMPEG Video Decoder [7], which runs over a separate WebSocket instance. Measurement data is returned to the web client at a constant rate.

## 4 Experiments

The following experiments can be performed with the experimental setup [4]:

- Projection of an object on the retina;
- Function of the iris;
- Normal sighted eye, near-sightedness, and far-sightedness;
- Eye accomodation while looking at nearby objects;
- Presbyopia (old people sightedness);
- Yellow spot and blind spot.

As shown in Fig. 6, at the lab controller each parameter of the eye model can be adjusted in defined steps, which of course cannot be the purpose and intention for the school experiments. Therefore, two of the experiments are described below as examples; they may also cover several of the basic experiments listed above.

### 4.1 Experiment 1: Object Projection and Function of the Iris

The eye fundus is set to the normal-sighted (emmetropic) eye distance, the iris is fully open. The object distance is at maximum; the silicone eye lens is initially set so that the object is out of focus. First ideas for students' tasks may be the following:

1. Vary the refraction of the eye lens so that the object will be focused on the retina. This will not work. Find an explanation why this does not work [Solution: Because the object display is too bright, but an approximate value of refraction can be found by a minimum size of the bright spot on the retina.] Describe your observation. What else can you find out?
2. Close the iris to an appropriate brightness so that the contour of the object display is shown sharply. Describe your observation and explain the function of the iris concerning the imaging process in our eye.
3. Formulate hypotheses what you have to adjust in order to get a sharp image on the display? Check your hypotheses with the remote lab. Explain the imaging process of the human eye [Possible hint: Fine-tune the lens refraction].

In the web interface or in accompanying documents, students receive additional information about the actual processes in the human eye, supplemented by explanatory illustrations. For this experiment, the processes are:

- The lens of the eye changes its refraction with the help of the zonula fibers (*zonula ciliaris*) and the ciliary muscle (*corpus ciliare*).
- The iris has the function of an aperture. It reduces its diameter by the *musculus sphincter iridis*, and dilates by the *musculus dilatator iridis*. However, this function is controlled involuntarily in the eye—in contrast to the model.

### 4.2 Experiment 2: Near-Sightedness and Far-Sightedness

In the beginning, the model setup is similar to Experiment 1. The eye is emmetropic and lens and iris are adjusted with autofocus algorithms to display a sharp image on the retina. To investigate near-sightedness and far-sightedness students have to do the following:

1. In the web interface, select near-sightedness (*myopia*) or far-sightedness (*hyperopia*) for the second part of the exercise, respectively. The eye fundus is set accordingly in the model.
2. By means of the turret wheel, an appropriate correction lens has to be found so that the object is again sharply displayed on the retina.

The accompanying explanations are as follows:

- A myopic eye has a longer eyeball compared to a normal-sighted eye, so the object is imaged in front of the retina.

- A concave corrective lens allows the image to be sharply reproduced on the retina (sketch of the ray path). The refraction of the lens depends on the refractive power of the lens in diopters (one diopter corresponds to a focal length of one meter) and on the refractive power of the own lens.
- A hyperopic eye has a shorter eyeball compared to a normal-sighted eye, so that the object is imaged behind the retina.
- In this case, a convex corrective lens has to be used.

## 5 Learning Arrangement

From a science education point of view concerning the use of remote labs, it is important to design evidence-based learning arrangements [8]. The model of educational reconstruction forms the basis for the development of the learning arrangements [9].

A first learning arrangement, which considers the above-mentioned experiments, will be developed for the upper secondary level in Austrian schools. The design of a first learning arrangement in this remote lab is following these basic scientific ideas:

- The emmetropic human eye can image objects sharply on the retina by means of its refractive media, consisting of the cornea, anterior chamber, iris, posterior chamber, lens, and vitreous body.
- In myopic and hyperopic eyes, a sharp image is not possible because the incoming visual rays meet in front of and behind the retina, respectively. Corrective lenses can be used to remedy this situation.
- Accommodation is the adjustment of the refractive power of the eye lens to the distance of the fixed object. With age, the width of accommodation decreases, which leads to presbyopia.

Based on these key ideas, we aim to meet the initially addressed learning outcomes:

1. Students are able to explain the function of the lens and iris and the mechanism of imaging objects on the retina.
2. Students can describe the proportions in near-sighted and far-sighted eyes and can apply their knowledge when explaining possibilities of correction by convex and concave lenses.
3. Students can explain the principle of accommodation of the eye when viewing objects at different distances, as well as the relationships in presbyopia.

## 6 Anticipated Outcomes

Beginning with the winter semester of 2023, the remote lab will undergo an initial series of tests with representatives of upper secondary school teaching staff. With these results, the learning arrangements will be further elaborated or revised and the respective exercises will be finalized.

**Acknowledgments.** This project is funded by the Innovation Foundation for Education and is carried out as part of the Innovation Labs for Education program.

## References


1. Schlichting, L.C.M., de Ferreira, G.S., de Bona, D.D., de Faveri, F., Anderson, J.A., Alves, G.R.: Remote laboratory: application and usability. In: Technologies Applied to Electronics Teaching (TAEE), Seville, Spain, pp. 1–7 (2016). https://doi.org/10.1109/TAEE.2016.7528355
2. Pester, A., Klinger, T.: Distributed experiments and distributed learning. Int. J. Online Biomed. Eng. (iJOE) **16**(6), 19–33 (2020). https://doi.org/10.3991/ijoe.v16i06.13661
3. Borah, R.R.: School-based eye health intervention: a lens for the future. In: Ophthalmology Times Europe, vol. 19, no. 4, pp. 28–30, May 2023
4. Cornelsen Experimenta [Online]. https://en.cornelsen-experimenta.de/shop/de/Sekundarstufe/Physik/Optik/47030-Demonstration+kit+Functional+human+eye+model.html. Accessed: 30 May 2023
5. Lachenmayr, B., Friedburg, D., Buser, A.: Auge – Brille – Refraktion. 5 th ed. Georg Thieme, Stuttgart/New York (2016)
6. RFC 6455—The Websocket Protocol. https://datatracker.ietf.org/doc/html/rfc6455. Accessed 30 May 2023
7. JSMpeg—MPEG1 Video & MP2 Audio Decoder in JavaScript. https://github.com/phoboslab/jsmpeg. Accessed 30 May 2023
8. Krumphals, I., Steinmetz, T., Kreiter, C., Klinger, T.: The development of a learning arrangement in a characteristic curve remote laboratory. In: The Learning Ideas Conference, New York, June 14–16, 2023 (paper accepted)
9. Duit, R., Gropengießer, H., Kattmann, U., Komorek, M., Parchmann, I.: The Model of Educational Reconstruction—A Framework for Improving Teaching and Learning Science, pp. 13–27. Unpublished (2012). https://doi.org/10.13140/2.1.2848.6720

# The Development of the Ukrainian Teachers' Digital Competence in the Context of the Lifelong Learning in the Conditions of War

Oksana Ovcharuk(✉)

Institute for Digitalisation of Education of the National Academy of Educational Sciences of Ukraine, Kyiv 04060, Ukraine
oks.ovch@hotmail.com

**Abstract.** Teachers' ability to use ICT is crucial in the modern conditions for the Ukrainian teachers. This is caused by many factors, including the long quarantine and transition to distance and blended learning in 2020–2022, and military actions caused by Russian aggression. In Ukraine, teachers are faced with the challenge of teaching online, as many schools cannot organize classroom learning, and part of the school infrastructure is destroyed. The digital skills of teachers today are extremely important and require constant updating due to the development of evolving technologies. For Ukrainian teachers, who felt not only the consequences after the long quarantine period of 2020–2022, but also the consequences of military actions that continue to this day, increasing the level of professional skills in the use of ICT is critical. To understand the specific needs of teachers, a public survey was conducted in 2022 and 2023 to identify needs and gaps in their digital skills. The study, based on a public survey of the needs and on the identification of the level of digital competence of Ukrainian teachers, allowed to make recommendations for professional development institutions and educational policymakers in view of the importance of developing digital competence in the context of lifelong learning.

**Keywords:** Teachers' digital competence · Self-assessment of digital competence · Teachers' professional development

## 1 Introduction

Teachers' digital competence refers to key competences according to the European Union Council Recommendation on Key Competences for Lifelong Learning [1]. European countries have adopted national strategies, policies and nationwide initiatives in the digital skills area. They are aimed at promoting and supporting the development of digital skills for citizens, IT specialists, and the workforce in the education systems of EU member states [2]. For Ukraine, the issue of developing the digital competence of teachers became especially acute during the period of military aggression of the Russian Federation, when all schools were not able to return to face-to-face education, and a


M. E. Auer et al. (Eds.): ICL 2023, LNNS 899, pp. 239–246, 2024.
https://doi.org/10.1007/978-3-031-51979-6_25

significant number of students were out of school. During this period, in February 2022, the Ukrainian government suspended education in schools and resumed it on March 14, 2022 in remote mode. Despite the fact that the experience of distance learning during the pandemic left many tools for obtaining an education, they were not effective for everyone. Therefore, the State Education Quality Service, in cooperation with the "Education Quality Assurance System" initiative, which is implemented within the framework of the "Government Reforms Support in Ukraine" (SURGe) project, and the reform support team of the Ministry of Education and Science of Ukraine have developed advice for school leaders on the organization of the educational process in the conditions of war [3]. According to this advice during online classes, teachers should conduct consultations with children, integrated lessons, classes-conversations, polylogues and creative classes. At the same time, in the conditions of war, it is necessary to optimize the educational load. All these recommendations require a good command of digital tools for the organization of distance learning.

It turned out to be especially difficult and important to find out what exactly teachers need, what tools they have for remote lessons, and whether they have enough skills in using new technologies to organize the educational process. Some of the teacher surveys on this topic were conducted even during the COVID-19 quarantine, including, in particular, the survey of the State Education Quality Service of Ukraine, which showed the willingness of teachers to teach remotely [4]. However, it was possible to show a complete picture of the digital competence of teachers thanks to a survey at the beginning of 2022 (January-March 2022), during which teachers self-assessed the level of their own digital competence through an online questionnaire, which will be carried out by the Institute for Digitalization of Education of the National Academy of Educational Sciences of Ukraine. In 2023, this institution also conducted a survey of teachers, and has now received preliminary results for comparison between 2022 and 2023. It should be noted that the level of digital literacy demonstrated by Ukrainian teachers in the spring of 2022 made it possible to ensure a continuous educational process in Ukrainian schools, and such experience during the war is unique for European countries.

## 2 Literature Review

The practical digital skills of teachers are of considerable interest, and many studies are conducted on this topic. Basilotta-Gómez-Pablos, Matarranz, & Casado-Aranda (2022) investigate the publications on the teachers' digital competence surveys in Scopus and WoS in the two decades of 2000's and concluded that despite the multiple efforts of the professional development institutions and universities teachers self-assess that they have a low or medium-low digital competence, also they indicate on the absence of certain competencies related to the application of the digital skills in practice [5].

In 2019 European Framework for Digital Competence of Teachers: DigCompEdu highlighted six competency areas that teachers have to achieve and apply in their everyday practice (Caena & Redecker, 2019). The authors emphasize that teachers are responsible for creating educational digital environments, they should create opportunities for their students to gain experience and develop the ability to use ICT in education and everyday life [6].

In Ukraine, the interest to the development of digital competences of teachers and students has been noted for the last two decades, and especially in recent years, when the whole world felt the COVID-quarantine pandemic. Researchers Shyshkina (2018), Bykov, Spirin, & Pinchuk (2020) touch on the issues of using digital competencies in the educational environment, as well as the need to master the digital skills in the system of teachers' professional development in the context of lifelong learning. In their opinion, designing a digital environment for learning is an important condition for the successful inclusion of students in the learning process, which creates equal opportunities for all participants (Bykov, Spirin & Pinchuk, 2020) [7, 8].

The issue of the development of digital competences of teachers became especially acute with the beginning of the transition to distance learning during the large-scale aggression of the Russian Federation, when schools and teachers were forced to find digital solutions for organizing the educational process. The need to increase the level of digital literacy of teachers in the conditions of war in Ukraine has been highlighted in the works of Ukrainian scientists quite recently: the latest research in this field (Budnyk &Nicolaescu, 2022) analyze social challenges that primarily concern the training of future teachers to use digital technologies in studying [9]. Ivaniuk (2022) highlights educational opportunities for Ukrainian students, parents and teachers who left Ukraine and continue to obtain education through various programs [10]. Also, in cooperation with the online education studio "EdEra", the Ministry of Education and Sciences of Ukraine with the support of the UNICEF Children's Fund, the public association "Osvitoria" launched a free online course on safe education during the war for school teachers, principals and administration [11].

## 3 Research Methods

The survey of teachers was based on public opinion. During the research, quantitative methods were used, which are based on surveys of a certain number of respondents and allow obtaining numerical values. In 2022, 54,254 respondents from all regions of Ukraine took part in the survey (Bykov et al., 2022) [12]. In 2023, 42,708 participants answered the questions of the online questionnaire. This is a fairly significant share of teachers who voluntarily and anonymously expressed their problems and conducted a self-assessment of digital competence. The results of 2023 are currently being processed. A digital competence self-assessment tool was developed based on the Digital Competence Framework for Citizens as a basic tool and standard for people to use in their daily lives (Carretero Gomez, Vuorikari, & Punie, 2017) [13].

The set of questions in 2022 and in 2023 revealed the teacher's digital competence in five areas, such as information and digital literacy, communication and cooperation, creation of digital content, security, and problem solving. At the same time, such levels of digital competence as basic user, independent user, and professional user were applied.

## 4 Research Results

The survey in February-March 2022 revealed that, according to teachers, the implementation of ICT and the use of digital tools in the general secondary education system of Ukraine is not effective enough. In the answers to the open question about the main problems in the implementation of distance learning in schools, the respondents indicated the following: improper access to digital devices, weak provision of high-speed Internet connection, improper management of access to IT infrastructure by educational institutions (Bykov et al., 2022) [13]. In 2023, Ukrainian teachers added the following to the obstacles on the way to the advancement of distance learning: lack of Internet due to intermittent power outages (65.8%), low level of self-organization and motivation of students (41.3%), lack of time due to teacher overload (25, 9%), psychological difficulties during remote initiation (14.5%), air alarms, the facility is located in the occupied territory, the facility is destroyed or damaged. Today, Ukrainian teachers emphasize that they need advanced training. They note that they need to improve the methodology of conducting online lessons, acquire skills for creating educational videos, recording and editing lesson videos, need to familiarize themselves with online tools and services for student creativity and create and maintain their own blog, take courses for elementary school teachers, etc.

An important part of teachers' surveys was their self-assessment of the level of digital competence in the conditions that have developed in the education system. The obtained results in 2022 showed the following. Respondents' self-assessment of their "Information and digital literacy" demonstrated that the majority of teachers are able to search for information at the level of an independent (48.1%) and professional (30.9%) user; assess the reliability of information at the level of professional (44.8%) and independent (24.9%) users; store the information found at the level of professional (41.8%) and independent (32.3%) users' level.

In the field of "Communication and cooperation" pedagogues are able to communicate using various means of communication at the level professional (61.7%) and independent (13.4%) users; create and manage content at the level of independent (53.2%) and professional (19.6%) users; use online services at the level of independent (37.1%) and professional (38.9%) users; know how to use online tools for collaboration at the level of professional (39.2%) and independent (22.8%) users.

In the field of "Creating digital content" the vast majority of respondents are able to create multimedia content in various formats using various digital tools and environments at the level of basic (62.9%) and independent (32.4%) users; can use content formatting functions and various tools at the level of independent (61.9%) and basic (30.2%) users; know the rules for using content in accordance with copyright protection at the basic (46.9%) and independent (39.6%) user levels; have programming skills at the level of basic (72.9%) and independent (21.3%) users. On average, only 8% of respondents have the level of a professional user in this category.

In the field of "Safety", the respondents noted the presence of skills to ensure device and program system protection (basic level–67.3%, independent level–21.1%) and protect personal information on their digital devices (basic level–47.1%, independent level–38%). As well as the ability to use ICT safely for one's own health (professional level-48.9%, independent level-30.4%) and knowledge about the impact of digital technologies on everyday life and the environment (independent level-39.3%, professional level–46.7%).

In the field of "Problem Solving", the professional level of the user is on average, 12% of respondents have, which indicates certain gaps in the system of professional development of pedagogues that need to be improved. Half of the respondents have a basic user level in this area, as evidenced by the following data: the ability to solve problems that arise when using digital technologies (basic level–56.7%, independent level–38.8%); the ability to choose and to use an appropriate digital tool or service to solve non-technical problems (basic level–43.4%, independent level–42.8%); the ability to choose and use an appropriate digital tool to solve technical problems (basic level–55.7, independent level–38.6%); awareness of the need to update skills in the field of digital technologies (independent level-40.4%, basic level-36.3%) [8].

Since the above indicators for 2023 have not yet been calculated, and the processing of the obtained results for this year is ongoing, we now rely on another indicator that allows us to find out whether the general teachers' awareness and competence to use digital tools has increased for the past period. Teachers were asked which online learning resources and platforms they use to provide distance learning with students. We present a comparative table of 2022 and 2923 years (Fig. 1). A comparison shows that teachers began to use these resources more, that is, we assume that their awareness of the possibilities of using educational online resources has increased. It should also be noted that the mentioned resources for the period of 2022 and 2023 were saturated with materials from various subject areas for teachers and began to be in greater demand. Here we should take into account that it was a war period in Ukraine and many of teachers faced with the professional, psychological and other war-related issues in that time.

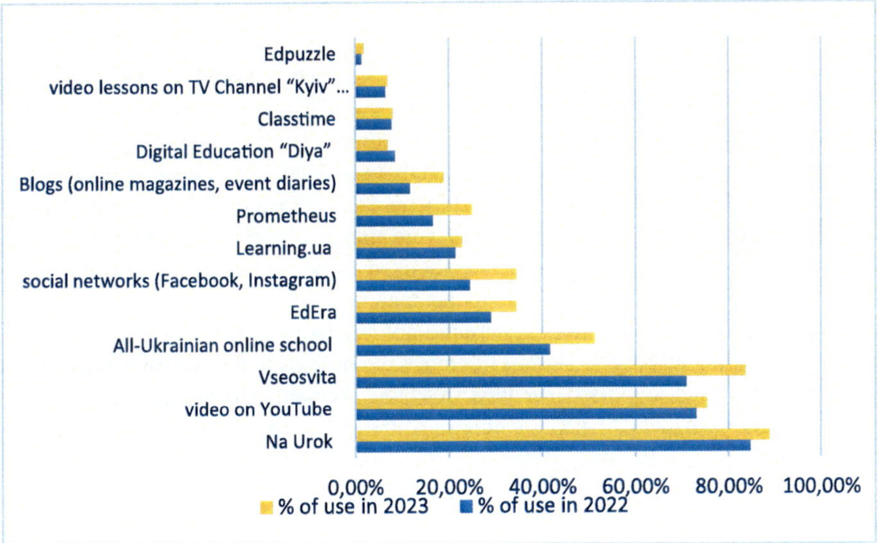


**Fig. 1.** Online resources and platforms that Ukrainian teachers use to conduct lessons (in comparison in 2022 and 2023).

At the same time, in the last years, opportunities for increasing the level of digital literacy have appeared for teachers. These are, in particular, initiatives from The

Ministry of Education and Science and professional education institutions in partnership with the civil society and scientific institutions. Such educational platforms as "Na Urok" (https://naurok.com.ua/), "Vseosvita" (https://vseosvita.ua/), "EdEra" (www.ed-era.com), "Digital Education.Diya" (https://osvita.diia.gov.ua/en) cover a wide range of topics and disciplines and provides practical techniques, workshops on the use of online applications, methods of working with parents, including in an inclusive classroom. Among the online courses and webinars offered to teachers by these platforms, many of them are dedicated to improving the level of digital literacy, for example: inclusive education in the conditions of distance and blended learning; how to create quality educational video content; educational online projects. "Vseosvita" offers about 800 webinars for teachers, including media literacy for teachers, using online tools for learning, workshops and examples of integrated lessons. "Digital Education.Diya" platform provides online handbook, guides, podcasts, tests for teachers aimed at improving their level of digital competence. The specified platforms are the most popular in Ukraine, and which teachers indicated during the surveys in 2022 and 2023.

Such resources create lifelong learning opportunities for teachers, the ability to choose the topics and practices they wish to improve, and the place and time of the learning itself. However, it should be noted that they have not yet become a generally accepted practice for teachers in Ukraine. After all, today there is a network of professional development institutes, which also provide the possibility of taking courses face-to-face and remotely, which remains the main one for teachers today.

## 5 Conclusions

Considering the results obtained during the teachers' survey, it should be noted that such factors as the COVID-quarantine and restrictions associated with the beginning of the Russian military aggression in Ukraine have a significant impact on the development of the digital competence of teachers. Given the global trend towards a decrease in the level of opportunities for teachers to work with students face-to-face in the classroom during 2020–2022, a significant number of teachers were able to master those digital skills that they had not used before. Ukrainian teachers realized the need to acquire new knowledge and competences in the field of ICT use in the context of lifelong learning. At the same time, the professional development system did not have time to fully meet the needs of teachers in acquiring new digital skills. IT companies and civil society organizations have joined the training and supported teachers, providing access to educational materials for teachers and students that they can use online. And this is a new, positive experience for Ukraine today.

It is crucial for Ukrainian teachers to constantly update the spectrum of distance learning tools. Creating conditions for improving their skills in the use of digital learning tools is an important condition for lifelong education. At the same time, revealing the opinion of teachers, determining their level of digital competence and digital readiness should become a strategic task of the Ukrainian school. Therefore, the constant conducting of surveys of teachers, the study of their experience in using digital tools, the introduction of innovations that contribute to the effective organization of the educational process in schools should be carried out consistently and become part of the monitoring of the quality of education as a whole.

It is positive that teachers continue to master new online tools and look for learning platforms that are convenient for themselves and their students, which will allow distance learning; they express readiness to improve their professional level in digital literacy, participate more actively in teacher online communities, and express their readiness to lifelong learning. Therefore, it allows us to express certain recommendations, among which, in particular:

- State authorities at various levels should support the development of action plans, programs, and events aimed at supporting teachers in mastering digital tools and creating conditions for their use in school education;
- It is important to support teachers during the war, where one of the tasks should be digitalization and the creation of a digital educational environment saturated with methodical materials, examples of lessons, instructions for using tools for organizing learning;
- The need for lifelong learning for teachers is an important task, but the psychological state and low willingness to use professional development opportunities during wartime is a problem and an obstacle, so it is important to consider this aspect during training for teachers. The creation of hotlines for counseling teachers can help to overcome daily issues related to the use of online resources by teachers for conducting lessons.

## References

1. European commission homepage. council recommendation on key competences for lifelong learning. https://education.ec.europa.eu/focus-topics/improving-quality/key-competences, Last Accessed 2023/05/03
2. An official website of the European Union. National strategies. https://digital-skills-jobs.europa.eu/en/actions/national-initiatives/national-strategies, Last Accessed 2023/05/03
3. How should a teacher organize his work during the war: recommendations of the State Education Quality Service. https://sqe.gov.ua/yak-vchitelyu-organizuvati-svoyu-robotu-p/?fbclid=IwAR27UOhBQXq3xjWyp6FCaUzt9Qc4e3_9HnmhoIpJxsJq3j-i68U-hgc2NgA, Last Accessed 2023/05/03
4. A third of students in wartime did not have permanent access to education—the results of the study. The official website of the State Education Quality Service of Ukraine. https://sqe.gov.ua/tretina-uchniv-v-umovakh-viyni-ne-mali-po/, Last Accessed 2023/05/03
5. Basilotta-Gómez-Pablos, V., Matarranz, M., Casado-Aranda, L.A., et al.: Teachers' digital competencies in higher education: a systematic literature review. Int. J. Educ. Technol. High. Educ. **19**, 8 (2022). https://doi.org/10.1186/s41239-021-00312-8
6. Caena, F., Redecker, C.: Aligning teacher competence frameworks to 21st century challenges: The case for the European digital competence. Framework for Educators (DigCompEdu). Eur. J. Educ. **54**(3), 1–14 (2019). https://doi.org/10.1111/ejed.12345
7. Shyshkina, M.: The general model of the Cloud-Based learning and research environment of educational personnel training. In: Auer, M., Guralnick, D., Simonics, I. (eds.) Teaching and Learning in a Digital World. ICL 2017. Adv. Intell. Syst. Comput., **715**. Springer, Cham. (2018). https://doi.org/10.1007/978-3-319-73210-7_94
8. Bykov, V., Spirin, O., Pinchuk, O.: Suchasni zavdannya tsyfrovoyi transformatsiyi osvity [Modern tasks of digital transformation of education]. UNESCO Chair Journal "Lifelong Professional Education in the XXI Century" (1) 27–36 (2020)

9. Olena Budnik, Inna Nicolaescu. Digital technologies in the training of future teachers: modern challenges of distance education. Stud. Educ. Didact.. 69, **1**(2), 69–78 (2022), https://czasopisma.marszalek.com.pl/images/pliki/ve/2/ve208.pdf
10. Ivaniuk I. V.: Creation of educational opportunities for Ukrainian students, parents and teachers in Warsaw during the war. Information bulletin No. 2. Kyiv: IDE NAESU (2022). https://lib.iitta.gov.ua/id/eprint/730437
11. A course on safe education during war has been created for teachers. Official website of the Ministry of Education and Science of Ukraine (8 December 2022), https://mon.gov.ua/ua/news/dlya-vchiteliv-stvoreno-kurs-pro-bezpechnu-osvitu-pid-chas-vijni Last Accessed 2023/05/03
12. Bykov B. Ю., Ovcharuk O. B., Ivaniuk I. B., Pinchuk O. П., Galperina B. O. (2022). The current state of the use of digital tools for organization of distance learning in general secondary education institutions: 2022 RESULTS. Information Technologies and Learning Tools, **90**(4), 1–18 (2022). https://doi.org/10.33407/itlt.v90i4.5036
13. Carretero Gomez, S., Vuorikari, R. and Punie, Y.: Dig comp 2.1: The digital competence framework for citizens with eight proficiency levels and examples of use. Publ. Off. Eur. Union, Luxemb. (2017)

# Framework for the Online Education with the Distributed Educational Resources

Galyna Tabunshchyk, Anzhelika Parkhomenko(✉), Sergey Subbotin, Iryna Zeleneva, Tetiana Holub, and Tetiana Kapliienko

National University "Zaporizhzhia Polytechnic", Zhukovskogo Str. 64, Zaporizhzhia 69063, Ukraine

parhom@zntu.edu.ua

**Abstract.** The paper presents solutions for organizing distributed educational resources based on the modern framework and innovative forms of ensuring the competencies of participants in the educational process. The approach to managing the work of distributed teams of developers of educational resources from two countries - Germany and Ukraine based on a flexible Scrum framework has been implemented. The results of the development of several e-learning modules and the organization of the infrastructure of virtual and remote laboratories, which were used by students through cases and problem-solving during pilot training, are shown.

**Keywords:** Distributed educational resources · Framework · Digital educational environment · Virtual and remote labs infrastructure · Virtual masters' network

## 1 Motivation and Purpose

It is known that distributed learning is a term for describing learning culture based on a very diverse collective educational experience of participants who, nevertheless, are united by one goal—the constant improvement of collective knowledge and skills and the desire to share them [1].

A feature of distributed educational resources involved in the process of such learning is that they are created by different specialists, in different spatial and temporal conditions, in different interactive environments, using different approaches and mechanisms. Distributed educational resources are not only educational materials that are concentrated in different places, but also the developers of these resources themselves. Therefore, it is important to support the developers of such resources in their development, continuous improvement, evaluation of their contributions and giving them the opportunity to share their achievements by creating effective integrated educational environments. As studies have shown, digital educational ecosystems have been successfully created in many higher education institutions today. Nevertheless, their fragmentation and closeness remain a problem, which does not allow students to use effectively all the educational resources of partner universities, as well as to overcome language, spatial and temporal barriers on the way to education [2].


M. E. Auer et al. (Eds.): ICL 2023, LNNS 899, pp. 247–254, 2024.
https://doi.org/10.1007/978-3-031-51979-6_26

Hence, an important task is to create an integrated learning ecosystem based on decentralized learning environments, which will allow developers to share learning resources successfully and students to access and effectively use educational content.

The "European Partnership for Project and Innovation Management" program is aimed at shaping the digital future of European and Ukrainian higher education institutions and includes the implementation of several projects, in particular ViMaCs (Virtual Master Cooperation on Data Science) and ViMUK (Virtual Master School Ukraine). These projects will ensure the formation of an integrated digital community of partner universities, further development and modernization of educational programs, digitalization of educational spaces of higher educational institutions [3].

The purpose of this work is to research and develop the framework for organizing digital educational environment based on distributed educational resources that will ensure the effective implementation of the online educational process and quality control of educational services.

## 2 Methodology

### 2.1 Structure and Key Components of the Framework

The framework for online education combines digital educational technologies, teaching methods and distributed educational resources that are created at Dortmund University of Applied Sciences and Arts, National University Zaporizhzhia Polytechnic, West Ukrainian National University, Kyiv National University of Construction and Architecture, National University Lviv Polytechnic. Digital educational environment is deployed on several platforms (Confluence, LMS Moodle) on servers in Germany and Ukraine [4].

To organize the work of distributed teams and structure distributed educational resources, several spaces were organized according to goals and features of two projects.

ViMaCs is the project aimed at creating virtual educational and laboratory infrastructure for digital learning and a portfolio of modules in the field of Data Science. The network unites Ukrainian and European partner universities, and the Dortmund University of Applied Sciences and Arts is its key component. The implementation of ViMaCs involves the development of new Data Science training modules, modernization of existing training programs, pilot training, implementation of IT cluster with the emphasis on the collecting and processing of business data and decision-making based on big data [5].

ViMUk is the project aimed at creating a joint master's program in the form of a virtual cross-border master's school in project management and digital transformation with a collaborative Digital Education Ecosystem (DEE) and collaborative e-learning modules. The central IT resource DEE is a distributed, mirrored back-end infrastructure that provides the distributed and connected LMS-based e-learning system with extra IT tools, in particular cloud-based project management tools.

To create a unified knowledge base and team workspace, the Confluence wiki was used, which provides information for distributed resources developers about module specification template and quality control policies.

The core of the educational environment is the DEE [6, 7], which provides developers and students with access to educational resources, activities and deliveries. The general structure of e-learning modules includes such sections as Didactic concepts, Module description, e-Learning materials, Guide-lines & Templates, Additional information (for example, Knowledge clips), Case studies, Assignments, Competence profile.

## 2.2 Distributed Teams Organization

The development of distributed educational materials provides greater flexibility, the ability to use the best practices of different countries and universities and it allows to create unique content that will be most useful to students. However, the organization of such a distributed structure can be complex, and many problems can arise when creating educational material for online learning with distributed educational resources (Fig. 1).

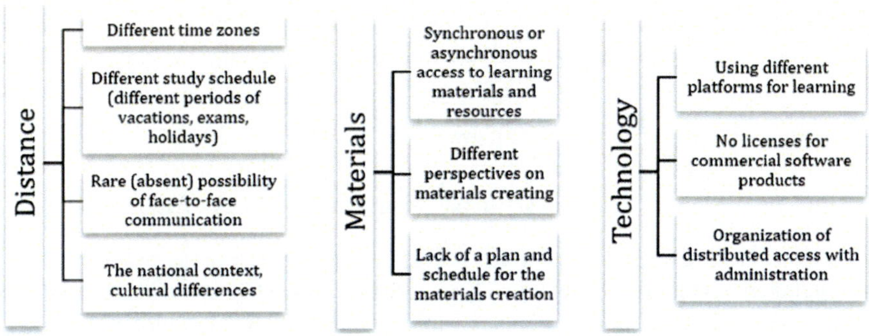


**Fig. 1.** Problems of distributed educational resources development.

To solve a part of these problems and to make the organization of the process of creating distributed learning materials more understandable, some of the project management frameworks for can be used, since project teams in this area are often distributed and include specialists from different countries with different levels of experience and capabilities. One of the most popular agile frameworks is Scrum. Scrum engages groups of people who collectively have all the skills and experience to do work and share or acquire needed skills [8]. Thus, it can be used to organize the efficient creation of distributed learning materials. Scrum uses elements that can help solve a huge part of the problems presented in Fig. 1 and it also offers clear formal events for adaptation and inspection. These events can be used to implement the following empirical Scrum pillars for creating distributed learning content:

- Transparency—each teacher, as a member of the team, knows what part of the materials' creation is his / her responsibility, knows the timing and sequence of tasks.
- Inspection—each teacher has the opportunity to check the current state of the created materials; in the presence of a formal/informal manager, a set of inspection measures is carried out to monitor compliance with the schedule for the development of materials, the quality and effectiveness of materials.

- Adaptation—based on the results of the inspection, it is possible to modify the development schedule, reassign tasks or conduct collaborative activities; this incremental approach helps optimize predictability and to control risk.

The following roles should be involved in the process of creating learning value with Scrum [8]:

- Product Owner(s)—stakeholders, representatives of universities and project managers for the development of educational courses, who are interested in the development and will evaluate the quality of the developed materials.
- Scrum Team—teachers from different universities and countries, who work together (not always geographically) to crate learning materials.
- Scrum Master—a person responsible for the effectiveness of the Scrum Team, coaching team members self-management and creating high-quality mate-rials, causing the removal of impediments to progress, and also organizing Scrum communication activities as productively as possible while respecting time constraints.

During Scrum communication events the Scrum Team and its stakeholders adjust the schedule for creating learning materials and inspect the results.

## 3 Main and Anticipated Findings

### 3.1 Implementation of Module Artificial Intelligence and Data Analytics

The decision-making tasks, as well as identification and forecasting the state of multidimensional objects and processes tasks, arise in various subject areas. In various applied areas such as engineering and economics, biology and medicine, military engineering, it becomes necessary to make decisions based on the data with the insufficient expert knowledge or absence of the analytical models. Despite the difference in the essence of such problems, they can be solved on the basis of the methods of Computational Intelligence, i.e., data-driven machine learning methods [9–11]. These methods don't depend on the applied tasks and can be used in practice for a wide range of problems of a different nature. The key object that Computational Intelligence methods work with is data sample, which is a set of collected observations describing objects or objects' state. Large samples allow researchers to achieve better accuracy and reliability of the models obtained on their basis, but at the same time they significantly slow down the process of model building. Therefore, the task of the sample reduction arises, which is solved by the methods of forming and selecting the samples [12]. On the other hand, the more features that characterize the instances of the sample, the more time it will take to build a model, the more complex the model will be, the more difficult it will be for a person to perceive it. Therefore, it is necessary to reduce the number of features, which requires the use of methods for evaluating and selecting features [13].

The Module "Artificial Intelligence and Data Analytics" aimed to address these needs for such target groups as Master students in various applied domains, practical specialists in programming and professionals in applied domains, applied researchers, university teachers who need new knowledge in data analytic for teaching students. This module aims to familiarize students with the basic concepts of data analysis and artificial intelligence, outline key techniques, and introduce the basics of Python programming with the focus on learning data analysis and machine learning libraries. The module provides an overview of the existing groups of data-driven methods in the context of tasks solving of constructing diagnostic and recognition models. After completing the module, the student will be able reasonably choose the most suitable method for solving a specific applied problem, to select and use methods of data dimensionality reduction, including feature selection, feature extraction (feature construction), instance selection and sampling. The module overall learning outcome is that the students will be familiarized with methods of data analytics, they will also receive both theoretical and practical knowledge to use the methods and to develop software for data analytics for various problem specific domain.

The module content is provided as a set of educational components (submodules) that can be studied selectively, taking into account the needs of the student and his/her previous knowledge and skills. It includes such submodules as Introduction into Python, Data analytics with Python, Natural language processing with Python, Computational intelligence for technical and medical diagnosis, Data samples for intelligent model building, Supervised machine learning basics in Python, Data hashing transformations for dimensionality reduction in diagnostic and recognition problems, Quality indicators of decision tree and forest based models, Thermal imaging as a data source for analysis in applications.

### 3.2 Implementation of Virtual and Remote Labs Infrastructure

The studying of the modules Cyber-physical systems and Design of IoT systems requires the organization of practical experiments with the components, protocols, software and hardware platforms used in the development of such systems. Therefore, the organization of the infrastructure of virtual, remote and hybrid laboratories based on technologies and tools of online engineering for prototyping and researching of reusable solutions has become an urgent task [14].

The analysis showed that when studying cyber-physical systems, a number of experiments with real control units and physical systems, as well as with their virtual models, can be carried out using the hybrid laboratory GOLDi (Grid of Online Lab Devices) [15]. For practical study of the development of hardware and software for cyber-physical and the IoT systems, a remote laboratory RELDES (Remote Laboratory for Development of Embedded Systems) can be recommended [16]. To study the issues of energy efficiency, security and cybersecurity of such systems, a number of experiments of the Smart House & IoT hybrid laboratory (Fig. 2) are available [17].

The cycle of laboratory works can be based on the usage of Simuli Virtual Lab and Wokwi online simulator, which are implemented as web services. These are powerful and convenient tools for creating prototypes of the IoT devices based on Arduino Uno, STM32 Nucleo F411RE, Raspberry Pi software/hardware platforms, as well as various sensors and actuators. They offer many ready-made examples, and also provide an opportunity to share ideas with other participants in the educational process.

It is also proposed to use free cloud platforms and the IoT services, in particular, Thinger.io and Adafruit.io to perform laboratory work. Thinger.io is a cloud-based IoT platform that contains all the tools students need to easily prototype, scale, and control

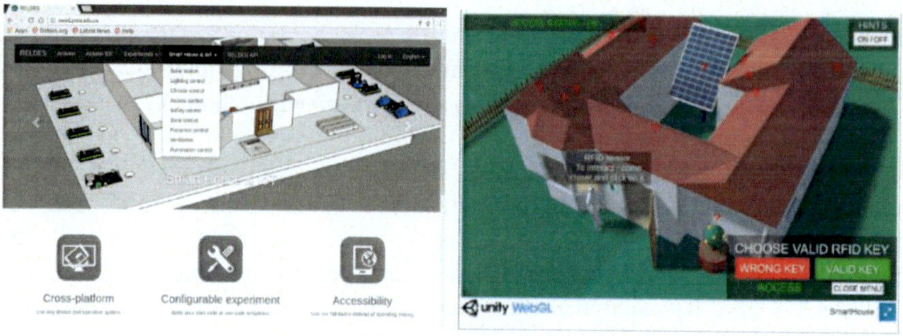


**Fig. 2.** Web interface of hybrid laboratory Smart House & IoT.

connected devices. Adafruit.io is a cloud service for receiving and storing data, as well as visualizing it in real time. It allows students to create a project, connect it to various web services (for example, weather services) and other devices.

Thus, the usage of online engineering tools will provide students with the necessary knowledge and skills for the implementation of their projects in the area of cyber-physical and the IoT systems, as well as for the successful application of acquired knowledge and skills in future professional activities.

### 3.3 Implementation of Hardware Design Module

The hardware design module is an important part of the overall curriculum. This module includes such courses as Designing Microprocessor Systems, Theory of Digital Systems Synthesis on FPGA, Reliability of Computer Systems. The set of materials is selected in such a way that students can master the basic methods of designing functional units of a computer, as well as methods for assembling a project from components.

Lectures are organized remotely, equipped with presentations and they are not monotonous, but in the form of a dialogue with the audience. From time to time, students participate in a discussion that activates their thinking and supports communication.

The laboratory workshop allows students to master all the main stages of designing modern digital equipment: theoretical preparation of the project, development of structural and functional diagrams, odelling and testing of the project. In the laboratory workshop, students learn how to develop projects using hardware description languages, such as VHDL and Verilog, which are widely used in practice. Simulation is performed on chips from top manufacturers such as Altera/Intel, Xilinx. At the same time, students learn how to work with the appropriate IDE Quartus II and Vivado environments. Attention is also paid to the problems of improving the reliability characteristics of software and hardware. Such tasks are taken into account when creating Cases, as well as when developing tasks for problem-based learning (PBL), which are performed in groups of 3–4 students. Students receive problem-based topics, such as how to improve circuit-on-a-chip topology; how to implement several independent devices on one chip; how to increase the reliability of the scheme by redundancy and at the same time find the minimum sufficient reserve.

When working on Cases, students can compete in achieving the best results. The competition is held between groups, with students choosing a project manager in each group. This approach contributes to the support of students' communication skills, the development of creative thinking, which is especially important in the context of distance learning. In addition, students learn to work in a team, form joint decisions, and find ways to exchange information effectively. This, in turn, makes it possible to form teams of students from different universities.

In the process of PBL research, students often form interesting solutions, which are used in the further research of masters and graduate students. Thus, students are involved in the scientific work of the department. The result of such work is, for example, a Hardware-Software System for accelerated text processing using FPGAs [18, 19].

All theoretical materials and presentations, as well as tasks for control and individual work, are available on the Moodle platform. This allows all students, regardless of circumstances and location, to master the techniques and solve the tasks. For discussions, consultations, and work on joint projects, such interactive platforms as Zoom, Google Meet, Google Team are used. Thus, the effect of being in a team is achieved, which is necessary for the quality work of students in distance learning.

## 4 Conclusion

The developed modern framework helped to cope with the challenges and intercultural differences in pedagogical approaches and methodologies for the development of educational resources. The implementation of the digital educational environment based on the developed framework will contribute to the organization of cross-border learning. Thus, students can study from almost anywhere in the world, and teachers can teach from any country in the world and provide high quality educational services.

**Acknowledgements.** This work is partly carried out with the support of DAAD projects ViMaCs (Virtual Master Cooperation on Data Science) and ViMUK (Virtual Master School Ukraine).

## References

1. Rahman, H.: Collaborative learning: an effective tool to empower communities. In: Handbook of Research on e-government readiness for information and service exchange: utilizing progressive information communication technologies. Information science reference, 588 p. (2010)
2. Otto, D., Scharnberg G., Kerres M., Zawacki-Richter, O.: Distributed learning ecosystems. Concepts, resources, and repositories, 304 p. Springer VS (2023)
3. EuroPIM, https://www.fh-dortmund.de/projekte/europim.php.
4. Wolff, C., Tabunshchyk, G., Arras, P., Otegi, J. R., Bushuyev, S., Verenych, O. et al.: Cross-border projects in digital education ecosystems. In: Mobility for Smart cities and regional development—challenges for higher education. Lect. Notes Netw. Syst. **389**, pp. 382-394. Springer, Heidelberg (2022)
5. ViMaCs, https://www.fh-dortmund.de/projekte/vimacs.php/

6. Wolff, C., Reimann, C., Mikhaylova, E., Aldaghamin, A., Pampus. S., Hermann E.: Digital education ecosystem (DEE) for a Virtual master school. In: 2021 IEEE International Conference on Smart Information Systems and Technologies (SIST), Nur-Sultan, Kazakhstan, pp. 1–7 IEEE (2021)
7. Mikhaylova, E., Aldaghamin, A., Ebberg, F., Tokanov, O., Wolff, C., Reimannn, C.: Digital education ecosystem (DEE): user-centred design of the student journey configurator. In: IEEE International Conference on Smart Information Systems and Technologies (SIST), Nur-Sultan, Kazakhstan, pp. 1–8 IEEE (2021)
8. The 2020 Scrum Guide TM, https://scrumguides.org/scrum-guide.html
9. Calin, O.: Deep learning architectures. A mathematical approach, 760 p. Springer, Heidelberg (2020)
10. Tabunshchyk, G., Subbotin, S., Arras, P., Trotsenko, E.: Intelligent data analysis for individual hypertensia patient's state monitoring and prediction. In: 2021 IEEE International Conference on Smart Information Systems and Technologies (SIST), Nur-Sultan, Kazakhstan, pp. 1–4. IEEE (2021)
11. Rabcan, J., Levashenko, V., Zaitseva, E., Kvassay, M., Subbotin, S.: Application of fuzzy decision tree for signal classification. In: IEEE Transactions on Industrial Informatics, **15**(10), 5425–5434 (2019)
12. Subbotin, S. A.: Evaluation of informativity and selection of instances based on hashing. In: Radio electronics, computer science, control, **3**, 129–137 (2020)
13. Ferri, F.J., Pudil, P., Hatef, M., Kittler, J.: Comparative study of techniques for large-scale feature selection, https://citeseerx.ist.psu.edu/viewdoc/download?doi=10.1.1.24.4369&rep=rep1&type=pdf
14. Parkhomenko, A., Gladkova, O.: Virtual tools and collaborative working environment in embedded system design. In: International Conference on Remote Engineering and Virtual Instrumentation (REV), Porto, Portugal, pp. 90–93. IEEE (2014)
15. Poliakov, M., Larionova, T., Tabunshchyk, G., Parkhomenko, A., Henke, K.: Remote laboratory for teaching of control systems design as an integrated system. In: 3th International Conference on Remote Engineering and Virtual Instrumentation (REV 2016), Madrid, Spain, pp. 339–346. IEEE (2016)
16. Parkhomenko, A., Gladkova, O., Kurson, S., Sokolyanskii, A., Ivanov, E.: Internet-based technologies for design of embedded systems. In: 13th International Conference on the Experience of Designing and Application of CAD Systems in Microelectronics (CADSM), Lviv, Ukraine, pp. 167–171. IEEE (2015)
17. Parkhomenko, A., Tulenkov, A., Sokolyanskii, A., Zalyubovskiy, Y., Parkhomenko, A.: Integrated complex for IoT technologies study. In: Online engineering & Internet of Things. LNNS, **22**(31), pp. 322–330. Springer, Cham (2017)
18. Golub, T., Zeleneva, I., Hrushko, S., Pavlishin, M.: A subsystem for hardware acceleration of text classification in the FPGA base. In: Academic notes of the Tavria National University in the name of V.I. Vernadskyi. **31**(70), 2, pp. 73–79 (2020)
19. Golub, T., Zeleneva, I., Hrushko, S.: Software method of preparing text data for their hardware processing using FPGA, In: International Scientific-practical Conference on Science, Engineering and Technology: Global Trends, Problems and Solutions, Prague, Czech Republic, pp. 16–20 (2020)

# Online Educational Courses Implementation in Technical Universities

Aigul Mendygalieva[1], Julia Lopukhova[2], Elena Makeeva[2,3](✉), and Natalia Strekalova[4,5]

[1] Makhambet Utemisov West Kazakhstan State University, Uralsk, Kazakhstan
[2] Samara State Technical University, Samara, Russian Federation
helenmckey2205@gmail.com
[3] Samara State University of Social Sciences and Education, Samara, Russian Federation
[4] Samara National Research University, Samara, Russian Federation
[5] Togliatti Academy of Management, Togliatti, Russian Federation

**Abstract.** This article focuses on the problem of rethinking the use of online education courses when teaching technical university students in social, humanitarian and major subjects. The authors examine both positive and negative aspects of introducing online educational courses (OECs) into today's educational process. The paper considers three types of online educational courses: Massive Open Online Courses (MOOCs), Shared Educational Online Courses (SEOCs) and University Internal Online Courses (UIOCs), which could successfully improve the existing system of teaching students both in Russia and Kazakhstan. The authors also rely on laws and orders regarding the introduction of online educational courses in university environment and approved by federal legislation and offer their vision for the implementation of these orders. The paper further describes the researchers' experiment on introducing online courses and blended learning into the educational process of two universities (Samara State Technical University—SamGTU and Makhambet Utemisov West Kazakhstan State University—WKSU). The authors draw the following conclusions: on the one hand, the use of online educational courses is mandatory for the successful existence of any educational organization in modern reality. On the other hand, OECs (MOOCs, SEOCs, UIOCs) used now as substitutes for some parts or modules of traditional face-to-face courses (in purely remote format) are not suitable for all students. Most students are ready to master online educational courses only under the guidance of teachers and prefer a blended learning format. When choosing between MOOCs, SEOCs, UIOCs, they give preference to UIOCs.

**Keywords:** Online Educational Courses (OECs) · Electronic Educational Recourses (EER) · Massive Open Online Courses (MOOCs) · Shared Educational Online Courses (SEOCs) · Distance Education · Online Learning

## 1 Introduction

Being a part of a broader international system of higher education, most universities both in Russia and in Kazakhstan are under pressure due to rapid technological changes and ongoing integration into the global economy. Information society development and


M. E. Auer et al. (Eds.): ICL 2023, LNNS 899, pp. 255–266, 2024.
https://doi.org/10.1007/978-3-031-51979-6_27

widespread dissemination of information technologies open up new opportunities for learning. Still, at the same time they challenge conventional views on teaching and learning and on developing students universal and professional competences. Most Russian and Kazakhstan university students attend traditional face-to-face lectures and seminars. Meanwhile, they are not always trained adequately, or do not always gain real experience and practical knowledge. To resolve this problem, university teachers and authorities are trying to find effective teaching tools that would help master theoretical knowledge and, at the same time, free up enough time for its practical application.

The paper considers three types of online educational courses: Massive Open Online Courses (MOOCs), Shared Educational Online Courses (SEOCs) [1] and University Internal Online Courses (UIOCs), which could successfully improve the existing system of teaching students both in Russia and Kazakhstan.

The authors make a special note that online educational courses (OECs) differ from online educational resources (OERs) in terms of their completeness. It means that OERs include "all digitized materials offered freely and openly to teachers, students and self-learners for use and reuse in teaching, learning and research" [2] and students and learners can use them from time to time, not even regularly. OECs, in their turn, are characterized by a clearly formulated goal, objectives and results, that is, by a certain completeness offered by this or that particular course. OECs, in their turn, are divided into Massive Open Online Courses (MOOCs), which are publicly available to everyone who wants to study, into Shared Educational Online Courses (SEOCs), which are closed and can be accessed only within the framework of network agreements between universities [1] and University Internal Online Courses (UIOCs). UIOCs are similar to MOOCs and SEOCs in structure and form but are not available from outside as such courses are developed within the university's online educational environment and are intended to be used as online substitutes for definite face-to-face courses traditionally taught by this certain university. Thus, the authors' OECs classification is based on the target audience of these online educational courses.

Thus, MOOCs are hosted on open educational platforms such as Coursera, EdX, Open Education, Lectorium, etc., while SEOCs are hosted on private university platforms and, as a rule, access to such courses is not open to a wide range of users, and in most universities, they are implemented only within the framework of a network agreement between universities. UIOCs, in their turn, are completely closed to external audiences. At the same time, both MOOCs and SEOCs have the format of exclusively remote online education, which is actively promoted by both the Republic of Kazakhstan and the Russian Federation. On February 8, 2019, The Ministry of Education and Science of the Russian Federation approved the "Action Plan of the Ministry of Science and Higher Education of the Russian Federation for the period from 2019 to 2024". In this Plan, Output cluster 2.1. Named "Ensuring the global competitiveness of Russian education" of the National Project "Education" includes a very important Subsect. 2.1.2 named "Development of online education" indicates the following expected result: "By 2024, at least 20% of students in higher educational programs should master their individual courses, disciplines or modules in the format of online courses, using the resources of other organizations" [3]. In turn, in the Republic of Kazakhstan, such documents as the Law of the Republic of Kazakhstan "On Education" dated July 27,

2007 (subparagraph 25, Article 5) [4], "Rules for organizing the educational process on distance educational technologies" (Order of the Minister of Education and Science of the Republic of Kazakhstan dated March 20, 2015 No. 137 with amendments, dated November 3, 2021, Order No. 547) [5] and "Model rules for the activities of educational organizations implementing educational programs of higher and (or) postgraduate education" Order of the Minister of Education and Science of the Republic of Kazakhstan dated October 30, 2018 No. 595; with amendments and additions (Order No. 207 dated 05/18/2020) [6] also strictly regulate the activities of educational institutions in the field of online education and technology. Thus, educational organizations have set clear tasks: to create and develop technological infrastructure of the online learning system, to implement measures to popularize the introduction of online technologies, to master online courses of other organizations. Thus, educational organizations have clear tasks: to create and develop technological infrastructure of the online learning system, to implement measures to popularize the introduction of online technologies, to master online courses of other organizations.

Another federal act, Order No. 816 of the Ministry of Education and Science of the Russian Federation dated August 23, 2017 "On Approval of the Procedure for the use of e-learning, distance learning technologies by organizations engaged in educational activities in the implementation of educational programs" [7] specify exactly how the introduction of e-learning should take place in an educational organization. E-learning, including the use of distance learning technologies, is implemented in universities in almost all areas of training. In correspondence courses or in part-time education e-learning is used to even a greater extent. Since recently there has been a trend of widespread coverage of educational organizations with e-learning, distance learning technologies have to be used not only in correspondence and part-time education, but in full-time education, too.

## 2 Research Background

The spread of the Internet and digital technologies partially erased the boundary between the concepts of online learning and distance learning. Online learning (synchronous and asynchronous) appeared long after distance learning and became its logical continuation with the development of the Internet and digital technologies. The researchers still see the difference and stress that both distance learning and online learning take place outside the classroom, but remote and asynchronous online learning is carried out according to the student's own rhythm and schedule without direct contact with the teacher. For synchronous online learning though, live communication between the teacher and students in real time is vital.

Thus, the authors are sure that the concept of "distance learning" is broader than the concept of "online learning". Distance learning includes online learning (synchronous and asynchronous). At the same time, modern distance learning can not exist without Internet technologies. Therefore, at the moment these two concepts have become as close as possible and in the very near future might merge completely.

Online learning is often used in the context and in connection with such concepts as e-learning and digital learning. They indicate the student's ability to acquire knowledge in various formats: audio, video, hyperlinked text, infographics, various programs, applications, platforms, games, tools and materials for acquiring knowledge through augmented reality, and much more. On the one hand, digital learning and e-learning can be considered synonymous. On the other hand, the authors believe that they differ from each other. Digital learning includes the use of any digital technologies that can help students comprehend knowledge and expand their skills. This may include the use of apps, games, social media, videos, and other digital learning tools. Digital learning exists in both online and offline education. E-learning, on the other hand, is a specific type of digital learning that involves the use of a learning management system (LMS) to deliver learning remotely in the form of online courses with different duration, including specific training materials, sometimes involving interactive elements to assess understanding. However, these terms are not always strictly limited as digital technologies can be used in various learning formats, including electronic. Online educational courses in the form that we all know and use now, appeared in the late 2000s, and the peak of their development came in 2012, when leading international platforms with open online courses were created. They still remain leaders in the market of educational services to this day: Coursera, EdX, Udacity, Khan Academy, etc. Russian educational platforms, such as Open Education, Lecture Hall, Universarium, etc., began to develop at about the same time, but their substantial and quantitative content lagged significantly behind their foreign counterparts at that time.

The authors realize that the transition of education to the Internet has generated a demand for new educational services. Nevertheless, mass online education is criticized both within and outside the academic community. Most discussions hold today on the use of open online educational resources, which include such forms as SEOCs and MOOCs, relate to the fact that there is no clear understanding of the role and place of SEOCs and MOOCs in the modern educational process.

MOOCs are also often criticized for the fact that this format does not provide educational motivation for students, and as a result, an extremely low percentage of students complete their online courses. The share of students who have successfully completed their MOOCs is below 10% [8]. MOOCs creators [9–11] realize that only a half of the students who signed up for a course actually study regularly, and only a small part of this half complete the entire course. One of the reasons for this situation is that some people sign up for the course out of curiosity, out of a desire to see the content of lectures and evaluate the forms of content presentation [12]. This part of the audience does not plan to complete the course at all.

This research reveals that the vast majority of Russian and Kazakhstan university teachers (from 100 involved) firmly believe that MOOCs are not capable of becoming a substitute for traditional education, because they form a so-called "commonplace thinking patterns» and give superficial knowledge as only full-time students have the opportunity to ask questions and clarify important issues while addressing to their teachers live. There is no such possibility when students attend an online course. Besides, the specific university atmosphere cannot be reproduced in open online educational courses.

As for SEOCs, the proportion of students who successfully complete courses hosted by other universities is much higher. This is explained by the format of the implementation of these courses into the educational process. This option involves conclusion of an agreement between universities, while students' mobility is only virtual in practice. In the further parts of the paper, the authors will describe the experience of implementing these courses in more detail.

A comparative study of MOOCs and SEOCS raises the question of the format of using online educational courses in the educational process. Thus, T.V. Semenova and K.A. Vilkova distinguish three main types of inclusion of online courses in educational programs: (1) embedding MOOCs in a mixed learning format, (2) replacing part of the full-time courses of the educational program with online disciplines, (3) creating an online master's program in which all courses are taught as MOOCs [13] and, in this research, as SEOCs, as well. N.V. Shcherbakov and S.S. Kirillova describe three main models of online education: "Model 1 "Blended learning using online course materials within the discipline": an online course is considered as an auxiliary material for a certain academic discipline. Model 2 "Online course accompanied by a teacher": an online course that was designed by teachers of the same educational organization which implements this academic discipline. Model 3 "Independent development of an online course with tutoring support": an online course that was developed by another educational organization and is implemented within the framework of the agreement on the network form of implementing the educational program" [14].

The authors believe that these points of view are complementary as they highlight the question of the effectiveness of using external (created by teachers of other universities) and internal (created by teachers of their own university) educational online courses. Based on this assumption, the paper further describes the experience of using external and internal educational courses.

## 3 Materials and Methods

In this work, the authors used such empirical research methods as observation and comparison, as well as theoretical methods including abstraction and analysis and synthesis.

Observation and comparison helped to go deep into modern educational process which involves traditional methods together with online educational courses and e-learning, to study its history and analyze existing problems and difficulties. Analysis and synthesis made it possible to decompose the process into its components and simulate the situation of using online educational courses. Abstracting, in its turn, made it possible to point out both positive and negative aspects of using online educational courses and e-learning.

## 4 Discussion

This part of the study examines the factors contributing to the creation of an optimal model for the use of external and internal online educational courses in the educational process of Samara State Technical University when they are integrated into a blended learning format or replace a part of face-to-face courses.

To create such a model, the authors developed an algorithm for using online educational courses designed by other universities into the educational process:

1. Defining goals and objectives: Before you start using online courses, it is necessary to determine which specific goals should be achieved with their help. For example, it can be a course supplement, additional material for students to work independently, some students can use the course as a basis for creating their own course, etc.
2. Search and selection of appropriate courses: Various resources can be used to search and select appropriate online courses, such as MOOC course catalogs, catalogs of courses from other universities, recommendations from colleagues, etc.
3. Course adaptation: After selecting courses, it is necessary to assess their compliance with the curriculum, program, and student needs. If necessary, the course can be adapted, for example, to remove unnecessary topics or add material that is necessary for the curriculum.
4. Integration of courses into the learning process: after adapting the courses, they can be integrated into the learning process. This can be both additional material and the basis for creating your own course.
5. Evaluation and analysis of results: after completing the use of online courses, it is necessary to evaluate the results and analyze their effectiveness. This can be done by interviewing students, analyzing test results, etc. In general, the model of using online educational courses of other universities in the educational process is a comprehensive approach to integrating such courses into the academic program, which allows to optimize the learning process as much as possible and increase the effectiveness of student learning.

Figure 1 presents the algorithm of using online educational courses of other universities in the educational process.

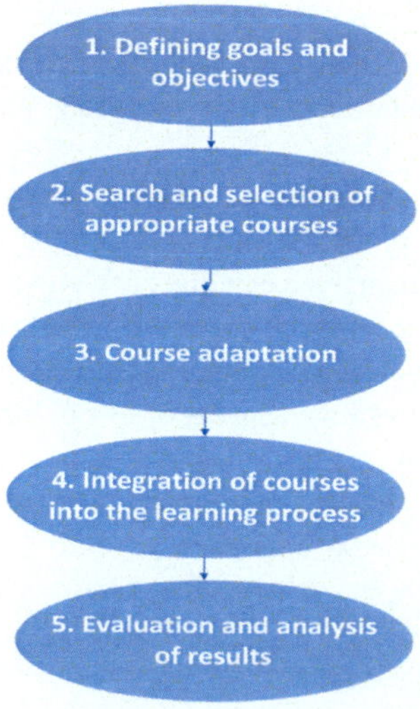


**Fig. 1.** The algorithm of using online educational courses of other universities in the educational process.

The authors have extensive experience in using online courses of varying types including those designed by other universities. In this regard, both practice and questionnaires show that teachers cannot just start simply using a certain ready-made course. There is a lot of additional work to be done, e.g. choosing the most suitable online courses, analyzing their content and assessing their compliance with curricula and programs, organizing students work with the chosen online courses, monitoring their progress and evaluating final results. All that is a labor-intensive process that requires additional effort and time on the part of teachers, which significantly reduces interest in using external online educational courses. And since a university teacher today is not only required to carry out educational and methodological work, but also to undertake scientific research there is practically no time left for the teacher to thoroughly analyze hundreds of ready-made online courses.

The experiment was conducted at Samara State Technical University (over 6 terms of 2020/2021, 2021/2022, 2022/2023) and at Makhambet Utemisov West Kazakhstan State University (over 2 terms of 2022/2023) and involved monitoring the use of educational online courses in social, humanitarian, and technical subjects. These subjects were traditionally taught as full-time courses, but some modules of these courses were replaced with MOOCs, SEOCs and UIOCs. Blended learning was also applied in some of these subjects. Such documents as "On the use and offset of the results of mass open educational courses at Samara State Technical University" [15] and "Rules for the development of mass open online courses at Makhambet Utemisov West Kazakhstan State University" [16] served as the legal basis for the experiment.

According to these regulations, any online course applied in these universities should meet the following requirements:

- MOOC/SEOC working hours coincide by at least 80% with the number of working hours of the corresponding subject (module).
- MOOC/SEOC content is aimed at building competencies developed by the corresponding subject (module). If these requirements are not met in full, the MOOC/SEOC can only be used to substitute a part of the corresponding subject (module) and combined with traditional in-class work. The list of MOOCs/SEOCs recommended for the use in SamSTU is approved annually by the decision of the university educational and methodological council.

The list of subjects in which some modules previously taught only in class were replaced by MOOCs / SEOCs / UIOCs (in purely remote format) during the experiment is as follows:

- MOOC: history (developed in St. Petersburg), life safety (Moscow), Russian language and speech culture (Yekaterinburg), descriptive geometry and engineering graphics (St. Petersburg).
- SEOC: business communications (Moscow), history of the oil industry (St. Petersburg).
- UIOC: business communication, philosophy, descriptive and computer graphics.

Blended learning was applied in teaching the following subjects.

- SEOC: life safety (75% of the course (lectures and practical training, developed in Yoshkar-Ola)—blended learning, 100% of laboratory works—in person), descriptive geometry and computer graphics (100% of lectures—blended learning, Yoshkar-Ola).
- UIOC: Russian language and culture of communication (80% of lectures—blended learning, 100% of practical training—in person), history, philosophy, psychology (50% of lectures—blended learning, 100% of practical training—in person).

Makhambet Utemisov West Kazakhstan State University used the only MOOC (developed in Moscow) as a part of Organic Chemistry course.

About 100 students of SamSTU and WKSU were trained at each of the above-mentioned courses. Table 1 summarizes the experiment results.

**Table 1.** Experiment results.

| Courses delivered using MOOCs/SEOCs/UIOCs at SamSTU and WKSU | Replacement model for part of full-time courses (in purely remote format) | | | Blended learning model | | |
|---|---|---|---|---|---|---|
| | MOOC | SEOC | UIOC | MOOC | SEOC | UIOC |
| History (Saint Petersburg) | + | | | | + | + |
| Life safety (Moscow) | + | | | | + | |
| Russian language and speech culture (Yekaterinburg) | + | + | | | | + |
| Business Communications (Moscow) | | + | + | | + | |
| Philosophy | | | + | | | |
| Psychology | | | | | | + |
| Organic chemistry | + | | | | | |

After completing the courses, their effectiveness was assessed according to such criteria as: the percentage of students who successfully completed their studies, students satisfaction with the materials presented, the amount of time required to study the material, its complexity, excessive theorizing or, conversely, ease, incompleteness, the quality of evaluation materials and transparency of evaluation, the presence / absence of feedback from the course teachers, the presence / absence of operational communication with technical support, the willingness of students to recommend the course to other students [17]. For this purpose, more than 1,000 students and 100 teachers of SamSTU and WKSU completed online questionnaires. Let's look at each indicator in more detail.

Percentage of students who successfully completed their studies. The analysis of the courses completed by students showed that in case when a face-to-face course was completely substituted with an inline course, the number of students who completed it in the MOOC format was 15%, in SEOC format—78% and in UIOC format—80%. In case this course was implemented in a blended learning format, the number of students who completed MOOCs increases to 35%, SEOCs—up to 90%, UIOC—90%. In order to understand this dependency, it is necessary to keep asking students (during the period when they study the course) what their goal is. If the goal of studying is only to get acquainted with the material, then the student is not interested in completing this OEC. When students realize that the course objective is not only learning new material, but also their mid-term assessment, then percentage of students who successfully complete their studies increases.

Students' satisfaction with the presented material. As a rule, the educational material of online courses consists of video lectures, presentations, a list of sources where students can get comprehensive information on the topic, quizzes and tests, etc. At the same time, questionaries showed that this information is not always enough when students want to study some course modules in more detail, for example, when it is necessary to clarify terminology, find a detailed solution to problems, etc. In this regard, UIOCs become more accessible to students, since the sources indicated in them have links directly related to the electronic library system of the university. The sources cited in MOOCs and SEOCs are not that easy to find and access. In addition, in a UIOC blended learning format, there is a possibility of direct communication between students and their teacher and students can discuss some issues in person. The advantage of UIOCs here is also manifested in the fact that students, being in direct contact with their teacher, are able to influence the content of the course. If the teacher sees that most of the students are not coping with some topics, tasks, tests, he can promptly change or correct the task in real time. SEOCs developers make such transformations of courses only if the statistics of several groups of different universities produce negative results.

The amount of time required to study the material, its complexity, excessive theorizing or incompleteness Not everyone is able to maintain the necessary pace of learning, complete the tasks provided for in the schedule on time and revise the training material. Students, as a rule, skim through the theory, mainly especially through presentations, and immediately begin to complete tasks. Students return to the theoretical material only when they face difficulties in performing any tasks or tests. And again, UIOCs the advantage here as university teachers can quickly make changes to the process of studying, extend deadlines (if possible), remove, or add materials, simplify or complicate their submission.

Quality of evaluation materials and transparency of evaluation. From the very beginning, students should know what they need to do to move on to mastering a new section of the course, what assessment tools are waiting for them and what the criteria for evaluating tasks are (minimal percentage required of passing the test, how many times it can be passed, how much theoretical materials need to be studied to move on to a new test, etc.). It is desirable that this information be presented at the beginning of the course. If this is not the case, then the evaluation materials may turn out to be "unpleasant" surprises when attending SEOCs and MOOCs, especially. With UIOCs, students can address such questions directly to their teacher. A good online course always contains a clearly defined assessment structure and different types of assessment materials. A significant disadvantage of all online courses is that their evaluation tools can only test students' knowledge. Practical skills can be evaluated only when students perform direct actions related, for example, to their professional activity.

Presence/absence of feedback from the course teachers. Thanks to feedback, students can interact with their teacher, who in turn must promptly monitor and analyze the information received, if necessary, using it to adjust the course. Of course, each online course provides feedback from its developer, which is true for MOOCs, SEOCs and UIOCs, as well. Students who mastered MOOCs and SEOCs, including SEOCs in a blended learning format, say that communication with the counsellor is quick, but the answers are mostly of a navigational or clarifying nature. UIOCs have more advantages in this regard since students can always meet their teacher and ask all necessary questions.

Availability / absence of operational communication with technical support. This indicator caused the same reaction of students of all courses, both in distance and blended formats since there were no complaints about technical support from students. All issues were resolved promptly and competently.

Students' willingness to recommend the course to other students. This indicator directly depended on whether students completed the course and how interesting and accessible course materials were. The survey showed that students recommend MOOCs to their fellow students only if they had no problems with the teacher responsible for this course in their university and with their mid-term assessment. In general, students do not think much of MOOCs and SEOCs and prefer live communication with teachers. Students unanimously choose and advise UIOCs in a blended learning format when they are sure that the material presented in the online course and presented by the teacher face-to-face is of the same quality. If this teacher is charismatic, intelligent and interesting as a person, students choose to study his course only in full-time format. Table 2 briefly summarizes these results.

## 5 Conclusion

The research yielded the following conclusions: on the one hand, the use of online educational courses is mandatory for the successful existence of any educational organization in modern reality, since they have the potential not as a substitute for conventional education, but as an addition to it. On the other hand, OECs (MOOCs, SEOCs, UIOCs) used now as substitutes for some parts or modules of traditional face-to-face courses (in purely remote format) are not suitable for all students. They are designed for effective independent learners who are trained to choose the content they need and can use OECs as a form of continuous learning and professional development. The vast majority of students are ready to master online educational courses only under the guidance of teachers and prefer a blended learning format. When choosing between MOOCs, SEOCs, UIOCs, they give preference to UIOCs as students can always meet their teacher and ask all necessary questions.

The authors plan to continue their research and develop a model of effective interaction of the educational process participants through the introduction of online educational courses on a large scale.

**Table 2.** Effectiveness of using external and internal online educational courses.

| Effectiveness of using external and internal online educational courses (number of students -100 per course) | Replacement model for part of full-time courses for MOOCs (in purely remote format) | | | Blended learning model | | |
|---|---|---|---|---|---|---|
| | MOOC | SEOC | UIOC | MOOC | SEOC | UIOC |
| Percentage of students who successfully completed their studies | 15 | 78 | 80 | 35 | 90 | 90 |
| Students' satisfaction with the presented material | - | ± | ± | ± | ± | + |
| The amount of time required to study the material, its complexity, excessive theorizing or, conversely, ease, incompleteness | ± | ± | ± | ± | ± | + |
| Quality of evaluation materials and transparency of evaluation | + | ± | ± | ± | ± | + |
| Presence/absence of feedback from the course teachers | - | ± | ± | ± | ± | + |
| Availability/absence of operational communication with technical support | - | ± | ± | ± | ± | + |
| Students' willingness to recommend the course to other students | - | ± | ± | ± | ± | + |

## References

1. Lopukhova, Y., Makeeva, E.: Creating Virtual Learning Environment: Shared Online Learning in University Education In:International Journal for Cross-Disciplinary Subjects in Education (IJCDSE), Volume 8, Issue 2, June, 2017. pp. 3046–3054. ISSN 2042–6364. (2017). https://doi.org/10.20533/ijcdse.2042.6364.2017.0412
2. Butcher, N.: A basic guide to Open Educational Resources (OER). Vanc.: Commonw. Learn. p. 134 (2011)
3. The activity plan of the Ministry of Science and Higher Education of the Russian Federation for the period from 2019 to 2024, approved by the Minister of Science and Higher Education of the Russian Federation M.M. Kotyukov on (February 08, 2019). http://fgosvo.ru/uploadfiles/prikaz_miobr/Plan_deyatelnosti_2019-2024.pdf, Last Accessed 2023/05/22
4. On education (The Law of the Republic of Kazakhstan, dated 27 July, 2007 No. 319-III) (as amended on 07/07/2020) https://iaar.agency/iaar/pravovye-akty-respubliki-kazahstan-v-oblasti-obrazovaniya/en, Last Accessed 2023/04/15
5. Order of the minister of education and science of the Republic of Kazakhstan, dated March 20, 2015, No. 137 https://cis-legislation.com/document.fwx?rgn=76276 Last Accessed 2023/04/25
6. Order of the minister of education and science of the Republic of Kazakhstan, dated October 30, 2018 No. 595; with amendments and additions (Order No. 207 dated 05/18/2020) https://iitu.edu.kz/en/articles/obrazovanie-en/normativno-pravovye-akty-v-sfere-obrazovania-en/ last accessed 2023/04/15

7. Order of the ministry of education and science of the Russian Federation No. 816 dated August 23, 2017 On Approval of the Procedure for the Use of e-learning, distance learning technologies by organizations Engaged in educational activities in the implementation of educational programs. http://ivo.garant.ru/#/document/71770012/paragraph/35:0, Last Accessed 2023/03/29
8. Daniel, J.: Making sense of MOOCs: Musings in a maze of myth, paradox and possibility. Journal of interactive Media in education, **3**. (2012)
9. Khalil, H., & Ebner, M.: MOOCs completion rates and possible methods to improve retention-A literature review. In: EdMedia+ Innovate Learning. pp. 1305–1313. Association for the Advancement of Computing in Education (AACE) (2014)
10. Barber, M., Donnelly, K., Rizvi, S.: An avalanche is coming: Higher education and the revolution ahead. London, Institute for Public Policy Research, available at: http://www.ippr.org/publication/55/10432/an-avalanche-is-coming-higher-education-and-the-revolution-ahead, Last Accessed 2016/12/23 (2013)
11. Giving Knowledge for Free: the emergence of open educational resources. In: Organisation for Economic Co-Operation and Development publications, France, available at: www.sourceoecd.org/education/ 9789264031746, Last Accessed 2023/02/23 (2007)
12. Mossley, D.: Open educational resources and open education. The Higher Education Academy, York Science Park, Heslington, p. 26 (2014)
13. Semenova T.V., Vilkova K.A.: Types of integration of mass open online courses into the educational process of universities. University management: Practice and analysis. **21**(6):114–126. (2017). https://doi.org/10.15826/umpa.2017.06.080, Last Accessed 2023/04/04
14. Shcherbakov N.V., Kirillova S.S.: On the introduction of online courses in the educational process of the university. In: Science and Education. **3**:1 (2020)
15. Regulation No. P-596 of 12/25/2020 On the use and offset of the results of mass open educational courses at the samara state technical university. https://samgtu.ru/uploads/documents/polojenie/P-596.pdf, Last Accessed 2023/03/22
16. Rules for the development of mass open online courses at Makhambet Utemisov West Kazakhstan State University, approved 2022/03/28, p. 13 Makhambet Utemisov West Kazakhstan State University, Uralsk (2022)
17. Lopukhova Y., Makeeva E., Gorlova E. Designing an academical online course for technical students: structure, content, assessment. In: Auer M.E., Rüütmann T. (eds) Educating Engineers for Future Industrial Revolutions. ICL 2020. Adv. Intell. Syst. Comput. **1328**. Springer, Cham. https://doi.org/10.1007/978-3-030-68198-2_63 pp. 682–689 (2021)

# The Method of Using EOSC Cloud Services for Math and Science Teachers' Training

Maiia Marienko(✉)

Institute for Digitalisation of Education of the NAES of Ukraine, M. Berlyns'koho St., 9, Kyiv 04060, Ukraine
popelmaya@gmail.com

**Abstract.** The article describes the essence of using EOSC cloud services for math and science teachers' training. The main components of the proposed learning method are target (target and target group), content, technology (methods, forms and learning tools) and results. A possible plan for training classes is given in the article. Teachers got acquainted with open science, the main structure of EOSC and its components, with specialized cloud services and tools to maintain international projects. The method of using EOSC cloud services for math and science teachers' training is experimentally tested, and the results are proved using Fisher's test. Experimental work on developing and implementing the method of using EOSC cloud services for math and science teachers' training occurred as a natural, cross-disciplinary pedagogical experiment in four stages: preparatory, ascertaining, formative and control. The methodological recommendations for using EOSC open science cloud services for teachers are formulated based on the experiment results. Specific components of learners' open science competence increased in the experimental groups compared to the control groups. The statistical analysis confirmed this conclusion. The proposed method is recommended to be implemented as part of teachers' training courses.

**Keywords:** Cloud technologies · Cloud-oriented systems · Teachers of natural and mathematical subjects · Scientific lyceums

## 1 Introduction

### 1.1 Problem Statement

The active implementation of the principles of open science in the system of teachers' training courses in the future perspective will guarantee the participation of Ukraine in the accession and further development of the European Open Science Cloud (EOSC), which is one of the priority tasks for the implementation of the Road Map of Ukraine's integration into the European research area (ERA). Considering the relevant specifics of work in a scientific lyceum, as the innovative educational establishments in Ukraine, the teacher should be able to manage the research activities of students using digital technologies. There is an urgent need to improve the qualifications of teachers for their further work in the scientific lyceum and to master the skills of using open science cloud services in the process of blended and distance learning.


M. E. Auer et al. (Eds.): ICL 2023, LNNS 899, pp. 267–274, 2024.
https://doi.org/10.1007/978-3-031-51979-6_28

### 1.2 The State of the Art

Many research infrastructures and European projects offer educational catalogues for storing and listing various educational materials, as stated in [4]. A catalogue of web applications consisting of educational materials is proposed in the EOSC-Pillar project. Research [4] briefly describes the scope and technical implementation of the EOSC-Pillar RDM training and support catalogue and how curation, quality assurance and sustainability are implemented.

The problem of designing an environment for supporting the work of a virtual team of scientists to exchange electronic resources and joint access was considered in studies [2, 8, 9]. These are environments of open science that can be used in the educational process.

A vast array of scientific research data can be used with benefit only if the user acquires specific skills in their processing and reuse. This problem is considered in the study [3]. The training material is customized for different communities, and participants are taught how to use Xarray, Dask, and in general, how to effectively access and nalyse large online datasets. Scientists in their work [3] suggest completing the training with group work, where participants can work with larger sets of scientific data: the classroom is divided into several groups. Each group works on different scientific questions and may use different datasets.

Research by the team of scientists L. Provost, F. Di Donato, E. Tóth-Czifra et al. [7] focused on online open science training held within the framework of the H2020 project "Transforming Research through Innovative Practices for Linked Interdisciplinary Exploration" (TRIPLE). The TRIPLE training toolkit is intended for training providers and research organizations that wish to develop and manage training activities as open educational resources.

Separate EOSC open science services were considered in a previous study [5]. However, the proposed classification was based on the main stages of scientific research. Open science cloud services for science and mathematics teachers need additional research.

### 1.3 Purpose

The purpose of the paper is to justify the model and develop the components of the methodology for using EOSC cloud services for training math and science teachers for work in a scientific lyceum in the graduating class. The effectiveness of the developed method of using EOSC cloud services for math and science teachers' training is to be experimentally proved.

## 2 The Conceptual and Terminological Body

The concept is based on the leading idea, according to which the methodically justified use of EOSC cloud services to improve the qualifications of teachers of natural and mathematical subjects will contribute to the increase of their professional competence and IR competence, the broader use of cloud-oriented tools and technologies of open science in the teaching process, the modernization of educational and scientific environment.

*Competence in open science* is the ability of a person to carry out successful research activities by the principles of open science based on knowledge, abilities, skills and personal attitude.

*Components of competence in open science* can be grouped into four main categories [6]: skills and experience required for open access publication; skills and experience in data research, management, analysis/use/reuse and dissemination; skills and experience of working in one's disciplinary community and outside it; skills and expertise arising from a general and broad conception of science as researchers interact with the wider public to enhance the impact of science and research.

## 3 Approach

The leading idea is expressed in *the research hypothesis*: methodically justified use of the method of using EOSC cloud services for math and science teachers' training will contribute to increasing their competence in open science, IR competence, more expansive use of cloud-oriented tools and technologies of open science in the teaching process, modernization of educational -scientific environment.

## 4 Actual Outcomes

*Structure of The Method of Using EOSC Cloud Services for Math and Science Teachers Training.*

*Target component.* Purpose: to increase the level and professional development of teachers due to the use of cloud-oriented systems of open science and to increase the level of competence in open science. Target group: teachers of natural and mathematical subjects.

*Content component.* The concept of open science and its significance for the teacher of natural and mathematical subjects. EOSC and its components. Specialized cloud services as means of implementing open science. International projects.

*Technology component.* Teaching methods: observation, demonstration, illustration, reproductive, search, research, educational discussion; the situation of cognitive novelty; the situation of interest, problem-heuristic. Forms of education: training, training courses, distance learning courses, seminars, webinars, master classes, individual consultations, lectures (traditional, problem-based) using cloud services and open science systems. Learning tools: EOSC toolkit (cloud services and resources), Google Classroom, Google Meet.

*Resulting component*: increasing the professional development of teachers due to the use of cloud-oriented systems of open science, increasing the level of competence in open science.

Minimum requirements for hardware and software on the user's device: browser and Internet connection (wired or Wi-Fi).

*Approximate plan of training classes.*

Topic 1. Registration in EOSC and project creation (4 h).

Topic 2. Selection and addition of cloud services/resources of open science (4 h).

Topic 3. Use of industry-wide cloud services/resources of open science (2 h).

Topic 4. Use of specialized cloud services/resources of open science (2 h).

Total: 12 h.

The use of cloud-based systems of open science by teachers in the process of learning and professional development leads to the introduction of these systems into school practice, wider the use of the best examples of cloud-based tools within the framework of school teaching, which will diversify the educational process and lead to an increase in its science orientation.

## 4.1 Implementation

To test the method of using EOSC cloud services for math and science teachers' training, a distance course, "Open science cloud services for educators", was developed, which was implemented within the framework of the certificate educational program "Information Systems and Cloud Technologies in the Educational Process". The distance course was taught during the introduction of quarantine restrictions due to the spread of COVID-19. All methods justified themselves very well. There were 921 pre-registered course participants, but 774 joined and started work. An even smaller number of participants completed the course - only 643. Since there were 921 registered participants of the distance course at the beginning, for convenience, they were divided into four groups of 230 people (the first group was 231 people) [10]. A list of all participants (in alphabetical order) was compiled based on the results of filling out the course application. In each group, this list, along with e-mail addresses, was made public to practice the skills of collective and group work. Distance course is developed based on Google Classroom. In a separate section of the course (Organizational issues), instructions for the Google Classroom user were posted.

Participants who completed the course were considered to be those who completed all practical tasks. Each task was evaluated as follows: 1 b. – credited, 0 b. - not counted. Each practical session was limited in time (it was calculated until the end of the day). The lecture and practical session were available to each course participant every day from 9:00 a.m. The deadline for their implementation was 15 h. This time was calculated taking into account individual and group counselling, possible technical problems, and the individual pace of completing the task by each course student.

In the course of the distance learning course, specific difficulties in the work were revealed [10]:

- At the registration stage, participants filled out the course application with errors, which subsequently made it difficult to perform individual practical tasks (where it was necessary to practice teamwork skills);
- Certain cloud services were not designed for the simultaneous work of several such users (in a reasonably short period). Therefore certain technical problems were subsequently resolved by correspondence with service developers;
- Non-localized cloud services caused the most difficulties for course participants;
- Due to technical problems, individual course participants could not physically pass individual practical tasks on time (in exceptional cases, they revised the material later than the deadline set for one or another topic).

In the final survey, the distance course participants thanked for the information, the acquired practical skills and the organizers' initiative. They expressed interest in such practical activities and needed more time to hold a master class. The participants noted that the presentation of the material was accessible, helpful and practically oriented. The listeners were determined to learn more about the tools of the European Open Science Cloud. The participants of the distance course expressed their wish for a more detailed and extensive study of EOSC's specialized cloud services of open science.

### 4.2 The Experimental Testing

Experimental work on the development and implementation of the method of using EOSC cloud services for math and science teachers' training took place as a natural, cross-disciplinary pedagogical experiment in four stages: preparatory (October 2019–December 2019), ascertainment (January 2020–February 2020), formative (March–December 2020), control (January–May 2021).

Control and experimental groups were formed as follows:

- The control groups included five groups of students of distance education courses at the Zhytomyr Polytechnic State University, Kryvyi Rih State Pedagogical University, a group of students who are members of the open Google group "Open Science in Education" (141 people). Students of the control groups took advanced training courses with a scientific component but without using the method of using EOSC cloud services for math and science teachers' training;
- The experimental groups included four groups of students of the distance education course at the Zhytomyr Polytechnic State University (395 people). The listeners of the experimental groups studied according to the method of using EOSC cloud services for math and science teachers' training.

To find out the state of formation of competence in open science and to evaluate the effectiveness of using the method of using EOSC cloud services for math and science teachers training, ascertaining sections of the following components of competence in open science were performed: skills and work experience in one's disciplinary community and outside it; skills and experience in data research, management, analysis/use/reuse, dissemination. Each component was considered separately and calculated according to the high, sufficient, medium and low levels.

Analyzing the ascertainment sections, we can see that the trainees have sufficient skills and experience in their disciplinary community and outside it but low–skills and experience in research data, management, analysis/use/reuse and dissemination.

Analyzing the results obtained after the formative stage of the pedagogical experiment, we can see that the percentage of a high level of skills and experience formation about data research, management, analysis/use/reuse and dissemination increased to 31%, and a sufficient level from 9% to 24%. At the same time, there is an increase in the number of course participants with sufficient skills and work experience in their disciplinary community and outside it: from 38% to 41%.

Let's check the reliability of the hypothesis about the existence, from a statistical point of view, of differences between the levels of skill formation and work experience in the own disciplinary community and outside it of the experimental and control groups based on the results of the final cut.

For this, we will use Fisher's criterion [1]. Let's formulate hypotheses:

$H_0$: The share of trainees who, according to the results of the study of the levels of skills formation and work experience in their disciplinary community and outside it, showed a high and sufficient level is more significant than in the control groups;

$H_1$: The share of trainees who, according to the study of the levels of skill formation and work experience in their disciplinary community and outside it, showed a high and sufficient level equal to in the control groups.

At the same time, only the fractions corresponding to the observations for which the effect occurs are used in the calculations. If the levels of formation of skills and work experience in one's disciplinary community and outside it are indicated to be high and sufficient, then "the effect takes place"; in the opposite case – "the effect is absent". The effect takes place. Control groups contained 52 participants (37%), and experimental groups contained 221 participants (56%). There is no effect. Control groups contained 89 participants (63%), and experimental groups contained 174 participants (44%). The experimental data fully satisfy the restrictions imposed by the angular Fisher transformation. The empirical value of the Fisher test is 3.9224; the critical value is 1.6449. According to Fisher's statistical criterion, the reliability of differences in the characteristics of the experimental and control groups is 95%.

### 4.3 Recommendations

The main methodical recommendations for using EOSC services for math and science teachers were formulated based on the research results.

1. First, consider the "Services" category from the EOSC catalogue. "Data sources" or "Data" can be used to conduct laboratory work on a specialized subject.
2. The "Scientific discipline" filter will help select cloud services depending on the academic subject.
3. Using most EOSC for math and science services for studying school subjects will be more appropriate for conducting practical and laboratory work.
4. EOSC cloud services will also be helpful in preparing scientific student work for participation in competitions of the Small Academy of Sciences of Ukraine.
5. Teachers can strengthen inter-subject connections using the EOSC toolkit in lessons (for example, mathematics and computer science).
6. The use of EOSC cloud services leads to an increase in the competence of open science not only among teachers but also indirectly among students. The level of IR competence among students also increases.
7. Introducing students to the EOSC portal, the teacher explains the principles of open science (using practical examples).
8. Particular attention should be focused on the "Training" resource category, as certain educational materials will be helpful to teachers in professional development.

## 5 Conclusions

The result of the experiment is the determination of the practicality and feasibility of possible forms for the large-scale introduction of innovative ICT tools, scientific-methodical and educational materials regarding the use of cloud technologies and open science tools and practices in the learning and research environment of pedagogical universities. Pedagogical approaches to using a cloud-oriented methodical system for improving the qualifications of teachers of natural and mathematical subjects for work in a scientific lyceum based on the State University "Zhytomyr Polytechnic" were tested. During the distance course, the participants studied specific cloud services for implementing open science in Ukraine. In addition, students learned how to work with certain cloud services to support joint work, organization of educational and research work, and distance learning.

The analysis of the results of the ascertainment stage of the pedagogical experiment on the introduction of the authors' method into the teachers' training showed that there is a need to improve the qualifications of math and science teachers required for work in a scientific lyceum. According to the statistical analysis of the obtained results, the characteristics of the experimental and control groups before the start of the experiment correspond to the significance level of 0.05, and, at the same time, the reliability of the differences in the characteristics of the experimental and control groups after the experiment is equal to 95%. The experiment proved the increase in the level of the open science competence of the participants, which indicates the effectiveness of the EOSC-based methodology implementation for science and mathematics teachers training for work in a scientific lyceum in a graduating class.

The methodology can be applied to teachers of scientific lyceums who plan to improve their qualifications for working with 11th-grade students. The main focus should be on implementing open research education and relevant skills and competencies into the training and professional development of teachers of scientific lyceums. It is recommended to conduct learning through distance professional training courses for teachers with the support of a community of scientists.

Further research will aim to create a classification of EOSC open science cloud services by subjects and learning objectives. It is planned to investigate their advantages and disadvantages; discover which types of lessons one can support with the use of cloud services of open science.

## References

1. Barot, T., Krpec, R.: Alternative Approach to Fisher's Exact Test with Application in Pedagogical Research. In: Silhavy, R., Silhavy, P., Prokopova, Z. (eds.) CoMeSySo 2018. AISC, **859**, pp. 50–59. Springer, Cham. (2019). https://doi.org/10.1007/978-3-030-00211-4_6
2. Bykov, V., Mikulowski, D., Moravcik, O., Svetsky, S., Shyshkina, M.: The use of the cloud-based open learning and research platform for collaboration in virtual teams. Inf. Technol. Learn. Tools **76**(2), 304–320 (2020). https://doi.org/10.33407/itlt.v76i2.3706
3. Fouilloux, A., Marasco, P.L., Odaka, T., Mottram, R., Zieger, P., Schulz, M., Coca-Castro, A., Iaquinta, J., Eynard Bontemps, G.: Pangeo framework for training: experience with FOSS4G, the CLIVAR boot camp and the science course. In: EGU General Assembly 2023, Vienna, Austria, EGU23–8756, pp. 24–28 (2023). https://doi.org/10.5194/egusphere-egu23-8756

4. Garcia, P.O., Berberi, L., Candela, L., Van Nieuwerburgh, I., Lazzeri, E., Czuray, M.: Developing the EOSC-Pillar RDM Training and Support Catalogue. In: Silvello, G. et al. (eds.) TPDL 2022. LNCS, **13541**, pp. 274–281. Springer, Cham (2022). https://doi.org/10.1007/978-3-031-16802-4_22
5. Marienko, M., Shyshkina, M.: The Design and Implementation of the Cloud-Based System of Open Science for Teachers' Training. In: Auer, M.E., Pachatz, W., Rüütmann, T. (eds.) ICL 2022: Learning in the Age of Digital and Green Transition. LNNS, **633**, pp. 337–344. Springer, Cham. (2023). https://doi.org/10.1007/978-3-031-26876-2_31
6. O'Carroll, C., Hyllseth, B., Berg, R., et al.: Providing researchers with the skills and competencies they need to practise Open Science. Publications Office (2017). https://data.europa.eu/doi/https://doi.org/10.2777/121253
7. Provost, L., Di Donato, F., Tóth-Czifra, E. et al.: Open Science Training in TRIPLE. Open Res Europe 2023, **3**:39 (2023). https://doi.org/10.12688/openreseurope.15430.1
8. Svetsky, S., Moravcik, O., Mikulowski, D., Shyshkina, M. The ICT Design for Modern Education Technology and Applications. In: Arai, K., Kapoor, S., Bhatia, R. (eds) FTC 2020, Volume 3. AISC, **1290**, pp. 281–291. Springer, Cham. (2021). https://doi.org/10.1007/978-3-030-63092-8_19
9. Svetsky, S., Moravcik, O., Shyshkina, M., Cervenanska, Z., Kotianova, J. The Knowledge-Based design of educational technology. In: Arai, K. (eds) FTC 2021, Volume 3. LNNS, **360**, pp. 759–775. Springer, Cham. (2022). https://doi.org/10.1007/978-3-030-89912-7_58
10. Vakaliuk, T.A., Marienko, M.B.: Experience of using cloud-oriented open science systems in the process of teaching and professional development of natural and mathematical teachers. Inf. Technol. Learn. Tools **81**(1), 340–355 (2021). https://doi.org/10.33407/itlt.v81i1.4225

# EFL Students' Engagement and Digital Transformation to Support Education in Difficult Times

Lorena Fernanda Parra Gavilánez(✉)

Universidad Técnica de Ambato, English Teaching Program, Faculty of Education, Ambato, Ecuador
lparra@uta.edu.ec

**Abstract.** The COVID-19 pandemic has influenced various dimensions of higher education systems globally, including the teaching and learning process of English as a Foreign Language. Pedagogical practice has been altered by the confinement of lockdown measures, which affected the level of student engagement. Consequently, this study aimed to analyze the level of student engagement during the pandemic. A realist evaluation guided the analysis of this study to answer the research question: what worked for the students in this context and why? Descriptive statistics were also used to analyze the results that emerged from a student engagement questionnaire. The results revealed that the level of student engagement varied according to the different dimensions analyzed using the questionnaire instrument. There was a strong connection between student capabilities and the teaching environment. Additionally, it can be concluded that distance learning during the COVID-19 pandemic lockdown was a successful teaching-learning medium for engaged students that are confident in using technology. It is essential to consider different engagement dimensions in different circumstances to help students achieve the desired learning outcomes during prolonged periods of imposed online learning.

**Keywords:** Students' engagement · EFL learners · Distance learning

## 1 Introduction

The COVID-19 outbreak originating in Wuhan, China, at the end of 2019, led to the temporary closure of great number of educational institutions at all instructional levels around the world. Indeed, most countries adopted strict social-distancing policies and lockdown strategies [1]. The resulting move to remote learning triggered a sudden change in the teaching and learning process [2]. The abrupt changes forced both students and teachers to move from classroom instruction to online classes or remote teaching. Ecuador maintained one of the longest educational confinement strategies in the world. For this reason, the aim of this paper is to analyse undergraduate English as a Foreign Language (EFL) student engagement during the COVID-19 pandemic in the Ecuadorian context.


M. E. Auer et al. (Eds.): ICL 2023, LNNS 899, pp. 275–286, 2024.
https://doi.org/10.1007/978-3-031-51979-6_29

In Ecuador, starting March 13, 2020, the government decreed that all public and private educational institutions had to work from home using different technological platforms and tools. This educational lockdown persisted in part into early 2022. Consequently, e-learning has become a new form of distance education that has allowed teachers to innovate technologies in order to create interactive classes that engage students in the learning process [3]. At the beginning of the pandemic, lecturers and teachers who were not familiar with the distance learning modality had to improvise their classes while new teacher training was offered to prepare teachers for this new paradigm with the purpose of guaranteeing the continuity and quality of education [4]. The lockdown has forced teachers to adapt their professional approaches, training, and abilities to fully integrate and utilize the information and communications technologies in the teaching process. Various technological platforms, such as Zoom, Teams and others, have become essential to online education. The result of this process has been to move the debate surrounding online education from an auxiliary role to that of one of the principal media of education.

In order to better understand student performance during the lockdown in Ecuador, this paper investigated Ecuadorian EFL learners' engagement in enforced online learning during the COVID-19 pandemic. To achieve this objective, the following research questions guided this study:

1. What is the level of engagement of a sample of Ecuadorian EFL undergraduate students in the implementation of online learning during the COVID-19 pandemic lockdown?
2. What worked for the students in this context and why?

## 2 Literature Review

### 2.1 Student Engagement

Student engagement is one of the main aspects of the teaching-learning process[5]. It is considered a multifaceted phenomenon that has been widely investigated in order to better understand the factors that influence student performance at the different levels of education[6]. In order to improve learners' ability to perform successfully, students must participate in the learning process and demonstrate creativity in the assignment creation process [7]. The term "student engagement" has evolved to refer to how involved or interested students appear to be in their learning, as well as their connections to their classrooms, institutions, and peers [8]. In other words, student participation or engagement determines their qualities in the classroom. However, guaranteeing a high degree of participation and ensuring the quality of learning during prolonged lockdowns, in which economic and emotional conditions affect performance, can be challenging [9]. Additionally, student participation in the classroom is about how students are involved in learning and how students communicate and interact with each other and with the teacher. Student participation has been significantly affected by the COVID-19 restrictions as learners have adopted a passive performance because of an obstructive learning atmosphere, a sense of self-isolation and negative learning attitudes [10]. The learning process depends on the conditions that institutions and professionals generate to stimulate and encourage student involvement [11]. From this perspective, the relationship between students and institutions generates an optimal environment for productive participation. However, optimal participation can be overshadowed by a series of limitations that emerge from extraordinary circumstances [12]. Moreover, institutions and staff are responsible for creating an environment that facilitates learning and provides learning opportunities. In all cases, the final responsibility for learning at the undergraduate level rests with students [6].

### 2.2 Online Distance Learning

Online learning is referred to in many terms in the literature, including tele-learning, virtual learning, computer-assisted learning, web-based learning, e-learning, networked learning, internet learning, and distance learning. For the purposes of this study, all of these terms are regarded as part of the same phenomenon. Online learning is a type of learning that allows students to use the Internet to access content and improve their educational opportunities [13]. The use of technology has helped to increase the quality of the teaching and learning process in general, as well as to facilitate this process in extraordinary circumstances such as in the COVID-19 pandemic [14]. Among the characteristics of e-learning is the novel approach to the conveyance of instruction by means of electronic data that upgrade learner aptitudes, information, or other learning practices [1]. Moreover, online learning refers to the learning condition that enables learners to obtain wider access to information and to carry out education without restrictions on place or time [15]. However, all the benefits that online learning offers can be overshadowed when this mode of education is forced on students by the circumstances, creating dissatisfaction for students that are not fluent users of technology [16].

Certain types of technology have been extensively adopted in the field of second or foreign language (L2) teaching and the discussion and study of technology is not new in language learning [17]. Moreover, online distance learning in general has had notable positive effects in widening the horizon of L2 learning and has influenced the nature of acquisition processes in two ways: by increasing the amount of L2 exposure and by broadening the scope of L2 input [18]. In a similar vein, [19] states that the use of technology may stimulate positive attitudes, for example, through an increased level of interest, motivation, interaction, and language production. The empirical data from extensive past studies has shown that technology-enhanced language learning provides a wealth of authentic information, as well as the opportunity to practice language skills and to participate more actively in class [20]. This digital openness could make education more accessible by allowing it to be adapted to individual needs. Because everyone absorbs content differently, information should be acquired in a variety of ways, whether through video, audio, text, or another medium [9]. Furthermore, individuals select what best matches their needs, which has an impact on how the information obtained is shown to demonstrate the knowledge gained [21].

# 3 Methodology

## 3.1 Participants

The participants of this study were 159 undergraduate students of the English Teacher Training Programme of a Faculty of Social Science of a public university in Ecuador. This is a public university that offers different types of graduate and postgraduate on-campus educational programmes. The sample is part of a face-to-face teaching programme that had to change unexpectedly to a distance learning programme due to the COVID-19 pandemic. The participants had been attending online classes continually for approximately 18 months at the time of the study. The ages of the students ranged from 19 to 25 years old. The students of the programme participated in this research voluntarily.

## 3.2 Instrument

To carry out this investigation, a questionnaire was used for data collection. Questionnaires offer the benefits of both standardized and open responses to a range of topics from a large sample or population [22]. The research used the Student Engagement Questionnaire (SEQ). This questionnaire has been widely used to investigate the level of student engagement in different types of research. The SEQ was first applied to taught courses as a graduate survey in Hong Kong in 2001, since then it underwent slight changes until its final version appeared in 2009 [23]. This instrument has been used to analyze the capabilities that university students have during the learning process and to assess the learning environment that teachers create in class. For example, this analysis has served to give feedback to teachers to design an adequate learning environment in different Spanish university populations [24]. Another study demonstrated that this questionnaire served to analyze the importance of student engagement to identify learning needs and enhance the quality of the teaching and learning environment in Moroccan universities [25]. The cited studies applied this questionnaire to learners of face-to-face programmes in normal circumstances unlike the current study that applied the questionnaire to students of online classes during a prolonged lockdown. The instrument used contained thirty-five statements focusing on student capabilities and the teaching and learning environment. These two general groups were further divided into seventeen sub-categories as follows:

**Student capabilities:**

– *Intellectual capabilities.*

- Critical thinking.
- Creative thinking.
- Self-managed learning.
- Adaptability.
- Problem solving.

– *Working together/Teamwork.*

- Communication skills.
- Interpersonal skills and group-work.

Computer literacy.

**Teaching and learning environment**:

- Active learning.
- Teaching for understanding.
- Assessment.
- Coherence of curriculum.

- *Teacher-Student relationship.*

- Relationship between teachers and students.
- Feedback to assist learning.
- Workload.

- *Student – student relationship.*

- Relationship with other students.
- Cooperative learning.

### 3.3 Study Case

This study employed a descriptive research design. To analyze the level of engagement of the EFL learners during the COVID-19 lockdown, descriptive statistics were used as this makes data visualization easy to understand through meaningful representation and leads to a simplified interpretation of the data collected [26]. Second, a realist approach was chosen as it offers answers to student engagement in relation to education and its issues—in this case EFL distance learning. The realist framework makes it easier to figure out what worked for the students in this case, and why [27]. This framework facilitated the identification of aspects of distance learning in the abrupt change from face-to-face learning to remote teaching as experienced in the EFL Ecuadorian context. By using the realist framework, this paper offers new insights into distance learning in the current situation.

The first component of realist evaluation is the context. The context reflects the reality into which an intervention is placed, as well as the mechanisms for achieving the desired outcomes [27]. In the same vein, the authors state that mechanisms, the second component of this framework, are the central tenet of realist evaluation as they form part of the design of an evaluation, which come to light during the evaluation process. Finally, the outcomes can be positive or negative, expected or unexpected. Pawson and Tilley argue that the outcomes, results or impact must be viewed as the complex outworking of multiple mechanism or context effects.

There are four stages to guide a realist evaluation. Stage 1 encourages the researcher to formulate the theory. This stage involves the researcher in outlining how a programme works and how the programme mechanisms will generate the desired outcomes. Stage 2 requires data collection to test the context, mechanism and outcomes (CMO) hypothesis, that is, what might work for whom in what circumstances and why. Any method that will open the "black box" of the programme mechanisms can be used [28]. Stage 3 entails analysing the data in terms of CMO settings, so confirming the initial premise and generating a transferrable understanding of what worked for whom and why in various situations. At this stage, a realist evaluation seeks to explore the casual relationship between the intervention's contexts, mechanism and outcomes [29]. Stage 4 results in the original hypothesis being refined for future testing. Realist evaluation reveals what did work for whom in what circumstances and why [30].

In the present study, each stage informed the research design as follows:

Stage 1- Programme theory.

The first step in this phase was to review the literature on implementation of online distance education during the COVID-19 lockdown and its influence on learner engagement. The literature guided the development of the initial program theory required. The programme theory for this study is articulated as follows.

- Change to online distance learning during Covid-19 pandemic will impact the level of EFL learners' engagement.

A more detailed breakdown of the CMO configurations for this initial programme theory is given in Table 1.

**Table 1:** CMO configuration of initial program theory

| Context | Mechanism | Outcome |
|---|---|---|
| C1 EFL students from the face-to-face programme changed abruptly to a distance learning programme | M1 Students will invest time to become familiarized with virtual tools in online distance learning | O1 Engagement in virtual learning environments |
| C2 Remote teaching during COVID-19 pandemic lockdowns | M2 Students will participate in synchronous and asynchronous learning activities | O2 Improvement of students' capabilities during the lockdowns |
| C3 Virtual teacher-student relationship | M3 Students will have virtual meetings with their tutors to clear up learning doubts | O3 Having effective feedback for better understanding of the course content |

Stage 2- Data collection to test the programme theory:

To test and refine the initial programme theory, the Student Engagement Questionnaire by Kember and Leung (2009) was applied. The questionnaire was used to find out the level of engagement of EFL learners in online learning as experienced during the extended lockdown of the COVID-19 pandemic.

Stage 3 – Analysis of data into CMO configuration:

The analysis of the data was intended to identify sequences and connections between CMO configurations in accordance with the Realist Evaluation Approach by exploring causal outcomes derived from mechanism in contexts [31]. For this, the results obtained from the questionnaire were analysed using a statistical descriptive analysis since this allowed the collection, analysis and characterisation of the data with the purpose of interpreting the results according to the study needs [32]. The analysis revealed whether the proposed CMO configurations were accurate. The contexts of the initial program theory were conserved as they were taken from the literature and represent the situation during COVID-19 pandemic lockdown. The mechanisms emerged from the statements of the Student Engagement Questionnaire, and the outcomes resulted from the most representative subcategories of the SEQ. The most representative subcategories emerged from the highest median values. Table 4 summarizes and explains the CMO re-configuration.

## 4 Results

For the first research question, the data collected in the questionnaire were analyzed using a descriptive analysis, which led to the following conclusions:

- The main category of the SEQ, Students' Capabilities, registered a median of 56 out of 70 possible points, with a range equal to 56, made up of a minimum score of 14 and a maximum of 70. This dispersion of data results in a standard deviation of 8.18 (Table 2). These results are made up of the scores obtained in the sub-dimensions, as shown in Table 3:
- The Intellectual Capabilities subcategory recorded a median of 40 out of 50 possible points, with a range equal to 40, made up of a minimum score of 10 and a maximum of 50. This dispersion of data results in a standard deviation of 5.96.
- The Working Together subcategory had a median of 16 out of 20 possible points, with a range equal to 16, made up of a minimum score of 4 and a maximum of 20. This data dispersion results in a standard deviation of 2.64.
- The main category, Teaching-Learning Environment, registered a median of 65 out of 85 possible points, with a range equal to 68, made up of a minimum score of 17 and a maximum of 85. This dispersion of data results in a standard deviation of 10.39 (Table 2). These results are made up of the scores obtained in the sub-dimensions, as shown in Table 3:
- The Teaching subcategory had a median of 36 out of 45 possible points, with a range equal to 36, made up of a minimum score of 9 and a maximum of 45. This dispersion of data results in a standard deviation of 5.59.
- The Teacher-Student Relationship subcategory recorded a median of 16 out of 20 possible points, with a range equal to 16, made up of a minimum score of 4 and a maximum of 20. This dispersion of data results in a standard deviation of 2.75.
- • The Student-Student Relationship subcategory had a median of 15 out of 20 possible points, with a range equal to 16, made up of a minimum score of 4 and a maximum of 20. This dispersion of data results in a standard deviation of 3.13.

**Table 2.** Descriptive statistics by dimensions

| Main Categories | Median | Maximum | Minimum | Rank | Standard deviation |
|---|---|---|---|---|---|
| Students' capabilities | 56.00 | 70.00 | 14.00 | 56.00 | 8.18 |
| Teaching—Learning environment variables | 65.00 | 85.00 | 17.00 | 68.00 | 10.39 |

**Table 3.** Descriptive statistics by sub dimensions

| Dimensions | Subdimensions | Median | Maximum | Minimum | Rank | Standard deviation |
|---|---|---|---|---|---|---|
| Students' capabilities | Intellectual Capabilities | 40.00 | 50.00 | 10.00 | 40.00 | 5.96 |
| | Working Together | 16.00 | 20.00 | 4.00 | 16.00 | 2.64 |
| Teaching—Learning environment variables | Teaching | 36.00 | 45.00 | 9.00 | 36.00 | 5.59 |
| | Teacher-Students Relationship | 16.00 | 20.00 | 4.00 | 16.00 | 2.75 |
| | Student-Student Relationship | 15.00 | 20.00 | 4.00 | 16.00 | 3.13 |

For the second research question, the data collected in the questionnaire and analysed using descriptive analysis also allowed for the determination, through the CMO reconfiguration, of what worked for whom, in what circumstances and why, as illustrated in Table 4:

**Table 4:** Summary of the CMO reconfiguration

| Context | Mechanism | Outcome |
|---|---|---|
| C1 EFL students from a face-to-face programme changed abruptly to a distance learning programme | M1 Students use knowledge to solve problems, consider different points of views, use own initiative and take responsibility for own learning | O1 Development of intellectual capabilities |
| C2 remote teaching during COVID-19 pandemic lockdowns | M2 Students are given the chance to participate in class, to be involved in a variety of teaching and assessment methods and receive orientation from teachers | O2 Good teaching and learning environment |
| C3 Virtual teaching–student relationship | M3 Communication between teaching staff and students is effective as both had access to internet and virtual tools | O3 Effective teacher-student relationship |

Stage 4 – Refining the programme theory.

At this stage, the realist evaluation framework encourages the evaluator to refine the initial programme theory according to what has emerged in stage 3. The refined programme theory is stated as follows:

Distance online learning during the COVID-19 pandemic lockdown had positive effects on students who developed their intellectual capabilities through the enhancement of critical thinking and management of self-managed learning. It also had beneficial effects for students who had a good teaching and learning environment through opportunities to experience active learning and appropriate feedback, and for students who maintained a good relationship with their teachers through effective communication.

## 5 Discussion

This study examined the effects of online distance learning on EFL learners' engagement as implemented during the COVID-19 lockdown in Ecuador. To this end, a realist evaluation approach guided the research to interpret and ehavio the results.

For the first research question, the findings demonstrated that the level of engagement in general was not affected by the implementation of online distance learning during lockdown. On the contrary, this finding is in agreement with the literature in providing evidence that online learning allows students to improve their educational opportunities [13]. According to the results that emerged from the Student Engagement Questionnaire, student intellectual capabilities were enhanced significantly. In other words, the learners demonstrated a high level of cognitive engagement. This dimension of student engagement requires forethought and willingness to understand complex concepts [6]. "Teamwork", another subcategory of students' capabilities, was also enhanced in online learning, meaning that students learn to participate in groups in a virtual learning environment. This is supported by other research that indicates that technology may stimulate positive attitudes and increase the level of group interaction [19]. Another relevant aspect that emerged from the SEQ is that the "Teaching" subcategory was rated highly. This means that teachers appear to have provided adequate mechanisms to include students in the teaching process, despite the fact that many were not familiar with distance-learning teaching strategies and were forced to improvise their classes, at least until some formal training could be offered [4]. The results also reported that there was a considerable improvement in the teacher-students' relationships, which also contributed to the greater level of student engagement in the learning process. From the ehaviourl perspective, both student ehaviour and teaching practice contributed to an appropriate environment for the development of engagement in the learning process [33].

For the second research question, a realist evaluation approach helped to determine what worked for whom, in what circumstance's and why. Distance online learning worked well in relation to this sample of EFL learners' engagement during the pandemic lockdown. The dramatic change that teachers and EFL students had to face during this process did not impede the progress of the learning process. Online education helped EFL students to be involved and interested in classroom activities, which agrees with the findings of Cheng [18], who states that online learning brings positive effects and broadens the horizon of L2 learning. The online modality worked well for EFL students who were exposed to activities that enhanced their critical thinking, such as solving problems with prior knowledge, considering different points of view, making judgements about different perspectives, and investing time in challenging activities. These types of activities promote the development of intellectual capabilities and are part of general cognitive engagement [6]. Another factor that helped EFL learners to stay engaged through online education was their exposure to a variety of teaching and assessment methods. Although the change from the face-to-face modality to online education meant that most teachers had to adopt methodologies that were new and unfamiliar, the learners felt that lessons were sufficiently inclusive to create an environment of engagement. The impact of the social context on the students can have both positive and negative effects that influence the socio-cultural engagement and, therefore, the learning process itself [34]. The quality of the relationship between teacher and students also affected the level of engagement during the pandemic lockdown. According to the results, learners benefited from effective communication with their tutors, which further facilitated the learning process. Students found it helpful that the teaching staff were available when they had doubts in relation to the subject content. Providing assistance and support during a challenging time appears to be essential as virtual learning may require greater flexibility [16].

## 6 Conclusion

In conclusion, the level of engagement of these EFL students did not suffer any appreciable negative impacts during the COVID-19 pandemic lockdown. The abrupt change of modality from face-to-face education to the online distance learning modality demanded the swift introduction of practices that ensured the level of student engagement. The mechanisms used in online learning education contributed through the development of students' intellectual capabilities and the promotion of a positive teaching-learning environmental and a strong teacher-student relationship. Pragmatically, this study illustrates how English language practices were conducted to engage students during the extended period of enforced remote teaching. Descriptive statistics and the realist evaluation approach helped to find robust results to answer the research questions. Ideally, future research should be conducted from the perspective of teachers to create a more comprehensive viewpoint in relation to successful student engagement in extended circumstances requiring online distance education.

**Acknowledgements.** The author would like to thank the Universidad Técnica de Ambato (UTA), and the Dirección de Investigación and Desarrollo (DIDE) for its financial sponsorship for the publication of the present research work. Thanks to the staff of the Educational Research program-Higher Education of Lancaster University for the suggestions on this paper.

## References

1. Octaberlina, R., Muslimin, A.: EFL Students perspective towards online learning barriers and alternatives using moodle/google classroom during COVID-19 Pandemic, Int. J. High. Educ., **9**:6, (2020)
2. Mabrur, A., and Suwartono, T.: Junior high school students' readiness to participate in E-Learning and Online EFL Classes during the COVID-19 Pandemic, Int. Soc. Sci. J., Jun. (2021)
3. Sevy-Biloon, T.: Virtual or face to face classes: ecuadorian university students' perceptions during the pandemic. English Lang. Teach. Educ. J. E-ISSN **4**(1), 15–24 (2021)

4. Navarro-Espinosa, J., Vaquero-Abellán, M., Perea-Moreno, A., Pedrós-Pérez, G., Aparicio-Martínez, P., Martínez-Jiménez, M.: The higher education sustainability before and during the COVID-19 Pandemic: A Spanish and Ecuadorian Case, Sustain. 2021, **13**:11, p. 6363, (2021)
5. Axelson and Flick, Defining Student Engagement, **43**:1, pp. 38–43, (2011), http://dx.doi.org.ezproxy.lancs.ac.uk/https://doi.org/10.1080/00091383.2011.533096, https://doi.org/10.1080/00091383.2011.533096
6. Fredricks, A., Blumenfeld, P., Paris, A.: School engagement: potential of the concept. State Evid. **74**(1), 59–109 (2004)
7. Sesmiyanti, S.: Student's cognitive engagement in learning process, Polingua Sci. J. Linguist. Lit. Lang. Educ. **5**(2), 48–51 (2016)
8. Axelson, R., Flick, A.: Defining student engagement. Chang. Mag. High. Learn. **43**(1), 38–43 (2010)
9. Tameemy, A., Alrefaee, Y.: Impact of COVID-19 on english language teaching in yemen: challenges and opportunities. TESOL Int. J. **16**, 238–252 (2021)
10. Bao, W.: COVID-19 and online teaching in higher education: a case study of peking university. Hum. Behav. Emerg. Technol. **2**(2), 113–115 (2020)
11. Krause, K., Coates, H.: Students' engagement in first-year university. Assess. Eval. High. Educ. **33**(5), 493–505 (2008)
12. Almarghani, E., Mijatovic, I.: Teaching in higher education factors affecting student engagement in HEIs-it is all about good teachin (2017)
13. Wang, N., Chen, J., Tai, M., Zhang, J.: Blended learning for Chinese university EFL learners: learning environment and learner perceptions. Comput. Assist. Lang. Learn. **34**(3), 297–323 (2021)
14. Rahman, H., Rajab, A. Wahab, A., Nor, Zakaria, W., and Badli, M., Factors Affecting Motivation in Language Learning, Int. J. Inf. Educ. Technol., **7**(7), pp 543–547, (2017)
15. Tanjung F., Utomo, A.: Investigating EFL Students' perception on online learning amidst covid-19 pandemic, ijiet (International J. Indones. Educ. Teaching), **5**(1), pp. 102–115, (2021)
16. Mulyani, M., Fidyati, F., Suryani, S., Suri, M., Halimatussakdiah, H.: University students' perceptions through e-learning implementation during COVID-19 pandemic: Positive or negative features dominate? Stud. English Lang. Educ. **8**(1), 197–211 (2021)
17. Amin, F., Sundari, H.: EFL students' preferences on digital platforms during emergency remote teaching: Video Conference, LMS, or messenger application? Stud. English Lang. Educ. **7**(2), 362–378 (2020)
18. Cheng, L.: Extended implications of technology in second language teaching and learning. Work. Pap. Appl. Linguist. TESOL **11**(2), 27–29 (2015)
19. Arslanyilmaz, A., Computer-assisted foreign language instruction: task based vs. form focused, J. Comput. Assist. Learn., **29**(4), pp. 303–318, (2013)
20. Alberth, A.: Technology-Enhanced teaching: a revolutionary approach to teaching english as a foreign language. Teflin J. **24**(1), 1–13 (2013)
21. Pomeroy, D.: Educational equity policy as human taxonomy: who do we compare and why does it matter? Crit. Stud. Educ. **61**(3), 329–344 (2020)
22. Cohen, L.: Research methods in education, Eighth edition. London, England ; New York, New Yok: Routledge (2018)
23. Kember, D., Leung, D.: Development of a questionnaire for assessing students' perceptions of the teaching and learning environment and its use in quality assurance. Learn. Environ. Res. **12**(1), 15–29 (2009)
24. Gargallo, B., Suárez-Rodríguez, J., Almerich, G., Verde, I., Cebrià, M.: The dimensional validation of the student engagement questionnaire (SEQ) with a spanish university population. Students' capabilities and the teaching-learning environment, An. Psicol., **34**(3), pp. 519–530, (2018)

25. Ben, I., Zerhouni, B.: Student engagement for quality enhancement and responding to student needs in the moroccan university: the case of the english studies track., Arab World English J., **10**(3), pp. 165–177, (2019)
26. Cohen, L., Manion, L., Morrison, K.: Research methods in education. Routledge, New York (2018)
27. Pawson, R.: Realistic evaluation. SAGE, London (1997)
28. McFau, H., FitzGerald, E.: A realist evaluation of student use of a virtual reality smartphone application in undergraduate legal education. Br. J. Educ. Technol. **51**(2), 572–589 (2020)
29. Harris, G.: The realist approach to evaluation research: an introduction. **19**(5), pp. 250–259, (2013). https://doi.org/10.12968/ijtr.2012.19.5.250
30. Graham, A, McAleer, S.: An overview of realist evaluation for simulation-based education, Adv. Simul. **3**(1), pp. 1–8, (2018)
31. Ryan, F.: A virtual law clinic: a realist evaluation of what works for whom, why, how and in what circumstances?, (2019)
32. Cohen, A.: Strategies in learning and using a second language, 2nd ed. Oxfordshire, [England] ; New York, New York: Routledge, (2014)
33. Kahu, E.: Framing student engagement in higher education. Stud. High. Educ. **38**(5), 758–773 (2013)
34. Dall'Alba, G., Barnacle, R.: An ontological turn for higher education, Stud. High. Educ., **32**(6), pp. 679–691, (2007)

# The Methodology for Using the Cloud-Based Open Science Systems in Higher Education Institutions

Mariya Shyshkina(✉)

Institute for Digitalisation of Education of NAES of Ukraine, M. Berlyns'koho St., 9, Kyiv 04060, Ukraine
marimodi@gmail.com

**Abstract.** The paper describes the methodology for using cloud-based open science systems in training students at the master's level in educational sciences. This methodology aims to increase the ICT competence of learners, modernise the university learning and research environment, and enforce collaborative, open, project-oriented learning. The new learning course "Smart Technologies in Education" was developed and introduced in several universities to train students in the educational sciences through the "ICT for Education" speciality. The cloud-based learning and research environment specially designed, developed and implemented for this purpose is to support the proposed approach. The course contains several modules and themes on adaptive learning and research design. The particular module is devoted to open science practices. The cloud services of different kinds deployed and designed to support collaborative and adaptive learning and research covered MS Office 365 (MS Teams and Power BI), MS Azure (virtual machines), and European open science cloud (EOSC). The practical tasks were to train collaboration skills in virtual teams, open learning and research materials exchange, adaptive data processing and visualisation, electronic learning resources selection and use and others. The practical part also included learning projects and creative tasks. The work substantiates that the implementation of the methodology for using open science systems in the educational process of pedagogical universities will contribute to the introduction of innovative forms and methods of learning, more active use of the emerging ICT, the development of ICT competence of learners, enforcing team working and research skills.

**Keywords:** Cloud technologies · Cloud-based systems · Open science · Digitalisation of education · Educational personnel

## 1 Introduction

### 1.1 Problem statement

The current trends in the learning environment design in higher education institutions cover a wide range of more flexible, personalised, open education and research systems, cloud-based tools and services. The emerging ICT are to support collaborative learning and research spaces, adaptive and big data processing, mobile and intuitive environments, remote and virtual laboratories and others.


M. E. Auer et al. (Eds.): ICL 2023, LNNS 899, pp. 287–294, 2024.
https://doi.org/10.1007/978-3-031-51979-6_30

The cloud-based tools transformed the university environment, enabling massive data processing, collaborative use and reuse of research results, greater scalability of computational capacities and service-oriented architecture. The open science approach enforces open, collaborative, project-oriented learning. The role of science and engineering education as the driving force of the information society is growing. Therefore, in the education of a modern teacher, not only open learning but also open science technologies should be presented, which are now an integral part of adaptive, personalised learning of a new age.

The most important trend is educational personnel training because they are the driving force for the digitalisation of education. The active implementation of open science services, being the cloud-based tools of a new generation, is one of the promising directions for increasing the ICT competence of learners and modernising the learning and research environment of higher education institutions.

## 1.2 The State of the Art

The launching of the European Open Science Cloud (EOSC) project in 2018 opened a new stage of the cloud-based systems of open science development. The adaptive power of cloud-based learning raised the urgent need for educational personnel training to tackle problems of using breakthrough ICT [3].

Special personnel must effectively introduce and use ICT in education to develop and implement practical learning tools and techniques in their subject areas. We are talking about training the digitalisation of education personnel. In 2010 at the Institute for Digitalisation of Education of NAES of Ukraine, a new speciality, "ICT for education", was launched for PhD training in the field of educational sciences; more than 50 dissertations considered the issues in this speciality (https://iitlt.gov.ua/eng/atestat/spetsializovana-vchena-rada/avtoreferaty-dysertatsiyi.php). A similar speciality introduced into master's training programs in educational sciences in 2018 ensured the cross-cutting integration of ICT training.

The emerging ICT introduced into the learning process of educators will ensure better results in bringing innovative practices into various fields of education during their professional activity. In particular, this applies to the latest technologies of cloud-based open science systems. We are talking about a new methodology of learning and research systems design, the foundations of which are the cloud-oriented approach [3] and the conception of open science [8]. This approach is to support collaborative learning and research, access to the learning environment at any time and from any place, the use of the most modern services for the collective, as well as for individual work, big data processing and so on. The methodology contributes to the general principles of open science, such as open access, open data, research methods, communication, and evaluation [8]. Cloud-oriented approaches, the main features of which are the flexibility and scalability of the IT infrastructure, are to provide open science principles in practice and achieve significantly higher efficiency in ICT training [3].

The main innovative features of a cloud-oriented open science learning environment are access to storage space and the most powerful cloud services and computer capacities for data processing and collaboration. The spectrum of educational and research activities is expanding due to a greater professional focus on the discipline's content and increased access to research tools [4]. Methodological features of the organisation of collaborative activity of students on educational projects using Microsoft Office 365 are considered in [4]. A scientific and methodological study of the problems of cloud-oriented learning and research environment formation in educational institutions requires further research in the context of the priorities of open science [3, 7]. In particular, it concerns the impact of robust platforms for storing and processing significant amounts of research data, data processing services with shared access to education, and the flexible selection and scaling of computer capacities [2]. Using cloud-based platforms in combination with communication services gives much more opportunities for the functioning of virtual research teams [2, 4]. These trends were significantly amplified by the impact of the pandemic period, with mixed and distance learning and research processes gaining mass adoption and use [5]. In this regard, we are talking about the methodology for forming a cloud-oriented learning and research environment in educational institutions using open science systems, and this methodology requires profound elaboration, implementation and use [3].

A research study of the problems of introducing open science services into the education system at its various levels is a promising trend. In particular, the advantages of introducing open science practices in the early stages of a scientific career are considered in [1]. The authors believe it is essential to form the ability to use ICT tools that facilitate the exchange and documentation of the scientific work of young scientists effectively and transparently [1]. Karen Maex, Rector Magnificus of Amsterdam's University, expresses concern about the independence of science in the modern digital age and advocates the European "Digital University Act" [6]. The author emphasises that researchers and teachers need access to platforms and data for educational and research purposes to ensure qualitative research.

The open science services and the importance of ensuring mass and equal access to them for scientists at different stages of their careers need careful attention. But these trends need to be introduced also into the training of educators and teachers [7]. It favours further research on creating and implementing appropriate learning methods for developing professional and ICT competencies of educational personnel in pedagogical universities. Educators are called upon to ensure the implementation of promising approaches and practices of open science in educational institutions.

The methodology for using the cloud-based services of open science for training the education digitalisation personnel is a need at various levels. The professionals in the digitalisation of education should have sufficient ICT competencies to understand and adopt new approaches and appropriate methods and tools and have sufficient management competencies to work in education management. They also should possess good investigative competencies to educate a new generation of researchers. The professional competencies of the education digitalisation personnel should cover open science competencies considered in detail during the previous pedagogical experiment concerning teachers' training [7], and ICT competencies, particularly in cloud technologies use [2; 3]. The educators are to be able to create, design and implement cloud-based open science systems in higher education institutions.

Given the trends of the growing role of scientific education of teachers and the introduction of emerging ICT, new learning methods for implementing the cloud-based systems of open science into the training process need consideration.

### 1.3 Purpose

The research idea is that the creation and use of cloud-based open science systems in the learning process in higher education institutions will positively affect the rganization of this process and conducting scientific research. It will create conditions for developing new forms, methods and technologies of learning. There is a need to ensure and increase the level of ICT competence of the education rganizationn personnel, which, in turn, will lead to positive qualitative changes in the rganization of learners' learning and research activities with a broader take up of modern ICT.

*The* research aims to develop and justify the methodology for using cloud-based systems of open science in the learning process of students at the master's level in educational sciences and to provide recommendations on the most promising ways of introducing new content, forms and methods into the educational process in HEI.

## 2 Actual Outcomes

### 2.1 The Analysis of the Learning Needs for the Education Digitalisation Personnel Training

As noted above, the problems of digital transformation of the educational sphere come to the fore, along with ensuring sustainable improvement of digital competence of the personnel in the educational sector, overcoming the shortage of highly qualified staff for the development of the digital economy and igitalization of society in general. Indeed, the success of the igitalization of education, including the trend of introducing promising cloud technologies of open science, depends on the personnel – specialists who are directly involved in the processes of digital transformation of education, as well as on the consistency, and coordination of actions of all participants.

In Ukraine, training in digitalisation of education is carried out at the level of magistracy, graduate school (field of study 011 "Educational, pedagogical sciences," speciality "Information and communication technologies in education"). The analysis of educational programs for this speciality, implemented in various higher education institutions, including higher pedagogical education, showed differences in approaches, sequence of content, and learning outcomes between the master's level and PhD. The existing programs need to cover the aspect of cross-cutting and continuity in the training system for the digitalisation of education personnel. The analysis also showed that the programs must reflect current trends, particularly adaptive learning systems, immersive technologies, open science technologies, etc.

At the beginning of our study, a pilot survey of master's students (field of study 011 "Educational, pedagogical sciences" speciality "Information and communication technologies in education") was conducted in the 2020–2021 academic year. The sample was 30 people. The survey focused on students' awareness of the current digitalisation trends in education and science, the most promising tools, etc.

It showed that the vast majority (97%) of respondents were not aware well of the basic concepts of open science, open data, etc. 76.7% of the respondents had only general ideas about these main terms, had heard or read something about certain aspects, but did not have a thorough understanding. Only 3% of the total number of respondents confirmed that they had a conscious understanding of the essence of the fundamental open science concepts, were sufficiently aware of European trends, knew the goals of the European open science cloud, could indicate specific types of services and give examples of their use.

### 2.2 The Implementation and Testing

The methodology for using cloud-based systems of open science in training Masters in Educational Sciences by the specialty "ICT for Education" is to satisfy better their educational and research needs and increase their ICT competence level.

Being aware of the need to ensure the cross-cutting training of personnel in the digitalisation of education (mainly based on institutions of higher pedagogical education), which implies the consistency and continuity of the formation of relevant competencies, in particular, the ability to apply cloud technologies of open science at different levels of education (Bachelor - Master - PhD) and integration of ICT learning content and relevant techniques, we undertook the training course "Smart Technologies in Education". The course is part of a master's degree program (field of study 011 "Educational, pedagogical sciences," speciality "Information and communication technologies in education") of the National University of Life and Environmental Sciences in 2020–2021. Among the aims of training within this course are the mastering and practical application of open science services and technologies and the ability to apply them in practice, educational and professional activities.

Among the learning methods of this course, there are such as explanatory-illustrative, practical, partial search, problem-search, and problem-heuristic. Among the forms of learning are lectures, seminars, laboratory work, independent work, and individual and group educational projects and creative tasks. Learning tools covered electronic resources and adaptive cloud services for open science systems and tasks (in particular, MS Office 365, MS Teams, MS Power BI, MS Azure, AWS and others).

Practical tasks were focused on creating educational projects "in the cloud," acquiring the skills of presenting and processing data in a cloud-oriented environment (Office 365), using adaptive data processing services (Power BI), creating and using virtual machines to use the computing power of the cloud servers (MS Azure). The result of the learning is: increasing the level of organisation of pedagogical research and increasing the level of ICT competence of learners.

The model of the cloud-based learning and research environment specially designed, developed and implemented to support the proposed methodology is depicted in Fig. 1.

*The brief content of training classes.*

*Module 1 covers adaptive systems for educational and research purposes.* The concept of adaptive learning systems is the main in this module. It is related to such trends in the modern cloud-based learning and research environment as "big data", "smart data", and "FAIR data". The scientific and educational community's awareness and broader

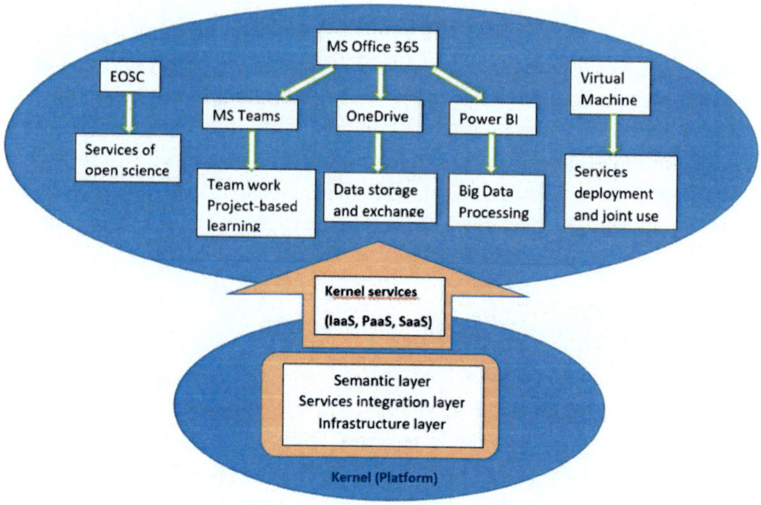


**Fig. 1.** The cloud-based learning and research environment model to support the proposed methodology.

take-up of these entities cause their introduction into the learning process as a basic need. From this point of view, the adaptive learning systems in the context of the latest trends in the formation of Society 5.0 are characterised by integrating physical and cyberspace based on intelligent digital technologies, virtual and augmented reality, and big data processing. The evolution of adaptive learning systems in education is considered to emphasise that the emergence and use of cloud technologies for educational systems gives a real opportunity to realise at a qualitatively new level their automatic adjustment to rapidly changing information and communication and computer processing needs of users. Practical work on this module was carried out using Microsoft Office 365 and Microsoft Teams services, enabling students to create group and individual educational projects. Cloud services for adaptive data processing, particularly Microsoft Power BI, require special attention within the open science data processing framework. Students used this service to perform practical work on downloading, visualisation and adaptive analysis of various data arrays.

*Module 2* covers designing an *adaptive cloud-oriented educational and research environment*. This module considered the basics of designing adaptive cloud-oriented systems, particularly virtual machine deployment technologies. There are two approaches to studying this material. Due to the first approach, each student creates a virtual machine in the account; for example, on MS Azure of AWS platform. The second, the teacher creates a virtual machine; in practical classes, students learn how to access it and perform specific actions with its services. For this course, we chose the second approach, which is more focused on group work of students and practising skills. It is more relevant for the "ICT for Education" speciality. The actual development of applications on virtual machines may be more appropriate for teaching students in computer science. At the same time, creating a virtual machine for deploying various types of services can be offered to the student as a creative task.

In the second part of this module, there is an in-depth study of the services of open science. These services form a new and promising class of cloud-based systems used as a tool for the digital transformation of learning and research processes. The central concept of the design is FAIR data processing principles (Findable, Accessible, Interoperable, Reusable). In this concern, the EOSC, a pan-European infrastructure that provides many data processing services available to scientists, is the focus of attention. Its mastery will allow students to use the most powerful cloud services for their pedagogical and research activities. Therefore, it is essential to consider the peculiarities of the selection and use of EOSC services within the content of this module.

The peculiarity of the learning course is the practical part that included learning projects and creative tasks.

Within the experimental group were four experimental groups of students (85 respondents), and the control group included two groups of students (20 respondents). The experimental group learned the new course "Smart Technologies in Education" due to the proposed learning methodology, as the control group was not taught this learning course but also studied Educational sciences at the master's level. In the experimental group, almost everybody was aware of the open science concept and tools after completing the learning course, and 70% of students showed high and medium literacy levels. The open science competencies are peculiar in this context, but they were not measured in this experiment. The quantity of students of the speciality "ICT for Education" at the master's level is not high. It is a new speciality, and it is introduced only in a few universities, and the survey covered almost all of the students of this speciality for the moment. It was impossible to get a statistically valid sample to test the difference in learning for open science competence formation. At the same time, the ICT competence of learners was measured in the aspect of the cloud-based systems use. The students in the control group also learned ICT during their study, and they also learned to use cloud services in their education and research activity. The percentage of students with a high level of ICT competence for cloud technology use increased from 20% to 56%, while in the control group, it increased from 18% to 32%. The statistical significance of the difference in the level of ICT competence after completing the learning course was confirmed by Fisher's test.

## 3 Conclusions/Recommendations/Summary

Implementing the methodology for using open science systems and services to support the learning process in higher education institutions proved effective. It positively impacted the organisation of students' collaborative work, improved their activity and motivation to study, and improved learning outcomes. Introducing the cloud-based systems of open science into the learning process raised the professional competence of students of the speciality "ICT for Education". It raised ICT competence, mainly for using cloud technologies and awareness of the open science concept, tools and practices. Research results and information about the existing opportunities, services and advantages of using cloud services, network tools and platforms of open science, e-infrastructures and the EOSC may be disseminated in the system of pedagogical universities for training master's students in educational sciences and in-service teachers. The results were implemented in the international project V4 Educational Academic Portal for Integrating IT into Education (EDUPORT), 2022–2023.

## References


1. Allen, C., Mehler, DMA.: Open science challenges, benefits and tips in early career and beyond. PLoS Biol **17**(5), (2019), https://doi.org/10.1371/journal.pbio.3000246
2. Bykov, V., Mikulowski, D., Moravcik, O., Svetsky, S., Shyshkina, M.: The use of the cloud-based open learning and research platform for collaboration in virtual teams. Inf. Technol. Learn. Tools **76**(2), 304–320 (2020), https://doi.org/10.33407/itlt.v76i2.3706, last accessed 2023/05/18
3. Bykov, V., Shyshkina, M.: The conceptual basis of the university Cloud-based learning and research environment formation and development given the open science priorities. Inf. Technol. Learn. Tools **68**(6), 1–19 (2018), https://journal.iitta.gov.ua/index.php/itlt/article/view/2609/1409
4. Glazunova, O.G., Kuzminska, O.G., Voloshyna, T.V., Sayapina, T.P., Korolchuk, V.I.: E-environment based on Microsoft Sharepoint for the organisation of group project work of students at higher education institutions. Inf. Technol. Learn. Tools **62**(6), 98–113 (2017), https://journal.iitta.gov.ua/index.php/itlt/article/view/1837
5. Krylova-Grek, Y., Shyshkina, M.: Online learning at higher education institutions in Ukraine: achievements, challenges, and horizons. Inf. Technol. Learn. Tools. ICT and learning tools in the higher education establishments **85**(5), 163–174 (2021)
6. Maex, K.I.J.: Digital University Act.pdf. University of Amsterdam/Amsterdam University of Applied Sciences. Presentation (2021), https://doi.org/10.21942/uva.13553825.v1
7. Marienko, M.V.: Tools and services of the Cloud-Based systems of open science formation in the process of teachers' training and professional development. Lect. Notes Bus. Inf. Process. Book Ser. (LNBIP) **429**, 108–120 (2021)
8. Open Science. Policy Brief, December 2015. ERA Portal, Austria (2015), https://era.gv.at/public/documents/2714/ERA_Open_Science_POLICY_BRIEF_December_2015.pdf

# Redesigning Digital Delivery of Postgraduate Programmes in the Post-Pandemic Era: A Sri Lankan Experience

Neelakshi Chandrasena Premawardhena(✉)

Department of Modern Languages, University of Kelaniya, Kelaniya, Sri Lanka
neelakshi@kln.ac.lk

**Abstract.** This paper presents the results of a new teaching approach applied to a postgraduate programme conducted at a state university in Sri Lanka. The study reviews the change of teaching approaches from the onsite delivery to virtual learning during the pandemic and how it evolved in the post-pandemic era. Albeit remote learning was imperative during the pandemic, the universities were given the flexibility to decide on the mode of delivery and redesign their teaching approaches once the world returned to new normal. With the novel experience gained through virtual teaching and learning during Covid-19, higher education institutions across the world did not fully return to analogue teaching as before. The numerous advantages of virtual delivery resulted in the undergraduate and postgraduate programmes continuing with the online teaching or adopting the blended mode. This study presents the results of conducting Master of Arts in Linguistics programme in different modes, how the teaching approaches changed over the years during pre-pandemic, pandemic as well as post-pandemic eras. The research design contained mixed method approach which included a survey and recorded feedback obtained during the sessions. The new learner centred approach adopted during the post-pandemic times yielded more benefits and positive feedback when analysing the survey results and the comments given by the students.

**Keywords:** Collaborative learning · Post-pandemic learning experience · Postgraduate programmes · Re-designing online delivery · Virtual learning environment

## 1 Introduction

The world went through an extraordinary time from early 2020 and innovative approaches were adapted by higher education sector to deliver their study programmes with minimum interruption to the academic activities. 'Home Office', 'Work from Home', 'Virtual Learning', 'Digital Delivery' were very common expressions and practices that were the order of the day. In the context of universities in Sri Lanka the Government supported online delivery of study programmes by providing free data access to the LMS and ZOOM platform to all the staff and registered students [1–5]. This provision is present even to date, thus encouraging the continuation of digital delivery of


M. E. Auer et al. (Eds.): ICL 2023, LNNS 899, pp. 295–306, 2024.
https://doi.org/10.1007/978-3-031-51979-6_31

undergraduate and postgraduate programmes. During the pre-Covid 19 times only a few of the postgraduate programmes in the country were conducted entirely online, while the face-to-face delivery was more common. Some programmes adapted the blended approach too, although the onsite delivery was the standard method. The present study illustrates how the transition from onsite to blended learning and transformation to virtual learning occurred over a period of five years of conducting the Master of Arts in Linguistics at the University of Kelaniya, Sri Lanka from 2018 to 2023. The programme of two year's duration including a research component leading to a MA thesis is conducted in English and Sinhala medium (local language) during the weekend since the majority of the students are employed.

**Table 1.** Mode of delivery of the study programme

| Year | Mode of delivery |
|---|---|
| 2018/2019 | Face-to-face |
| 2019/2020 | Face-to-face |
| 2020/2021 | 75% face-to-face, 25% online |
| 2021/2022 | Online |
| 2022/2023 | Online |

As presented in Table 1 the first two batches experienced only onsite teaching whereas the third batch was confronted with mobility restrictions imposed due to Covid-19 towards the latter part of the taught programme which resulted in the remaining course content being delivered online after March 2020. The fourth and fifth batches experienced only online delivery.

With the world adapting to new normal in the latter part of 2022 the higher education sector was expected to return to traditional mode of delivery. Nevertheless, it was encouraged to continue with virtual delivery or adapting a blended approach and many universities in Sri Lanka took advantage of the experience gained and lessons learnt during the digital transformation to redesign the study programmes to suit online delivery or blended teaching. By the time the 2022/2023 programme of MA in Linguistics commenced, the restrictions due to the pandemic were no longer present. However, due to the numerous advantages of online delivery experienced during the previous two years it was decided to continue with the virtual sessions.

## 2 Purpose

The aim of this study is to review the experience of both purely onsite or virtual delivery of a postgraduate programme which later adapted new teaching strategies during the new normal, how the sessions were designed to be more student-centred than in the previous years and what the teacher and student perspectives were. Despite being given the choice of reverting back to face-to-face delivery with all mobility restrictions imposed during the pandemic being lifted, as illustrated in Table 1 the MA in Linguistics programme for the 2022/2023 batch also followed the mode of delivery of the previous year by resorting to online delivery of the entire taught programme comprising five course modules and the thesis. The majority of the students requested online delivery due to several reasons. The high cost of transport, time factor and the ability to follow the programme from the familiar home environment were some of the reasons cited by them. The Sinhala medium students were summoned during the first eight weeks for onsite lessons but resorted to online delivery subsequently due to student requests. Considering all the feedback from students and staff of the course conducted in the previous four years and further lessons learnt, the course was given a face-lift with more hours being allocated for teaching sessions of each module to integrate a more interactive and collaborative learning approach. How the new approach was received by the students could be analysed through the present study albeit the survey was conducted specifically with relevance to the course module in Applied Linguistics. It is expected that the results obtained from this study will contribute to deciding the future mode of delivery as well as the re-designing of the sessions relevant to other course modules in the study programme.

Numerous studies conducted relevant to higher education during and after the pandemic across the world bring to light that the traditional mode of delivery is no longer the order of the day [5–8]. Many a scholar strive to find the best possible direction for delivery of study programmes in the higher education sector in the post pandemic era [9–13]. It is expected that the results of this study will contribute to re-think and re-design the postgraduate programmes in the future in a more student friendly and student centred approach.

## 3 Approach

The subject of this study is the course module in Applied Linguistics in the two-year MA in Linguistics programme conducted by the researcher in 2022/2023 bearing 5 credits equivalent to 10 ECTS which had only 10 sessions of two hours each in the previous years. While redesigning the course modules the number of sessions allocated for Applied Linguistics was increased to 20 in 2022, thus allocating 40 teaching hours over a period of 20 weeks. In addition, a seminar of two hours in preparation for the examination was allocated as in the previous years. The increased number of sessions provided the researcher ample opportunities to redesign the course module into a more practical oriented series of sessions which facilitated the students to use the content learnt during the interactive group activities using breakout rooms of the Zoom platform. In the previous years, several studies were conducted on the experience of transformation of delivery of postgraduate programmes [1, 3], the results of which were also a deciding factor when redesigning the present course module. The batch comprised over 100 students in the English medium programme and 11 in the Sinhala medium. The English medium batch was of a multi-lingual, multi-ethnic and multi-religious nature, thus representing a cross section of the multiethnic, multi-cultural Sri Lanka which also provided the students to learn from each other about unique cultural and linguistic aspects. The Sinhala medium group consisted of Sinhala native speakers who also had a sound knowledge of English.

The newly designed structure of the Applied Linguistics course module comprised 20 sessions of 2 h each as stated above. During every session the first hour was allocated to delivery of content planned for the day using PowerPoint slides with embedded audio, video material. Images to support the content and a handout. The second hour was dedicated to collaborative learning via group activities in breakout rooms to use the learnt content practically and to present the results to the participants. The attendance was around 80% at each session, which was a very positive sign. The researcher conducted the 16 initial sessions of the course module while two other academic staff members were given 2 sessions each to cover the content of Translation and Second Language Teaching. Thus, the data analysed in this study only represent the feedback and responses given by the students related to the sessions conducted by the researcher.

## 3.1 Sample Sessions

The content of the 16 sessions conducted by the researcher contained selected topics of relevance due to the vastness of the field of Applied Linguistics. The content included among others an introduction along with the scope of Applied Linguistics, language and culture, cultural diversity and harmony, reflection of own culture, intercultural communication, cultural heritage (customs and traditions, folklore) first and second language acquisition, second and foreign language teaching, lesson planning, testing and evaluation, discourse analysis, language policies and language planning. The Postgraduate Learning Management System of the university was utilised to upload the content, additional reading material, handouts, outcomes of the group activities and lecture recordings as well as recordings of the group activities. As stated above, the lesson plan for each session comprised delivery of theoretical knowledge using PowerPoint presentations with relevant videos and examples. The second part of the session comprised of group activity in breakout rooms usually comprising 4 groups where the participants were assigned to the rooms automatically. Once the group activities were completed the students presented their work during the third session. Finally, feedback was obtained from the students regarding the lesson, content, and delivery as well as the outcome of the group activities presented by their peers which could be submitted orally or via chat facility. The following sample lessons will provide an insight into the structure of a typical 2-h session conducted for the 2022/2023 batch of students.

### 3.1.1 Session on First Language Acquisition (Day 2)

Deliver the content via PowerPoint presentation including milestones of first language acquisition supported by relevant videos (50 min).

Students are divided into 4 groups and sent to breakout rooms. 4 Worksheets are already uploaded on the LMS carrying following instructions and relevant images for each group activity to write a dialogue (40 min):

Group 1—A child refusing to sleep. Parents trying to put the child to sleep

Group 2—A group of children playing together

Group 3—A child in a store demanding a toy and refusing to leave

Group 4—Parents trying to feed a child who is refusing to eat (a very common occurrence in Sri Lankan households)

Medium: English for the English medium group and Sinhala for the Sinhala medium group of students.

Each group presents the dialogue (20 min)
Comments and feedback (10 min).

### 3.1.2 Session on Discourse Analysis (Day 10)

Deliver the content via PowerPoint presentation on discourse analysis including theories with relevant examples supported by video clips (60 min).

Students are divided into 4 groups and sent to breakout rooms. 4 Worksheets carrying authentic text/transcript are already uploaded on the LMS.

Task: Note down information related to the participants in the discourse (age, gender, social status, educational background, profession etc.). What is the setting, theme? Comment on the language used (35 min).

Medium: English for the English medium group and Sinhala for the Sinhala medium group of students.

Each group presents the findings (15 min).

Comments and feedback (10 min).

In comparison to the lesson plans for the previous batches, whether it was onsite or online delivery, group activities were never or rarely included due to time constraints since only 10 sessions of 20 h' duration were allocated for each course module. Due to the restructuring of the sessions with more hours allocated for each module (double the amount for Applied Linguistics) interactive sessions could be designed facilitating collaborative learning for the 2022/2023 batch of students.

Furthermore, a similar method was adopted during the seminar session of two hours conducted to prepare students for the written examination. The seminars for each module were conducted at the completion of the delivery of all five course modules. In the previous years revision of the ten sessions conducted for Applied Linguistics was carried out using slides to go through the content briefly to refresh the knowledge of the students. However, the session was designed with a more learner-centred approach for the present batch of students. After briefly presenting the main topics discussed during the 16 sessions, students were made aware of how to attempt the questions given at the examination. The paper would carry 8 questions and the students are required to answer only 4 questions within three hours. Instructions were given on how to structure the answers supported by examples. The group activity contained 4 questions from past papers where each group was assigned to present the structure of the answer for one question. This task gave the students an opportunity to share the knowledge with each other, discuss at length and provide a summary of what a possible answer would be. The groups presented their summaries, and all four groups could share the information, give their feedback and comment on the others' presentations. During the final comments and feedback session all the students unanimously agreed that this group activity prepared them very well for the final examination. It was a rewarding experience for the students as well as the researcher due to the active involvement of the participants. Thus, the seminar was not a passive session containing two hours of merely listening to content of 16 sessions being repeated once again as in the previous years.

### 3.2 Research Design

Mixed method was applied to obtain the results of the study using qualitative and quantitative data. Firstly, daily feedback of each session was obtained via chat and oral comments retrieved through the video recording and Zoom chat. Secondly, a questionnaire was administered at the end of the module on Applied Linguistics comprising data obtained through Likert scale based responses and open-ended questions via Google form. Thirdly, the results obtained at the examination for this module were analysed to ascertain the level of performance of the students. It is noteworthy that the previous batches also responded to online questionnaire surveys at the end of the course module on Applied Linguistics barring the 2018/2019 batch where only oral feedback was obtained at the end of face-to-face sessions. The advantage of online delivery is the provision to record all the sessions including oral feedback. The recordings of lectures as well as presentations of group activity were made available to the students immediately on the Postgraduate Learning Management System (PGLMS) of the university.

The online survey conducted at the completion of the 16 sessions by the researcher was modeled on a similar survey administered for the students of the previous batch in 2021/2022 [1]. Due to the changes introduced to the present batch new questions were introduced to the present survey i.e., questions 11, 12 and 13 in Section two. The Google form comprised 22 questions in three sections. The first segment contained five questions on demographic data including gender, age group, district of residence, employment status and occupation. As illustrated in Table 2 the second section of the questionnaire contained thirteen questions regarding the learning experience based on Likert scale responses ranging from 1) strongly agree to 5) strongly disagree. The responses to questions in Section two provided data for a statistical analysis of the perceptions of the students on the new approach of content delivery of the course module in Applied Linguistics.

Table 3 indicates the four open ended questions given in section three of the survey.

The questions given in Section three aimed at having access to qualitative data from the students on what appealed to them mostly during the sessions on Applied Linguistics, the advantages and disadvantages of solely learning online and suggestions for improvement. The questions in this section focused on the overall opinion on the mode of delivery, accessibility issues, regular participation as well as the learner centred approach adopted during the sessions and online participation enabling to obtain a qualitative analysis of the study programme.

## 4 Actual Outcomes

The Masters programme continues to date with online teaching, remote supervision of thesis and online viva-voce examinations. The demographic data obtained from Section one of the questionnaire survey revealed that only a quarter of students of the subject of this study reside in the two districts close to the location of the university, namely Colombo and Gampaha. Therefore, continuation of online delivery of the programme is a blessing in disguise for the students as stated by them in the open-ended questions of Section three of the survey. Furthermore, 95% of the students are employed fulltime, thus having only the weekend to devote to their personal matters. The majority of them

**Table 2.** Section two of the survey

| Component | Question | Scale |
|---|---|---|
| 1 | My overall online learning experience during the MA programme was positive | 1. Strongly agree<br>2. Agree<br>3. Neither agree nor disagree<br>4. Disagree<br>5. Strongly disagree |
| 2 | I participated in the majority of the lectures conducted online | |
| 3 | I had no difficulties in adjusting to following the entire course work of the MA programme online | |
| 4 | The Learning Management System (PGS LMS) was helpful to access lecture handouts and preparatory material including videos before the lecture was conducted online | |
| 5 | The Learning Management System (PGS LMS) was helpful to access lecture recordings and additional material uploaded by the lecturer to revise the content | |
| 6 | It was more convenient than travelling to the university every Saturday for onsite lectures | |
| 7 | The lectures were clear and easy to follow | |
| 8 | The recordings of the lectures helped to understand the lecture content better | |
| 9 | I had internet issues and could not join the online sessions | |
| 10 | I prefer online mode of learning against face-to-face- sessions | |
| 11 | The group activities helped me to understand the content delivered during the lecture better | |
| 12 | The group activities also helped me to get to know my batch mates and interact with them online | |
| 13 | Through the group activities I could learn from my peers and enhance my knowledge | |

are between the ages of 30 and 45. Hence, many have family and work commitments that leave them limited time to focus on their studies. Thus, digital delivery is ideal for working students or students with young children as they can focus on the home front as well as learning. There were comments from some students that their children also join the sessions if any topics of their interest are discussed i.e., reflection on own culture, language acquisition, folklore. Thus, interactive sessions focusing on practical use of the content learnt during the lectures have been extremely beneficial. The analysis of qualitative and quantitative data reveal that the new teaching and learning approach has been well received by the students.

**Table 3.** Section three of the survey

| Component | Question |
|---|---|
| 1 | What did you most like in the Applied Linguistics lecture sessions? |
| 2 | In your opinion what were the advantages of following the MA in Linguistics programme entirely online? |
| 3 | In your opinion what were the disadvantages of following the MA in Linguistics programme entirely online? |
| 4 | Please give any suggestions for improvement |

## 4.1 Analysis of the Quantitative Data

The data presented in the section two of the questionnaire reveal the positive attitude towards digital delivery and new approach to teaching as depicted in Fig. 1.

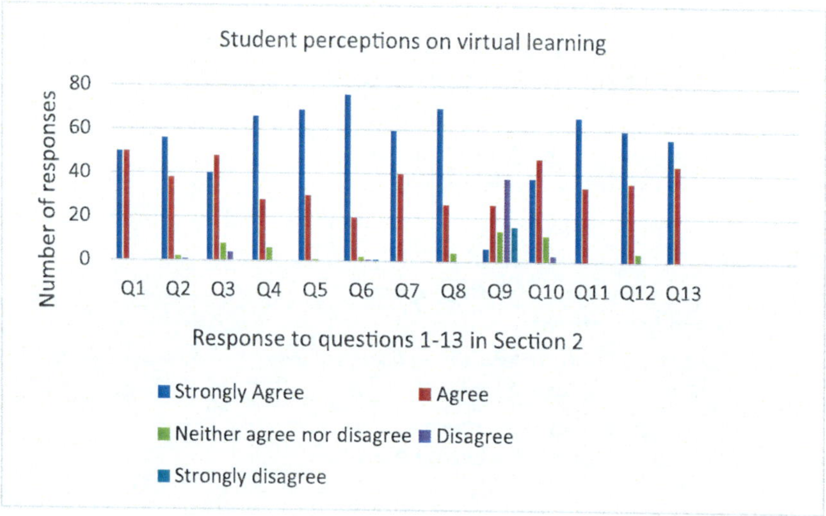


**Fig. 1.** Postgraduate student perceptions on online learning

Apart from a very few indecisive or non-committal students who responded with 'neither agree not disagree' all the others showed their appreciation of the online sessions. For Questions 1, 7, and 13 on the overall positive experience, clarity of lectures and ability to easily follow them as well as learning from peers through group activities and enhancing the knowledge, all the responses were stated as 'strongly agree' or 'agree'. With regard to the rest of the questions too, the majority responded very positively, thus appreciating the new approach adapted to delivering the sessions in Applied Linguistics.

## 4.2 Analysis of the Qualitative Data

All the respondents mentioned that the sessions were extremely useful, and that they highly appreciated the student-centred approach. The fact that there was ample opportunity for interaction between Sinhala and Tamil native speakers and to learn from each other was a mutually beneficial experience. Sinhala is the language spoken by the majority in Sri Lanka whereas Tamil is the next major language spoken in the country by the Tamil and Muslim communities [14]. As mentioned above, the English medium group consisted of multi-cultural and multi-linguistic participants that provided many opportunities to interact with each other, which would not have been possible if no interactive group activities were added to the sessions. None of the respondents mentioned the preference for any onsite sessions as against the previous batch where 20% of them wished for face-to-face delivery [1]. The four open-ended questions in the survey presented similar views of the students. A summary of the responses is given below:

**Q1 What did you most like in the applied linguistics lecture sessions?**

Methodology used, interactive sessions, collaborative learning, group activities, content, presentations, student centered learning, inclusion of relevant videos, using examples from day-to-day life, sharing of practical experiences with other participants, learning from each other.

**Q2 In your opinion what were the advantages of following the MA in Linguistics programme entirely online?**

No issues with long distance travelling, saving transport cost, travel time and energy, easy to follow sessions, use of recordings, easy to focus and concentrate from the comfort of home.

**Q3 In your opinion what were the disadvantages of following the MA in linguistics programme entirely online?**

Connectivity issues, power cuts and power failures, lack of physical presence of peers and lecturers, noise from home surroundings disturbing during group activities when microphone is on.

**Q4 Please give any suggestions for improvement.**

Including a revision session onsite, conducting a mock examination.

The feedback at the end of each session as well as the questionnaire survey demonstrated the success of the new structure of the sessions. Extremely positive feedback was evident on Likert scale responses as well as the views expressed on open ended questions. The virtual delivery was not felt by the students as a disturbing factor since all the sessions contained interactive group activities which was a most rewarding experience according to their statements. With the re-designing of the sessions during the academic year 2022/2023 it is evident that the students found that the advantages and benefits of online learning outweigh the disadvantages. The group activities and the learner-centred approach during the sessions were mentioned by 98% of the students as what they most liked about the sessions. Saving of travel time, cost and energy was mentioned by the majority of the students as the advantages. Since at least 65% of the students come from faraway locations virtual delivery was a blessing for them saying time and money. The unavoidable connectivity issues, power cuts and power failures were the disadvantages mentioned by most students while a few mentioned the lack of physical proximity to the lectures and peers as a drawback of virtual learning. Nevertheless, all acknowledged the fact that under the present circumstances of rising cost of living and high inflation resulting in an economic crisis in the country, virtual learning is the best option available. Thus, this study illustrates how digital delivery can be rewarding even when the circumstances are not the same as during Covid-19 and where no one was forced into 100% virtual teaching. The fallacy of the success of purely face-to-face teaching was proven otherwise by the positive student feedback on virtual delivery. Hence, the traditional view on the success of programmes in Humanities only if they are delivered onsite could be proven wrong through the feedback given. Furthermore, the analysis of results prove that the student performance was far better than during the times of face-to-face delivery. In the module for Applied Linguistics only around 7% of the students scored less than 70% marks which earned the majority a grade 'A'.

Moreover, the interactive sessions brought the hidden talents and creativity of the students to light. There were opportunities to present their acting, singing, recitation and creative abilities. The group activities presented results of collective efforts and contributions of all the participants involved barring a few who experienced connectivity issues.

The challenges faced during face-to-face sessions were no longer present. For instance, the technical difficulties faced by the lecturers in setting up the sound systems, laptop computer and multimedia projector were no longer a concern. Having all the required material at hand saved in one's own computer avoids issues of being confronted with unknown slow computers or ones that suddenly shut down at their own will that are available in the lecture halls. The full two hours could be utilised to deliver the lesson planned for the day without waiting for late-comers, ones who take time to partake a meal between the lectures or those leaving early to get the last long-distance buses to their destinations. Thus, the time allocated was more than sufficient to deliver the lesson, conduct group activities, present them in the main room and obtain feedback from the peers as well as the lecturer. Albeit the lecturers and students never met physically there was ample opportunity to know their strengths and weaknesses. Some students also got the chance to meet each other during a cultural exchange programme conducted onsite, enabling them to further practice the knowledge on culture and intercultural communication in real-life situations. The new method of facilitating the students to practically apply the knowledge gained during each session made them more self-confident. The lecturers also admitted that they are more relaxed during digital delivery since they only have to focus on the session and not be concerned with peripheral issues i.e., slow or malfunctioning computers and projectors, power failures in lecture halls.

## 5 Conclusions and Recommendations

The analysis of the feedback during the virtual sessions and the questionnaire survey reveals that the new approach to teaching was much appreciated by the students due to numerous advantages it provided. The student-centred approach provided opportunities for deeper understanding of the content delivered with interactive collaborative learning activities integrated during the second hour of each session. The improvements to the structure and organisation of the weekly sessions, inclusion of group activities was among the highlights mentioned by the students. Multicultural. Multi-religious background of participants paved way for broadening one's horizons, learn from each other and respect each culture, language, and religious beliefs. The composition of the group was a mirror on Sri Lanka. The sessions were an example not only for effective delivery of subject knowledge but also to demonstrate how effective it is to broaden one's horizons in a multi-cultural, multi-religious, multi-ethnic setting to learn from each other about different customs and traditions, lifestyle and beliefs, which was clearly an ideal setting for Applied Linguistics. The creativity of the students and their numerous hidden talents including reciting, acting, singing (folk songs) could be highlighted during the group activities that gave them self-esteem and pride of one's ethnic and cultural background. Due to the extremely positive feedback and the rewarding experience gained not only by the students but also by the researcher, it is highly recommended to design future programmes too in a similar student-centred approach facilitating collaborative learning.

**Acknowledgements.** The author wishes to acknowledge the support of the Department of Linguistics, University of Kelaniya, Sri Lanka and the contribution of students of MA in Linguistics by way of feedback and participation in the questionnaire surveys conducted.

## References

1. Premawardhena, N.C.: The impact of virtual learning on undergraduate and postgraduate programmes: a Sri Lankan Experience. In: Auer, M.E., Pachatz, W., Rüütmann, T. (eds) Learning in the Age of Digital and Green Transition. ICL 2022. Lecture Notes in Networks and Systems, vol 633. Springer, Cham, pp. 312–323 (2023)
2. Premawardhena, N.C.: New dimensions in online teaching and learning of foreign languages: proximity at a distance. In: Auer, M.E., Hortsch, H., Michler, O., Köhler, T. (eds) Mobility for Smart Cities and Regional Development - Challenges for Higher Education. ICL 2021. Lecture Notes in Networks and Systems, vol 389. Springer, Cham, pp. 622–633 (2022a)
3. Premawardhena, N.C.: Remote supervision: a boost for graduate students. In: Auer, M.E., Hortsch, H., Michler, O., Köhler, T. (eds) Mobility for Smart Cities and Regional Development - Challenges for Higher Education. ICL 2021. Lecture Notes in Networks and Systems, vol 389. Springer, Cham, pp. 633–644 (2022b)
4. Udara, S.P.R., Arachchige, K.L.T., Sathsara, P.P. et al.: Immediate actions to minimize the impact of COVID-19 on education. Intern. J. Multidiscipl. Res. Publicat. (IJMRAP) **4**(2), 25–34 (2021)
5. Hayashi, R., Garcia, M., Maddawin, A., Hewagamage, K.P.: Online Learning in Sri Lanka's Higher Education Institutions during the COVID-19 Pandemic. ADB Briefs No 151 (2020)
6. Bashir, A., et al.: Post-COVID-19 adaptations; the shifts towards online learning, hybrid course delivery and the implications for biosciences courses in the higher education setting. Front. Educ. **6**, 711619 (2021). https://doi.org/10.3389/feduc.2021.711619
7. Singh, J., Steele, K., Singh, L.: Combining the best of online and face-to-face learning: hybrid and blended learning approach for COVID-19, post vaccine, & post-pandemic world. J. Educ. Technol. Syst. **50**(2), 140–171 (2021)

8. García-Morales, V.J., Garrido-Moreno, A., Martín-Rojas, R.: The transformation of higher education after the COVID disruption: emerging challenges in an online learning scenario. Front Psychol. (2021)
9. Carrillo, C., Assuncao, F.: Online teaching and learning practices in teacher education after COVID-19: lessons learnt from the literature. J. Educ. Teach. (2022). https://doi.org/10.1080/02607476.2022.2153018
10. Martin, U.: Education after COVID (2021) Education After COVID -- THE Journal https://thejournal.com/articles/2021/05/12/education-after-covid.aspx Last accessed 2023/06/02
11. Rasli, A., Tee, M., Lai, Y.L., Tiu, Z.C., Soon, E.H.: Post-COVID-19 strategies for higher education institutions in dealing with unknown and uncertainties. Front. Educ. **7**, 992063 (2022). https://doi.org/10.3389/feduc.2022.992063
12. Rapanta, C., Botturi, L., Goodyear, P., et al.: Online university teaching during and after the Covid-19 crisis: refocusing teacher presence and learning activity. Postdigit. Sci. Educ. **2**, 923–945 (2020). https://doi.org/10.1007/s42438-020-00155-y
13. Leal Filho, W., Lange Salvia, A., Abubakar, I.R. et al.: Impacts of the COVID-19 pandemic on routines of higher education institutions: a global perspective. Sustainability **14** (2022). https://doi.org/ https://doi.org/10.3390/su142114105
14. Jayasundara, N.S.: Language education policy planning in Sri Lanka: concern for unity, reality and rationality. Intern. J. Sci. Res. Public. **9**(2), 199–206 (2019). https://doi.org/10.29322/IJSRP.9.02.2019.p8626

# The Plausibility of Personalizing Interfaces Using the Big Five Personality Traits

Dina A. Zekry(✉) and Gerard T. McKee

The British University In Egypt, Cairo, Egypt
dinazekry@hotmail.com, Gerard.McKee@bue.edu.eg

**Abstract.** The research described in this paper reports the plausibility of personalizing interfaces based on the Big Five Personality Traits to create an enhanced user experience. Three datasets are used in this research: the ICS-BUE dataset, the AraPersonality dataset, and Open Psychometrics dataset. The data analysis consists of clustering the participants' data and finding trait combinations. HDB-Scan and K-means methods are used to create clusters. The clusters produced did not have distinct dominant traits nor varying trait combinations. The results show that openness is the most dominant trait across all datasets, neuroticism is the least dominant trait across all datasets, and extraversion and conscientiousness alternate in dominance in the third and fourth positions. Accordingly, the predominant trait combinations are OACEN and OAECN. The AraPersonality dataset has agreeableness as the most dominant trait while the other two datasets have openness as the most dominant trait. The Open Psychometrics dataset has conscientiousness as the second dominant trait. Based on these findings, interface designers should not depend on a single dominant trait and should consider trait combinations. Accordingly, interface designs should cater to the predominant trait combination.

**Keywords:** The big five personality traits framework · Human-computer interaction · Clustering · Interface design · Personalizing interfaces

## 1 Introduction

A Learning Management System (LMS) is a web-based software application that is used to plan, implement, and assess a specific learning process [1]. A user interface (UI) is the medium through which the users interact with the system [1]. Interface design focuses on anticipating what users might need to do and ensuring that the interface has elements that are easy to access, understand, and use to facilitate those actions. Personalised interface design is a way that tailors the interface design to suit the user's preferences and needs [1]. Personalisation increases user engagement, satisfaction, and productivity. This research focuses on role-based personalisation where the users are grouped together based on a set of pre-determined and similar characteristics [1]. Personality analysis is used as the data for the user characteristics. The Big Five personality traits test is a framework to assess an individual's personality, dividing an individual's personality along five main traits: conscientiousness, agreeableness, openness, extraversion, and neuroticism [2]. An individual scores varying percentages on each trait and the trait with the highest percentage is referred to as the dominant trait.


M. E. Auer et al. (Eds.): ICL 2023, LNNS 899, pp. 307–318, 2024.
https://doi.org/10.1007/978-3-031-51979-6_32

### 1.1 The Big Five Personality Traits

The Big five personality traits test is a framework to assess an individual's personality. The framework divides an individual's personality to five main traits. Namely, conscientiousness, agreeableness, openness, extraversion, and neuroticism [3, 4, 5, 6, 7]. An individual scores varying percentages on each trait and the most dominant trait is the trait with the highest percentage.

The framework relies on being a continuum where an individual has each trait that occurs along the spectrum. Meaning, conscientiousness ranges from impulsive and disorganized traits to disciplined and careful traits, agreeableness ranges from suspicious and uncooperative traits to trusting and helpful traits, openness to experience ranges from a preference for routine and practical traits to imaginative and spontaneous traits, extraversion ranges from reserved and thoughtful traits to sociable and fun-loving traits, and neuroticism ranges from calm and confidence traits to anxiety and pessimism traits.

Accordingly, each trait can be said to measure certain characteristics. Openness measures acceptance of new ideas, adventures, and innovation. Conscientiousness measures an individual's sense of consciousness, duty, discipline, and responsibilities. Extraversion measures an individual's social activity, source of joy, and integration with the community. Agreeableness measures how adaptive the person is he is a collaborating person, trusts and satisfies others needs at the expense of him/herself. Neuroticism measures the emotional instability of the individual, temper, negative emotions, and lack of self-confidence.

### 1.2 Clustering

Clustering is a machine learning technique that divides groups of objects into similar categories based on similar features [8, 9]. Clustering enables splitting the data into subsets of data where the data objects have high inter-similarity and low intra-similarity [8]. Clustered data enables pattern detection and helps in insight identification [8, 9]. Also, clustering helps divide the market into subgroups which allows customized/personalized marketing [9, 10].

The Big Five Personality Traits clustering space is a five-dimensional space. Two techniques are investigated in this paper. Namely, HDBScan and K-means.

HDBScan is a method that is based on the DBScan algorithm that is used to group like data together and can deal with high dimensional data [10]. HDBScan can handle real-world data with varying densities and is a fast algorithm. It works by finding the distance between each point and its farthest neighbour defined by the minimum samples parameter [10]. HDBScan is chosen for two main reasons, it can handle high dimensional space data and it does not require previous knowledge on the number of clusters to be produces or the centroids of the clusters [10].

K-means clusters the data by separating samples in K groups of equal variances and measuring how internally coherent the clusters are [11]. K-means clustering assigns N data points into K clusters so that similar data points can be grouped together. The algorithm checks each data point and calculates the distance from all centroids then assigns its membership to the nearest centroid [9] [11]. The number of clusters is determined by the K value. If the K value is unknown the Elbow and Silhouette method can be used to find the optimum value of K [9, 11].

### 1.3 Research Aims

The research aims to report on the plausibility of personalising interfaces based on the Big Five Personality Traits framework. This is to promote personalising the LMS interfaces based on the students' personality to enhance the students' experience. Each student personality test result produces a trait combination that expresses the student's personality. The research aims to cluster students' personality test results so that students with similar personality are grouped together. Accordingly, the LMS can be designed to cater for each of the distinct clusters produced. This led to the study of three research hypotheses:

H1 Well separated clusters with distinct dominant personality traits will be formed from a student personality dataset collected from ICS-BUE students.
H2 Egyptian personality consists of the same dominant traits and trait combination. This was explored using the AraPersonality dataset.
H3 Different nationalities would have varying dominant traits and trait combination. This was explored using the Open Psychometric dataset.

## 2 Datasets

The ICS-BUE dataset is gathered by the paper authors. Authors in [13] gathered the AraPersonality dataset from Egyptian twitter users. The Open Psychometrics dataset is an open-source dataset available on Kaggle and gathered by an open psychometric website that allows online users to take the Big Five Personality traits test [12, 14].

### 2.1 ICS-BUE Dataset

The dataset is gathered on the campus of the British University in Egypt (BUE). All participants are Informatics and Computer Sciences (ICS) students. The batches chosen for the research are prep year, year one after prep, and year two after prep. The participants have the same computer and domain expertise. Participants age from 17 to 21 years old. Participants are males and females. However, the data has not been divided based on gender. Participants are Egyptian citizens who mostly live in Cairo. Some participants used to live in neighbouring governments.

All students took the Big Five Personality Traits test using Truity website [12]. The participants submitted the results in pdf format. The pdf contains the percentage of each trait for the participant. The pdfs were transformed to csv format where the columns represent the Big Five Personality Traits, rows represent each participants' data, and each cell contains the percentage of the trait.

An email was sent to students informing them with the experiment aim, location, and asking them to participate. Two incentives were used to attract the students. First, the students were promised to win a surprise gift. The gift would be given to 5 randomly chosen students from each batch. Second, the students would be given candy after they completed the questionnaire. Also, the experiment was setup on the university e-learning with an instructions pdf to enable the students to complete the questionnaires online.

The following are the steps in which the experiment was conducted:

(1) The student comes into the lab.
(2) The researcher explains the experiment.
(3) The researcher hands the student a consent form.
(4) The student takes his/her time to read the consent form.
(5) The student decides to participate then he/she signs the consent form.

The participant takes the Big Five Personality Traits test through Truity website [12]. When the student completes the test, he/she saves the results in PDF form and uploads the PDF on the e-learning in a link specified for the experiment.

The average time to complete the experiment was 15 min. Students/participants with weak English vocabulary or slow readers take an average of 20 min. The following are terminologies and questions that almost more than half of the participants required assistance to understand "I often feel blue", "I seldom feel blue", the term "abstract ideas", "chores", "flourishing", "inquisitive", and "vivid imagination". The experiment was setup for 4 days and the total number of participants was 208. 28 participants completed the questionnaires online while 180 took the questionnaires on campus. Table 1 shows the total number of participants according to batches.

**Table 1.** Number of participants per batch

| Batch | Number of participants |
|---|---|
| Prep year | 30 |
| Year 1 after prep | 107 |
| Year 2 after prep | 71 |
| Total | 208 |

The prep year batch has the lowest number of participants with only a total of 30 participants. Also, the participants struggled to understand the correlation between the big five personality traits test and interface design. Prep year students are taught introductory modules that discuss the basics of Informatics and Computer Sciences. However, in preparatory year there are no modules that discuss human computer interaction, or user experience. Accordingly, the students had just enough information to participate in the experiment.

Year one after prep batch has the highest number of participants with 107 participants. Year one after prep study a module named Human computer interaction. The participants have clear understanding of the experiment and were intrigued to complete it. After completing the questionnaires, multiple students commented that they enjoyed the study and were pleased with the Big Five Personality Traits test. Additionally, multiple students wanted to understand their personality test results in depth.

Year two after prep batch has a total of 71 participants. The participants had clear understanding of the experiment and would complete it with fewer questions in comparison to other batches. The participants showed interest in the next phase of the research where they wanted to know how the data will be used.

## 2.2 AraPersonality Dataset

The AraPersonality dataset is provided by authors in [13]. The authors gathered the data through a questionnaire that focused on Egyptian participants. The participants are twitter users, and the questionnaire is presented in Arabic. The paper aimed to provide a dataset for the Egyptian personality traits curated using the Arabic language. This is due to the lack of gathering personality traits from Arabic region and the lack of presenting the questionnaire in Arabic dialect. The paper randomly selected 5 questions to evaluate each trait from Costa and McCrae's (1992) NEO-PI-R Domains (10-item scale). Accordingly, the questionnaire consisted of 25 questions in Arabic. The authors produced classification models to classify users' personality according to twitter feeds. However, this research focuses on the questionnaire dataset results. The dataset consists of 92 participants. The questionnaire results are saved in csv file in the same format as the ICS-BUE dataset.

## 2.3 Open Psychometric Dataset

The Open Psychometrics dataset is an open-source dataset available on Kaggle and gathered by an open psychometric website that allows online users to take the Big Five Personality traits test [14]. The dataset contains 1,015,342 questionnaire answers. The website recorded the location of the submissions. The dataset contains submissions from 217 countries. The greatest number of submissions is 443,320 from the United States of America (USA). There are 788 submissions from Egypt and 57,371 from Great Britain. The total number of submissions from different countries is 822,054. Table 2 shows the country with its corresponding number of participants and the total number of participants from all countries.

**Table 2.** Number of participants per Country

| Country | Number of participants |
|---|---|
| United States of America | 443,320 |
| Great Britain | 57,371 |
| Egypt | 788 |
| All countries | 822,054 |

The format of the dataset is in a.csv file. The dataset consists of 50 question answers on the Likert scale. Accordingly, there was a need to calculate the percentage of each participant's personality traits. For the trait calculation the Eqs. 1, 2, 3, 4, and 5 were used [4]. Q stands for question followed by the number of questions. For each variable the Likert score corresponding to the question number is substituted in the equation. The scores at the beginning of the equation are fixed scores for each trait. Each equation calculates a trait, and each trait is denoted by its first letter. The total score calculated ranges from 0 to 40 for each trait. Accordingly, a trait that has score equal to 40 is equivalent to 100%.

$$E = 20 + Q1 - Q6 + Q11 - Q16 + Q21 - Q26 + Q31 - Q36 + Q41 - Q46 \quad (1)$$

$$A = 14 - Q2 + Q7 - Q12 + Q17 - Q22 + Q27 - Q32 + Q37 + Q42 + Q47 \quad (2)$$

$$C = 14 + Q3 - Q8 + Q13 - Q18 + Q23 - Q28 + Q33 - Q38 + Q43 + Q4 \quad (3)$$

$$N = 38 - Q4 + Q9 - Q14 + Q19 - Q24 - Q29 - Q34 - Q39 - Q44 - Q49 \quad (4)$$

$$O = 8 + Q5 - Q10 + Q15 - Q20 + Q25 - Q30 + Q35 + Q40 + Q45 + Q50 \quad (5)$$

## 3 Results and Discussion

The following paragraph walks through the research steps/cycle. In the beginning the research focused on students from the British University in Egypt. After clustering and analysing the data a need to compare findings with a more general dataset emerged. The initial hypothesis was that well separated clusters with distinct dominant personality traits will be formed from the data collected from ICS-BUE. However, the clusters formed indicated that openness and agreeableness are the dominant traits followed by extraversion, conscientiousness, and neuroticism. Moreover, the clusters formed had the same dominant traits which meant that there was no distinction in the dominant traits. The clusters produced similar personality trait combinations.

Accordingly, AraPersonality dataset was utilized. AraPersonality was chosen because it focuses on Egyptians' personality traits. The second hypothesis was made that the Egyptian personality consists of the same dominant traits and there is a high chance that the same trait combinations are discovered. The clustering discovered that openness and agreeableness remain to be the dominant traits. However, the order of the other traits varies slightly where the third dominant trait is conscientiousness followed by extraversion and neuroticism. The acronym for AraPersonality dataset is AOCEN and for the ICS-BUE dataset is OAECN. The clusters produced show that openness and agreeableness are the dominant traits, neuroticism is the least dominant/recessive trait, and conscientiousness and extraversion alternated in arrangement.

Before formulating a conclusion, it was decided to explore an Open Psychometrics dataset and compare the results. The third hypothesis was made that different nationalities will have different dominant traits and trait combination. The Open Psychometrics dataset consists of participants from different countries such as Unites States of America (USA), Great Britain (GB), Egypt (EG), and other countries. The Open Psychometrics dataset produced the following combination of dominant traits openness, conscientiousness, agreeableness, extraversion, and neuroticism (OCAEN). Each of the following countries Egypt, USA, and GB were clustered separately. However, the same combination of dominant traits was discovered (OCAEN). The following was observed across all datasets, neuroticism is the least dominant trait in all datasets, openness is the most dominant trait in all datasets, agreeableness is the second most dominant trait, and extraversion and conscientiousness alternate in the third and fourth position of dominance.

There are two main steps to find the dominant traits in each dataset. First, clustering the data using HDBscan and/or K-means. Second, calculating the highest average score for each trait and arranging the traits from highest to lowest average score in each cluster. Accordingly, each cluster will produce an acronym that indicated the order of its dominant traits. The dataset dominant trait can be identified from the acronyms produced.

### 3.1 ICS-BUE Dataset

Table 3 shows the 3 methods used for clustering, the total number of clusters produces, each cluster number, and the acronym given. Each acronym is a combination of the first letter from the Big Five Personality traits. The two dominant traits in all the clusters are openness and agreeableness. The least dominant trait is neuroticism. Extraversion is the third dominant trait followed by consciousness.

**Table 3.** ICS-BUE dataset acronym

| Method | Total number of clusters | Number of data points | Cluster label | Acronym |
|---|---|---|---|---|
| HDBScan-Euclidean | 17 | 26 | 17 | OACEN |
| | | 6 | 7 | ECOAN |
| | | 6 | 9 | AOECN |
| HDBScan-Manhattan | 3 | 3 | 0 | OANCE |
| | | 8 | 1 | AOCEN |
| | | 106 | 2 | OACEN |
| K-Means | 2 | 97 | 0 | OACEN |
| | | 100 | 1 | AONEC |
| | 3 | 53 | 0 | AONEC |
| | | 71 | 1 | OACEN |
| | | 73 | 2 | OAECN |
| | 4 | 65 | 0 | OACEN |
| | | 53 | 1 | AENOC |
| | | 55 | 2 | OAECN |
| | | 24 | 3 | AENOC |

Table 4 shows each year with its corresponding clusters and acronyms. The most dominant trait in each batch is openness and agreeableness.

When all the data is clustered together openness and agreeableness are the dominant traits. Also, when the clusters are formed for each batch separately, openness and agreeableness are the dominant traits. Accordingly, the most dominant trait is openness followed by agreeableness. The dominant trait is calculated by the following method. Each column represents a trait of the big five personality traits. Each row corresponds

**Table 4.** ICS-BUE dataset acronyms per year

| Year | Cluster Label | Number of data points | Acronym |
|---|---|---|---|
| prep | 0 | 9 | AOCNE |
| | 1 | 11 | EONAC |
| | 2 | 8 | OCEAN |
| Year 1 | 0 | 60 | OACEN |
| | 1 | 40 | ANOEC |
| Year 2 | 0 | 13 | OANCE |
| | 1 | 23 | OECAN |
| | 2 | 33 | OAENC |

to an individual and each cell represents the percentage the participant scored on a personality trait. Each cluster produced consist of a set of records/rows that represent the participants who belong to that cluster. The average score is calculated for each trait. Accordingly, the highest average score indicates the dominant trait in the cluster. The dominant trait is calculated for each cluster produces. Additionally, in each cluster the traits are ordered from the highest average score to the lowest average score. This is used to represent the cluster in terms of initial letters of the big five personality traits. For instance, AOCNE represents cluster 0 produced from prep year dataset where the dominant trait is agreeableness (highest average score) and the recessive trait is extraversion (lowest average score).

To further investigate the dominant traits in the datasets 60 records of participants who scored the highest percentage of openness and agreeableness were discarded then the clusters were formed again. It was expected that after removing those participants the clusters will indicate the dominance of other traits. However, openness and agreeableness remained the dominant traits. This was tested for each batch as well as the dataset in general.

Accordingly, there is no doubt that the dominant traits for the participants are openness and agreeableness. The other traits are found with varying percentages. Neuroticism is the lowest/ most recessive trait. Consciousness appears to be the third most dominant trait more than extraversion.

In conclusion, KNN produces better clusters than HDBscan. HDBscan using Manhattan distance produces better clusters than Euclidean distance. The use of elbow method with silhouette method enables verification of the optimum K value. Openness and agreeableness are the dominant traits in the collected dataset. The interfaces to be designed should cater for the dominant traits.

### 3.2 AraPersonality Dataset

K-means produced two clusters of the following acronym OACEN and OACNE. Table 5 shows the acronyms produced for the AraPersonality dataset. The clusters show that openness and agreeableness are the dominant traits. Conscientiousness is the third dominant trait. Neuroticism and extraversion alternate in cluster one and two for the fourth and fifth position.

**Table 5.** AraPersonality acronyms

| Method | Cluster label | Number of data points | Acronym |
|---|---|---|---|
| K-Means | 0 | 48 | OACEN |
| | 1 | 44 | OACNE |

## 3.3 Open Psychometric Dataset

In the Open Psychometrics dataset, the dominant trait is openness. Table 6 shows the clusters produced and their corresponding acronym.

**Table 6.** Open Psychometric dataset

| Country | K value | Cluster label | Number of data points | Acronym |
|---|---|---|---|---|
| Egypt | K = 2 | 0 | 498 | OCEAN |
| | | 1 | 290 | OCANE |
| | K = 3 | 0 | 290 | OCANE |
| | | 1 | 236 | OCNEA |
| | | 2 | 262 | OEACN |
| Great Britain | K = 2 | 0 | 28789 | OECNA |
| | | 1 | 28582 | OCANE |
| | K = 3 | 0 | 28582 | OCANE |
| | | 1 | 13438 | OEACN |
| | | 2 | 15351 | ONCEA |
| USA | K = 2 | 0 | 28789 | OECNA |
| | | 1 | 440,099 | OCANE |
| | K = 3 | 0 | 170899 | OCANE |
| | | 1 | 163687 | OCAEN |
| | | 2 | 134302 | OCNAE |
| Open Psychometric Dataset | K = 3 | 0 | 333314 | OCANE |
| | | 1 | 287068 | ONCEA |
| | | 2 | 245902 | OEACN |

The Egyptian participants from the Open Psychometrics dataset have openness as the dominant trait. However, consciousness is the second dominant trait. In the other two datasets agreeableness was the second dominant trait. In the USA, the second dominant trait is consciousness as well. In Great Britain the second dominant trait is split between consciousness and extraversion.

### 3.4 K = 1 for All Datasets

Each dataset is clustered with K-means into one cluster. This is to arrange the traits in terms of dominance with relevance to all datasets. Table 7 shows the produced clusters. The ICS-BUE dataset trait combination is openness, agreeableness, extraversion, conscientiousness, neuroticism. The AraPersonality dataset trait combination is agreeableness, openness, conscientiousness, extraversion, and neuroticism. The Open Psychometrics dataset trait combination is openness, conscientiousness, agreeableness, extraversion, neuroticism. The three countries that are chosen from the Open Psychometrics dataset namely United States of America, Great Britain, and Egypt produce the following trait combination openness, conscientiousness, agreeableness, extraversion, neuroticism.

**Table 7.** Acronyms per dataset

| Dataset | Acronym |
|---|---|
| ICS-BUE | OAECN |
| AraPersonality | AOCEN |
| Open psychometrics | OCAEN |
| EGY-Open Psychometrics | OCAEN |
| US-Open psychometrics | OCAEN |
| GB-Open psychometrics | OCAEN |

Openness is the most dominant trait in all datasets. The locally gathered datasets (gathered in Egypt) have agreeableness as the second dominant trait. The Open Psychometrics dataset has consciousness as the second dominant trait. Extraversion is mostly the fourth dominant trait. Neuroticism is the least dominant trait.

The datasets used all apply the Big Five Personality Traits test. However, in the ICS-BUE dataset the Truity website is used to produce the trait scores in percentages. The authors of the AraPersonality dataset chose 5 questions to represent each of the five traits and calculated the score of the traits out of 100 [13]. The Open Psychometrics dataset used the psychometric website to calculate the personality traits score out of 40 [14]. Accordingly, each dataset gathered the data in a different method and calculated the scores in a different method as well. However, all represent the user personality in terms of the Big Five Personality Traits. The variation in the scores was disregarded since the research reported in this paper focuses on the arrangement of the traits from the dominant trait to less dominant trait. This is not affected by the variation in scores.

## 4 Conclusion

First, a dataset collected from individuals that have similar characteristics will not produce distinct clusters, where distinct clusters are defined by differences in the dominant personality trait. On the contrary, clusters will have the same dominant trait but varying trait combination arrangement (H1, H2, H3). Second, a nationality can have similarity in the dominant trait but varying trait combination (H2). Third, different nationalities have varying dominant traits and trait combination (H3). It can be assumed that participants willing to take a personality test are generally open to new experiences, therefore have an openness or agreeableness trait. All datasets accumulated the data by self-reported individuals who wanted to take the test. Accordingly, the dominant trait of those individuals aligns with their willingness to participate in the studies and accounts for the similarity in the results across the three datasets. Finally, what do these results mean for interface design, specifically role-based personalization of interfaces. Most importantly, interface designers should design for the dominant trait combination rather than the dominant trait alone. Designing for the trait combination gives better scope for personalizing interfaces by taking into account the aggregated representation of the user.

## 5 Future Work

To ensure a representative sample, it is important to collect data randomly from individuals who were not initially interested in taking the personality test. This will help discover new insights and is expected to reduce the dominance of the openness and agreeableness traits. Including individuals who did not have a specific interest in the test will allow a broader exploration of personality traits. To understand the correlation between individuals with different personality traits and design elements, individual's reaction to different design choices such as colour schemes, layout structures, typography, or interactive elements should be analysed. The objective is to utilize the understanding of personality traits and design preferences to create personalized learning management systems to provide users with an enhanced user experience.

## References

1. Al-Emran, M.: in Proceedings of International Conference on Emerging Technologies and Intelligent Systems. ICETIS 2021 (2021)
2. Xie, I., Matusiak, K.K.: Discover digital libraries: theory and practice. Elsevier, Amsterdam, Netherlands (2016)
3. Luketina, P.: Big five personality traits. Kaggle (2021). https://www.kaggle.com/code/petarluketina/big-five-personality-traits/data
4. Goldberg, L.R.: The development of markers for the Big-Five factor structure. Psychol. Assess. **4**, 26–42 (1992)
5. McCrae, R.R., Costa, P.T., Jr.: A five-factor theory of personality. In L.A. Pervin, O.P. John (Eds.), Handbook of personality: Theory and research (pp. 139–153). Guilford Press (1999)
6. Introductory Psychology. ER Services, p. Chapter 11 (2021)
7. Lim, A.: Simplypsychology.org (2020). https://www.simplypsychology.org/big-five-personality.html

8. Han, J., Kamber, M., Pei, J.: Data mining. Morgan Kaufmann, Waltham, Ma (2012)
9. Han, J., Kamber, M., Pei, J.: Data mining: concepts and techniques. Elsevier Science, Burlington (2012)
10. Stewart, G., Al-Khassaweneh, M.: An implementation of the HDBSCAN* clustering algorithm. Appl. Sci. **12**(5), 2405 (2022)
11. K-means clustering and related approaches: Chapman & Hall/CRC Computer Science & Data Analysis, pp. 87–132 (2012)
12. The Big Five Personality Test: Truity (2022). https://www.truity.com/test/big-five-personality-test
13. M. S. Salem, S. S. Ismail, and M. Aref, "Personality traits for Egyptian twitter users dataset," Proccedings of the 2019 8th International Conference on Software and Information Engineering, 2019
14. Big five personality test. https://openpsychometrics.org/tests/IPIP-BFFM/

# Digital Technologies in a Modern University: Current Experience and Critical View

Svetlana V. Barabanova[1](✉), Nataliya V. Nikonova[1], Natalya N. Gazizova[1], Maria A. Khvatova[2], and Irina I. Romashkova[3]

[1] Kazan National Research Technological University, Kazan, Russia
sveba@inbox.ru
[2] Bauman Moscow State Technical University, Moscow, Russia
[3] Financial University, Moscow, Russia

**Abstract.** The article considers features of e-learning in the process of teaching various disciplines to students of modern Russian university. The authors present positive and negative aspects of the introduction of e-learning elements into the traditional educational process. The features of the use of information technologies in traditional and electronic learning and possibilities of their combination are considered. The authors propose different types and forms of digital technologies for introduction into the educational process.

**Keywords:** Information technology · Online course · E-learning

## 1 Context

Global informatization processes also require changes in higher education. The development, implementation and application of e-learning technologies are worldwide practices and tend to evolve. Digital technologies and digital devices are increasingly penetrating our lives and most students have grown up in the context of their widespread use, and now it is impossible to imagine a student who does not know how to work with a computer, smartphone and other gadgets. Information and communication technologies influence the choice of the future profession and accordingly determine to a large extent the choice of direction of training.

It appears that e-learning in higher education is primarily about teaching students knowledge and skills using information technology and e-learning resources. The transition to open education is a modern approach to resolving the contradictions between traditional education and the constantly evolving culture of society, which makes it possible to ensure the continuity of the educational process and the formation of the foundations of the information society. The main purpose of study in a higher educational institution, in a modern university is to ensure the competitiveness of its graduates, their ability to build a successful career or their own business. Training at the university is mainly focused on the development of professional competencies, but employers recently also require future professionals to have IT competencies and competencies Soft skills, which include: communication, ability to work in a team, Skills of critical thinking, leadership and responsibility, knowledge of work ethics and organization.


M. E. Auer et al. (Eds.): ICL 2023, LNNS 899, pp. 319–325, 2024.
https://doi.org/10.1007/978-3-031-51979-6_33

## 2 Approach

The specialization of students in technological fields requires extensive and in-depth knowledge in basic disciplines such as mathematics, chemistry, physics and consideration of their interconnectedness. The rapidly changing conditions of our life, technical and technological improvements in production, widespread digitalization require the future graduate to have a broad knowledge for work in the future specialty, the ability to navigate in science andtechnical and economic information, independently update and upgrade their professional knowledge and master modern digital technologies professional in their professional field. Therefore, the learning process should contribute to the formation of a person with a high common culture, having fundamental knowledge and skills in the future professional activity and the ability to learn new knowledge and new technologies independently. Modern higher education institutions can offer forms of training that will combine traditional educational technologies and modern digital technologies, which include the technical equipment of classrooms, author online courses, modern electronic textbooks that take into account both the directions of training, and the training of specific students, as well as video materials, educational programs, simulators, etc. The training of modern specialists is not only the transfer of educational material, but also the opportunity to teach students to learn throughout life. With the global digitalization of society, digital technologies are also transforming the educational process. At the same time, their use is limited by the technical equipment of audiences, the presence or absence of accompanying electronic materials, as well as the ability of teachers to apply them in the educational process, as well as from many other things. At the same time, because of the changing environment, teachers have to adapt to changes in the educational process [1] and must also be well versed in the subjects taught and be specialists in their subject areas, at the same time be able to use modern information technologies in their classes [2].

In addition, traditional learning largely ignores the individual abilities of students, such as how quickly students learn information or whether additional material is required to develop skills. After all, to better understand the rules in many basic and special disciplines, such as «Mathematics» or course «Jurisprudence», as for example, when considering complex texts of a legal nature, it is recommended to solve a large number of problems and examples. And the time limit of the courses of disciplines leads to the fact that many students do not develop the necessary skill and understanding in solving basic examples and, accordingly, this leads to the fact that in the future they can not solve more complex examples and applied tasks. To solve the problems arising in the educational process when applying digital technologies in traditional learning, we offer the introduction of such modern learning technologies as mixed learning, which is the integration of traditional and electronic, as well as distance education technologies. The combination of traditional face-to-face training with modern technology is the main advantage of mixed learning. The possibility of using a mixed e-learning model reduces the amount of time spent on theoretical presentations by creating video lectures and making more efficient use of time in classroom sessions [3].

When constructing the course of higher mathematics, we used a modular approach to the presentation of the material, in which each section is an independent unit. In this way, each module is accompanied by independent and monitoring work. In the Moodle system, the control work can be replaced by a test. To determine the level of preparedness of students used rating system, which allows to objectively assess the knowledge and skills received by students.

## 3 Actual Outcomes

With the introduction of digital technologies in the educational process, the role of the teacher not only does not diminish, but on the contrary becomes even more significant, and its functions—broader. Teachers have to work in constantly changing situations, their work becomes more and more difficult. Teachers who work with students according to the same standards belong to different generations and have different competencies accordingly. Therefore, they use different resources for their training programmes, depending on their age and degree of e-ownership.

In the meantime, e-learning and the creation of e-learning resources are not simply the translation of traditional methods and pedagogical technologies into an electronic environment. First of all, serious changes in the mentality, psychology, professionalism of the teachers themselves are needed. The psychology of digitalization is a new and very topical topic in preparing teachers for work in modern conditions. Secondly, there is a need to introduce modern technologies to the educational process for all participants in the educational process with the participation of professionally trained professionals. The scope of the teacher's activities is significantly expanding, but its dependence on technical support measures is also increasing. In addition to being the author or co-author of the e-learning course, if possible, the author must still be up to date with the latest developments, have an idea of the new e-learning courses in his discipline that are being developed in other higher education institutions, as well as at adjacent departments; to make changes in the traditional educational process—to have a course of video lectures, for example, on the most complex issues of the course; to upgrade skills in terms of mastering new methods and methods of teaching, the use of distance learning technologies; to be able to work with the group in an online format. The authors' use in the educational process of elements of the interactive course «Mathematics—Electronic Course» showed that mixed learning using both traditional methods and electronic educational resources is positively perceived by students, not only increases students' interest in performing various types of tasks using electronic resources, but also in the educational process in general. An anonymous survey of students showed that they are interested in tasks that can be performed using electronic resources. The opportunity to listen to and view lectures, will stop at difficult incomprehensible moments, make corrective notes in the notes—all this was also noted by students as a positive aspect.

So the e-learning course can be used by teachers in different ways. For example, to consolidate and systematize knowledge as a simulator or used to test the knowledge gained, or as a device that simulates the actions of students and the environment of learning and communication.

All this allows you to reduce the distance, save time on transfer and verification of tasks, make it possible to monitor the results of training and analyze them, allow to make transparent the educational process itself. But, unfortunately, we are not talking about a qualitative change in education. This can be done within the framework of traditional learning. However, their joint use with traditional training technologies will improve the quality, lead to a new level of interaction of the subjects of the educational process [4].

The proposed complex contributes to solving the problem of pedagogical stimulation of creative self-development of students, their active self-managed cognitive activity and is divided into three main parts: didactic materials, which are the core of the kit, universal didactic kit for the student, as an information model of the didactic system, as well as an electronic complex, which can include electronic textbooks, computer system for testing, access to remote libraries and much more. The electronic complex has the following functions:

- to be a full tool of the learning process;
- contain all modules and parts of paper;
- integrate, supplement and systematize various teaching tools;
- to disclose the basic requirements to the content of the studied discipline;
- meet the principles of science;
- implement information and procedural and practical content;
- submit material systematically and consistently;
- Implement continuity of training at all stages;
- to satisfy the principles of modularity, «compression» of educational information, individualization, accessibility, combination of breadth and depth of presentation, rigour and clarity;
- to visualize the educational material with the necessary minimum of text and visualization, facilitating the understanding and assimilation of new concepts;
- meet all requirements of work programs;
- to meet the goals and objectives in the development of core competencies.

It achieves mobility and flexibility of its arrangement, optimal combination of interests of teachers and students, consideration of interdisciplinary connections, requests of individual directions and, in particular, specialties. The Basic Training Manual, which is a reference and reference system, provides theoretical information for solving problems and detailed illustrations with examples. Intensive training technology is based on the application of the kit and the use of the cabinet for the organization of the process of self-learning, and the rating system allows to obtain an objective and comprehensive assessment of educational achievements and to carry out pedagogical monitoring of the quality of training. As a result of the rating control, the levels of competence development are determined by the areas of training:

(1) Threshold; (2) Advanced; (3) Excellent.

At the same time, the electronic educational and methodical complex should not completely replace teaching aids and classroom classes with a teacher, but should be an assistant, both a teacher and a student, is a supplement to classical forms of training, To be a conduit in the ocean of information and a guide for choosing a learning path. The digitization of training manuals is insufficient for the successful use of e-learning materials. Their use is only an initial condition for the further development of teaching, the criterion for which is its usefulness to the student. The main objective of electronic teaching materials should be to make maximum use of the capabilities of the computer and other electronic means—video materials, access to various resources, the ability to build different graphs, Diagrams, diagrams, based on the data obtained during the course of training [5].

A survey of 20 first-year students, 18–19 years old, suggested analyzing lectures by the same teacher in different formats—online and offline, showed that students perceive different formats. In addition, the teachers themselves often conduct classes using different techniques. So in the lecture in a face-to-face format, teachers more use gesticulation, monitor the reaction of the audience, adjusting their pace to the audience. More often than not, the narrative style is replaced by the conversation format, where the teacher asks questions on the understanding of the material and formulates questions for the audience to comment the answers. In the classroom teachers more often give examples from life, personal experience in solving certain issues, can tell a joke or a funny case from life, if it is appropriate in the context of a lecture, give examples of scientific practice, Professional experience in relation to the topic. In the online format, lectures are more often held in a more concise form, most often it is a lecture—a monologue, without deviation from the given topic. Most often, online students notice that online lectures are not appropriate examples from private life, funny cases and the like. Increasingly it is said that lectures in the online format are good in the form of concentrated material that is convenient to listen to and make the necessary notes. In this format, the lectures should be concise, informative, than in the audience, where deviations from the strict presentation of the material allow the listeners a little rest and return to the study of the material. Detailed explanations of the material are more welcome in person. This is due to the fact that a better understanding, you need feedback from the teacher, an opportunity to discuss what you have heard, to apply in practice and to discuss the results in a group of students. An e-course may thus be recommended:

(a) students who need to revisit selected topics for the best possible learning;
(b) students in preparation for independent or control studies, examinations;
(c) students who are in debt or are in debt for a semester on selected topics;
(d) students who, for various reasons (illness, competition, personal circumstances) were unable to attend classes;
(e) students in other specialties who have not studied the subject to this extent or who have not covered certain topics or sections;
(f) for remote work. This course can be used when teaching students in a distance format. And teachers' participation here is indirect.

Teaching a specific subject, such as mathematics, applied mathematics, algebra and geometry, can be divided into several stages. The preliminary stage checks the presence of residual knowledge of the mathematics course.

The first stage consists in the acquisition of lecture material and skills to solve the basic problems of mathematics by given formulas. During the lecture the basic concepts and definitions are introduced. Their practical application is considered. The lectures give examples explaining the use of mathematical concepts, give visual interpretation of mathematical results. The given examples make the presentation more accessible and allow the listener to better learn the material.

The second stage—practical exercises, detailed solution of the main tasks under each section.

In the process of learning, a student needs to be able to assess his or her level of knowledge to understand whether he or she has mastered enough of the material passed or to some topics need to return for more detailed study. It is not always possible to assess your own capabilities objectively. Since the teacher cannot control and direct the actions of each student at all times, it is necessary to organize ways of self-control. Thus, the control of module uptake can be carried out in the rating system through control points on the module material: standard calculation task, training test tasks, intermediate testing.

The third stage—the execution of the calculation tasks for each section and the decision of the control work presented in the form of tests.

The fourth step is to perform the final test for all sections of the course.

## 4 Conclusions

Digital technologies are being actively developed and are finding their place in many areas of the educational process. Along with the development of e-learning, new scientific directions connected with pedagogical technologies of e-learning and distance education are emerging. Considering the issues of transforming educational activity in the digital reality, it is necessary to emphasize the primary need for the development of a normative act, reflecting the rapid introduction of distance learning and e-learning into the real educational process (taking into account the traditions of legal regulation in Russia). This is necessary not only for the regulation of educational activities, educational process with the use of distance learning technologies and electronic educational resource, but also to protect the rights and legitimate interests of teachers and students while increasing the burden on the former and the need for appropriate material and technical guarantees for the latter. Local, university standard-setting should also be actively developed. As a result of the analysis of the results of the use of digital technologies in mathematical training, we have obtained that the application of the electronic course together with the traditionally conducted classes individualizes the learning process, It enables a more visible presentation of educational material and motivates its assimilation, develops cognitive activity and contributes to a deeper understanding of new concepts and methods. There has also been an increase in the number of successful students in experimental groups compared to control groups under equal initial conditions. At the same time, the development of the organization of the educational process in face-to-face training using e-learning still remains one of the main tasks in higher education.

## References

1. Aleksandrov, A.YU., Vereshchak, S.B., Ivanova, O.A.: Cifrovizaciya rossijskogo obrazovatel'nogo prostranstva v kontekste garantij konstitucionnogo prava na obrazovanie. Vysshee obrazovanie v Rossii **28**(10), 73–82 (2019)
2. Antamoshkina, E.N., Saakyan, M.A.: Podgotovka kadrov dlya industrii turizma v usloviyax cifrovizacii obrazovatel'nogo prostranstva. Turizm: pravo i e'konomika. №3, S. 27–29 (2021)
3. Barabanova, S.V., Kajbiyajnen, A.A., Krajsman, N.V.: Cifrovizaciya inzhenernogo obrazovaniya v global'nom kontekste (obzor mezhdunarodny'x konferencij). Vy'sshee obrazovanie v Rossi (1), 94-103 (2019)

4. Gazizova, N.N., Nikonova, G.A., Nikonova, N.V.: Uchebno-metodicheskij komplekt po matematike dlya studentov tekhnologicheskogo universiteta. Vysshee obrazovanie v Rossii **2**, 56-61 (2018)
5. Galikhanov, M.F., Khasanova, G.F.: Podgotovka prepodavateley k onlayn-obucheniyu: roli, kompetentsii, soderzhanie. Vysshee obrazovanie v Rossii **28**(2), 51–62 (2019). https://doi.org/10.31992/0869-3617-2019-28-2-51-62
6. Gazizova, N.N., Nikonova, N.V., Suntsova, M.S., Barabanova, S.V., Strelnikova, I.A.: Special aspects of organizing teaching activities by simultaneous learning. Lect Notes Netw Syst **390**, 726–736 (2022)

# Use of Digital Tools Contributing to the Digital Transition in Engineering and Data Science Courses

Alberto Cardoso(✉) and Jorge Henriques

University of Coimbra, Department of Informatics Engineering, Centre for Informatics and Systems, University of Coimbra (CISUC), Coimbra, Portugal
{alberto,jh}@dei.uc.pt

**Abstract.** Digital technologies represent important tools for achieving quality education, which is one of the fundamental components of the United Nations' sustainable development 2030 agenda. Teaching and learning processes can benefit from the use of digital tools, improving teachers' productivity and giving students flexibility and support for collaborative work, contributing to their engagement. Particularly in engineering courses, open-source digital tools, such as Jupyter Notebook and Moodle, among many others, provide a suitable environment to develop and share educational materials, combining different types of resources and enabling the use of different programming languages. These tools can be used in a classroom context as well as for assessment purposes. This article presents examples of the use of technology to enhance teaching and learning in Engineering and Data Science courses, seeking to contribute to the development of innovative teaching methodologies in engineering, promoting the digital transition in education.

**Keywords:** Digital tools · Teaching and learning · Digital transition · Engineering education

## 1 Introduction

Sustainable development depends considerably on quality education and information technology has emerged to disseminate shared knowledge and can be seen as a primary driving force to support education development, which is one of the key components of the United Nations' sustainable development 2030 agenda. Technology-assisted learning tools, such as mobile devices, digital tools, and remote and virtual laboratories, have been broadly used to change and improve the quality of education in schools and institutions [1].

The Internet of Things (IoT) has proven to be a robust mechanism to integrate a worldwide learning experience, providing pervasive access to data. Thus, the Internet and the Information and Communication Technologies offer an adequate environment to improve the teaching and learning processes, supported by innovative teaching methodologies, digital tools, and online resources.


M. E. Auer et al. (Eds.): ICL 2023, LNNS 899, pp. 326–334, 2024.
https://doi.org/10.1007/978-3-031-51979-6_34

Collaborative work is of great importance, especially in engineering and data science courses, where the teamwork and interaction between teachers and students are very frequent. Therefore, there are tools available to develop and share online resources, providing teaching and learning conditions that respond to current societal challenges [2] and the necessary digital transition in education [3].

Thus, teachers should feel challenged to tailor their classes to address the new behavior of students, taking advantage of existing tools and technologies to prepare educational materials that can contribute to the development of collaborative resources and enhance interaction between different actors.

In this context, this article aims to present some examples of the use of digital tools in different subjects of engineering and data science courses, showing how these tools can contribute to the digital transition in engineering education, as well as to improve teaching and learning processes, supporting also collaborative activities.

The remaining of the article is organized as follows: Sect. 2 describes the methodology followed. Section 3 presents an example of an Engineering and Data Science curricular unit, and Sect. 4 presents some examples of using the referred digital tools and discusses some validation results obtained with the activities carried out with students. Finally, Sect. 5 concludes the article by highlighting the main findings that resulted from this work.

## 2 The Approach

Teaching and learning processes can benefit from the use of digital tools, improving teachers' productivity and giving students flexibility and support for collaborative work, contributing to their engagement. Particularly in engineering courses, open-source digital tools, such as Jupyter Notebook and Moodle, among many others, provide a suitable environment to develop and share educational materials, combining different types of resources and enabling the use of different programming languages. These tools can be used in a classroom context as well as for assessment purposes.

This article presents examples of the use of technology to enhance teaching and learning in Engineering and Data Science courses, seeking to contribute to the development of innovative teaching methodologies in engineering, promoting the digital transition and the improvement of the quality of education.

### 2.1 The Jupyter Notebook

The Jupyter Notebooks are digital tools that can be open with a web browser, making practical the use of the same interface running locally like a desktop application, or running on a remote server. Therefore, notebooks provide an environment for programming, offering several advantages for students and teachers as these are free and open-source software, as well as it can be used to promote teaching/learning based on innovative approaches and reproducible research [4].

For example, a teacher can make available the notebooks on a web server and easily give students access. Each notebook file is documented in the JSON format with the extension '.ipynb', making easier the process to write, manipulate and share these files.

Therefore, notebooks can be used to record a computational piece of code to explain it, or a given subject in detail to others, and a variety of tools help users to conveniently share notebooks [5].

In recent years, the use of Jupyter Notebooks has increased considerably due to its simplicity, active support community and availability of numerical and plotting libraries, being widely used for teaching purposes, especially in higher education, and in particular in the areas of Science, Technology, Engineering and Mathematics (STEM) [6].

### 2.2 Moodle

Another digital resource that can be considered is the Moodle platform, which can be used as a repository to store tutorial information and guidelines to carry out the practical works. Moreover, this platform can also be used to build quizzes for self-evaluation and assessment of students. Each quiz can contain different types of questions (multiple choice, matching, true/false, numerical, and short answers, etc.), obtained in a random manner from a database with questions grouped by topics.

Moodle can be used to support different engineering subjects, even to carry out remote and virtual experiments using, for example, existing laboratory experimental setups [7].

There are also many examples of using Moodle to create an overall framework for innovative transdisciplinary academic environments in engineering, making it possible to close efficiently, effectively, and practically the gap between modern engineering education and the practical skills needed [8].

## 3 Engineering and Data Science Courses

Engineering and Data Science has become a very important area in terms of research and in engineering formation. Data Science has become a very important area in terms of research and engineering training. It is an interdisciplinary field that uses scientific methods, processes, algorithms, and systems to extract knowledge and insights from data in various forms, both structured and unstructured, like data mining [9–11].

Engineering and Data Science employs techniques and theories, drawn from many fields within the context of mathematics, statistics, information science, databases, and computer science, to analyze and understand phenomena with data, trying to unify diverse areas, such as statistics, data analysis, machine learning and their related methods [9].

The resources referred in this article were considered in the curricular unit "Introduction to Data Science and Engineering" (IDSE) of the 1$^{st}$ year (2$^{nd}$ semester) of the Degree in Engineering and Data Science of the University of Coimbra, in Portugal.

The main goal of the Degree in Engineering and Data Science is to provide the future engineer with the basic foundations for the various areas that constitute Engineering and Data Science, guarantying the training of professionals with a significant set of theoretical knowledge, methods, and practical skills. Additionally, it also enables the practice of the profession in contexts that involve the operationalization of fundamental concepts in the design of solutions for small and medium complexity/scale problems and support for the implementation of complex solutions, whether in the component linked to Data Engineering, whether in the field of Data Science.

This curricular unit intends to introduce the areas of data science and data engineering, providing student with an overview of the area, its methodological principles, its challenges, and its main applications. One of the goals is that student create sensitivity to the set of technical, scientific, and methodological challenges that a Data Science Engineer will experience in his/her professional practice, allowing him/her to create sensitivity for choosing the appropriate methodologies in the analysis and design of solutions as well as for the creation of value in Data Sciences.

Therefore, this curricular unit serves as a link to the other disciplinary curricular units that make up the curricular plan, enabling the student to always integrate disciplinary knowledge in a more comprehensive perspective of Engineering and Data Science.

It is also intended to foster autonomous learning and collaborative work, interpersonal relationships, and oral and written communication, taking advantage of digital technologies.

### 3.1 The Syllabus of the Curricular Unit

The syllabus of the IDSE curricular unit includes the following topics:

Introduction: What is Data Science and Data Engineering?

- Big Data and Data Science
- Why now? – Datafication
- Current landscape of perspectives
- Skill sets needed

Problems and Applications.

- Essential Concepts of Data
- The data science life cycle and pipeline
- Typical problems: regression, classification, clustering and association rules
- Popular Data Analytics Applications

Data Engineering landscape.

- Common challenges in data engineering
- Data, memory and storage
- Operating Syst.
- Database Syst.
- Networking and the Internet
- S/W Engineering
- HPC & Cloud computing

Data Science landscape.

- Why do we need different methods?

- Common Challenges in data science
- Exploratory Data Science
- Data preparation and cleaning
- Feature Engineering and curse of dimensionality
- Regression, classification, clustering
- Fusion
- Validation

Basic Machine Learning Algorithms.

- Linear Regression
- k-NN

The teaching activities comprise theoretical and practical classes, in which students must carry out various practical works. These woks involve several practical exercises using Python as programming language.

### 3.2 Examples of Practical Works

The practical works of the IDSE curricular unit includes, for example, the following topics:

- Structured Data: NumPy Structured Arrays
- Structured Data: Pandas Structured Objects
- Structured Data in Pandas: Indexing and Selection; Time Series
- Database creation using Pandas objects and sqlite3.

These practical works are very important to promote students' contact with various modules of the Python programming language, such as NumPy or Pandas modules, which are essential tools in the field of Engineering and Data Science.

## 4 Use of Digital Tools

Digital technologies are powerful tools that can help improve teaching and learning processes in many ways, making it easier for instructors to generate instructional materials and providing new methods for people to learn and collaborate. In a context where the Internet and IoT favor the interconnection of smart devices and collaboration between people, it is important to use the potential of advanced digital technology to transform the educational process in an effective and efficient way. Technology fosters and encourages creativity and gives students a sense of success, encouraging further learning by thinking outside of traditional techniques [1].

Proper use of digital tools to enhance teaching and learning activities is challenging but improving the quality of lectures and the motivation and benefits for students to learn must be taken into account.

Digital technologies, such as the Jupyter Notebook, offer an environment and resources that facilitate student learning, allowing them to learn at their own speed and explore content independently.

In addition, the existence of tools such as Moodle, which provides means to develop exercises and assessment tests, allows students to provide self-assessment means that give them a perception of their state of knowledge on the subjects dealt with.

### 4.1 Examples of Using Jupyter Notebooks

In this context, Jupyter Notebooks are provided to students with the necessary information to implement the resolution of each practical work of the curricular unit. This procedure is very appreciated by the students because they can develop the code collaboratively, as well as get feedback from the teachers easily. Practical classes are very interactive, and most students are able to carry out their work in an adequate and timely manner, promoting the use of digital tools to support learning and training processes.

As an example, Fig. 1 presents the main steps to create a Python Database using the module sqlite3.

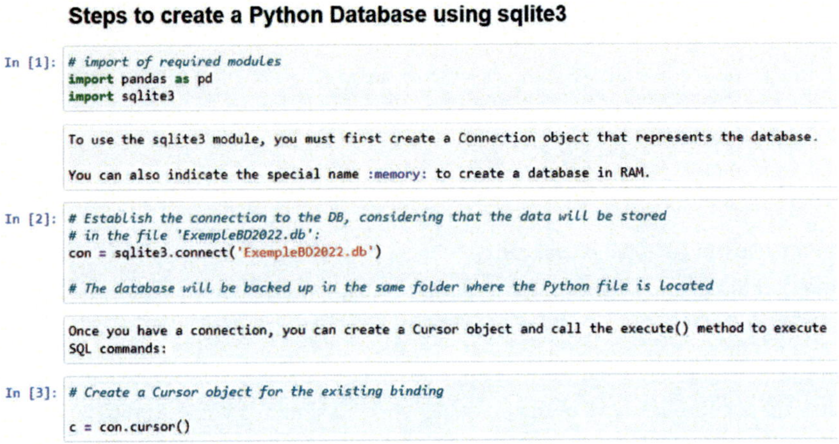

**Steps to create a Python Database using sqlite3**

```
In [1]: # import of required modules
        import pandas as pd
        import sqlite3
```

To use the sqlite3 module, you must first create a Connection object that represents the database.

You can also indicate the special name :memory: to create a database in RAM.

```
In [2]: # Establish the connection to the DB, considering that the data will be stored
        # in the file 'ExempleBD2022.db':
        con = sqlite3.connect('ExempleBD2022.db')

        # The database will be backed up in the same folder where the Python file is located
```

Once you have a connection, you can create a Cursor object and call the execute() method to execute SQL commands:

```
In [3]: # Create a Cursor object for the existing binding

        c = con.cursor()
```

**Fig. 1.** Cells of a Jupyter Notebook to create a Python database.

In Fig. 2, the cells of a Jupyter Notebook to create relationships/tables of a Python database are also shown.

### 4.2 Methodology Description

The methodology followed and implemented in the practical classes of the IDSE curricular unit has proven over the last three years that it motivates students to participate in the classes and, more importantly, support and enhance the students' learning process in an important curricular unit for their curriculum in Engineering and Data Science.

Being students of the 1st year, it is important to encourage them to use the available digital tools to their benefit, whether in the classroom or to support the learning process, as well as for self-assessment.

**Creating Relationships/Tables**

Consider the Relationships/Tables `CIDADAOS`, `MUNICIPIOS` and `ESTADO_DIARIO`

```
In [4]: # Using the execute() method, create the relation/table - CIDADAOS
        # (ID, integer Primary Key; Cidadao_Nome, text; Municipio_ID, integer; Data, date)

        c.execute('''CREATE TABLE CIDADAOS
                    ([generated_id] INTEGER PRIMARY KEY,
                    [Cidadao_Nome] text,
                    [Municipio_ID] integer,
                    [Data] date)
                    ''')

Out[4]: <sqlite3.Cursor at 0x23283028f80>

In [5]: # Create the relationships/table - MUNICIPIOS
        # (ID, integer Primary Key; Municipio_ID, integer; Municipio_Nome, text)

        c.execute('''CREATE TABLE MUNICIPIOS
                    ([generated_id] INTEGER PRIMARY KEY,
                    [Municipio_ID] integer,
                    [Municipio_Nome] text)
                    ''')

Out[5]: <sqlite3.Cursor at 0x23283028f80>

In [6]: # Create the relationships/table - ESTADO_DIARIO
        # (Cidadao_Nome, text; Municipio_Nome, text; Data, date)

        c.execute('''CREATE TABLE ESTADO_DIARIO
                    ([Cidadao_Nome] text,
                    [Municipio_Nome] text,
                    [Data] date)
                    ''')

Out[6]: <sqlite3.Cursor at 0x23283028f80>
```

**Fig. 2.** Cells of a Jupyter Notebook to create relationships/tables of a Python database.

In addition to the contribution of digital tools, such as Jupyter Notebook and Moodle, to allow the accomplishment of the practical works, a significant checkpoint is the final project that students have to specify and develop.

This project intends to design and develop a small Database Management System to manage some institutional tasks, involving the three main ETL steps (extraction, transformation, loading) for data integration, in a data engineering context.

The application scenario can be chosen by the students, being an example the patient database management in an emergency hospital unit.

For the implementation of the project, it is suggested the use of the Python programming language and the use of modules such as NumPy, Pandas or SQLite3, among others, as well as the use of a relational database.

### 4.3 Results Assessment

Even though a systematic evaluation of the impact of the use of digital tools in the curricular unit was not carried out, it can be stated that the methodology followed in the curricular unit was attractive to the students and motivated their active and collaborative participation in the classes and in carrying out activities of learning.

At the end of the semester, it was clear that the digital tools fulfilled the objectives of motivating students to learn concepts and programming methodologies in the context of Engineering and Data Science, and of promoting more adequate and effective learning processes.

The use of digital tools in the 1st year of the course has also boosted their use in other subjects of the course, contributing to the increase of students' skills and their performance in general.

Analyzing, in particular, the resulted projects, even being with small complexity, it is important to highlight that the students were able to specify the database system and implement in Python the managing system with a simple interface.

These results reveal the importance of using digital tools during the semester for project development and to raise students' awareness of the relevance of the digital transition.

In order to assess the pedagogical effect of using digital tools, it will be necessary to ask students to answer a questionnaire about the activities carried out and the learning process, considering, for example, the following questions [6]:

- Can the content of the proposed activities using digital tools be easily understood?
- Is the way in which the contents are presented in the activities attractive?
- Did the activities help me to understand the theoretical contents?
- Do the activities help me study for exams?
- Working with digital tools, does my interest in the course increase?
- Was the time allocated to the activities sufficient?

Even without carrying out this survey, it was possible to informally verify that the feedback from the students was very positive, demonstrating a strong satisfaction with the use of digital tools to develop the practical works, as well as regarding their own impression about how the different activities helped them to understand concepts, prepare for exams, and increase interest in the course.

Therefore, it can be considered that the use of digital tools, such as Jupyter Notebooks and Moodle, is useful to improve the teaching process and contributes to increase students' motivation for disciplines, improving their performance and their awareness of the relevance of the digital transition in the most diverse contexts of application.

## 5 Conclusion

This article presents a methodology supported on the use of digital tools, such as Jupyter Notebook and Moodle, contributing to the digital transition in engineering courses, and describes the use of these tools in the context of an Engineering and Data Science course.

Considering the benefits of digitalization, which offers many opportunities to develop innovative teaching and learning processes, it is crucial to use technologies and digital tools to improve the teaching and learning processes in the era of digital transition.

Therefore, this article describes the main aspects of the methodology followed by the authors in subjects of engineering and data science courses and presents some examples of the use of digital tools that can support teachers and motivate students, as well as contribute to their engagement in the learning process.

Given the positive feedback from students and the perception that the digital tools have helped them understand concepts, prepare for tests and increase interest in the course, it can be concluded that the use of this methodology and digital tools in engineering and data science courses can contribute to promoting the digital transition in a collaborative and integrated way.

In addition, it is expected that the proper use of these technologies in education, especially in engineering courses, will improve the digital learning environment and student performance.

**Acknowledgment.** This work is funded by the FCT - Foundation for Science and Technology, I.P./MCTES through national funds (PIDDAC), within the scope of CISUC R&D Unit - UIDB/00326/2020 or project code UIDP/00326/2020.

## References

1. Haleem, A., Javaid, M., Qadri, M.A., Suman, R.: Understanding the role of digital technologies in education: a review. Sustain. Operat. Comput. **3**, 275–285 (2022)
2. Abdulrahaman, M.D., Faruk, N., Oloyede, A.A., Surajudeen-Bakinde, N.T., Olawoyin, L.A., Mejabi, O.V., Imam-Fulani, Y.O., Fahm, A.O., Azeez, A.L.: Multimedia tools in the teaching and learning processes: a systematic review. Heliyon **6**(11) (2020)
3. Kustec, S., Vestager, M.: Digital transition in education is already taking place but it needs to be accelerated, Article of the Slovenian Presidency of the Council of the European Union 2021 (2022)
4. Raju, A.B.: IPython notebook for teaching and learning. In: Natarajan R. (Eds) Proceedings of the International Conference on Transformations in Engineering Education. Springer, New Delhi (2015)
5. Kluyver, T., Ragan-Kelley, B., Pérez, F., Granger, B., Bussonnier, M., Frederic, J., Kelley, K., Hamrick, J., Grout, J., Corlay, S., Ivanov, P., Avila, D., Abdalla, S., Willing, C.: Jupyter development team: Jupyter notebooks—a publishing format for reproducible computational workflows. In ebook "Positioning and Power in Academic Publishing: Players, Agents and Agendas", pp 87–90 (2016)
6. Bascuñana, J., León, S., González-Miquel, M., González, E.J., Ramírez, J.: Impact of Jupyter Notebook as a tool to enhance the learning process in chemical engineering modules. Educ. Chem. Eng. **44**, 155–163 (2023)
7. Cardoso, A., Teixeira, C., Henriques, J., Dourado, A.: Internet-based resources to support teaching of modelling, simulation and control of physiological systems in biomedical engineering courses. In Proceedings of the 11th IFAC Symposium on Advances in Control Education (ACE2016), pp. 332–337 (2016)
8. Ralph, B.J., Woschank, M., Pacher, C., Murphy, M.: Evidence-based redesign of engineering education lectures: theoretical framework and preliminary empirical evidence. Eur. J. Eng. Educ. **47**(4), 636–663 (2022)
9. Cardoso, A., Leitão, J., Teixeira, C.: Using the Jupyter notebook as a tool to support the teaching and learning processes in engineering courses. In Proceedings of ICL2018 (2018)
10. Dhar, V.: Data science and prediction. Communic. Assoc. Comput. Mach. (ACM). **56**(12), 64–73 (2013)
11. Leek, J.: The key word in "Data Science" is not Data, it is Science. Simply Statistics (2013) (https://simplystatistics.org/)

# Assessing the Behavioural Component of Team Competence in the Digital Educational Environment

Tatiana Shaposhnikova, Alexander Gerashchenko(✉), Tamara Bus, Dmitry Romanov, and Kristina Khoroshun

Kuban State Technological University, 2 Moskovskaya Street, Krasnodar 350072, Russia
alexander_gerashchenko@mail.ru

**Abstract.** The readiness to work in a team is becoming increasingly important. This competence should be formed in the student at all levels of the system of continuing education (in a broader sense, develop throughout life). The task of objective assessment of the readiness to work in a team must be considered in the context of a larger task—assessing the professional orientation of the individual. The object of the research is the assessment of the student's readiness to work in a team, the subject of the research is the methods of the specified assessment implemented in the digital educational environment. Following the modern models of competences and personal and professional qualities, as well as models of social interaction, the authors of this article have substantiated the criteria for the behavioural component of readiness to work in a team. The authors have taken into account the fact that (at the highest levels of such readiness) an individual can work productively in different teams. The proposed method can be implemented both for students of general educational institutions and for students of higher educational institutions, provided that the educational environment allows the implementation of monitoring technologies.

**Keywords:** Team competence · Assessing the behavioural component · Digital educational environment

## 1 Introduction

In the modern world, such an individual competence as the readiness to work in a team is becoming increasingly important. Currently, the tasks facing all spheres of human activity are so time-consuming and complex that their solution is possible only with the combined efforts of professionals [1, 2]. However, this pooling of efforts is possible if there is a proper development level of key competences required for solving professional problems and such personality traits as the ability to interact with other people and the readiness to work in a team. Uniting individuals into a team will not be helpful enough without the proper level of key competences required to solve professional problems. Success from combining efforts depends on both the management (organization of the work of the team) and the ability of each team member to interact with other people,


M. E. Auer et al. (Eds.): ICL 2023, LNNS 899, pp. 335–342, 2024.
https://doi.org/10.1007/978-3-031-51979-6_35

their readiness to work in a team. Thus, the readiness of individuals to work in a team is a fundamentally important factor for any sphere of human activity, as well as a key competence required to solve professional problems.

The readiness to work in a team is a component of the individual's professional orientation, which includes project, research, creative and communicative components, and the communicative component is the readiness to work in a team [3]. Therefore, without assessing such a personality quality as the readiness to work in a team, it is impossible to assess professional orientation, as well as the success of professional education (in a broader sense, the quality of education).

This competence should be developed in the student at all levels of the system of continuous education, not only in the system of vocational education. Therefore, (for example, in the most advanced institutions of additional education of Russia) schoolchildren receive their early professional education, and one of its didactic tasks is to form the readiness to work in a team. The digital transformation of educational environments presents new opportunities for solving didactic problems, learning activities of students, as well as their interaction. However, students should be taught direct interaction in a team. Modern information technology is crucial, but its use should be a way to increase the success of solving didactic problems, not a purpose in itself.

However, due attention is not paid to such a metrological task as an objective assessment of this competence, probably because universal models and methods to assess competences (personal and professional qualities) have been developed. Competences include the following standard components: operational (knowledge and skills), motivational-value, reflective, emotional-volitional, and behavioural (personal experience in relevant activities). The behavioural component is the most important, since it gives meaning to the rest, as well as competences in general [3–5].

The most interesting of the methods developed to assess the components of competences and personal and professional qualities is the diagnostic assessment of the behavioural component. Its criteria are conventionally divided into two groups: (1) parameters specific to a given competence; (2) parameters that reflect the relationship between the operational and behavioural components, i.e. to what extent the knowledge and skills are used in personal experience in the relevant activities.

The task of objective assessment of the readiness to work in a team must be considered in the context of a larger task—assessing the professional orientation of the individual. The research problem is the question: "How to assess the individual readiness to work in a team objectively?" The research purpose is to develop a method for the assessment of the behavioural component of readiness to work in a team. The object of the research is the assessment of the student's readiness to work in a team, the subject of the research is the methods of the specified assessment implemented in the digital educational environment. The parameters of the first group (the criteria specific to team competence as the readiness to work in a team) are to be highlighted.

## 2 Methods

The methodological foundations of the research are the systemic, sociological, competency-based, qualimetric, activity-based and probabilistic-statistical approaches. The research methods are as follows: the analysis of scientific literature, normative documents and advanced pedagogical experience, modelling, methods of qualimetry, methods of mathematical statistics, method of expert assessments, methods of set theory. The research also involves the pedagogical experiment, which has revealed a high level of differentiating abilities of the proposed criteria.

The research base is Kuban State Technological University (Krasnodar, Russia).

## 3 Results

The allocation of criteria for the behavioural component of team competence should be based on the models of the individual's personal experience in social interaction (teamwork). If $Z$ is the set of teams in which the analyzed individual fulfilled his/her role, $S_i$ is the set of members of each team (beside the analyzed individual), then the set of individuals with whom the analyzed student worked in teams is as shown in Eq. (1).

$$C = \bigcup_{i=1}^{z} S_i \tag{1}$$

The number of such individuals is as shown in Eq. (2).

$$c = card(C) \tag{2}$$

Here *card* is the power of the set, U is the symbol of the union of sets, and the number of teams where the analyzed individual participated is as shown in Eq. (3).

$$z = card(Z) \tag{3}$$

Accordingly, the number of individuals in the *i*-th team, beside the analyzed student, is as shown in Eq. (4).

$$s_i = card(S_i) \tag{4}$$

If the number of work cases of the analyzed individual in the *i*-th team equals $n_i$, then the total number of work cases in different teams is as shown in Eq. (5).

$$N = \sum_{i=1}^{z} n_i \tag{5}$$

If the work time in the *j*-th case was $t_j$, then the total work time of the analyzed student in teams is as shown in Eq. (6).

$$T = \sum_{j=1}^{N} t_j \tag{6}$$

If $\eta_i$ is the set of individual work cases in the *i*-th team, then the total set of individual work cases in teams is as shown in Eq. (7).

$$\eta = \bigcup_{i=1}^{z} \eta_i \tag{7}$$

The number of individual work cases in the $i$-th team is as shown in Eq. (8).

$$n_i = card(h_i) \tag{8}$$

The total number of work cases in different teams is as shown in Eq. (9).

$$N = card(h) \tag{9}$$

If the set of competences (in addition to readiness for social interaction and the implementation of one's role in the team) that the analyzed individual shows in the $k$-th case is $D_k$, then the total set of manifested competences is as shown in Eq. (10).

$$D = \bigcup_{k=1}^{N} D_k \tag{10}$$

When assessing such a competence as readiness for social interaction and the implementation of one's role in a team, other competences are also taken into account, because the meaning of social interaction and the realization of one's role in the team is to show the already formed competences, to increase the success (efficiency) of the team's work. The implementation of one's role in the team cannot be better (but can be worse) than the level of already formed competences allows. An individual who is a true professional and truly ready for social interaction must work successfully in a wide variety of teams. The absolute degree of similarity of the $i$-th and $j$-th teams is the following value presented in Eq. (11).

$$\varepsilon_{i,j} = card\left(S_i \bigcap S_j\right) \tag{11}$$

The relative degree is the following value presented in Eq. (12), where ∩ is the symbol of the intersection of sets.

$$\gamma_{i,j} = \frac{card\left(S_i \bigcap S_j\right)}{card\left(S_i \bigcup S_j\right)} \tag{12}$$

For the teams that are completely dissimilar (in relation to the analyzed individual), Eq. (13) is applicable.

$$\varepsilon_{i,j} = 0 \tag{13}$$

The absolute degree of difference between the two teams (in relation to the analyzed individual) is as presented in Eq. (14).

$$\phi_{i,j} = \left(S_i - S_j\right) \bigcup \left(S_j - S_i\right) \tag{14}$$

The relative degree of difference between the two teams (in relation to the analyzed individual) is as shown in Eq. (15).

$$\varphi_{i,j} = 1 - \frac{card\left(S_i \bigcap S_j\right)}{card\left(S_i \bigcup S_j\right)} \tag{15}$$

The analysis of the primary models of the analyzed individual's work in teams made it possible to single out the criteria for the behavioural component.

The first criterion for the behavioural component of team competence (readiness to work in a team) is the number of teams in which the analyzed individual took part as presented in Eq. (16).

$$K_1 = z \tag{16}$$

Accordingly, the second criterion is the number of individuals with whom the analyzed student was in the same team is as shown in Eq. (17).

$$K_2 = c \tag{17}$$

The third, fourth, fifth and sixth criteria, which are the total number of work cases in different teams N, the total work time of the analyzed student in teams T, the number of incoming information flows w and the number of outcoming information flows q respectively, are shown in Eqs. (18), (19), (20) and (22).

$$K_3 = N \tag{18}$$

$$K_4 = T \tag{19}$$

$$K_5 = w \tag{20}$$

$$K_6 = q \tag{21}$$

The role in the team can be implemented in different ways (and it should be kept in mind that "as hundred cases of bad work is no substitute for two cases of good work", the ninth criterion (K9) should be more correct (adequate) than the eighth one. Accordingly, the seventh criterion takes into account the level of work (implementation of the role) of the analyzed student in teams, as it is shown in Eq. (22).

$$K_7 = 1.2\ n_{hst} + n_{vh} + 0.8\ n_h + 0.6\ n_d + 0.4\ n_a + 0.2 n_l \tag{22}$$

Here $n_{hst}$, $n_{vh}$, $n_h$, $n_d$, $n_a$, $n_l$ are, respectively, the number of when an individual worked in a team at the highest, very high, high, due, average and low levels (cases with the lowest level of work are not taken into account). It is shown in Eq. (23).

$$N = 1 n_{hst} + n_{vh} + n_h + n_d + n_a + n_l \tag{23}$$

If the implementation of the role in the team is assessable on a linear scale (preferably on the most perfect, i.e., the scale of relations), then Eq. (24) is applicable.

$$K_8 = \frac{\sum_{j=1}^{N} R_j}{B} \tag{24}$$

$R_j$ is the level of implementation of the role in the team in the $j$-th case on a $B$-point scale (division is used for normalization).

As "a hundred cases of bad work is no substitute for two cases of good work", the ninth criterion (K9) should be more correct (adequate) than the eighth one. Finding the empirical average (i.e. dividing by $N$) will create an absurd situation, because less successful cases of an individual working in a team will "level out" more successful ones. From the point of view of the authors of this article, it is necessary to apply the statistical scree plot method (in scientometrics, the $h$-index is calculated on its basis). In accordance with this method, the success of the analyzed individual in teams is equal to $H$, if in at least $H$ cases (out of $N$) the level of his work on the $B$-point scale was at least in each case. The suggested function type is presented in Eq. (25).

$$f(H, B) = \frac{B}{5} \cdot (1 + 0.1 \cdot H) \tag{25}$$

An even better form of the ninth criterion is the number of cases of teamwork with the level of work of an individual not lower than predetermined, for example, $0.4 \cdot B$.

Let the level of implementation of the individual's role in a team be assessed on a hundred-point scale. Let there be 10 cases of work, and the implementation of the role was 55, 35, 45, 30, 50, 60, 55, 55, 50, 55. The number of cases where the level of the individual's work in a team is not lower than 40 points on a hundred-point scale is 8.

The tenth criterion, presented in Eq. (26), reflects the diversity of the individual's social interaction.

$$K_{10} = \sum_{i=1}^{z} \sum_{j=i+1}^{z} \left( \frac{1}{1 + \varepsilon_{i,j}} \right) \tag{26}$$

This criterion does not correlate with $z$, since the degree of difference between pairs of teams is very different. The generalized criterion takes into account both the difference between teams and the level of work of an individual as shown in Eq. (27).

$$K_{11} = \frac{\sum_{i=1}^{z} \sum_{j=i+1}^{z} \sqrt{r_i \cdot r_j \cdot \varphi_{i,j}}}{B} \tag{27}$$

Here $r_i$ and $r_j$ are, respectively, the average level of work (on a linear $B$-point scale) of the analyzed individual in the $i$-th and $j$-th teams; division is necessary to get rid of the affiliation to the dimension of the linear scale.

The twelfth, thirteenth and fourteenth criteria, presented in Eqs. (28), (29) and (30) respectively, reflect the extent to which the individual's work in a team allowed him/her to demonstrate his/her competences (in addition to readiness for social interaction and the implementation of his/her role in the team).

$$K_{12} = card(D) \tag{28}$$

$$K_{13} = \sum_{k=1}^{N} card(D_k) \tag{29}$$

$$K_{14} = \frac{\sum_{i=1}^{card(D)} F_i}{B} \quad (30)$$

Here $F_i$ is the degree of manifestation of the $i$-th competences on a linear $B$-point scale; division is necessary to "get rid" of the dimension of the scale. The thirteenth indicator reflects that a student can demonstrate the same competence several times. The meaning of the ability to work in a team is to fully demonstrate the competences (in addition to team competence) required to solve problems.

The digital transformation of the educational environment creates favourable prerequisites for the implementation of monitoring technologies, including the methods of qualimetric diagnostic assessment [6]. Assessing the student's competences, including team competence, is possible due to the digital footprint. Typical examples of teamwork in the digital environment include the implementation of an educational network project, involving WebQuest [7], and the teamwork at a digital laboratory, which can also be assessed using the introduced criteria.

The analysis of the educational process, both undergraduate students and high school students in the regional school technology park at Kuban State Technological University (Krasnodar, Russia) shows the proper differentiating ability of the proposed criteria (Table 1), which meets the requirements of qualimetry. The sample included 2083 undergraduate students (986 of the enrolment year 2018 and 1097 of the enrolment year 2019) and 37 high school students. The maximum value of the second criterion is less than two times less than the maximum value of the first criterion because teams joined by an individual did not always differ in composition.

As it can be seen, the twelfth criterion has the least differentiating ability, i.e. the number of competences (others than team competence) demonstrated in joint work. The reason is that the competences are affiliated with training courses as part of the educational process so that the need to manifest competences is determined by the educational process and the success of such manifestation depends on the student.

## 4 Conclusion

The methods for assessing team competence still have to be developed. However, it is obvious that the teamwork models should be the scientific basis for identifying criteria and choosing methods for assessing this competence. Accordingly, sociological and competence-based approaches should be the leading methodological foundations of this assessment. The team is an integral social system, and the implementation of the individual's role depends on both key competences and readiness for social interaction. Research prospects include the development of a method for the integrative assessment of the success of an individual in a team, as well as for the integrative assessment of this competence (both on the relationship and on the nominal scales).

**Table 1.** Values of criteria for the behavioural component of team competence.

| Criterion | Minimum | Maximum |
|---|---|---|
| $K_1$, units | 2 | 14 |
| $K_2$, units | 5 | 25 |
| $K_3$, units | 8 | 41 |
| $K_4$, academic hours | 16 | 82 |
| $K_5$, units | 21 | 178 |
| $K_6$, units | 19 | 143 |
| $K_7$, units | 2.23 | 32.7 |
| $K_8$, units | 2.86 | 36.2 |
| $K_9$, units | 1 | 28 |
| $K_{10}$, units | 0.23 | 1.59 |
| $K_{11}$, units | 0.15 | 1.37 |
| $K_{12}$, units | 21 | 29 |
| $K_{13}$, units | 37 | 55 |
| $K_{14}$, units | 3.78 | 43.63 |

**Acknowledgments.** The research was carried out with the financial support of the Kuban Science Foundation in the framework of the scientific project No. APS-21.1/44.

## References

1. Klaic, A., Burtscher, M.J., Jonas, K.: Fostering team innovation and learning by means of team-centric transformational leadership: the role of teamwork quality. J. Occup. Organ. Psychol. **93**(4), 942–966 (2020)
2. Klavans, R., Boyack, K.: Research portfolio analysis and topic prominence. J. Informet. **11**(1), 1158–1174 (2017)
3. Boonsri, S., Pupat, P., Suwanjan, P.: Dual vocational students' competency: a second order confirmatory factor analysis of occupational competency in enterprise. Mediterr. J. Soc. Sci. **10**(1), 105–115 (2019)
4. Shaposhnikova, T., Gerashchenko, A., Romanov, D.: Information competency as a success factor in distance learning. In: Nazir, S., Ahram, T.Z., Karwowski, W. (eds) Advances in Human Factors in Training, Education, and Learning Sciences. AHFE 2021. Lecture Notes in Networks and Systems, vol 269, pp. 307–314. Springer, Cham (2021)
5. Yu, T.: Examining construct validity of the student online learning readiness (SOLR) instrument using confirmatory factor analysis. Online Learn. J. **22**(4), 277–288 (2018)
6. Khaperskaya, A.V., Minin, M.G.: E-Learning platform and pedagogical monitoring in the context of digital transformation. Higher Educ. Russia **30**(4), 131–138 (2021)
7. Shaposhnikova, T., Gerashchenko, A., Shabanova, T., Vyazankova, V., Romanova, M.: WebQuest as a factor of teaching teamwork. In: Nazir, S. (eds) Training, Education, and Learning Sciences. AHFE (2022) International Conference. AHFE Open Access, vol 59, pp. 206–212. AHFE International, USA (2022)

# E-Student in the Mozambican Context: An Analysis of Higher Education Students' Challenges Regarding to E-learning Implementation

Domingos Luis Rhongo(✉) and Bonifacio da Piedade

Catholic University of Mozambique (FGTI), Beira, Mozambique
{drhongo,bpiedade}@ucm.ac.mz

**Abstract.** Between 2020 and 2021, Mozambican higher education was affected by the COVID-19 pandemic, and in this scenario, the institutions resorted to the hybrid model; but e-learning in developing countries has not been successfully implemented because many of these countries have a very high digital divide, with low digital literacy and socio-economic problems actively contributing to the failure of the introduction of e-learning in higher education in Mozambique; however, higher education institutions had no choice unless migrating from conventional to e-learning; in this sense, this article aims to analyze the main challenges that students encountered in the learning process in the face of the introduction of e-Leaning, in five Mozambican higher education institutions. The methodological approach proposed in this paper was Quanti-qualitative insofar as the obtaining of the main data was through a survey of 345 students. In addition to the information collected through questionnaires. Given the above, results point to serious difficulties arising from the digital divide. The results show that online learning platforms such as Moodle, WhatsApp, Skype, Moocs, Google Meet, MS Teams and Zoom have been introduced. One of the gaps pointed out by the students was the lack of motivation for online classes, lack of ICT skills, and computer, and mobile resources with internet access, thus contributing to poor performance in the learning process; given these results, we can conclude that if higher education institutions have no urgent structural intervention, the government, parents and guardians to improve ICT access, the country could verge a digital divide.

**Keywords:** ICT skills · e-Leaning · E-Student · High Education

## 1 Introduction

The information society has become a reality as the massification of Information and Communication Technologies (ICTs), plus the continuous use of computer and mobile devices connected to the internet has contributed to the development and economic growth in developed and developing countries [1, 2]. In the area of education, the use of ICTs leverages growth with new dynamics and new models in teaching. Indeed, higher


M. E. Auer et al. (Eds.): ICL 2023, LNNS 899, pp. 343–354, 2024.
https://doi.org/10.1007/978-3-031-51979-6_36

education experiences new models of teaching, integrating technologies at the centre of learning. But this integration of ICTs was neither homogeneous nor compulsory, depending on the socioeconomic and cultural conditions in each reality.

With the recently experienced scenario characterised by the urgent computerisation, accelerated by COVID-19, due to the need to reduce the spread, the face-to-face classes were converted to a remote model (e-Learning) to materialise the social distance. This transformation was experienced almost everywhere in the world and at all levels of education; in higher education in particular there were more effects of COVID-19, according to [3], as most Education Institutions tried to keep the classes. Due to this social distancing, the continuation of the teaching activities required a financial effort of the higher education institutions, a didactic change of the teachers and students that in some situations would be for the first time using digital platforms [4]. In most cases, not all institutions were prepared and in equal conditions to face these changes of adopting ICTs in the Pedagogical process [5]. In the year 2020 data from UNESCO pointed to more than 290 million students who were left without classes because of the pandemic COVID-1, and in this context, a movement began around the world that the teaching and learning process would move to the remote modality [3].

## 1.1 E-Learning

The development of the Information Society has allowed cultivating a modern education, in this sense, it has been improved with various technologies incorporated in both distance and face-to-face education, this has been added to the internet and other platforms related to social networks that by the ease of use, increase the possibilities of effectively implementing teaching mediated by technologies; This process began to take shape and discussion within the academics and in 1999 Eliot Masie presents in his speech a context related to the concept of the teaching process based on web technologies, and was exactly in this speech that taken the concept of e-learning as learning based on the Web or that is the use of technology to create, manage and share training (Rhongo et al., 2020). In what is understood e-learning is the process of virtual or remote learning supported by information and communication technologies such as computers, tablets, mobile devices and the internet. In the perspective of Ugolini "e-learning should not be understood only as of the use of ICTs in educational activities, but also as an exploration of new opportunities that the pedagogical use of ICTs makes possible" (p. 9) [7]. There is a sharing of responsibility between the teacher and the students, without disregarding, or disrespecting the teacher's role as organiser and orchestrator of lessons and guardians of the teaching process.

## 1.2 Context of Education and E-Learning in Higher Education in Mozambique

The article [8] shows that the history of education in Mozambique had three phases or moments namely: (a) Indigenous Education, which was intended for the transmission of cultural values; (b) colonial education for the children of assimilated; and (c) post-colonization education, with a major objective was the introduction of the National Education System (SNE).

But with the advent of ICTs, a fourth moment begins that we can call the digital era education period with a strong impact on the teaching modality in which e-Learning reaches Higher Education Institutions in an administrative way and in a forced way by the pandemic as previously alluded.

As for the role of HEIs in informatization, it followed the informatics policy that from early on recognized the role of higher education in the process of implementation and expansion of ICTs, as well as in the improvement of citizens' living conditions, therefore, among several responsibilities assigned to HEIs, under that policy, the following stand out: use of ICTs to expand access to higher education [5]. The strategy approved by the Council of Ministers also frames e-learning as a learning model supported by ICTs [9].

Between 2020 and 2021, Mozambican higher education, as in many countries around the world, was affected by the COVID-19 pandemic, and in this scenario, the institutions resorted to the hybrid model that is aided by ICTs to carry out the teaching and learning process. According to [10], Mozambique has more than 50 higher education institutions with more than 230,000 students, and the fact, the institutions have adapted the teaching model by introducing e-Learning.

However, e-learning in developing countries has not been successfully implemented because many of these countries have a very high digital divide, with low digital literacy and socio-economic problems that have also actively contributed to the failure of the introduction of e-learning in higher education in Mozambique. However, higher education institutions had little choice but to migrate from conventional to virtual classrooms, so they all implemented various teaching platforms.

As mentioned, some higher education institutions were not successful in implementing the e-Leaning model, mainly in remote teaching where classes are conducted through the Internet, associated with the fact that students do not have computers, Internet, smartphones and even less the skills to effectively use the teaching platforms that were forcibly implemented.

Because in Mozambique, this phenomenon was little investigated from the student's point of view, before the implementation and after the implementation, it is pertinent and important to question: (i) whether or not the students were prepared for these transformations of teaching taking into account that the universities did not have electronic equipment for the students and that the students had to have their resources to attend the classes, it is also questioned: (ii) whether the students had enough skills to attend effectively in the online model? (iii) Which platforms were most commonly used and which had the best command and most comfortable model for students? and (iv) What were the biggest challenges in this process in the face of so many insufficient resources?

In this sense, this article aims to explore the main challenges that students encountered in the learning process in the face of the introduction of e-Learning, in ten Mozambican higher education institutions.

## 2 Methodology

### 2.1 Data Collection, Procedures and Analysis

This research had a Quantitative approach, where a survey was conducted exclusively targeted at higher education students to explore the demographic characteristics, experience and challenges of using the learning platforms. This modality was defined due to the genre of study that was important to cover a large number of students. According to CNAQ and Rhongo & Piedade, the country has more than 230,000 students distributed among higher education institutions [10, 11], thus initially we defined several 400 students from five high educational institutions.

A convenient way to collect data was through the Google Form, which facilitates the sending, collection, and monitoring of responses and data analysis and this firm can reach more people cost-effectively and sustainably [10].

After the creation of the questionnaire on the platform was forwarded by email and WhatsApp to the course coordinators, pedagogical directors and directors of the faculties mentioned so that it was forwarded to students.

The research had a total of 12 questions, closed type, divided into 4 categories namely: (a) Identification; (b) History of platform use; (c) ICT use skills, and (d) Challenges of using technology platforms in teaching and learning.

The consent of the participants was via a text on the form and was further legitimized because it was self-participation in the online modality, from the students' computer or mobile device.

For data analysis, we had two phases, the first in which the Google form is used to collect and analyze the answers, and the second in which a systematic review is made to promote the bibliography to support the data collected and provide a position and a field for discussion.

## 3 Results and Discussion

### 3.1 Demographic Data

The study was conducted in five Higher Education Institutions namely: Catholic University of Mozambique, ISCED University, Rovuma University, Lúrio University, and Samora Machel Military Academy, the selection criterion was accessibility at undergraduate and postgraduate levels. A total of 345 students responded voluntarily, of which 208 (60.3%) were from public institutions and 137 (39.7%) from private institutions respectively. Most universities in Mozambique are located in urban areas, while students live in the surrounding neighborhoods between 3 and 20 kms.

### 3.2 Gender Distribution

In terms of gender, 243 males corresponding to 70.4% and 102 females corresponding to 29.6% of the total participated (see Fig. 1).

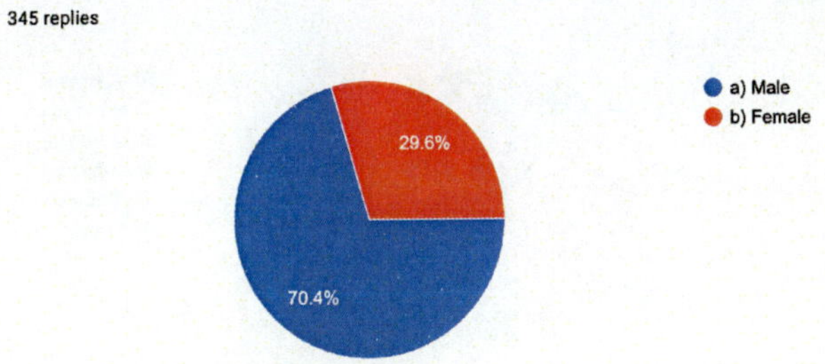


**Fig. 1.** Distribution of respondents by gender

### 3.3 Distribution of Respondents by Age Group

The majority of the respondents are in the age group 20–25 years with 34.8% (120 persons), followed by 30–45 years (107 persons), then 25–30 years with 21.2% (73 persons), then 17–19 years with 10.4% (36 persons), over 45 years participated with 1.7% (6 persons) and at the end those less than 17 years with 0.9% (3 persons) (See Fig. 2).

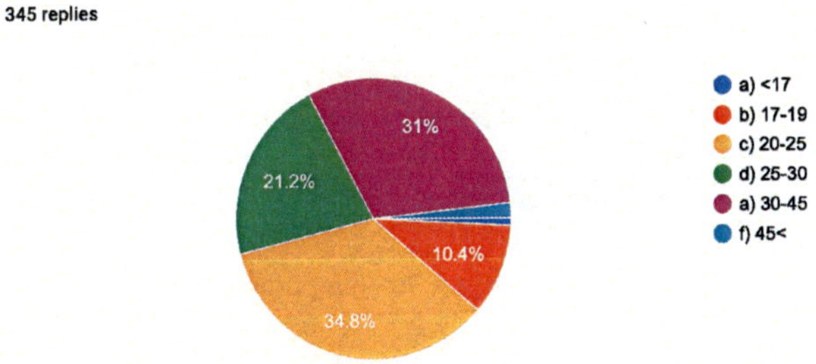


**Fig. 2.** Distribution of respondents by age group

### 3.4 Distribution of Respondents by Academic Level

In terms of academic level, 317 undergraduates graduated with a Bachelor's Degree, 24 Master's Degrees and 4 PhDs. In specific order, 4th-year undergraduates accounted for 29%, 1st years for 28.7%, 3rd years for 18%, 2nd years for 15.7%, Master's Degrees for 7% and PhDs for 1.2% (See Fig. 3).

### 3.5 Timing of Use of Remote Learning Technologies

The use of technologies in the classroom context peaked between 2020 to 2021 when COVID-19 was peaking in the world. In this context, students were asked about the moment they started using technologies and platforms and the internet before or at the time of the COVID-19 pandemic. Indeed 233 students who account for 67.5% responded that they started taking classes through technological means on particular platforms and

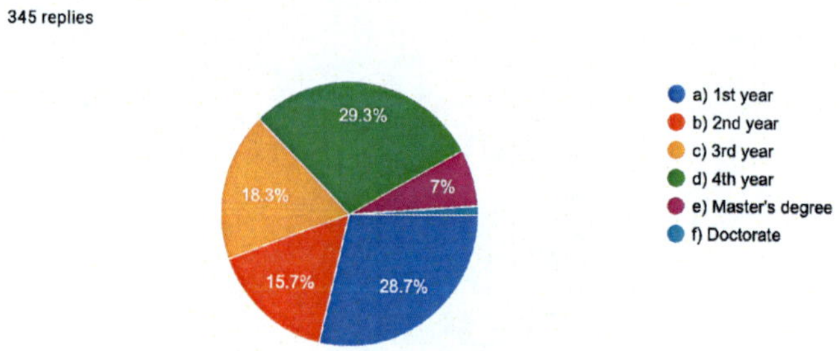


**Fig. 3.** Distribution of respondents by the level of education

the internet at the time of the pandemic and 112 (32.5%) before [10] the pandemic (See Fig. 4); it is an opinion almost consensual on par with a study conducted with teachers in the same circumstances and the same universities presented in the study by, where most report having started using technological resources when the COVID-19 pandemic began.

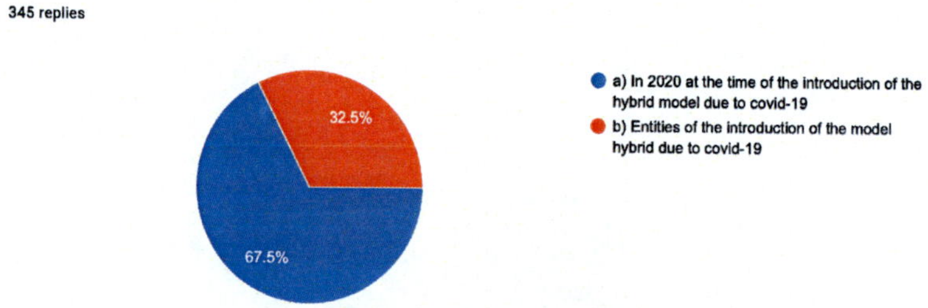


**Fig. 4.** Use of technologies in the teaching and learning process before and after the pandemic

### 3.6 Preparing Institutions to Introduce Educational Technologies

Were the institutions prepared in terms of technology and equipment to deal with remote learning, the majority of the responses refer to the position that institutions urged prepared (53.3%), and (24.1), state that does not see (Fig. 5); this position is reinforced in the results obtained in [10], where teachers expressed the same feeling, however, there are reservations in our point of view because the fact of having a computer laboratory, equipped with computers connected to the internet or even having an introduction to computer science subject in the curricular component does not mean preparation to implement and successfully face the remote learning model, e-learning requires more methodological and pedagogical components beyond the administrative ones.

### 3.7 Skills in Using the Platforms Used in Teaching

In terms of skills in using the teaching platforms, 58.3% corresponded to 201 students, followed by 32.5% (112) of students who reported that they possess these skills minimally, then we have the percentage of students who do not possess skills corresponding

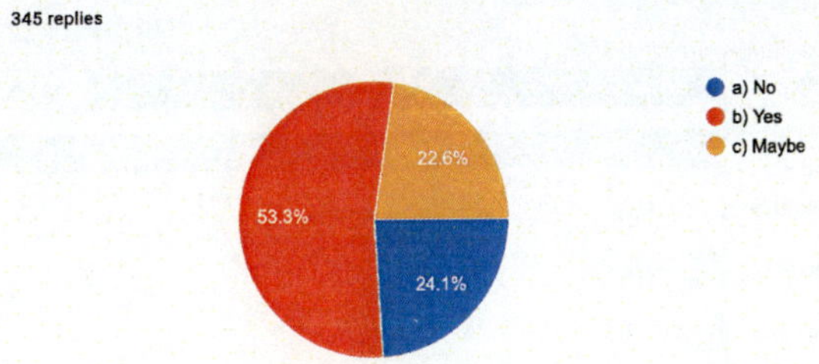


**Fig. 5.** Prior preparation of the institutions in the use of the platforms

to 6.1%, that is, a total of 21 students and lastly, we have that have difficulties in situating whether or not they possess the skills with a total of 3.2% corresponding to 11 students (See Fig. 6).

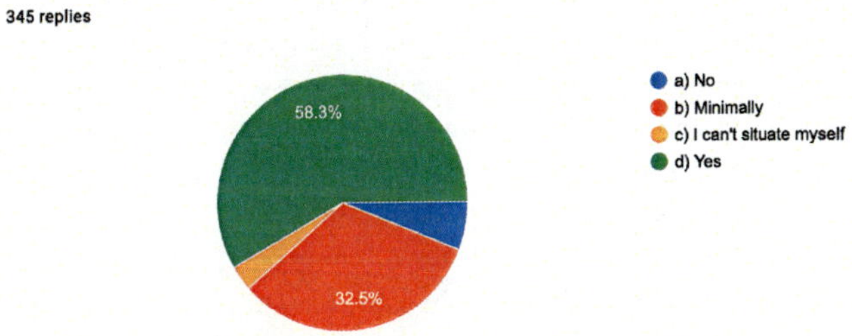


**Fig. 6.** Level of skills in using teaching platform

### 3.8 Platforms Used in Pedagogical Management at the Institution

According to Fig. 7, the platforms most used by students in COVID-19 pandemic was the Google Classroom with 64.1% followed by the Moodle with 55.4% concerning the others, corresponding to the same opinion expressed by teachers in the article by Rhongo & da Piedade [10]. The fact that these two platforms (Google Classroom and Moodle) were the highlights, confirms the simplicity of these platforms with an excellent evaluation where the aforementioned evaluation made at the Catholic University of Mozambique [12], showing a high simplicity of use, and free of charge; in addition, the Google Classroom platform is integrated into Google Suite, thus providing the possibility of its use in for education by the academic community [10].

### 3.9 Platforms Used in Synchronous Classes

The teaching activities in a remote model require the use of certain platforms that allow asynchronous communication and interaction in real-time to suppress the presence of physical space in a conventional classroom, so students responded that the most used platforms were Google Meet and Whatsup with almost 53% followed by Zoom with 46.4%, and further down in Microsoft Teams 7.5%, Skype with 6.4% and finally WebEx

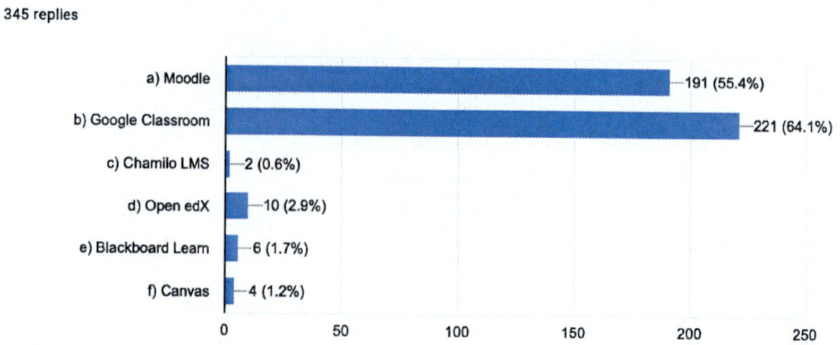


**Fig. 7.** Platforms used for teaching and learning in the Institution where you study

with 2% (Fig. 8). These platforms were used strategically in several educational institutions. Google Meet and Zoom have an equal rating of 4.5 stars out of 5 in Gartner and in a study by James Mordy [13], where 88% of users recommend Google Meet while 94% recommend Zoom. The report presented by CNAQ shows that Mozambique in the context of higher education initially stood out when the state of emergency decreed the use of WhatsApp and Skype in 163 organic units field field field[11]. It is perceived this trend because there was more familiarity with them mainly in the urban environment, where students have some electronic devices, Salimo states that in some contexts in a classroom, between 40 to 60% of students have laptops and almost 100% have mobile phones with internet access [14]. While Whatsup has an evident worldwide popularity for social use, being pre-installed in most smartphones, and given this acceptance in some university contexts as a pedagogical support tool in communication, feedback synchronously and asynchronously [10, 15, 16].

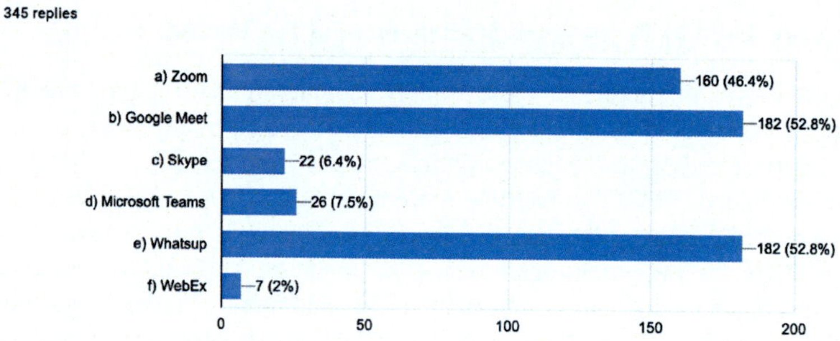


**Fig. 8.** Platform used in Teaching and Learning concretely in the interaction

### 3.10 Teaching Model of Preference

In terms of comfort with the three models of teaching, around 70.1% of students prefer the face-to-face model, corresponding to 242 students, 23% around 78 students feel comfortable with the online model, and 7.2% corresponding to 25 students prefer the hybrid model (see Fig. 9). In this sense, an important point is observed and we will highlight it about the preference of the online model, the result of almost 78 students in the universe of 320 students who prefer face-to-face, shows a transition and acceptance that in the long term may improve by national policies and particularise in HEIs, which does not take away the possibility of moving towards higher education effectively implemented in the e-Learning environment.

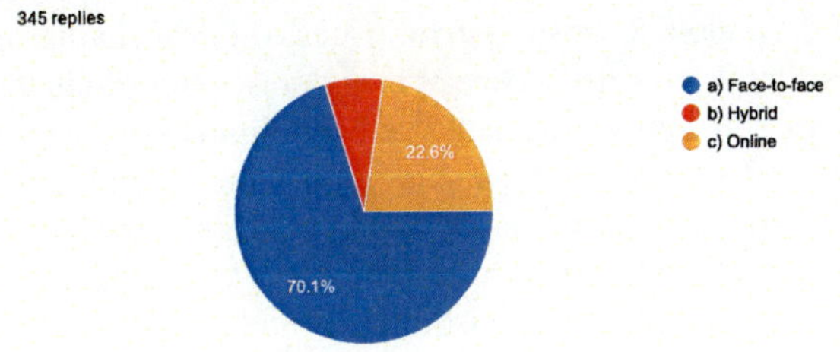


**Fig. 9.** Respondents' preferred teaching model

## 3.11 Challenges Facing E-Learning Teaching

About the challenges faced by students in the face of this new reality and which, for the most part, constitute the greatest challenge, the lack of computer resources and internet connection stands out, making up 59.4% of students, 23% point to issues of motivation, with 19.7% associating the lack of competences and lack of experience in this environment. Didactic and pedagogical issues were also highlighted with approximately 16.2% and, lastly, with less than 10%, were related to digital literacy and culture (Fig. 10).

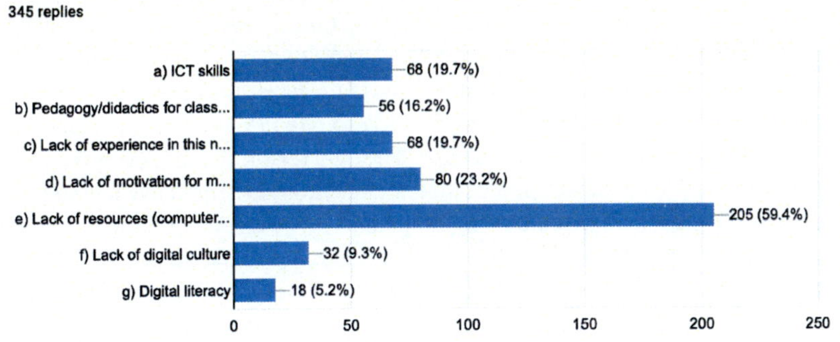


**Fig. 10.** Challenges encountered in the teaching and learning process

These results confirm the stage of developing countries that do not have for economic reasons the penetration rates of technologies and the internet is slow (Rhongo et al., 2018). The results show that student doesn't like hybrid teaching due to a lack of computer resources including the internet, lack of ICT skills, lack of experience in this new environment, and issues with teachers' e-pedagogy problems.

In higher education, teachers seek to promote active learning, and this process requires ICT skills [10], on the part of the teacher and the student, so it must be well defined what resources are missing, and what skills are needed in the transformation of face-to-face classes into online classes. To have e-Students properly speaking, we have the conviction that the HEIs should give answers to the students about the means and the necessary conditions for learning should be created by the institution and not rely on the fact that they have smartphones connected to the Internet to conclude that a teaching modality can function in such an environment. Mozambique has a poverty rate of 87.5% and the majority of the population resides in rural areas (66.5%), living below the poverty line this situation makes it classified as one of the poorest countries in the world [5]. The government has made some efforts as mentioned in Salimo about the project MoRENet (Rede de Educação e Pesquisa de Moçambique) of national ambit with the objective of interconnecting academic institutions of higher education and research, developing non-profit activities [5, 14], this project was quite important in its first years, but it was discontinued as it was removed in some institutions and then they were without internet.

## 4 Conclusions

As presented, a total of 345 mostly male students participated in this research in which most of them (300) are in the age group between 20–45 years of age, mostly undergraduate (317) and the rest post-graduates. The research was able to bring out the picture that most universities started with the integration of educational technologies at the peak of the covid-19 pandemic. The same ones affirmed to have competencies in e-Learning technologies with a major focus on Google Classroom and Moodle for the pedagogical management, and Google Meet and Zoom for the synchronous interaction, which is justified by being free, multiplatform and easy to use respectively. Before the end, they were asked about their preference between face-to-face, online and hybrid teaching and they replied that they are more confident with face-to-face teaching, although there was a surprise in the percentage of those who mentioned online teaching where the rate was 23%. To conclude, they responded about the challenges they have faced in the process of electronic teaching and learning, and most pointed to the lack of computer resources, lack of resources to connect to the internet, motivation to attend online classes, and lack of appropriate skills for this model and didactics. The data resulting from the research, lead to the conclusion that if the institution of higher education does not have an intervening role in the structure of the curriculum, accompanied by government policies, and even definition of the role of parents to improve access and use of ICTs, the country could create a digital divide. It would be important to implement a digital literacy policy in basic education so that higher education can receive students capable of studying with such resources; higher education institutions should make some effort to equip their laboratories with computers with Internet access and new technologies. The government should create a policy of diffusion and massification of access to new technologies to respond to current problems. According to Salimo for effective use of Digital in Mozambican HEIs, an ICT plan (Plano TICs) should be introduced and it should be an integral part of the strategies of the Universities so that the objectives can be achieved [14]. One of the ways forward would be to materialise the project announced in 2021 which mentioned that higher and technical-vocational education students in Mozambique will now benefit from the "one student, one computer" initiative given the difficulties of access to ICTs in higher education, especially with the challenges of hybrid education [18], and associate state partnerships with operators to support higher education through subsidised rates for unlimited Internet access, under very low payment conditions, as happened in 2020, when the price of Internet access for students was reduced to 100 MZN(abbreviation of new mozambique metical, the currency of Mozambique, abbreviated) and others were free of charge [11]. Technologies provide numerous possibilities to enhance student learning [19]. In higher education, the role of the teacher is relevant in the introduction of technology, mainly in the provision of programs, instructional resources, learning strategies, stimuli and the creation of a learning community in which students feel safe, confident and connected [20], but the reality and especially in Mozambique is visible that most teachers were not prepared to assume these responsibilities of creators of the e-learning process [12].

## References


1. da Educacao, M.: Plano Tecnológico da Educação (2011)
2. Rhongo, D.L., De Almeida, A., David, N.: Analysis of the adoption and use of ICT for e-government services: the case of Mozambique. In IST - Africa 2019, no. IST-Africa Week Conference (IST-Africa), p. 10 (2019)
3. Sobral, S.R.: O impacto do COVID-19 na educação Sónia Rolland Sobral. Res. Gate, no. July (2020)
4. UNSECO: COVID-19 and higher education: Today and tomorrow (2020)
5. Samussone, L.B., de F. R. Silveira, S., Brunozi Júnior, A.C., Alexandre, D.C.S., Reis, A.O.: Fatores condicionantes para a tendência de uso de tecnologias de informação e comunicação (TICs) no ensino superior em Moçambique. Res. Soc. Dev. **10**(6), e56910616053 (2021). https://doi.org/10.33448/rsd-v10i6.16053
6. Rhongo, D., Mura, S., Da Piedade, B.: The e-learning environment in developing countries : case of Catholic University of Mozambique, vol. 1, no. 25–28 September. Springer International Publishing (2020)
7. Ugolini, F.C.: Esperienze di e-learning nell'istruzione superiore in Europa, 1st ed. Roma, Italia: Aracne (2009)
8. Rhongo, D., Mura, S., Da Piedade, B.: The e-learning environment in developing countries : case of Catholic University of Mozambique. In ICL2019 – 22nd International Conference on Interactive Collaborative Learning, no. 25–28 September, p. 1871 (2019)
9. Conselho de Ministros, *Estratégia da Educação à Distância 2014–2018*. Mozambique, pp. 1–68 (2013)
10. Rhongo, D., Piedade, B.: E-Teaching in higher education : an analysis of teachers ' chal- lenges facing e-learning in Mozambique. In: Auer, M.E., Hortsch, H., Michler, O., Kohler, T. (Eds.) Mobility for Smart Cities abd Regional Development - Challenges for Higher Education. ICL2021. Lecture Notes in Networks and Systems. Springer, no. 389, pp. 1461–1472 (2022)
11. CNAQ: Resultados preliminares do inquérito sobre e-learning em tempos de COVID-19," Maputo (2020)
12. Mura, S., Rhongo, D.: Análise da utilização da plataforma Moodle nos cursos de Doutoramento : estudo de caso da Universidade Católica de Moçambique. Educ. a Distância 8(1), 25–46 (2018). https://claretiano.edu.br/revista/169/revista-educacao-a-distancia
13. Mordy, J.: Zoom vs. Google Meet: Which is the best video conferencing tool? (2020)

14. Salimo, G.I., Mondlane, U.E., De Engenharia, F.: Contributos Para O Ensino Superior Em Moçambique : OS, pp. 4–8 (2017)
15. Paiva, F.: WhatsApp alcança presença recorde em 99% dos smartphones no Brasil. *Mobile Time* (2020). https://www.mobiletime.com.br/noticias/27/02/2020/whatsapp-alcanca-presenca-recorde-em-99-dos-smartphones-no-brasil/ (accessed Aug. 22, 2020)
16. Susilo, A.: Exploring Facebook and Whatsapp as supporting social network applications for English learning in higher education, pp. 10–24 (2008)
17. Rhongo, D.L., De Almeida, A., David, N.: eGovernment in Mozambique : Past , Future and New Prospects. IST Africa 2018(E-Government), 1–8 (2018)
18. Opais: Estudantes do ensino superior vão receber computadores a partir de Novembro. *O PAIS - SOICO*, Maputo (2021)
19. Hoskins, B.J.: The art of e-teaching. J. Contin. High. Educ. **58**(1), 53–56 (2010). https://doi.org/10.1080/07377360903524641
20. Triyono, M.B.: The indicators of instructional design for E-learning in Indonesian vocational high schools. *Procedia - Soc. Behav. Sci.*, vol. 204, no. November 2014, pp. 54–61 (2015). https://doi.org/10.1016/j.sbspro.2015.08.109

# Online Laboratory Lessons: A New Era of Science Education

Werner Beyerle(✉)

WIFI Niederösterreich, St. Pölten Karat EDV Ltd, Wien, Austria
wbeyerle@karat.at

**Abstract.** This article discusses the possibilities, advantages, and drawbacks of moving previously on-site mechatronics laboratories online. Almost overnight, the Corona pandemic made it necessary to find, evaluate and implement software packages that allowed this shift without decreasing lab quality. Since nobody knew at the time, how long it would be needed to use online labs, it was necessary to achieve a smooth transition between online and on-premises labs. It was also necessary to switch between both possibilities at any time without delay for preparation. So, it was necessary to use the same books and documents within the online labs, which were used before in the traditional labs. So, the exercises were well defined, and the software products had to deal with this issue. This work describes the criteria of this evaluation in detail. For one year students participated in online laboratories under lockdown conditions. The end of the lockdown situation was met with consistent feedback that the online laboratories should remain an option, which indicates a strong popularity among students. Because of this the institute decided to keep two thirds of the training opportunities online with every third appointment being held on-site to allow room to encounter and tackle everyday problems.

**Keywords:** Laboratory · Online · Online laboratory

## 1 Introduction

With the advent of technology and the internet, the education sector has undergone significant changes. Online learning has become a common and accessible mode of education and laboratory lessons are no exception. Due to lockdowns during the Corona pandemic online learning suddenly became the only viable mode of knowledge transfer rather than just an option. It might not be true for some lab exceptions in Austria but if many labs simply didn't happen because of restrictions.

The open-source project VISIR is well known for its ability to conduct online labs. (Garcia-Loro et al., 2021; Gustavsson, I., Zackrisson, J., Håkansson, L., Claesson, I., & Lagö, T. L.) However the approach used in VISR could not be used in our case. This article explores the feasibility of carrying out laboratory lessons online, its benefits and limitations, and how it can impact science education in the future. Being a lecturer within a preparation course for the master exam in mechatronics, I had to carry out laboratory


M. E. Auer et al. (Eds.): ICL 2023, LNNS 899, pp. 355–362, 2024.
https://doi.org/10.1007/978-3-031-51979-6_37

lessons for control technology supporting the theory lessons. The theory lessons are concentrating on the design of logical circuits. One important task students have to learn is how to simplify these circuits. They learn how to use Karnaugh-diagrams to achieve simplification. It is very important, that students learn that design of digital circuits and their actual realization can freely be chosen at any time during the implementation. (*Steuern und Regeln für Maschinenbau und Mechatronik*, 2013). So it is very important, that nearly all lab-experiments can be conducted using all available technologies. Therefore, the main topics were:

- relay circuits
- electronic logic circuits
- pneumatic circuits
- controller and control circuits

The realization with PLCs (Programmable Logic Controllers) is missing in this lab because a special lab and theory course is dedicated to this topic. However even in this separate lab the same exercises are used to show that design and implementation are different tasks and the decision for one or the other way of realization can be freely chosen at any time. The design process does not assume a special form of realization. So, one can easily understand that the selection of lab-exercises must be made carefully in order to achieve the learning goals. For this reason, it was necessary to select such software systems, where all the necessary lab-exercises could be carried out without any limitations.

## 2 Advantages of Online Laboratory Lessons

*Convenient and accessible*: Online laboratory lessons allow students to access virtual experiments from anywhere, at any time. This is especially beneficial for students in remote areas or for those who may face practical difficulties in visiting physical labs, especially during the lockdowns when social distancing guidelines made personal attendance impractical.

*Cost-effective*: Online laboratory lessons are less expensive compared to traditional laboratory lessons as they eliminate the need for expensive equipment and materials. Additionally, students can repeat virtual experiments as many times as they want, thus saving time and money. While our institute already has the needed facilities and equipment to teach on-site laboratories, we can see the benefit for facilities that are planning to add new experiments.

*Safe and risk-free*: Online laboratory lessons eliminate the risks associated with traditional laboratory lessons such as handling expensive equipment and dealing with hazardous currents and voltages. This makes it a safer option for students and teachers alike.

*Customizable and interactive*: Virtual laboratory lessons can be designed to cater to the individual learning pace and style of each student. Additionally, students can interact with simulations, animations, and 3D models to better understand complex concepts.

*Transparency and interaction*: In an online environment it is very easy and comfortable to discuss examples and problems using screen-sharing. Students can help each other and discuss their problems without disturbing their colleagues as they would have done by having audible conversations in the on-site.

The VISIR approach assumes, that there exists a physical lab with measuring instruments, which can be under remote control and therefore could be used by students in an online setting. However due to the lockdown conditions nobody could be on-premises and therefore nobody could set up the measuring equipment. So, it was necessary to find a solution which did not need any on-site personnel or components. In this course students had to carry out lab exercises parallel to the theory lessons in order to strengthen their understanding.

## 3 Implementing the Online Laboratory

To start with an online laboratory a set of possible software solutions had to be evaluated.

The key measures during the evaluation were:

- Ease of installation and handling
- Price
- Functionality especially in regard of the planned lab exercises.
- The possibility of switching between online and on-premises at any time

To have a smooth transition from labs in presence to online labs, we tried to use the same exercises with the proven manuscripts the students were used to.

After the evaluation was carried out, the following products were suggested for use in the four different types of online labs:

- relay circuits software:" Relays-Simulator 2012.1 for Windows" freeware

  Students liked this software very much for the possibility to show which current paths are active. A situation which can be verified in on-premises lab only with a high measuring effort. In the simulation phase all contacts can be operated only by double-clicking.
- electronic logic circuits

  software: "LogikSim 0.64" GNU general public license
- pneumatic circuits

  Software: "Festo Fluidsim". The decision for this software was easy because there was an education version already available to the institute. This version is not the latest one, however the functionality of the available version completely fulfilled our needs for the given exercises.

  The Software Fluidsim is capable of handling standard pneumatic circuits as well as electro-pneumatic circuits. During the simulation it is very easy to evaluate the functionality of all components. Students reported that it was very easy for them to find conceptual errors in their designs. For documentation purposes the software was also very well suited. In standard labs the complete documentation must be done manually, which is much more time consuming.
- controller and control circuits

  Software: BORIS (block oriented simulation) by WINFACT. This software package comes as an add-in with the book "Simulation of technical processes" (Kahlert, 2004), which was used as course-book.

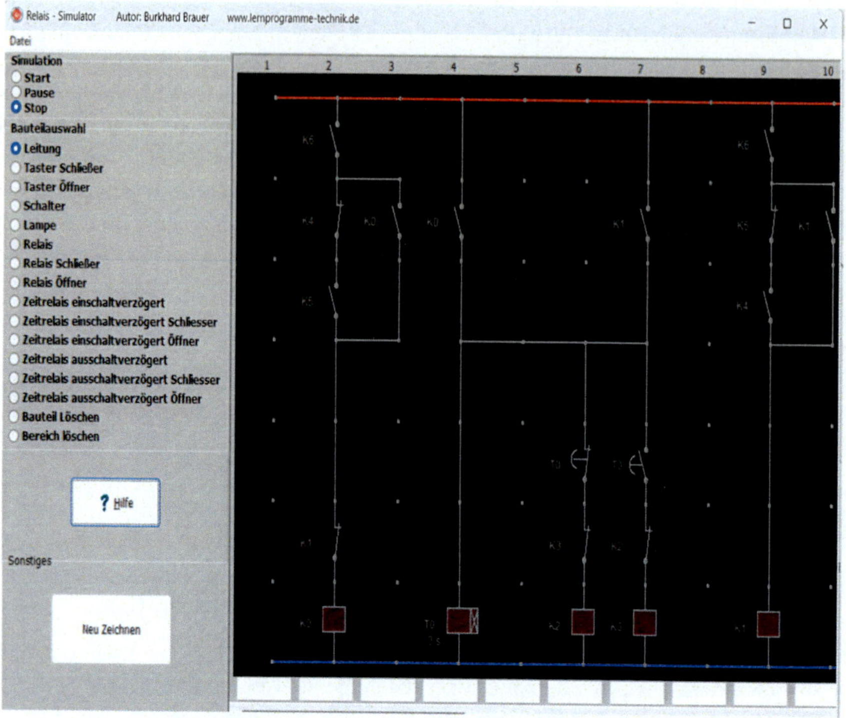


**Fig. 1.** Relays Simulator 2012.1 reversing contactor circuit

Students liked the software BORIS especially because of the possibilities to play around with several parameters. The results could immediately be seen in the diagrams. The software is offering a comparison mode, where the diagram with the old parameters is kept. So, the effect of altering some parameters can be easily seen. A functionality which can hardly be realized in on-premises labs.

At HEAD17 I presented another approach using a raspberry pi for PLC training at home. (Beyerle, 2017). We investigated this possibility together with a few volunteers. We sent them breadboards together with the necessary wires. For every type of lab-exercises we prepared a box with the necessary components. Additionally, a power supply and two multimeters were needed. One of the multimeters had to have oscilloscope capabilities. So, we ended up in a quite big box with a value of more than € 500, - each. Since the box was quite big and heavy it was costly to send it via mail or courier service. Although the volunteers were quite happy with this solution and their possibilities, we could not realize this solution for all online labs. It was too costly and needed a large effort for preparation. When switching back to labs on-premises all the materials were obsolete because it was not compatible with the existing training environment. All labs were equipped with components of the German company HPS. (Hps SystemTechnik GmbH, 2023) and mixing existing components with the contents of the boxes would not have made much sense.

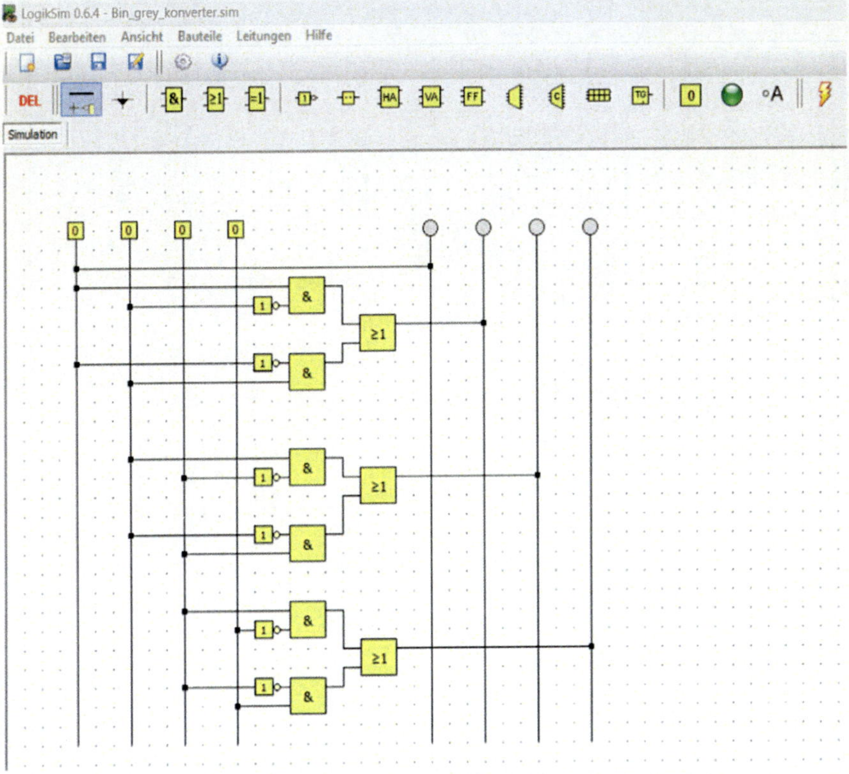


**Fig. 2.** LogikSim 0.6.4 Binary-Grey-converter

Although the preparation for the theory necessary for the lab does not depend on the way the lab is delivered—on premise or online—it is worthwhile to consider using the concept of the flipped classroom (Pfennig, 2017). Before starting with the online labs, the students were explicitly informed, that it is their task to prepare the necessary theory at home and that there will be no spare time to catch up with missing theoretical knowledge.

## 4 Limitations of Online Laboratory Lessons

*Limited hands-on experience*: While online laboratory lessons are convenient and safe, they lack the hands-on experience that is crucial for science education. This may hinder the development of practical skills and decrease a student's ability to apply their knowledge to real-life situations.

*Dependence on technology*: Online laboratory lessons are dependent on a stable internet connection and sufficient bandwidth which can often be an issue in remote areas. This can diminish a student's ability to access and participate in virtual laboratory lessons.

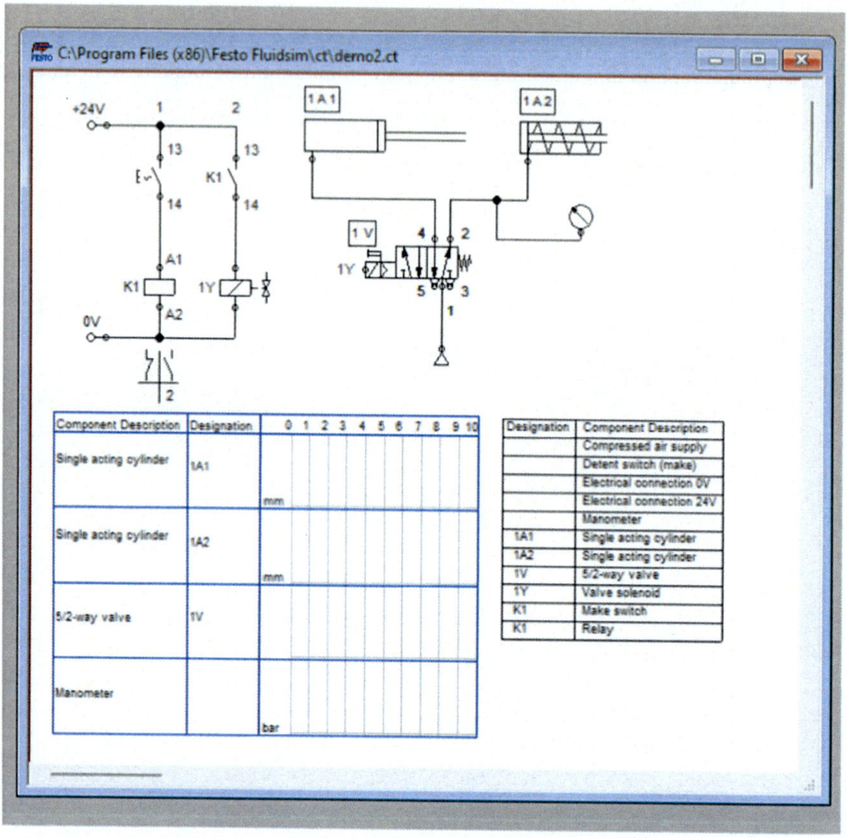


**Fig. 3.** Festo Fluidsim Example for a circuit with E-pneumatic

*Limited availability of virtual labs*: Currently, the availability of virtual laboratory lessons is limited, and not all experiments can be carried out online. This can limit a student's exposure to a range of laboratory experiments and techniques. During our evaluation process we only chose software packages that were suited to carry out all planned experiments. So, we could achieve, that traditional labs and online labs were 100% identical. Students attending online labs had no drawbacks over their colleagues attending in presence. This was also necessary to get comparable and fair evaluations.at the end of the course.

From all the mentioned limitations the first one—limited hands-on experience—was rated the most serious. During the final examinations students must build up several circuits. Of course, in this situation a lot of common problems come up. If you never had to deal with these problems, it is very difficult to find out why a circuit does not perform in the way one would have expected.

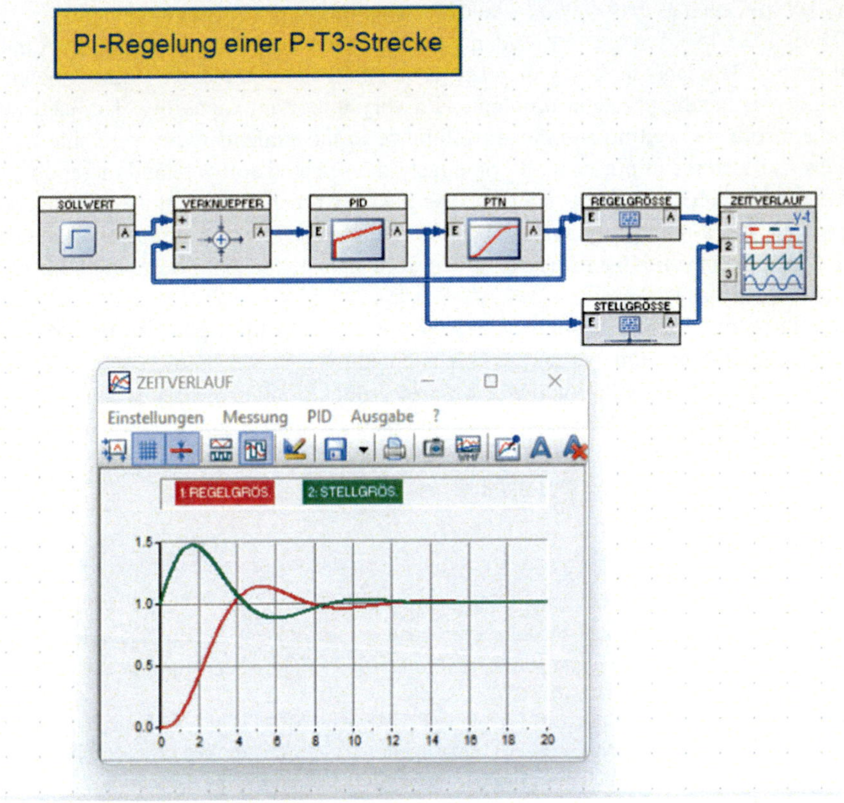


**Fig. 4.** Lab dealing with a PT3-controlled system.

## 5 Conclusion

Online laboratory lessons have the potential to revolutionize science education and make it more accessible, convenient, and safe. However, it is crucial to understand its limitations and work towards bridging the gap between traditional laboratory lessons and virtual laboratory lessons. After one year the lockdowns ended and labs could again be conducted in presence. We all were very astonished, that the students overwhelmingly wanted to keep the online labs. They brought their personal laptops to the laboratories and carried out their exercises using the software. We found out that the main reason for this behaviour was the students strong wish to be all together during the labs. For security reasons it is not allowed to have more than 8 participants in a traditional laboratory. By incorporating a balanced approach, combining both traditional and virtual laboratory lessons, students can get the best of both worlds. Our first assumption was, that it might be a good idea to change every week between online and onsite. During the first two months we tried several intervals. At the end we decided that every third lab should be carried out in presence. With this solution both students and lecturers were happy.

This allows the students to get a good feeling for all possible problems occurring in everyday life, such as broken wires, cold solder points or defective components.

During the lockdown a few students finished their courses and had do take their final exams. The labs are a key element to be able to pass the exams, since showing the ability to handle practical problems is a very important capability. The pass rate and the results showed no significant difference to the students, who were able to do all labs and lectures in presence. So obviously the selected software and methodology was the right decision to help students in successfully finishing their studies. At the end of each course, a mandatory survey must be completed. The return messages showed overall satisfaction with the measures which were taken to make the best out of this very complicated situation.

In conclusion, online laboratory lessons are a step in the right direction towards making science education more accessible and inclusive. However, more work needs to be done to overcome its limitations and make virtual laboratory lessons a more comprehensive learning tool. Due to the positive feedback and the good results in the final exams, lecturers and students were inspired to find new applications, where a complete or partial transition to online will make sense.

## References


Beyerle, W.: An affordable and modular development environment for PLCTraining. In *Proceedings of the 3rd International Conference on Higher Education Advances* (S. 179–188). Universitat Politècnica València (2017)

Garcia-Loro, F., Macho, A., Castro, M., Alves, G.R., Marques, M.A., Lima, N., Rodriguez-Gil, L., Najimaldeen, R., Fernandez, R.M.: Remote laboratory VISIR: recent advances, initiatives, federation, limitations and future. In *2021 IEEE Global Engineering Education Conference (EDUCON)* (S. 1754–1757). IEEE (2021). https://doi.org/10.1109/EDUCON46332.2021.9453961

Gustavsson, I., Zackrisson, J., Håkansson, L., Claesson, I., Lagö, T.L.: The visir project–an open source software initiative for distributed online laboratories. In *REV* (2007)

*hps SystemTechnik GmbH.* (2023, 28. Mai). https://hps-systemtechnik.com/

Kahlert, J.: Simulation technischer Systeme: Eine beispielorientierte Einführung (1. Aufl.). Vieweg Praxiswissen. Vieweg (2004). https://doi.org/10.1007/978-3-322-80247-7

Pfennig, A.: Flipping the classroom and turning the grades—a solution to teach unbeloved phase diagrams to engineering students. In *Proceedings of the 3rd International Conference on Higher Education Advances* (S. 73–81). Universitat Politècnica València (2017)

*Steuern und Regeln für Maschinenbau und Mechatronik: Die beigefügte CD enthält die Bilder des Buches und die Lösungen zu den Aufgaben und Übungen* (13. Aufl.).: *Bibliothek des technischen Wissens.* Verl. Europa-Lehrmittel Nourney Vollmer (2013)

# Embedding AI into LMS and eLearning Platforms

Eleni Ioannou Sougleridi[1], Spyros Kopsidas[2](✉), Denis Vavougios[1], Aggelos Avramopoulos[1], and Athanasios Kanapitsas[1]

[1] Department of Physics, University of Thessaly, 3rd Km Old National Road Lamia–Athens, 35100 Lamia, Greece

[2] Department of Informatics and Telecommunications, University of Thessaly, Papasiopoulou 2, 35131 Lamia, Greece

spyros@uth.gr

**Abstract.** Learning management systems (LMS) have revolutionized education by removing restrictions on time, space, and boundaries. The platforms are now essential for delivering instruction and training in universities, businesses, and other institutions. However, learning requirements are quickly evolving. A training program participant today expects to receive programs that are tailored to his needs, competence, and ability. To meet the learners' hyper-personalized needs, LMS platforms must be reimagined. Massive amounts of data are being generated and gathered by learning platforms, and when these data are processed using cutting-edge data science tools like AI, machine learning, etc., they can produce insightful data intelligence that will help to further enable efficient, adaptive, and personalized learning and training programs. AI and machine learning can assist us in processing these data to enhance the functionality of an eLearning or LMS platform and modernize it in accordance with new training methodologies and learner expectations. Moreover, it can help us obtain in-depth, accurate, and insightful intelligence on how people are interacting with an LMS down to the micro level, in order to improve the effectiveness of the system and assist us in delivering more personalized programs for higher participation and retention rates. In this paper, we present a structure and methodology for integrating AI into an LMS platform, in an ethical and effective way.

**Keywords:** Artificial Intelligence · e-learning · LMS

## 1 Introduction

Nowadays, education is taking a turn towards e-learning and LMSs [1]. Learning Management Systems (LMSs) gained a prominent role during the pandemic [2]. LMS is a software application that "provides administration, documentation, tracks, reports, automates and delivers educational courses and learning programs" [3]. Among the most popular LMSs are Blackboard, Moodle, Web CT and Canvas [4]. LMSs provide the user with progress tracking, higher learning retention. Addition, they are easy to use, cost-efficient and present flexibility and scalability [5]. Furthermore, LMSs can provide the


M. E. Auer et al. (Eds.): ICL 2023, LNNS 899, pp. 363–368, 2024.
https://doi.org/10.1007/978-3-031-51979-6_38

instructor with data analysis, trend analysis and action. LMS data is generated automatically and thus being cost-effective and relatively easy to analyze [6]. Moreover, LMSs make automated recommendations for courses to students by collecting and extracting metadata. They can also be used as part of blended learning techniques, as well as be part of both synchronous and asynchronous learning. Also, LMSs are currently being used for self-learning by allowing the student to self-assess, as well as alternative forms of learning such as collaborative learning.

## 2 Big Educational Data

Big data is the term referring to the vast amount of information that is created by learners and is collected daily during the learning process [5]. Big data includes student data such as demographics (age, ethnicity, gender, geographical data, etc.) and course data such as enrolment headcounts, grades and completion rates. These include the LMS log data, the task completion rates, the online learner performance and progress and the e-assessment scores. Educators can keep track of each learner's preferred activities [9]. LMSs collect this big data, sort, filter and examine it for the improvement of the learning experience and the prediction of students' performance by identifying areas of weakness.

Educational Data Mining (EDM) examines the educational data for the upgrade of teaching. Gaining popularity in the recent years, EDM offers dropout prediction and prevention through early intervention. The most commonly used Data Mining techniques include classification, clustering, visual data mining, statistics, association rule mining, regression, sequential pattern mining, text mining, correlation mining, outlier detection, causal mining and density estimation [7].

EDM contributes to evaluating the learning material by providing feedback supported learning, evaluating the task complexity, planning strategies and offering pedagogical support to both the student and the teacher. By predicting dropout and measuring retention, it gives the educator a glimpse into the motivation, satisfaction, reflection and awareness of the student. Furthermore, it monitors students' learning performance by overseeing skills, grades, deficiencies, elapsed time, correctness, engagement, participation and competency. EDM offers a modelling of learning behavior by supporting action, pattern and knowledge modelling [8].

However, LMSs have a hard time adapting to the diverse profiles and learning backgrounds of students [9]. Furthermore, motivation and engagement issues while using LMSs have been observed. This results in a passive attitude towards learning, due to boredom and poor engagement [10]. Another noteworthy issue is that the volume of Big Data that is produced is too large and too complex.

## 3 AI in Education

Lately there has been talk about the use of Artificial Intelligence (AI) in Education. A growing worldwide interest in AI and how it can enrich learning methods and techniques by offering a diverse approach [11]. Educational learning platforms implementing AI are becoming increasingly popular. AI can render a LMS more accurate, accessible, actionable and accountable. The processing of Big Data is often strenuous. This is where

AI jumps in to reap the benefits of real-time data analysis, by extracting, analyzing and structuring the Big Data produced by students and clustering their log files, simplifying thus the learning process for the educator. Moreover, it puts forward the classification of learning materials and grading. An Intelligent LMS evaluates skills, knowledge and overall performance.

For example, rule-based systems, which are used for educational purposes store and manipulate knowledge to interpret information in a useful way. Currently LMSs provide all students with an "one-size fits all" approach. Intelligent agents can detect the different learning styles, each student's level and attention rates [12]. By taking into account all the previous factors and the system logs, the Intelligent Agent produces the appropriate learning material for each student separately [13–18]. AI measures engagement by varied methods e.g. Emotional measurement methods, automatic facial expression recognition and measuring electrodermal activity. Intelligent Tutoring Systems provide assessment of the student before and after student-system interaction. Machine Learning algorithms predict academic performance by running diagnostics of learners' performance. Personalization takes into account the different learning aims, learning approaches, content, learning pathways, cognitive models, preferences, context and individual learning groups, making academic material more relevant to each learner. In such manner, there is an enhancement of academic performance and an improvement of the learning process as the educator pays equal attention to all students.

Machine Learning algorithms analyze each student's demographic, socioeconomic and academic data, that are generated through the student-system interaction. LMSs systems using AI are more dynamic, allowing for an interchange between the system and the user. AI looked encouraging when it comes to adaptive learning [19, 20]. Adaptive learning dynamically adapts, based on a learner's characteristics and needs. Learning becomes adaptable to the learner's evolving skills and knowledge [39, 40]. Learners with physical disabilities or learning difficulties could benefit from adaptive learning systems such as Intelligent Agents. Adaptive learning maximizes learner satisfaction by continuously modifying the learning efficiency and effectiveness. It takes into account each student's unique conditions, preferences, abilities, background knowledge, interests, goals, learning style, behavior, engagement and effort, performance, thinking process, emotions, working memory capacity, personality traits and response to learning activities. Furthermore, it adapts to student's needs as the student progresses. Pedagogically and research-based intelligent adaptive learning technology accesses and stays in the Zone of Proximal Development (ZPD) for each learner. That means that it provides the right next lesson at the right level of difficulty at the right time. Intelligent LMSs are dynamic systems and update frequently the learning model offering adapting guidance such as adaptive feedback, hint and recommendation generation.

With the help of AI, interchange between the user and the system is possible [21], making thus the learning experience more goal-oriented. As an Intelligent Agent gets immediate feedback from the user [21], it enables interaction between the user and the system and the identification of at-risk students is prompt.

The different AI approaches in adaptive education systems include Condition-Action Rule-based Reasoning, Fuzzy Logic (FL), Decision Trees, Bayesian Systems, Neural

Networks, Genetic Algorithms, Hidden Markov Systems, Random Forest, K-means, Case-based reasoning, Naive Bayes, Support Vector Machine, and Data Mining [22–24].

## 4 Enhancing LMS with AI

This study suggests a methodology for enhancing an LMS with AI power. Our methodology for enhancing an LMS with AI, is based on educational data utilization and exploitation and focuses on:

(1) enabling customized and adaptive training,
(2) gaining a deeper understanding of learners' needs, goals, and expectations,
(3) acquiring a thorough comprehension of the learning programs' effectiveness,
(4) enhancing retention and engagement rates,
(5) improving the educational experience,
(6) automating the management of learning,
(7) creating intuitive content that meets the needs of the learner.

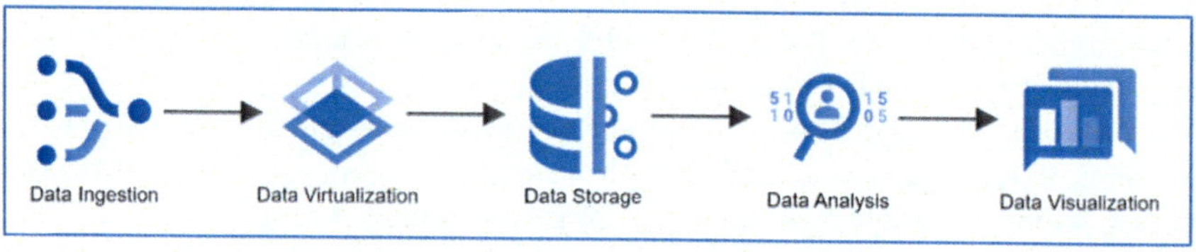


**Fig. 1.** Data flow diagram of an AI enhanced LMS

In order to achieve the above functionalities, the main processes and components that have to be implemented and be embedded into an LMS, are the following.

### 4.1 Data Ingestion

Data ingestion is the process of importing massive, diverse data files from various sources into a single, cloud-based storage medium like a data warehouse, data mart, or database, where they can be accessed and analyzed.

### 4.2 Data Virtualization

Data virtualization is an approach to data management that enables an application to retrieve and manipulate data without requiring technical information about the data, such as how the data is formatted or where it is physically located.

### 4.3 Data Storage

Data storage, also known as information preservation, is the process of keeping information and making it as accessible as possible using technology.

### 4.4 Data Analysis

Data analysis is the methodical application of statistical and/or logical techniques for describing and illustrating, condensing and summarizing, and evaluating data.

### 4.5 Data Visualization

The graphic representation of information and data is known as data visualization. Data visualization offers an accessible way to see and understand trends, outliers, and patterns in data by utilizing visual elements like charts, graphs, and maps. Additionally, it offers a great way to clearly present data to non-technical audiences.

## 5 Conclusions and Future Work

This study proposes a methodology for embedding artificial intelligence in LMS and eLearning platforms, through a technological and data management/analysis approach. More work has to be done, in order to ensure the efficiency, usability, user-friendliness, and ethical management of learner data, without disrupting the pedagogical aspect, since issues like teaching methods, curriculum, and instructional design are strongly considered.

## References


1. Dogan, M.E., Goru Dogan, T., Bozkurt, A.: The use of artificial intelligence (AI) in online learning and distance education processes: a systematic review of empirical studies. Appl. Sci. **13**, 3056 (2023)
2. Raza, S.A., Qazi, W., Khan, K.A., Salam, J.: Social isolation and acceptance of the learning management system (LMS) in the time of COVID-19 pandemic: an expansion of the UTAUT model. J. Educ. Comp. Res. **59**(2), 183–208 (2021)
3. Retrieved from https://en.wikipedia.org/wiki/Learning_management_system
4. Ushakov, A.: Learning content management systems in flt: Canadian experience. Intern. J. Eng. Lang. Literat. Stud. **6**(1), 25–32 (2017)
5. Dahdouh, K., Dakkak, A., Oughdir, L., Messaoudi, F.: Big data for online learning systems. Educ. Inf. Technol. **23**, 2783–2800 (2018)
6. Liu, M., Yu, D.: Towards intelligent E-learning systems. Educational and Information Technologies (2022)
7. Retrieved from IEDMS. Educational Data Mining. Available online: https://educationaldatamining.org/ (accessed on 27 December 2022)
8. Aldowah, H., Al-Samarraie, H., Fauzy, W.M.: Educational data mining and learning analytics for 21st century higher education. Telematics Inform. **37**, 13–49 (2019)
9. Oliveira, M., Barreiras, A., Marcos, G., Ferreira, H., Azevedo, A., Vaz de Carvalho, C.: Collecting and analysing learners data to support the adaptive engine of OPERA, a learning system for mathematics. In Proceedings of the 9th International Conference on Computer Supported Education - Volume 1: A2E, ISBN 978–989–758–239–4; ISSN 2184–5026, SciTePress, pp. 631–638
10. Ahmad, M., Ismail, S.N., Hamid, S., Alaboudi, A.A., Jhanjhi, N.Z.: Exploring students engagement towards the learning management system (LMS) using learning analytics. Comput. Syst. Sci. Eng. **37**(1), 73–87 (2021)

11. Holmes, W., Bialik, M., Fadel, C.: Artificial intelligence in education. Promise and implications for teaching and learning, Publisher, Center for Curriculum Redesign (2019)
12. Ahmad, W.F.W., Hashim, A.S.: Enhancement of learning management system by integrating learning styles and adaptive courses, advances in intelligent systems and computing, pp. 211–218 (2017)
13. Herlocker, J.L., Konstan, J.A., Terveen, L.G., Riedl, J.T.: Evaluating collaborative filtering recommender systems. ACM Trans Inform Syst **22**, 5–53 (2004)
14. Klašnja-Milicevic, A., Ivanovic, M., Nanopoulos, A.: Recommender systems in e-learning environments: a survey of the state-of-the-art and possible extensions. Artif. Intell. Rev. **44**, 571–604 (2015)
15. Srisa-An, C., Yongsiriwit, K.: Applying machine learning and AI on self-automated personalized online learning. In Fuzzy Systems and Data Mining; Tallón-Ballesteros, A.J., Ed.; IOS Press: Amsterdam. The Netherlands **5**, 137–145 (2019)
16. Moreno-Guerrero, A.-J., López-Belmonte, J., Marín-Marín, J.-A., Soler-Costa, R.: Scientific development of educational artificial intelligence in web of science. Fut. Intern. **12**, 124 (2020)
17. Smutny, P., Schreiberova, P.: Chatbots for learning: a review of educational chatbots for the Facebook Messenger. Comput. Educ. **151**, 103862 (2020)
18. Mousavinasab, E., Zarifsanaiey, N., Niakan Kalhori, S., Rakhshan, M., Keikha, L., Ghazi Saeedi, M.: Intelligent tutoring systems: a systematic review of characteristics, applications, and evaluation methods. Interac. Learn Environ **29**(1), 142–163 (2018)
19. Chetyrbok, P.V., Shostak, M.A., Alimova, L.U.: Adaptive learning using artificial intelligence in distance education. In Proceedings of the Distance Learning Technologies, Yalta, Crimea, 16–21 September 2021 (2021)
20. Adnan, M., Al-Saeed, D.H. Al-Baity, H.H., Rehman, A.: Leveraging the power of deep learning technique for creating an intelligent, context-aware, and adaptive m-learning model. Complexity, vol. 2021, Article ID. 5519769 (2021)
21. Potode, A., Manjare, P.: E-learning using artificial intelligence. Intern. J. Computer Sci> Inform. Technol. Res. **3**(1), 78–82, Month: January - March 2015 (2014)
22. Colchester, K., Hagras, H., Alghazzawi, D., Aldabbagh, G.: A survey of artificial intelligence techniques employed for adaptive educational systems within e-learning platforms. J. Artif. Intell. Soft Comput. Res. **7**(1), 47–64 (2017)
23. Mehta, P., Saroha, K.: Analysis and evaluation of learning management system using data mining techniques, 5th ICRTIT (2016)
24. Manhiça, R., Santos A., Cravino, J.: The use of artificial intelligence in learning management systems in the context of higher education: systematic literature review, 17th Iberian Conference on Information Systems and Technologies (CISTI), Madrid, Spain, 2022, pp. 1–6 (2022)

# Pedagogical Value of Educational Technologies in the COVID-19 Pandemic: EdTech Experts' Perspectives from Hungary, Kazakhstan, and Poland

Assel Csonka-Stambekova(✉)

Kazinczy, Hungary
assel.stambekova@ppk.elte.hu

**Abstract.** The purpose of this qualitative study was to explore EdTech experts' perceptions about innovative practices introduced in the development of educational technologies to meet teaching and learning needs during the COVID-19 pandemic. The author of this study inquired how EdTech experts viewed technological developments and their pedagogical value in local EdTech that delivered emergent remote education (ERE) in Hungary, Kazakhstan, and Poland from March 2020 to autumn 2021. The following research question guided the study: What are EdTech experts' perceptions on the pedagogical use of technology associated with this mode of instruction? Semi-structured interview data with eight participants was collected and analysed with MAXQDA software. Findings indicated that EdTech experts viewed the pedagogical value of EdTech as challenges and opportunities for teaching and learning in online learning environments. The article suggests further avenues for research and contributes to the knowledge base in educational technology research.

**Keywords:** COVID-19 pandemic · Pedagogical value · EdTech

## 1 Introduction

The effects of educational technology (EdTech[1]) use in teaching and learning have received considerable attention in the last 30 years in secondary school context [1, 2]. The role and the impact of EdTech on teaching and learning illustrate somewhat contradictory results. Some scholars recognise the beneficial effect of EdTech integration into teaching and learning [3]. Others, however, have made calls to identify critical areas where educational technology research could improve [4, 5]. Moreover, throughout the global COVID-19 pandemic EdTech was a chief tool to provide education remotely during the crisis. Teachers overnight have been forced to digitalise instructional materials, teacher and learner interactions, and deliver learning materials and assessments via Zoom, Google Classroom, MS Teams. However, the field of educational technology

[1] In this article EdTech is used interchangeably with technologies, digital and ICT tools primarily used for educational purposes during the COVID-19 pandemic.


M. E. Auer et al. (Eds.): ICL 2023, LNNS 899, pp. 369–380, 2024.
https://doi.org/10.1007/978-3-031-51979-6_39

research needs more scholarship [6] on how these communication platforms, initially developed for business, meet teaching and learning needs. Furthermore, the field challenges pedagogical practice [7]. Evidence shows that most "teaching" during emergency remote education (ERE) [8] was operational. That is, teaching via Zoom or MS Teams was teacher-led [9, 10] had a nature of "broadcasting", and lacked interaction [11, 12]. Nevertheless, an integration of the pedagogical, technological, organisational, and personal aspects in EdTech without compromising student learning (SL) needs further clarifications. Despite this seeming interest in investigating the interplay of EdTech and pedagogical practices to promote SL, these have been limited to studies in higher education with weak research designs [13]. Hence, this qualitative study aimed to explore and understand what pedagogical value EdTech can bring to teaching and learning practices. In this paper, I argue that when considering education online in K-12, educators and teachers need to recognise and embrace pedagogy, learning, technology, and the human aspect of education to deliver effective teaching and learning.

This paper captures the diverse voices of EdTech experts[2] during the pandemic. This qualitative study is part of a larger mixed-methods convergent parallel multiple-case study. The qualitative data was gathered independently from the quantitative data in the larger mixed-methods study and published separately. The researched context and participants- EdTech experts from Hungary, Kazakhstan, and Poland- share several similarities and differences. On one hand, education systems in these research sites are characterised by the past Soviet legacy. On the other hand, given the differences in the development of economies and democracies in these countries, these contextual factors shape participants' epistemological and ontological beliefs. As such, the results of this study should be treated with care in interpretation. Furthermore, cultural differences need to be considered when discussing the results of this qualitative-oriented and designed study. The following research question guided this study: What are educational technology experts' perceptions of the pedagogical use of technology associated with this mode of instruction?

## 2 Reviewing Empirical Literature

Technological developments brought opportunities for improved experiences in teaching and learning. For example, how collaborative work fosters knowledge construction and task achievement [14] and how technologies help build a stronger sense of student-centred learning [15]. Student-centred learning is evident in teacher technology practices employing authentic learning using digital tools [16]. Interestingly, research shows that when it comes to technology integration in the classroom, teachers choose practicality over pedagogy regardless of their beliefs [17, 18].

The recent COVID-19 pandemic has accelerated the adoption of EdTech for teachers and students' use during ERE. This abrupt shift to ERE has highlighted the need for

[2] EdTech experts are professionals who were in close contact pre-COVID-19 and during-COVID-19 with EdTech product development or provided professional development on the technology use for learning purposes to K-12 schools or technical equipment to schools. Additionally, these are professionals who participated in the evolution of educational technologies responding to challenges in remote teaching and learning caused by the novel virus COVID-19.

schools to invest in infrastructure and training to support effective use of EdTech [19]. Because technology affordances [20–22] and limitations differ from country to country, some researchers refer to these countries as low-tech and high-tech [23]. Low-tech countries experienced unavailability of power supply, internet connection, and technology access. Thus, these countries such as Libya and Yemen did not introduce ERE. In these countries, radio or TV served as emergency solutions [23]. Teachers from high-tech and developed countries experienced technological- and pedagogical-related challenges. While high-tech countries in general do not lack technology infrastructure, during ERE these countries struggled with slow internet connection, slow pace of institutions to adopt virtual learning [23], students' and teachers' digital illiteracy, and issues with student engagement and assessment. Another distinguishing feature of high-tech countries, according to one study [7] is that teachers sought support from their schools' ICT teams (p. 16). In contrast, in low-tech countries teachers had to call a helpline or access a national website to resolve technical issues [7]. This evidence shows binary distinctions between those who had technology access and who did not cause barriers to SL and teacher effectiveness during ERE. Furthermore, evidence illustrates that teachers lacked technological, pedagogical, and content knowledge [24] and tried to replicate classroom experience via EdTech. Considering these gaps in teacher knowledge, technology affordances and limitations, and different pace of technology integration in classrooms, the author of this article attempted to understand the following: what pedagogical value, as perceived by EdTech experts, EdTech brings into teaching and learning.

## 3 Methods

### 3.1 Sampling Procedure and Sample

According to the design of this qualitatively-oriented study, the sampling procedure was criterion- and convenience-based [25]. Using a criterion-based sampling form [26], the sample of participants included eight EdTech experts from Hungary, Kazakhstan, and Poland. Table 1 below provides the demographical information of the study participants. Four respondents were collaborating with schools and ministries of education in each research site during the pandemic. These respondents distributed EdTech tools to schools, including software and hardware, learning materials, and open educational resources, and provided teacher professional development to mitigate the challenges in remote teaching and learning caused by the COVID-19. Three participants were highly involved with a language mobile application to provide language learning solutions to K-12 students. One participant co-launched a support group for families and children facing digital inequality during ERE in one of the research sites. None of the participants received remuneration for their participation. Prior to data collection, I obtained digitally-signed informed consent from all participants. All interviews were audio-recorded via Zoom and participants verbally reaffirmed their consent before the interview. The semi-structured interview protocol with EdTech experts consisted of 13 questions addressing the research questions of this study and demographic questions. The interview protocol was guided by the theoretical framework on student engagement online, student learning, and the digital divide. The interviews ranged from 50 to 110 min (M = 71,43, SD = 20.25).

**Table 1.** Participants' demographic information

| Participants' ID | Country of origin | Years of experience | Industry representation | Gender |
|---|---|---|---|---|
| P1 | Hungary | 8 | Language start-up | Male |
| P2 | Hungary | 5 | IT start-up | Male |
| P3 | Hungary | 5 | Computer Science education start-up | Male |
| P1 | Kazakhstan | 10 | Big Tech | Male |
| P2 | Kazakhstan | 10 | Education leader in EdTech | Male |
| P3 | Kazakhstan | 10 | Language IT start-up | Male |
| P1 | Poland | 16 | Big Tech | Male |
| P2 | Poland | 3 | EdTech foundation | Female |

*Note* Big Tech refers to Apple, Amazon, Microsoft, Google/Alphabet, and Facebook/Meta Instrumentation[30]

### 3.2 Data Analysis

Interview data was transcribed with an AI-powered tool otter.ai, verbatim- proofread and member-checked to ensure the trustworthiness of the findings. I used thematic analysis (TA) as the main method to analyse and report themes within data [27]. MAXQDA version 2022, release 22.4.1 software (Verbi Software, 2022) was used to analyse and group segments of codes (see Table 2) to develop the themes (see Fig. 1).

## 4 Findings

### 4.1 Opportunities in Technological Innovations

Through a careful analysis of participants' interview data, distinct sub-themes were developed in exploring the opportunities in technological innovations. These sub-themes included (a) new quality of teaching and learning, (b) collaborative learning, and (c) student engagement online. Overall, the EdTech experts' perspectives were based on their remote lesson observations during the pandemic, collaboration with teachers and schools in the implementation of ERE in their respective countries, and conversations with educational authorities about the role of EdTech in education. Table 2 offers selected supporting participants' quotes due to page limitations.

**Table 2.** Data extract, with codes applied using a blended approach in TA

| Alignment between Research question (RQ) and Interview questions (IQ) | Data extract | Coded for |
|---|---|---|
| RQ: How educational technology can be used to help students gain cognitively?<br>IQ: How do technologies help students learn English? | My personal feeling is that there is always much better interaction when we see each other, the teacher and the student in person (P3, Poland) | Importance of social interactions in learning |
| RQ: How educational technology can be used to help students gain cognitively?<br>IQ: Could you describe how current major educational technology tools support teacher-learner interactions? And learner-learner interactions? | So if we look at the way teachers have used technology, it's been very transmission-based. Teachers have gone to simple lecture and demonstration tasks and the majority of lessons I've seen is that …they'll have a quick talk, quick questions and answers, they'll throw either a prompt on the whiteboard or they will show a quick video. And then people just do it, and then answer and that's your 40 min finished. (P2, Kazakhstan) | Broadcasting the lesson |

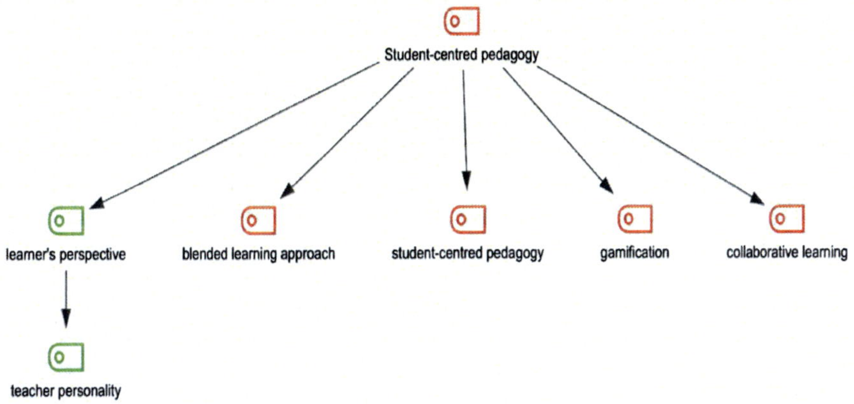


**Fig. 1.** Developed a thematic plan, showing one of the eight identified themes from the interviewees' data concerning the research questions of this study. Themes have been captured by MAXQDA 2022 (Release 22.4.1). *Source* Author

**New quality in teaching and learning.** In this study, all participants unanimously agreed. The overwhelming majority of the participants asserted that technology integration into teaching and learning can greatly benefit the pedagogy given the technology affordances. Four participants provided detailed descriptions about how EdTech personifies education. A common view amongst these participants was that EdTech empowers learners with a choice in how to learn. Consequently, some participants stated that this change in the learning paradigm could lead to improved economies and advancements in the 21st century teaching and learning.

Regarding the specific role of EdTech in various aspects in teaching and learning, participants unanimously emphasized the need for technology integration use in curriculum design, lesson planning, teacher professional development, and learner assessment. Most participants believed that teachers should incorporate EdTech into their pedagogy. They were quite vocal about providing a handful of valuable insights. For example, a participant from Kazakhstan observed the pressing need for a significant change of the existing national curriculum in school education. According to this participant, transformation of secondary education is necessary, and technology should play a crucial role in driving the transformation. Having observed 40 to 50 remote lesson observations this participant highlighted an orientation toward low thinking skills in classroom teaching practices as the weakest side in the current school curriculum. The participant went further and explained how teachers could use technologies to empower students for content creation, skills enhancement, and enabling students' higher-order thinking skills. Building upon this viewpoint, another participant further voiced the potential of technology in facilitating the preparation of the 21st century teachers.

**Collaborative learning.** With regards to this sub-theme, a recurrent sense among the participants was in considering EdTech as a tool for collaborative learning. Five participants commented on the useful features such as breakout room, sharing files, conversing with students via a chat, managing students' assignments in accessible folders in video conference tools such as Zoom or MS Teams. Four participants acknowledged that these tools had notable enhancements in their affordances during the COVID-19 pandemic. As such, offering teachers and students an opportunity to utilise familiar classroom tools in online settings is one of the technology innovations. In other words, a significant surge of third-party applications emerged for educational purposes in what formerly was known as business videoconferencing tools. One participant, representing a Big Tech, emphasized continuous learning beyond the boundaries of the classroom. As a result, features such as file storage and various communication tools between teachers and students had to be introduced in one of the well-known EdTech platforms to promote collaborative learning.

Interestingly, out of all the participants, only one, identified as an educational leader, observed a minimal use of EdTech for collaboration in 40 to 50 ERE lessons in Kazakhstan. Namely, a use of break out rooms or the synchronous nature of ERE during the pandemic failed to fully illustrate the potential of EdTech for fostering collaborative learning. This participant stressed that current approach to collaborative learning lacks the level of engagement that could be achieved using business videoconferencing tools such as Zoom, MS Teams, or Google Classroom.

**Student engagement online.** EdTech experts, who developed mobile applications and represented an EdTech foundation, were vocal about how EdTech can engage students and create authentic learning online. These participants talked about how they diversified the content in the mobile language and computer science applications they developed to increase student engagement. Student engagement, as they said, was a valuable indicator of the interaction between the technology and the learner. Therefore, it was increasingly important that the content they provided in the mobile applications was matching students' skills, knowledge, and introduced some level of challenge.

Furthermore, these participants talked at length about various approaches embedded in EdTech to increase student engagement. One of them is gamification that is purposefully built-in to EdTech. Beside the pure pedagogical purpose of engaging students into learning, some of these participants honestly said that they integrated gamification into EdTech for business purposes. In other words, the higher the engagement with the application and the content is, the longer the student uses the application, and the more money comes to developers. When questioned about how pedagogically valuable this approach was to them, some of them responded that the purpose of any business is to make profits. EdTech is not an exception and the better it is pedagogically equipped, the more profit a company makes.

Other participants, who were not involved into mobile language or programming applications, viewed gamification differently. According to these participants, gamification should be aimed at SL regardless the environment. For example, a participant from Poland emphasised how secondary and high school students, who are deeply immersed in technologies, challenged teachers skills to adapt their pedagogy accordingly. Overall, the participants viewed EdTech as a valuable pedagogical tool that can enhance teaching practices, strengthen learner engagement, and equip students with essential 21st century skills and knowledge.

## 4.2 Constraints in Technological Innovations

A divergent theme developed from the data set with a focus on constraints that occur as technologies become more advanced. The theme of constraints in technological innovations included the following sub-themes (a) digital divide, (b) lack of teachers' technological pedagogical knowledge, and (c) social connections. Table 3 illustrates some of the participants' supporting quotes.

**Table 3.** Participants' supporting quotes for the theme of opportunities and constraints in technological innovations

| Theme | Participant quote | Participant pseudonym |
|---|---|---|
| Opportunities in technological Innovations:<br>(a) new quality of teaching and learning<br>(b) collaborative learning<br>(c) student engagement | '…imagine … a school where students are learning how to fix cars, and they probably do not have every single type of car in the school to check…in virtual reality there are applications prepared by the car manufacturers which help to understand this car, showing this in VR, engaging students to be more deeply in this particular topic' | P3, Poland, September 22, 2021 |
| | 'Learning does not happen only during the video or in the online meeting. Learning has to take place before the meeting…' | P1, Kazakhstan, January 15, 2021 |
| | '…gamification and interactive elements make the class more engaging…frankly, we will move into this space a little bit as well in terms of helping actually the teacher-learner collaboration…' | P3, Hungary, March 29 2021 |
| Lack of teachers' technological pedagogical knowledge | There's a huge resistance which is also a generational question partially, I believe. Around "I don't like technology. I don't want to get used to it. I don't want to change. I don't want to use Zoom. What is Zoom? It's for young people | P3, Hungary, 29th March 2021 |
| | You don't want to change for a while… teachers in those regions are older than the average. And the average teacher age in Hungary is old already | P2, Hungary, February 16, 2021 |

**Digital Divide.** Four participants out of eight expressed their concerns about increasing digital inequalities in education revealed during the pandemic. They have explicitly used phrases such as "lack of access to technologies" (P2, Hungary, February 16, 2021), "challenge was in the infrastructure of the schools…" (P1, Kazakhstan, January 15,

2021), "so even if you are wanting to do distance learning, how do you do it when there is no Internet?" (P2, Kazakhstan, February 5, 2021), "poor Internet speed or device capabilities or finance access to devices" (P3, Kazakhstan, February 19, 2021).

Four participants noted some teachers possessed basic ICT skills which were insufficient for ERE. The participants referred to it as a disadvantage for students because teachers were unable to access and integrate various available technological solutions to their remote lessons. For instance, teachers used messenger and WhatsApp's functionality of sending audio messages for homework assignments, which deteriorated the quality of learning. Participants expressed concern that teachers did not increase their abilities to use technologies as means of self-learning.

**Lack of teachers' technological pedagogical knowledge.** In sharing their views on the constraints of technology use for teaching and learning, most participants firmly stated teachers' weak preparation on knowing about technology for educational purposes. Specifically, the participants shared how they observed teachers' functional level of using technology during ERE. That is, some other participants described teaching during the pandemic that they have observed as broadcasting of traditional lessons via Zoom or MS teams. Another aspect that three participants noted was teachers' resistance. Two participants felt that teachers were not willing to change because of their retirement age and lack of time to prepare technologically based lessons. Another participant connected it to the typical teachers' behaviour and to the country contextual factors.

Furthermore, almost all participants voiced their concerns that teachers lacked knowledge how to integrate pedagogy and subject knowledge with EdTech for the benefit of student learning. That is, teachers did not really know how a technology could diversify their pedagogy and promote students' higher order thinking skills. Some participants were questioning the quality of initial teacher education and professional development based on their remote lesson observations and conversations with teachers during the pandemic. These participants emphasised that teachers were surviving while transitioning to ERE. The participants felt that teachers' 'survival pedagogy' limited students' learning needs and the potential of technologies was not exploited fully during the pandemic lessons.

**Social connections.** Another topic that the participants had similar views of was about missing human connection in EdTech. Participants provided different examples about how communication and social interactions were important for learning, and especially for children of school age. Participants believed that despite a wide variety of tools to organise and manage learning in OLEs, EdTech lacked providing social connections. For instance, absence of in-person peer-to-peer interactions impacted the way ERE was carried out in the researched context. Half of the participants firmly stated that the most effective lesson was the one with human interaction. Another half of participants brainstormed on technological substitutes of peer-to-peer interactions online. Examples included holograms, augmented, and virtual reality. However, two participants emphasised the advantageous position of EdTech for improving pronunciation or listening skills in language education over actual human presence.

## 5 Discussion and Conclusion

The current study attempted to understand what EdTech experts considered as pedagogical value in EdTech. The study stemmed from a larger mixed-methods convergent parallel multiple-case study conducted online during the pandemic in Hungary, Kazakhstan, and Poland. This study contributes to the educational technology research knowledge-base a somewhat different than usual point of view on technology integration in teaching and learning. Specifically, the study draws on EdTech experts' perspectives about the opportunities and constraints in technological innovations for teaching and learning.

It is noticeable that the sample of participants in this study was highly aware of the affordances and limitations of EdTech. Consequently, the participants have shared their valuable insights about various nuances of the pedagogical value that EdTech bring into post-pandemic education. Given the characteristics of the study sample, the findings focused on major opportunities and constraints of technological innovations as perceived by the participants. As such, the participants acknowledged technology advancements that bolster new quality in teaching and learning, opportunities for collaborative learning, and student engagement online.

Participants commented on how student-friendly some of the technologies have become. For instance, how MS Teams offered other ways to maintain teaching and learning beyond familiar synchronous business communication tools such as chats and video calls [22] has shown how current technology offers higher quality of audio and video reducing the time lag between sound and image to 300ms. Moreover, because of the pandemic active users of [28] grew from 10 million to 300 million from December 2019 to April 2020.

Furthermore, the idea of personalised learning, expressed by some participants in this study, can help us further understand how students learn best when technology is involved. This finding supports evidence from previous studies [21]. It further supports the idea of enriching learners' online learning environments (OLEs) by increasing the effectiveness of technology affordances.

Collaborative learning was seen as another advantage in marrying EdTech with pedagogy. This study finding emphasised that EdTech provides learner autonomy in the types of interactions with peers, teachers, and technology. Because the use of EdTech depends on an end user's abilities and knowledge, EdTech allows multiple variations for students to co-create and co-construct knowledge.

The findings of this study show that one strong conclusion can be drawn: technologies represent immense opportunities for high quality in teaching and learning given their affordances [20] such as (a) connectedness; (b) sharing; (c) collective intelligence; (d) empowerment; (e) multimodality. Considering technological disruptions in teaching and learning caused by the pandemic, adequate technology integration in education can aid teachers in meeting students' needs and diversify their pedagogy. It can therefore be assumed that EdTech will be further used in education despite the known constraints and increasing digital inequalities. This assumption is widely shared in other studies [29]. However, it should be stressed that contextual factors such as school context, teachers' and students' demographics shall be considered in drawing this conclusion for larger populations. Therefore, the methodological approach and conclusions might differ from the current study.

At the same time, there are apparent disparities in experts' professional experience and how they viewed the pedagogical value of EdTech in learning and teaching. For example, participants who had been involved in school leadership or subject teaching in the past have given in-depth views on teachers' pedagogical knowledge, the priority of pedagogy over technology, and the essence of learning. In contrast, participants who came from industry or business viewed education transmitted to the technological domain from the perspective of skills development for future jobs. This view suggests that the ways EdTech is promoted for use in educational reforms in these countries is enacted through disruptions in learning and teaching, and innovation.

Most participants recognised that there is currently little understanding and knowledge of how technologies can be fully used to foster student learning. Specifically, the participants felt concerned about the lack of teachers' understanding of technologies and technological pedagogical knowledge. Participants described how teachers' functional use of technologies for practising lower order thinking skills, for instance, limited students' knowledge acquisition during ERE. Furthermore, more significant aspects of the digital divide such as a lack of or absence of EdTech access, strong internet connection and bandwidth, and limited use of online tools for students to create content restricted student learning. As mentioned in the literature review, lack of technology access and connectivity were issues found around the globe. The findings of this study are in line with previous research [23].

The aspect of social presence will remain essential in OLEs. To this end, teachers will have to develop technological and pedagogical knowledge to be able to support student learning. The evidence of this study suggests that while technological developments occur rapidly, social presence impacts learning extensively in OLEs. Future research is needed to understand how else EdTech is able to provide social presence using qualitative or mixed-methods research design.

## References

1. Al-Qallaf, C.L., Al-Mutairi, A.S.: Digital literacy and digital content supports learning: The impact of blogs on teaching English as a foreign language. The Electronic Library (2016)
2. Dashtestani, R., Hojatpanah, S.: Mobile-assisted language learning in a secondary school in Iran: discrepancy between the stakeholders' needs and the status quo. In Handbook for Online Learning Contexts: Digital, Mobile and Open: Policy and Practice pp. 157–174. Cham: Springer International Publishing (2021)
3. Yu, A., Trainin, G.: A meta-analysis examining technology-assisted L2 vocabulary learning. ReCALL **34**(2), 235–252 (2022)
4. Hennessy, S., Mavrikis, M., Girvan, C., Price, S., Winters, N.: BJET Editorial for the 50th anniversary volume: looking back, reaching forward. British J. Educat. Technol. **50**(1), 5–11 (2019)
5. Cowling, M.A., Crawford, J., Vallis, C., Middleton, R., Sim, K.: The EdTech difference: digitalisation, digital pedagogy, and technology enhanced learning. J. Univ. Teach. Learn. Pract. **19**(2), 1–13 (2022)
6. Renz, A., Krishnaraja, S., Schildhauer, T.: A new dynamic for EdTech in the age of pandemics. In CONFERENCE Paper presented at ISPIM innovation conference virtual (2020)
7. Gouseti, A.: 'We'd never had to set up a virtual school before': opportunities and challenges for primary and secondary teachers during emergency remote education. Rev. Educ. **9**(3), e3305 (2021)

8. Bozkurt, A., Sharma, R.C.: Education in normal, new normal, and next normal: observations from the past, insights from the present and projections for the future. Asian J. Dist. Educ. **15**(2), i–x (2020)
9. Ding, A.C., Ottenbreit-Leftwich, A., Lu, Y.H., Glazewski, K.: EFL teachers' pedagogical beliefs and practices with regard to using technology. J. Dig. Learn. Teac. Educ. **35**(1), 20–39 (2019)
10. Hall, T., et al.: Education in precarious times: a comparative study across six countries to identify design priorities for mobile learning in a pandemic. Inform. Learn. Sci. **121**(5/6), 433–442 (2020)
11. Utomo, M.N., Sudayanto, M., Saddhono, K.: Tools and strategy for distance learning to respond COVID-19 pandemic in Indonesia. Ingénierie des Systèmes d'Information **5**(3), 83–90 (2020)
12. Mu'awanah, N., Sumardi, S., Suparno, S.: Using Zoom to support English learning during Covid-19 pandemic: Strengths and challenges. Jurnal Ilmiah Sekolah Dasar **13**;5(2), 22–30 (2021)
13. Xu, Z., Banerjee, M., Ramirez, G., Zhu, G., Wijekumar, K.: The effectiveness of educational technology applications on adult English language learners' writing quality: a meta-analysis. Comp. Ass. Lang. Learn. **2**; 32(1–2), 32–62 (2019)
14. Schellens, T., Valcke, M.: Collaborative learning in asynchronous discussion groups: What about the impact on cognitive processing? Comput. Human Behav. **21**(6), 957–75 (2005)
15. Henderson, M., Selwyn, N., Aston, R.: What works and why? (2015)
16. Ertmer, P.A., Glazewski, K.D.: Essentials for PBL implementation: Fostering collaboration, transforming roles, and scaffolding learning. Essen. Read. Probl. Based Learn. **58**, 89–106 (2015)
17. Kimmons, R., Hall, C.: Toward a broader understanding of teacher technology integration beliefs and values. J. Technol. Teach. Educ. **24**(3), 309–335 (2016)
18. Vongkulluksn, V.W., Xie, K., Bowman, M.A.: The role of value on teachers' internalization of external barriers and externalization of personal beliefs for classroom technology integration. Comput. Educ. **118**, 70–81 (2018)
19. UNESCO. ICT Competency Framework for Teachers, version 3 (2018). Available at: https://en.unesco.org/themes/ict-education/competency-framework-teachers
20. Merchant, G.: The trashmaster: literacy and new media. Lang. Educ. **27**(2), 144–160 (2013)
21. Correia, A.-P., Liu, C., Xu, F.: Evaluating videoconferencing systems for the quality of the educational experience, Distance Education (2020)
22. Smith, E.: Why conference call technology never works. Motherboard Tech by Vice (2020). https://www.vice.com/en_us/article/y3mxyw/why-conference-call-technology-never-works
23. Hazaea, A.N., Bin-Hady, W.R., Toujani, M.M.: Emergency remote English language teaching in the Arab league countries: challenges and remedies. Comput. Assis. Lang. Learn. Electr. J. **22**(1), 201–222 (2021)
24. Hughes, J.: The role of teacher knowledge and learning experiences in forming technology-integrated pedagogy. J. Technol. Teac. Educ. **13**(2), 277–302 (2005)
25. LeCompte, M.D., Schensul, J.J.: Designing and conducting ethnographic research: An introduction. Rowman Altamira (2010)
26. Patton, M.Q.: Qualitative research and evaluation methods. Thousand Oaks. California (2002)
27. Braun, V., Clarke, V.: Using thematic analysis in psychology. Qual. Res. Psychol. **3**(2), 77–101 (2006)
28. Zoom. Dashboard statistics and graphs (2020). https://support.zoom.us/hc/en-us/articles/360039656511-Dashboard-Statistics-and-Graphs
29. Jaskulska, S., Jankowiak, B.: Teachers' attitudes towards distance learning during the COVID-19 pandemic. Educ. Stud. **57**, 47–65 (2020)
30. Birch, K., Bronson, K.: Big tech. Sci. Cult. **31**(1), 1–14 (2022)

# AI and Learning Analytics in Engineering Education

# Designing IoT Introductory Course for Undergraduate Students Using ChatGPT

Abdallah Al-Zoubi(✉) and ChatGPT

Princess Sumaya University for Technology, Amman, Jordan
zoubi@psut.edu.jo

**Abstract.** For the Internet of Things (IoT), a rapidly growing field, it is necessary to have engineers with solid hardware and software backgrounds. Designing an introductory course for a new bachelor's degree program in IoT Engineering can be difficult due to the diversity of the subject and the need to strike a balance between theoretical concepts and practical applications. ChatGPT's natural language processing capabilities can be used to design the course content, learning outcomes, material, and interactive tests, quizzes, and assessments that provide students with a tailored educational experience. With proper prompt engineering approaches, ChatGPT can assist instructors in developing engaging, dynamic lessons that incorporate real-world examples and use cases. The proposed approach might improve student engagement and understanding while requiring less time and effort to design and deliver the course. The proposed approach offers a promising solution for designing effective engineering courses that equip students with the theoretical knowledge and practical skills to design and implement IoT systems.

**Keywords:** IoT · Undergraduate education · Course design · ChatGPT · Bloom's taxonomy

## 1 Introduction

The Internet of Things (IoT) has transformed several facets of human life, including smart homes, healthcare, agriculture, transportation, and energy management [1]. IoT facilitates effective communication and data exchange by tying commonplace items to the internet, enhancing resource management, security, and quality of life [2]. IoT enables real-time monitoring, remote access, and decision-making, promoting sustainable development, accessibility, and improved user experiences across domains. Due to the rising demand for qualified professionals in this quickly evolving field, IoT is becoming increasingly important in higher education [3]. The inclusion of IoT in academic curricula helps students acquire the knowledge, abilities, and competencies required to succeed in smart homes, healthcare, agriculture, transportation, and energy management. Students can assist in creating original solutions to a variety of problems by grasping the fundamentals of IoT technologies, protocols, and applications. Integrating IoT in higher education institutions improves students' readiness for the digital age and their capacity to participate in the Industry 4.0 transformation [4].


M. E. Auer et al. (Eds.): ICL 2023, LNNS 899, pp. 383–394, 2024.
https://doi.org/10.1007/978-3-031-51979-6_40

Bing Du et al. reviewed undergraduate university education in IoT Engineering in China, including its history, status, curriculum, and problems encountered [5]. The authors found that the IoT Engineering curriculum was an unsystematic patchwork and was deficient in practical platforms. They then provide suggestions for further development and exploration of IoT education by presenting a Technical Knowledge Map of IoT Engineering. The problem of IoT talent shortage and how to encourage universities and institutions to establish a major in IoT Engineering to train high-level IoT talent and prompt educators to reflect on its achievements and problems, contributing to IoT talent cultivation worldwide. Al-Zoubi et al. designed a bachelor's degree program in IoT Engineering at Princess Sumaya University for Technology to bridge the ICT market gap in Jordan and to meet demands for specialized engineers in this rapidly evolving area. The university adopted an optimization design process for the program, including a cycle of consultation with stakeholders, followed by a formal procedure of rectification and accreditation. The program was successfully launched at the beginning of 2021/2022 with an intake cohort of 50 students. The article also discusses the definition and hype surrounding IoT and its applications in various industries [6]. The first course in this program was launched in the second semester of the academic year 2022/2023 that carries the title: "Introduction to IoT."

In this research paper, we propose the incorporation of ChatGPT, an advanced AI-powered language model, in the design of the course as this offers several benefits, including the ability to generate relevant and up-to-date content, facilitate personalized learning experiences, and improve student engagement. By leveraging the capabilities of ChatGPT, educators can create interactive course materials that cater to various learning styles and foster a deeper understanding of IoT concepts. Furthermore, the integration of ChatGPT into the course design can help students develop essential IoT skills through collaborative learning and problem-solving activities, ultimately preparing them for a successful career in IoT engineering.

## 2 IoT Education

The number of universities offering courses in iot has increased significantly, with china setting the bar as the first nation to offer iot engineering as a standalone undergraduate program. In China alone, Over 700 universities provide a bachelor's program in IoT engineering [5]. This innovative initiative demonstrates China's dedication to developing talent in the quickly developing IoT industry. Other universities that offer bachelor's degree programs in IoT engineering worldwide include Florida International, USA; Waterford Institute of Technology, Ireland; Asia Pacific University, Malaysia; and Sydney University, Australia, while Huddersfield University and De Montfort, UK, offer IoT bachelor degree program under computer science discipline [6].

Even though it is impossible to predict the future with absolute certainty, a significant growth in the number of bachelor's degree programs in IoT engineering offered by universities worldwide is expected over the next decade. The increasing demand for skilled IoT professionals, driven by the rapid expansion of IoT applications across various industries, such as smart homes, healthcare, agriculture, transportation, and energy management, will encourage higher education institutions to develop comprehensive IoT engineering programs. These programs will focus on providing students with a strong foundation in IoT technologies, protocols, and applications while incorporating interdisciplinary studies in computer science, engineering, data analytics, and information security. Consequently, the academic landscape will see a proliferation of IoT engineering programs essential for meeting the industry's needs and fostering innovation and collaboration among students and researchers.

In addition to the growth in the number of IoT engineering programs, the future will witness a transformation in how these programs are delivered. extension, interdisciplinary approach, emphasis on practical experience, focus on security and privacy, industry collaboration, online and flexible learning options, and specializations within IoT engineering are some potential directions for the future of IoT engineering programs. As universities embrace digital technologies and new teaching methodologies, we can expect a shift towards a more blended learning approach, combining traditional classroom instruction with online courses, virtual labs, and real-world projects. This approach will provide students with a more flexible and interactive learning experience, enabling them to develop practical skills and engage in collaborative problem-solving activities. moreover, emerging technologies, such as AI-powered chatbots, virtual and augmented reality, and remote IoT labs, will play a crucial role in enhancing the quality of IoT education, making It more accessible and engaging for students globally. Consequently, these advancements in IoT engineering education will contribute to a highly skilled workforce capable of driving the future of IoT innovations and implementations across various sectors

## 3 AI-Enabled Models and ChatGPT in Education

There are numerous instances of universities implementing AI and machine learning generally for various applications, including grading, student support, and learning analytics. To help students in their online courses, the Georgia Institute of Technology developed Jill Watson, an AI-powered teaching assistant [7], built on IBM's Watson platform. The platform was created to respond to student inquiries, offer feedback, and support discussion of course-related subjects. With the help of this AI teaching assistant, the environment for online learning has seen improvements in student support and engagement. On the other hand, the university of michigan created the ecoach system. This personalized learning tool uses machine learning algorithms to give students individualized feedback and support based on their needs. To help students succeed in their courses, ECoach provides personalized study plans, educational materials, and inspirational messages [8]. In addition, the university of technology sydney uses a learning analytics platform powered by AI to monitor students' engagement, behavior, and feelings as they engage with online course materials. Furthermore, a connected, adaptive, and innovative learning platform was created to suggest the best study materials based on the changing needs of students [9]. This system continuously incorporates new learner data through learning and big data analytics, enabling it to adjust and advance. these illustrations show how AI and machine learning technologies are used in various university-related contexts, such as learning analytics, personalized learning, and student support.

AI plays three main roles in education supporting individual students, entire classes, and large groups of learners. At the individual level, the emphasis is on developing teaching techniques and approaches tailored to a student's requirements. This prompted the development of "Intelligent Tutoring Systems," which have turned out to be just as successful as human tutors. AI at the class level aims to help teachers manage entire classes rather than specific students. Tutoring, grading, and virtual reality-based learning are notable examples of how AI is used in the classroom to improve teaching and learning. Lastly, AI aims to analyze learner interactions with systems at the cohort level and modify the educational system in response to these interactions' results. At this level, important applications include identifying at-risk students, predicting potential dropouts, and analyzing learners' interests, behaviors, and performance [10].

ChatGPT is a new innovative AI language model which employs deep machine learning to process and generate natural language text. ChatGPT was trained on various text data, including books, articles, and online conversations, to have the capacity to understand the context of a given prompt and generate appropriate responses. As a result, it can hold detailed discussions, impart precise knowledge on various subjects, and pick up on the subtleties and complexities of human language. Comparatively to earlier language models, which frequently required assistance in figuring out the meaning and intent of a given text, this represents a significant advancement. Another crucial feature of ChatGPT is its ability to carry out complex cognitive tasks and generate text of a high standard that is hard to distinguish from human writing. Due to its capacity to elicit knowledge and address complex academic problems, chatgpt can offer precise and reliable answers to questions that would otherwise be challenging to find through web searches. By offering support to students, teachers, and administrators in various ways, ChatGPT Can make a significant contribution to collegiate education. It can act as a research assistant, language learning tool, on-demand tutor, and content creator for course materials. Additionally, ChatGPT can support grading, administrative duties, and student engagement by integrating with learning management systems and discussion forums. When integrating AI language models into university curricula, it is crucial to consider ethical issues, data privacy concerns, and potential biases. This is because they should be supplemental to conventional teaching methods.

## 4 Prompt Engineering

Prompt engineering may be defined as a step-by-step procedure for producing the inputs that determine the output of an AI language model. [11]. It entails carefully designing and modifying input prompts to get precise, pertinent, and context-specific responses from an AI language model like ChatGPT [12]. Prompt engineering necessitates creating queries, propositions, or instructions that direct the AI model to produce the desired output or data. It relies on context, task description, specificity, and iterations to produce successful models. Accordingly, prompt engineering is essential to maximize the performance of AI language models and increase their usefulness across various applications, including content generation, problem-solving, and decision-making support. Users can fully utilize the capabilities of AI models and ensure that the responses are closely aligned with their requirements and objectives by skillfully constructing prompts.

For an AI language model like GPT-4, prompt engineering carefully designs and structures the input query or prompt to elicit precise, pertinent, and helpful responses. Prompt engineering aims to improve the user's interaction with the AI model and produce the desired results by framing queries or statements in a particular way, providing context, or directing the AI's response. This is particularly significant because the quality of the input directly influences the output produced by AI. Advisory and information-seeking, creative, instructional, and opinion-based prompts can interact with ChatGPT for various purposes depending on the context and the desired information.

With its adaptability and versatility in mind, ChatGPT can handle a variety of prompts. Several broad categories of prompts that can be used with the model are outlined in the list produced by ChatGPT itself below:

**Table 1.** Type of prompts and corresponding examples for IoT course design.

| Prompt type | Example |
|---|---|
| Information-seeking | What are the key components of an IoT system, and how do they interact with each other? |
| Advice-seeking | What are some best practices for ensuring security and privacy in IoT devices and networks? |
| Creative | Design a smart home system that uses IoT devices to improve energy efficiency and enhance residents' quality of life |
| Opinion-based | What are the most significant challenges in implementing large-scale IoT solutions in urban environments? |
| Instruction-based | Explain how to set up and configure a basic IoT sensor network to monitor greenhouse environmental conditions |
| Comparison | Compare and contrast the advantages and disadvantages of using Zigbee and Bluetooth Low Energy (BLE) as communication protocols in IoT networks |
| Hypothetical | Imagine a city with a fully integrated IoT infrastructure. How would this impact daily life, transportation, and resource management? |
| Personal (adapted) | Based on your training data, what are people's common misconceptions about IoT and its applications? |
| Language Translation | Translate the following sentence about IoT from English to Spanish: 'The Internet of Things connects everyday objects to the internet, allowing them to send and receive data.' |
| Data analysis | Given the data on energy consumption in a smart home, identify patterns and provide recommendations for reducing energy usage and costs |

This list needs to be more comprehensive because ChatGPT can handle many prompts, but it gives a general idea of the different kinds of prompts that can be used when communicating with the AI. Designing a complete and exciting learning experience for an introductory IoT engineering course using prompt engineering is possible.

We can create a curriculum that spans a variety of subjects, from fundamental ideas to complex applications, by fine-tuning the prompts for ChatGPT. This method also enables the creation of engaging lectures that use case studies, real-world examples, and visual aids to illustrate key concepts more effectively. To expose students to real-world challenges in IoT engineering, prompt engineering can help design engaging tutorials and workshops. One can also ask for step-by-step instructions, sample code snippets, or troubleshooting advice for various projects. Additionally, this approach permits a variety of exams and quizzes that gauge students' knowledge of theoretical and practical IoT engineering concepts.

Through prompt engineering, students can receive individualized feedback and guidance based on their performance and advancement. This specialized strategy attends to their needs and fosters a deeper comprehension of the subject. Instructors can also improve the prompts as the course progresses based on student feedback, performance metrics, and emerging trends to keep the content current and pertinent. In essence, prompt engineering improves the design of an introductory IoT engineering course, enabling a dynamic and effective learning experience that covers key concepts and offers practical, hands-on exposure to the IoT domain.

## 5 ChatGPT-Based Course Design

The conversation with ChatGPT regarding the course design started at the beginning of February 2023 and at the start of the spring semester. ChatGPT-3.5 was available then, and later ChatGPT-4 Was utilized upon its release in April 2023. Prompt engineering was utilized to create a comprehensive and engaging curriculum and design effective, specific prompts to guide ChatGPT in generating meaningful and useful content. In the context of course design, prompt engineering can be employed in various stages, as depicted in Table 2.

**Table 2.** Aspects of course design and corresponding prompts used

| Design aspects | Description | Prompt |
|---|---|---|
| Course Objectives and Outcomes | Using prompts like the ChatGPT can help identify and refine the goals and objectives of the course. This will ensure that the course content aligns with these objectives and adequately prepares students for subsequent coursework and professional success | List the key learning objectives and outcomes for an introductory IoT engineering course |
| Syllabus and Content Creation | Prompts can help create a structured and balanced syllabus covering essential topics in IoT engineering. AI-generated suggestions can be reviewed and adjusted to ensure the syllabus meets the learning objectives and offers a strong foundation in IoT concepts | Generate a detailed syllabus and topic list for a 16-week introductory IoT engineering course |

*(continued)*

**Table 2.** (*continued*)

| Design aspects | Description | Prompt |
|---|---|---|
| Lesson Plan Development | Create detailed lesson plans for each syllabus topic and generate learning activities, examples, and discussion questions. The AI-generated lesson plans can be further customized to fit the needs and learning styles of the students | Design a lesson plan on IoT network protocols for a 60-min class session |
| Assignment and Assessment Creation | Develop assignments and assessments that measure students' understanding of the course material | • Create an assignment that requires students to design a simple IoT system using Arduino and sensors<br>• Design a multiple-choice quiz on IoT security and privacy concerns |
| Supplementary Resources and Materials | Curate supplementary resources and materials to support students' learning. This can help me generate diverse assignments and assessments that effectively evaluate students' knowledge and skills | • List relevant books, articles, and online resources for learning about IoT networking<br>• Identify case studies showcasing real-world IoT applications across different industries |

The first step in the design process concerned the course learning outcomes, their alignment with the ABET program learning outcomes, and bloom's taxonomy. The prompt used and the course learning outcomes produced at this stage are shown in the snapshot of Fig. (1) after several trials since the chatbot initially generated the old ABET learning outcomes until it was made aware of the new seven outcomes and the proper format of Bloom's taxonomy.

ChatGPT was then asked to "act as a professor designing the course to devise the content over the 16-week semester, one lecture per week, and to give the title of the lecture, its description, and examples of content, bearing in mind that there is one lecture for orientation and introduction to the course in week 1, one lecture for the mid-term exam in week 8, one lecture for the final exam in week 16". The results are given as shown in Table (3), which includes a practical project in week 14 and week 15.

Another fundamental part of the conversation was concerned with the assessment tools. ChatGPT was asked to "act as a professor designing the course to prepare a list of questions for mid-term and final exams, four assignments, four worksheets, and what course learning outcome each question tend to measure and finally to give the key answer.

Microcontrollers like the Arduino family were also investigated as part of the course content, particularly in IoT projects. ChatGPT suggested the coverage of topics such as how to program an Arduino, interface with sensors and actuators, and use it to connect

**Table 3.** Course content and distributions as designed by ChatGPT.

| No | Title | Description | Examples |
|---|---|---|---|
| 1 | Orientation and Introduction to IoT | Overview of the course, IoT fundamentals, and the role of IoT in today's world | Course outline, IoT history, and applications |
| 2 | IoT Components and Architecture | Introduction to IoT components and their roles, along with IoT system architectures | Sensors, actuators, gateways, IoT architecture |
| 3 | IoT Connectivity and Protocols | Overview of various communication technologies and protocols used in IoT | Wi-Fi, Bluetooth, Zigbee, MQTT, CoAP |
| 4 | IoT Device Programming and Platforms | Introduction to IoT development platforms and basics of IoT device programming | Arduino, Raspberry Pi, ESP32, Node-RED |
| 5 | Physical Layer in IoT Systems | Introduction to the physical layer, including IoT systems' sensors, actuators, and hardware interfaces | Sensor types, ADC, GPIO, I2C, SPI |
| 6 | Network Layer in IoT Systems | Overview of the network layer in IoT systems, including communication technologies, protocols, and routing | OSI model, IPv6, 6LoWPAN, RPL |
| 7 | Application Layer in IoT Systems | Exploration of the application layer in IoT systems, focusing on data processing, analytics, and user interfaces | REST, MQTT, Web Sockets, APIs, user interfaces |
| 8 | Mid-term Exam | Assessment of students' understanding of the course material covered in Weeks 1–7 | - |
| 9 | IoT Security and Privacy | Introduction to security and privacy concerns in IoT and best practices for ensuring security | Encryption, authentication, secure boot |
| 10 | IoT Standards and Interoperability | Introduction to IoT standards and the importance of interoperability in IoT systems | IEEE, IETF, OCF, Zigbee Alliance |
| 11 | IoT Applications: Smart Home and Cities | Overview of IoT applications in smart homes and cities, including the technologies and use cases | Home automation, traffic management, smart lighting |

*(continued)*

**Table 3.** (*continued*)

| No | Title | Description | Examples |
|---|---|---|---|
| 12 | IoT Applications: Healthcare and Industry | Introduction to IoT applications in healthcare and the industrial sector, focusing on wearables and IIoT | Telemedicine, fitness trackers, predictive maintenance |
| 13 | Data Management and Analytics in IoT | Understanding data storage, processing, and analytics in IoT systems | Cloud storage, data processing, machine learning |
| 14 | Practical Project (Part 1) | Students design and implement a simple IoT system, working on a chosen domain's hardware and software components | Prototyping, coding, project documentation |
| 15 | Practical Project (Part 2) | Students refine and test their IoT projects, focusing on system integration, data management, and security | Testing, debugging, presentation |
| 16 | Final Exam | Assessment of students' understanding of the course material covered in Weeks 9–13 and their ability to apply it in the project | - |

to the internet or other devices in an IoT system. An example of how to incorporate Arduino into the course content included the following material:

- Basic syntax of Arduino programming language
- Programming structure: setup() and loop()
- Digital input and output using Arduino
- Analog input and output using Arduino
- Using serial communication to communicate with a computer

An Arduino program was given that reads a digital signal from a button and toggles an LED on and off accordingly. The practical project in weeks 14 and 15 was dedicated to building an IoT system around Arduino.

*This approach to content design* provides several advantages over traditional course design methods. First, it allows for a dynamic, interactive, and engaging learning experience by incorporating various techniques, such as hands-on activities, group projects, online discussions, gamification, and interactive simulations. Students can also use external resources from the internet to give them a different perspective on the subject matter and help them grasp complex concepts more easily. However, it is crucial to ensure that the online resources are reputable and reliable and to provide guidance and support to students on using them effectively, asking questions, and receiving answers in real-time. These methods will encourage collaboration, critical thinking, and real-time application

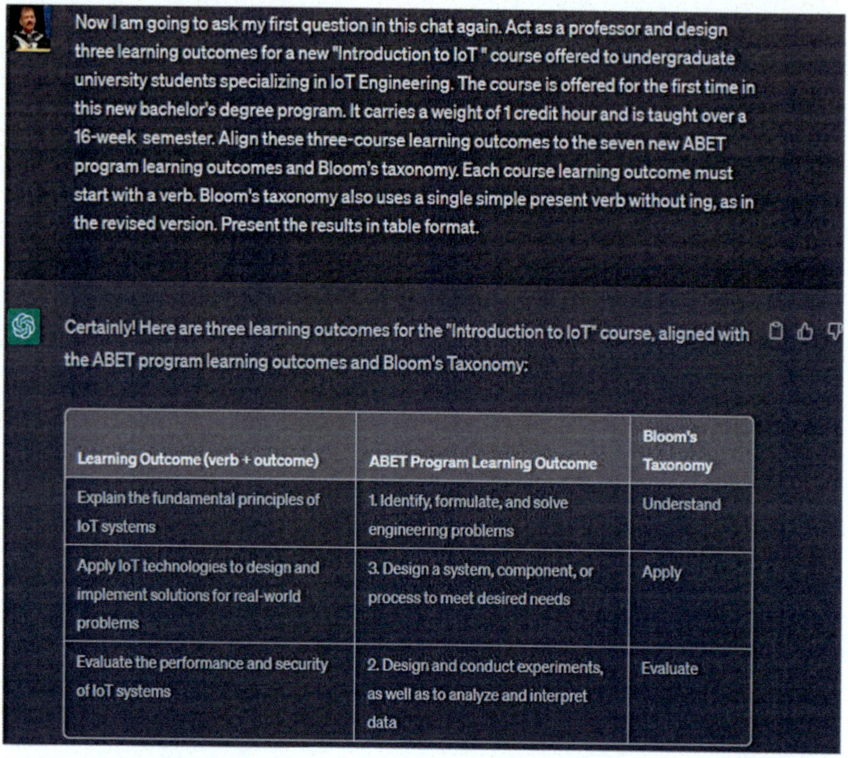


**Fig. 1.** Snapshot of the prompt to generate the course learning outcomes and results.

of concepts. Using these techniques, the IoT course can become an exciting and engaging learning experience for students. Second, it enables personalized learning, as AI technology can adapt to the needs of individual students. Finally, it provides a scalable and cost-effective solution for delivering high-quality IoT education to undergraduate students.

However, one of the main limitations of ChatGPT is that it may generate responses that need to be more accurate and appropriate in specific contexts. Therefore, it is important to use ChatGPT cautiously and verify its responses before using them in any context, especially when quoting references or dealing with the most recent data. Additionally, ChatGPT's responses may be biased or reflect the biases of its training data. While ChatGPT can offer precise and reliable answers to questions that would otherwise be challenging to find through web searches, it may only sometimes provide the most comprehensive or nuanced answers.

## 6 Conclusions

The use of ChatGPT in designing an engineering course for undergraduate students provides several advantages over traditional course design methods. The course is designed to provide students with a solid foundation in the fundamental concepts of IoT and the practical skills needed to design and implement IoT systems. By leveraging AI technology, an interactive and personalized learning experience was created that balances theoretical concepts and practical applications while providing a scalable and cost-effective solution for delivering high-quality IoT education. The proposed approach emphasized the importance of prompt engineering in course design to ensure comprehensive and engaging content that equips students with the knowledge and skills needed for success in the field. However, it is important to use ChatGPT cautiously and verify its responses before using them in any context, and that the online resources are reputable and reliable in providing guidance and support to students on using them effectively, asking questions, and receiving answers in real-time.

Future work could explore the potential of combining ChatGPT with other AI technologies or incorporating feedback mechanisms to enhance the proposed approach's effectiveness. The approach, however, offers a promising solution for addressing the challenges of designing effective IoT courses and preparing students for careers in this rapidly growing field. The new features of ChatGPT4, including its ability to analyze and produce images, diagrams, flowcharts, and other visual means, may be utilized further to improve course design and delivery.

## References

1. Al-Fuqaha, A., Guizani, M., Mohammadi, M., Aledhari, M., Ayyash, M.: Internet of things: a survey on enabling technologies, protocols, and applications. IEEE Commun. Surv. Tutor. **17**(4), 2347–2376, Fourth quarter (2015). https://doi.org/10.1109/COMST.2015.2444095
2. Kaur, K.: A survey on internet of things – architecture, applications, and future trends. 2018 First International Conference on Secure Cyber Computing and Communication (ICSCCC), Jalandhar, India, pp. 581–583 (2018). https://doi.org/10.1109/ICSCCC.2018.8703341
3. Yusuf Kalınkara, Tarık Talan: Role of the Internet of Things in Education System. 1st International Conference on Innovative Academic Studies, September 10–13, Konya, Turkey (2022). https://www.icias.net
4. Gairola, A.K., Kumar, V.: Role of Internet of things and cloud computing in education system: a review. In: Mehra, R., Meesad, P., Peddoju, S.K., Rai, D.S. (Eds.) Computational Intelligence and Smart Communication. ICCISC 2022. Communications in Computer and Information Science, Vol. 1672. Springer, Cham (2022). https://doi.org/10.1007/978-3-031-22915-2_5
5. Du, B., Chai, Y., Huangfu, W., Zhou, R., Ning, H.: Undergraduate university education in internet of things engineering in China: a survey. Educ. Sci. **11**(5), 202 (2021). https://doi.org/10.3390/educsci11050202
6. Al-Zoubi, A.Y., Tahat, A., Wahsheh, R., Taha, M., Al-Tarawneh, L., Hasan, O.: A bachelor degree program in IoT engineering: accreditation constraints and market demand. Int. J. Eng. Pedagogy. **12**(4), 17–34 (2022). https://doi.org/10.3991/ijep.v12i4.31429
7. Skrebeca, J., Kalniete, P., Goldbergs, J., Pitkevica, L., Tihomirova, D., Romanovs, A.: Modern developments trends of Chatbots using artificial intelligence (AI)," 2021 62nd International Scientific Conference on Information Technology and Management Science of Riga Technical University (ITMS), Riga, Latvia, pp. 1–6 (2021). https://doi.org/10.1109/ITMS52826.2021.9615258
8. Rebecca L. Matz, Kyle W. Schulz, Elizabeth N. Hanley, Holly A. Derry, Benjamin T. Hayward, Benjamin P. Koester, Caitlin Hayward, Timothy A. McKay.: Analyzing the efficacy of ECoach in supporting gateway course success through tailored support. In LAK21: 11th International

Learning Analytics and Knowledge Conference (LAK21), April 12–16, 2021, Irvine, CA, USA. ACM, New York, NY, USA, p. 10 (2021). https://doi.org/10.1145/3448139.3448160
9. Shawky, D., Badawi, A.: Towards a personalized learning experience using reinforcement learning," in Machine learning paradigms: Theory and application. Springer, pp. 169–187 (2019). https://doi.org/10.1007/978-3-030-02357-7_8
10. Ahmad Kashif, Junaid Qadir, Ala Al-Fuqaha, Waleed Iqbal, Ammar El-Hassan, Driss Benhaddou, Moussa Ayyash.: Data-driven artificial intelligence in education: a comprehensive review (2020). EdArXiv. June 19. doi:https://doi.org/10.35542/osf.io/zvu2n
11. Wang Shuai, Harrisen Scells, Bevan Koopman, Zuccon, G.: Can ChatGPT write a good boolean query for systematic review literature search? (2023). ArXiv abs/2302.03495
12. Abid Haleem, Mohd Javaid, Ravi Pratap Singh: An era of ChatGPT as a significant futuristic support tool: A study on features, abilities, and challenges. Bench Coun. Trans. Benchm. Stand. Evaluat. **2**(4) (2022). https://doi.org/10.1016/j.tbench.2023.100089

# Monitoring Student Performance Based on Educational Measurements

Vira Liubchenko(✉), Nataliia Komleva, and Svitlana Zinovatna

Odesa Polytechnic National University, 1 Shevchenko av., Odesa 65044, Ukraine
{lvv,komleva,zinovatnaya.svetlana}@op.edu.ua

**Abstract.** This paper presented an approach to monitoring student learning performance based on their assessment. The proposed information technology consists of two phases: modeling and diagnostics. During the modeling phase, each student's average grades for each semester are calculated, and then the dataset is clustered to identify patterns of student learning performance. The diagnostic phase uses the set of patterns and the student's current grade set to search for the closest pattern to the student's average grade set. The proposed information technology was tested on the success rate of undergraduate students in the Software Engineering program. The results showed that the proposed "dynamic" prediction method improves the work of supervisors with students by providing them with additional information.

**Keywords:** Student performance · Monitoring · Data analysis · Information technology

## 1 Introduction

Data analysis provides the means for solving different tasks, particularly diagnostic ones. The approach extracts informative features from various available data and builds a model based on these features. The model is then used to diagnose new entities with the same features. Usually, such an approach is applied under the condition of precision measuring. This approach can be adapted to work with educational measuring.

A significant problem in the learning process is the monitoring of student performance. At a particular moment, students' interest or motivation may disappear, difficulties with the complexity of learning materials may arise, or other events may affect performance in the learning process. However, tracking such events and recommending adequate corrective actions is not always possible because of the many influencing factors. Therefore, adapting digital diagnostic methods for the learning process can be helpful.

The paper aims to develop information technology to predict students' learning performance based on their assessment. The result of using the information technology is designed to support the decision-maker in monitoring students' learning performance.

The paper is structured as follows. Section 2 provides an overview of the published research on which our work is based. In Sect. 3, we described the information technology


M. E. Auer et al. (Eds.): ICL 2023, LNNS 899, pp. 395–402, 2024.
https://doi.org/10.1007/978-3-031-51979-6_41

of student performance prediction. Section 4 describes the application of the proposed information technology for the particular case. Finally, the general conclusions of the work are collected in Sect. 5.

## 2 Ways to Deal with the Issue

The problem of the quality of higher education is relevant for any country. Its assessment is especially important for public higher education institutions. In [1], the issue of defining a common understanding of quality for US universities was studied, for which "collaborative effort across states and with other stakeholders with an interest in and responsibility for aspects of quality assurance will be needed to arrive at a shared understanding." In [2], the need to develop a performance measurement system in higher education that is result-oriented rather than input-oriented for US universities was justified.

In [3], it was shown that there is no single criterion for constructing indicators for determining the quality of education. There is a "methodological gap in the statistical processes, the theoretical pieces of evidence, and the number of investigations in every level of education." There was proposed to consider the issue in the future "design and construction of multidimensional models for educational management quality measurement needs of educational institutions."

One of the aims of evaluating higher education performance based on grades is to decrease the rate of student failures. To identify and reduce the number of students who do not meet the required standards, it is necessary to consistently monitor the learning activities and behaviors of students in the classroom. However, monitoring a substantial number of students poses significant challenges.

The paper [4] aimed to thoroughly comprehend student performance prediction by examining approximately 260 studies conducted over the past 20 years. The authors specifically focus on two aspects: significant factors that strongly impact student performance prediction; various data mining techniques, encompassing prediction and feature selection, were used in the studies.

With the abundance of educational data accessible both online and offline, there exists an unprecedented opportunity to examine learning anomalies through a data-centric approach. The research [5] concentrated on five specific areas that have garnered significant attention: predicting underachievement, predicting dropout rates, identifying mental health issues, predicting challenges with graduation, and forecasting employment difficulties.

In [6], machine learning was applied to processing many student surveys to determine satisfaction predictors, revealing the degree of impact of questions on the result, particularly "survey items relating to course management and teaching being consistently most influential."

However, in [7], the authors found that with multiple data-driven ways to analyze higher education performance, there may be a problem with the lack of indicators to measure the quality of education or their simplification. The reason is that different stakeholders have different views on higher education goals.

In [8], different metrics were considered for evaluating the student model (Mean Absolute Error, Area Under Curve, Root mean squared error, log-likelihood). The conclusion was made about the applicability of various indicators for assessing models of affective states and evaluating models of skills.

The authors of [9] proposed a predictive method to detect students at risk of failing in a blended learning course. They utilize a neural network and a set of prediction variables derived from the online learning activities of students in a learning management system. The findings suggest that a neural network-based approach enables early identification of students prone to failure.

In another work, the influence of heterogeneous user behavior on predictive analysis is studied. The authors propose "an unsupervised way to construct a social behavior graph based on spatiotemporal data and to model social influences" [10].

In [11], it is shown that if students do not achieve the expected indicators of progress, learning can change "through (a) measurement that is sensitive to student's skill proficiency and how it changes following instruction, (b) the quality of instructional materials and contexts in which students are taught, (c) the quality of practice time, and (d) arrangement of reinforcement to support maintenance and application (i.e., generalization) of basic skills". In [12], dropout forecasting for underachieving students is solved as a problem of labeling sequences or predicting time series. The authors of [13] proposed a group classifier that combines the results of several generally accepted classifiers to identify students at risk of failing the studied course. This made it possible to evaluate the progress of academic performance and track the optimal time when this progress falls.

In [14], for recommend tests in e-learning systems, it is proposed to conduct data analysis based on creating several stacks of models using information from tests passed by former students. At the same time, training data was used for classification and regression problems.

The paper [15] proposes to consider the educational process as a system with negative feedback, where the deviation of the values of the learning process features from normal ones diagnoses the type of problem that interferes with the stable state of the educational process.

Based on an analysis of the publications, we can conclude the applicability of classification and regression methods for describing and predicting students' learning performance.

## 3 Information Technology Description

The input data for the proposed information technology are datasets containing the history of students' grades. It is known in which semester each grade was received. We do not consider which course grades are received, as students in the same program can study different courses each semester. We should pay attention to the small size of the datasets, which does not allow depth learning methods. Therefore, information technology is developed based on machine learning methods.

The proposed information technology can be divided into modeling and diagnosis phases.

The modeling phase aims to identify meaningful patterns of student learning performance. This is done using datasets on students' grades who have completed their studies, i.e., over an entire cycle of studies in a particular program. Clustering techniques are used to identify patterns. The procedure consists of four steps.

1. Calculate average grades in each *j*th semester for each *i*th student $A_{i,j}$, i.e., move to the *N*-dimensional feature space, where *N* is the number of semesters in the program.
2. Perform clustering of the dataset in the transformed feature space.
3. Frame the centroids of representative clusters as a set *PS* of learning performance patterns for the curriculum under analysis.
4. Analyze the low-number clusters. If an expert considers them a necessary performance pattern, then the centroid of the corresponding cluster is added to the set of patterns *PS*.

The diagnostic phase uses the set *PS* and the student's current grade set. The procedure consists of two steps only.

1. Calculate the average semester grades, i.e., transform the initial feature space to the *M*-dimensional feature space, where *M* is the number of full semesters the student has studied.
2. Search for the closest to the vector of average grades for *M* first values of the pattern from the set *PS*. The found pattern with its description is provided to the decision maker for corrective action.

The logical schema of "dynamic" prediction is shown in Fig. 1.

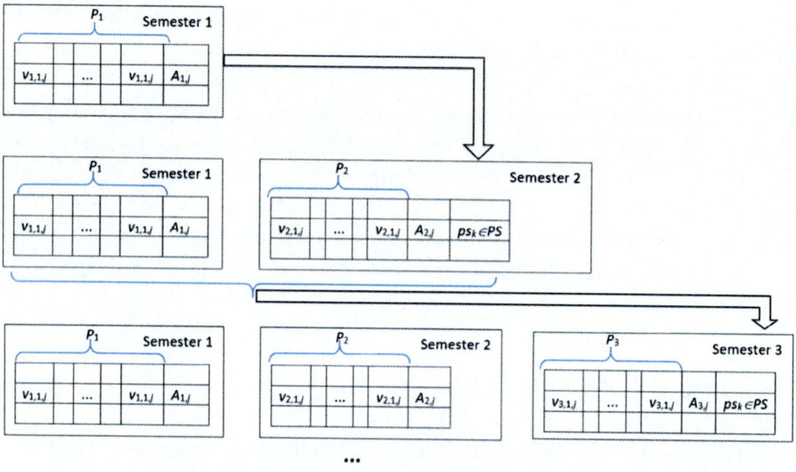


**Fig. 1.** Logical schema of the prediction process

It can be assumed that the classification results in the initial semesters will need more quality to predict. In this case, we use a regression model to predict performance.

Information technology results become the basis for monitoring student performance and identifying corrective actions. Figure 2 shows the schema of the proposed information technology. The decision-maker uses the information technology results to select corrective actions for feedback to the student.

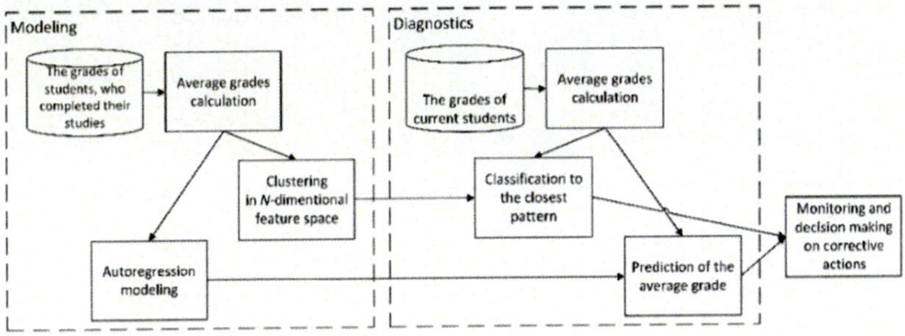


**Fig. 2.** Information technology schema

## 4 Case Study

We have tested the proposed information technology on the success rate of undergraduate students in the Software Engineering program. The dataset contained 88 records of the grades during eight semesters of study of the 2021 and 2022 graduates. The dataset contained 18% of A-student records, 23% of B-student records, 32% of C-student records, and 27% of D-student records. The distributions of the 2021 and 2022 graduates' grades had no statistically significant difference. Before the research began, all the records were anonymized.

In the first stage of the study, the average semester grades were calculated for each entry. This allowed a transition to describing objects in 8-dimensional space, each dimension corresponding to a particular semester.

Then we clustered the data into four clusters using the $k$-means method to find patterns of success. The number of clusters was chosen based on the interest in describing the patterns for the A-, B-, C-, and D-students. Figure 3 shows the resulting centroids of the clusters. The number of objects in each cluster was approximately the same.

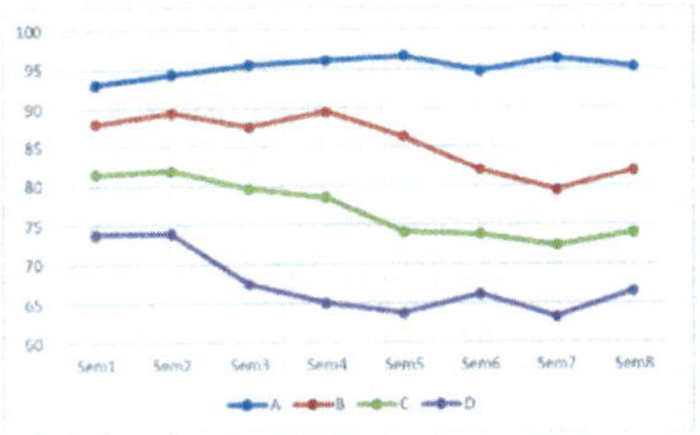

**Fig. 3.** Centroids of found clusters

Next, we evaluated the quality of the object classification in spaces whose dimensions varied from 1 (only the first semester completed) to 7 (7 full semesters completed). For the classification, we applied the $k$-nearest neighbor algorithm. Figure 4 shows the results of the evaluation. We used the accuracy and F1-measure for performance evaluation.

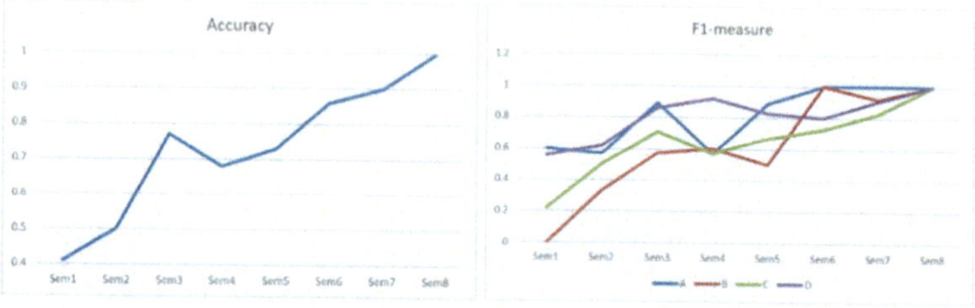


**Fig. 4.** Performance evaluation of the "real-time" classification

As we expected, classification quality in the initial semesters is poor. Therefore, for the first three semesters, consider using a regression model. Figure 5 shows the correlation matrix for semester grades, and we can note the significant dependencies between close semesters.

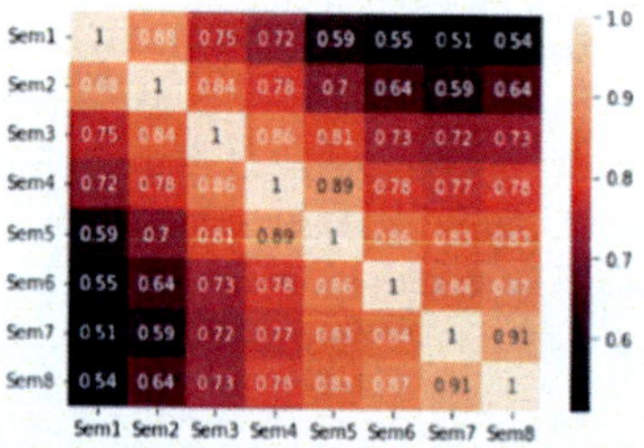


**Fig. 5.** Correlation matrix for semester grades

We studied the application of linear regression to predict students' success in the second, third, and fourth semesters. The study results are summarized in Table 1 ($m_i$ in the equations denotes the mean grade in $i$th semester).

We can see that using multiple regression does not significantly affect the prediction. It is also interesting to note that most of the records are predicted quite accurately; the decrease in the value of the coefficient of determination of a small number of records with significant deviation causes $R^2$. For example, Fig. 6 shows the relationship between the grades of the pairs first–second and second–third semesters and the linear regression models for them.

Let us summarize the result. As we expected, the prediction accuracy in the initial semesters based on patterns is low. Therefore, using a linear regression model from the second to the fourth semester is appropriate. Switching from paired to multiple regression does not yield significant improvement; hence is not applicable.

**Table 1.** Linear regression models for predicting the students' success

| # Semester | Prediction model | $R^2$ (training) | $R^2$ (test) |
|---|---|---|---|
| 2 | $m_2 = 13.99 + 0.84m_1$ | 75.8 | 82.8 |
| 3 | $m_3 = -1.3 + 0.98m_2$ | 69.4 | 75.3 |
| 3 | $m_3 = -0.98 + 0.95m_2 + 0.02m_1$ | 71.6 | 66.4 |
| 4 | $m_4 = -1.49 + 1.01m_3$ | 80.4 | 45.8 |
| 4 | $m_4 = -5.87 - 0.02m_3 + 0.18m_2 + 0.02m_1$ | 77.3 | 61.5 |

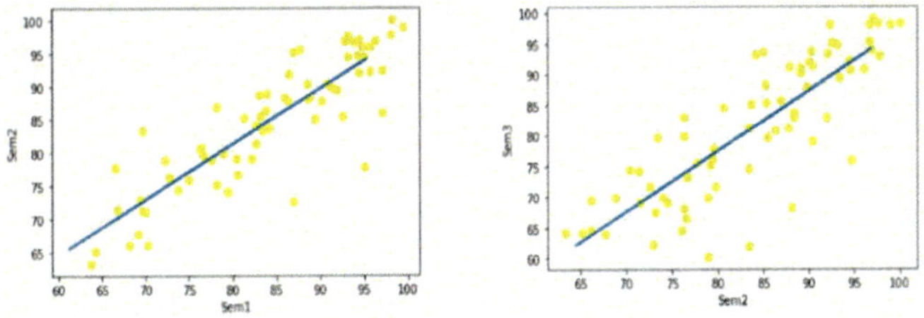


**Fig. 6.** Dependencies between the grades first–second and second–third semesters

The proposed "dynamic" prediction method improves the work of supervisors with students by supporting them with additional information. Using semester average grades provides the possibility to "cleanse" learning dimensions from the impact of choosing different courses, working with different teachers, etc.

## 5 Conclusions

In this work, we described an information technology that uses averaged semester grades to predict student performance in the learning process. The approach is based on clustering and $k$-nearest neighbor algorithms.

The case study demonstrated the effectiveness of data analysis in predicting student success. We have identified the clusters of successful students and explored ways to predict their success in future semesters. For the initial semester, we showed that prediction is achievable using linear regression models. The prediction quality evaluation showed that the prediction accuracy is acceptable for decision-making.

It can be argued that digital prediction methods are applicable to learning measures. This information technology will help improve decision-making support in academic supervision. The next step of our experimental research will be assessing the feasibility of normalizing students' assessments and extracting multi-curricula patterns.

## References

1. Tandberg, D.A., Martin, R.R.: Quality Assurance and Improvement in Higher Education: The Role of the States. State Higher Education Executive Officers Association, Washington (2019)

2. Weingarten, H.P., Hicks, M., Kaufman, A., Chatoor, K., MacKay, E., Pichette, J.: Postsecondary Education Metrics for the 21st Century. Higher Education Quality Council of Ontario, Toronto (2019)
3. Crissien-Borrero, T.-J., Velásquez-Rodríguez, J., Neira-Rodado, D., Turizo-Martínez, L.-G.: Measuring the quality of management in education. Review article. El profesional de la informaci6n **28**(6), e280604 (2019)
4. Batool, S., Rashid, J., Nisar, M., Kim, J., Kwon, H.Y., Hussain, A.: Educational data mining to predict students' academic performance: a survey study. Educ. Inf. Technol. **28**(1), 905–971 (2022)
5. Guo, T., Bai, X., Tian, X., Firmin, S., Xia, F.: Educational anomaly analytics: features, methods, and challenges. Front. Big Data **4**, 811840 (2022)
6. Langan, A.M., Harris, W.E.: National student survey metrics: where is the room for improvement? High. Educ. **78**(2), 1–15 (2019)
7. Loukkola, T., Peterbauer, H., Gover, A.: Exploring Higher Education Indicators. European University Association, Belgium (2020)
8. Pelanek, R.: Metrics for evaluation of student models. J. Educ. Data Min. **7**(2), 1–19 (2015)
9. Sukhbaatar, O., Usagawa, T., Choimaa, L.: An artificial neural network based early prediction of failure-prone students in blended learning course. Int. J. Emerg. Technol. Learn. **14**(19), 77–92 (2019)
10. Liu, H., Zhu, Y., Wang, C., Ding, J., Yu, J., Tang, F.: Incorporating heterogeneous user behaviors and social influences for predictive analysis. IEEE Trans. Big Data **9**(2), 716–732 (2023)
11. Daly, E.J., Martens, B.K., Barnett, D., Witt, J.C., Olson, S.C.: Varying intervention delivery in response to intervention: confronting and resolving challenges with measurement, instruction, and intensity. Sch. Psychol. Rev. **36**(4), 562–581 (2007)
12. Mubarak, A.A., Cao, H., Zhang, W.Z.: Prediction of students' early dropout based on their interaction logs in online learning environment. Interact. Learn. Environ. **30**(8), 1414–1433 (2020)
13. Alcaraz, R., Martinez-Rodrigo, A., Zangroniz, R., Rieta, J.: Early prediction of students at risk of failing a face-to-face course in power electronic systems. IEEE Trans. Learn. Technol. **14**(5), 590–603 (2021)
14. Mihaescu, M.C., Popescu, P.S., Mocanu, M.L.: Building and using multiple stacks of models for the classification of learners and custom recommending of quizzes. Electronics **11**(9), 1316 (2022)
15. Komleva, N., Liubchenko, V., Zinovatna, S.: Improvement of teaching quality in the view of a resource-based approach. In: 16th International Conference on ICT in Education, Research, and Industrial Applications, vol. 2740, pp. 262–277, Kharkiv, Ukraine (2020)

# Analysis and Improvement of Engineering Exams Toward Competence Orientation by Using an AI Chatbot

Thomas Fuhrmann(✉) and Michael Niemetz

OTH Regensburg, Seybothstr. 2, 93053 Regensburg, Germany
thomas.fuhrmann@oth-regensburg.de, http://www.oth-regensburg.de

**Abstract.** ChatGPT is currently one of the most advanced general chatbots. This development leads to diverse challenges in higher education, like new forms of teaching and learning, additional exam methods, new possibilities for plagiarism, and many more topics. On the other side with the development of advanced AI tools, pure knowledge will be less and less important, and demands from industry will change toward graduates with higher competencies. Education has therefore to be changed from knowledge-centered toward competence centered. The goal of this article is to use ChatGPT for analyzing and improving the competence orientation of exams in engineering education. The authors use ChatGPT to analyze exams from different engineering subjects to evaluate the performance of this chatbot and draw conclusions about the competence orientation of the tested exams. The obtained information is used to develop ideas for increasing the competence orientation of exams. From this analysis, it is visible that ChatGPT gives good performance mainly where knowledge is tested. It has, however, much more problems with transfer questions or tasks where students need creativity or complex insights for finding new solutions. Based on this result, exams and also lectures can be optimized toward competence orientation.

**Keywords:** Engineering education · Competence orientation · Exam analysis · AI chatbot · ChatGPT

## 1 Introduction and Motivation

AI applications are developing rapidly; ChatGPT is currently one of the most advanced general chatbots. It got high recognition in all areas of life, also in higher education. This development leads to diverse challenges in higher education, like new forms of teaching and learning, additional exam methods, new possibilities for plagiarism, and many more aspects. There are several highly discussed topics using chatbots in higher education. King wrote an editorial about higher education by doing a conversation with ChatGPT [10]. Lee and Soylu state that the advent of AI can give impulses for new forms of teaching and discuss ideas on how exams can be developed for a new learning experience [11].


M. E. Auer et al. (Eds.): ICL 2023, LNNS 899, pp. 403–411, 2024.
https://doi.org/10.1007/978-3-031-51979-6_42

There are several studies regarding the impacts of ChatGPT on all areas of higher education [12,17]. Several types of exams are solved by using ChatGPT, like in different areas of computer science [3,9,13,16,18]. It passed the national fundamentals of engineering exam in the US [15], passed the national high school English reading comprehension exam in the Netherlands better than an average student [5], passed a law exam [4], and a medical exam [7]. ChatGPT is used for different tasks during exam preparation and correction [15], and student support [16].

But the main long-term challenge due to the development of advanced AI systems that is not addressed by these short-time studies about ChatGPT is the question of further graduate competencies in their later working life. It is expected that many jobs will be replaced by AI, especially where routine work can be done better or cheaper. The consequence is that graduates should have high competencies in areas where they are difficult to be replaced by AI. Therefore, study programs should focus on teaching and examining these competencies. This article does a first investigation of the competence orientation of engineering exams using ChatGPT. The authors took several of their exams from courses that are taught in the Faculty of Electrical Engineering and Information Technology. These exams are from courses in different study programs and different semesters. The topics of the exams are very diverse. ChatGPT is used to solve these exams and the results are analyzed regarding performance. An in-depth analysis of ChatGPT's answers compared to typical student answers is done. This leads to ideas on how to improve existing exam concepts toward competence orientation.

Section 2 describes the competence orientation that is important for a study program. The analyzed exams are described in Sects. 3 and 4; all conversations with ChatGPT were done in German. Results of the analysis are given in Sect. 5, and suggestions for further exam development are made in Sect. 6. The conclusion and outlook are given in Sect. 7.

## 2 Competence Orientation in Future Working Life

There are numerous publications about the predicted future of work when AI is strong enough to replace many routine jobs. "Jobs with a lower risk of automation rely on social and creative skills. Therefore, the most important skills needed for these jobs are collaboration, self-regulation, knowledge construction, communication, real-world problem-solving, and the use of technology for learning" [1]. "The need for manual and physical skills, as well as basic cognitive ones, will decline, but demand for technological, social and emotional, and higher cognitive skills will grow" [6]. All these publications say that competencies on the lower levels of Bloom's Taxonomy [2] can be done by AI to replace humans in working processes. The lowest level of the taxonomy "Knowledge" is due to the large amount of memory capacity very easily done by AI, and also the second level "Comprehension" can be imitated due to a large amount of knowledge and algorithms to re-arrange this knowledge when generating answers. On the other side, people with competencies on higher levels of Bloom's taxonomy are urgently needed in working life. A large language model is incapable to create really new or original work, described in the highest levels "Synthesis" or "Evaluation" in Bloom's taxonomy. ChatGPT can support people by offloading work on the lower levels of Bloom's taxonomy so people can focus on the higher levels and can thus develop important competencies [8].

The logical consequence of this development is that education has to be adjusted toward the higher levels of Bloom's taxonomy. Lectures and also exams have to be adjusted toward competence orientation. In this article, the hypothesis is tested that ChatGPT can support in achieving this goal.

## 3 Exam: Algorithms and Data Structures

The course "Algorithms and Data Structures" (AD) is a part of the Bachelor Program "Intelligent Systems Engineering" in the third semester and is considered a part of the fundamental content. It consists of a lecture with four weekly hours and four ECTS credits and a lab course of two weekly hours and two ECTS credits. The lecture covers classical topics typically addressed in the basic education of computer science study programs. Compared to a typical AD lecture, which is treating algorithms and their properties like runtime complexity or optimality from a rather theoretical and mathematical point of view, here the focus is slightly shifted toward the selection and application of algorithms instead of mathematical proofs regarding their properties. In addition, this lecture contains a larger section on the parallelization of algorithms and their relation to aspects of the underlying computer hardware.

However, even with this slightly non-standard focus, the content is expected to be very well covered by textbooks, scripts, and exams publicly available via the internet and therefore it is to be expected that the training material for ChatGPT contained a considerable amount of beneficial input on AD topics.

The exam (written test with a duration of 90 min) intends to check the students' competencies to apply the acquired knowledge in a software development context. Also, there is a fraction of knowledge-targeting questions in order to measure the knowledge independently from the competence of the application. The exam of Winter Semester 2022/23 was fed into ChatGPT 3.5 (Version: 3. May 2023) and the obtained answers were rated according to the same expectations as the answers of the students. This test consists of four tasks, each of these structured into several sub-tasks with different complexity and sometimes including hints that are intended to help the students in finding the right solution.

An overview of the results is shown in Fig. 1. It compares the results obtained using ChatGPT to the average result[1] achieved by the 15 students participating in this exam. Each data point in the diagram represents one question of

[1] Comparing an individual result to an average is always problematic. We are following an old tradition in grading, here.

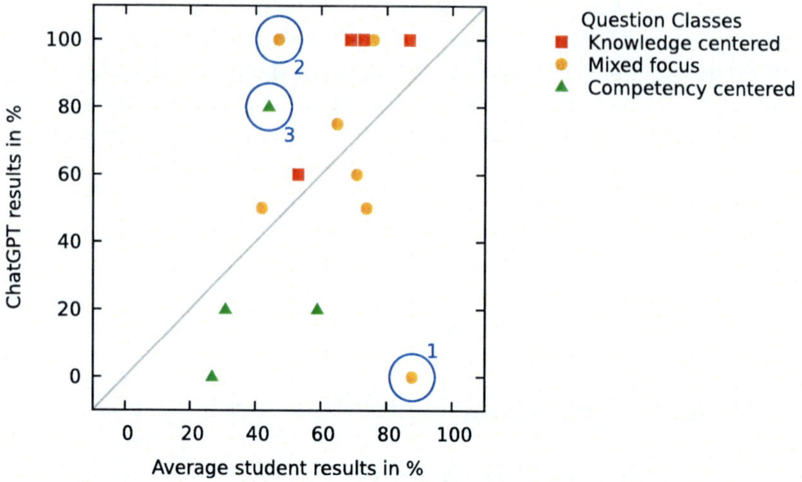


**Fig. 1.** Comparison of the results ChatGPT achieved in the AD exam to the average students' results. The questions are categorized into classes according to their degree of competence orientation anticipated by the authors of the exam. The expected trend of ChatGPT being stronger on knowledge-centered topics is visible with some exceptions

the exam with the questions differing in size (i.e. points) and scope (i.e. competence orientation vs. knowledge orientation). On a majority of questions (9:6), ChatGPT achieved better results than the average of the students. On questions supposed to be strongly knowledge-oriented, ChatGPT was always above the student average. On questions supposed to be strongly competency-oriented, ChatGPT was almost always below the student average with remarkable exceptions. On questions supposed to combine knowledge with some easy transfer application, the expected mixed picture was obtained. However, there were some surprises where both over- and underperforming of ChatGPT were observed. One question that nearly all students answered correctly but ChatGPT gave a very general and misleading answer with zero points (marked with circle "1" in Fig. 1) is the question to develop a rank determination algorithm for an arbitrary element in an unsorted list. Question "2" is about the two possibilities to execute a conditional jump in a processor pipeline that was perfectly answered by ChatGPT while students had higher problems. Question "3" is to implement a medium value search algorithm. The authors assume that question "1" is not a part of the training data set of ChatGPT while questions "2" and "3" are included.

## 4 Exams: Transmission Systems

The course "Transmission Systems" is an elective module with four weekly hours and five ECTS credits in the Bachelor Program "Electrical Engineering and Information Technology" in the sixth semester. The lecture covers basic topics related to the physical layer of communication systems, it includes communication channels, modulation formats, network topologies, error correction and cryptography, and the basics of optical communication systems.

Normally, it is a 90 min written exam with approximately 15 questions. During COVID-19, alternative exam formats were encouraged with reduced infection probabilities. Therefore, this exam was switched to seminary homework. Both types of exams are tested using ChatGPT 3.5 (Version: 12. May 2023).

### 4.1 Seminary Home Work

The seminary homework of the Summer Semester 2020 was to develop a fast internet connection for a rural village. Basic parameters for the village were given like the number of people, distances between houses, and necessary internet applications. Students should decide on a network technology and develop a scenario.

In Summer Semester 2021, students should develop an optical communication system using polymer optic fibers. They should decide on the parameters, should describe and calculate electronic circuits, and discuss the limits.

Both exams are seen to be fully competence oriented, as it is expected that no solution for this specific scenario can be found on the internet, but students have to develop their own solutions according to their knowledge. In both scenarios, ChatGPT failed completely in the exams. All answers remained on the general level that can be read on the internet. Also with very detailed questions to ChatGPT, it was not possible to get concrete solutions for the exam tasks.

### 4.2 90 min Written Exam

In Summer Semester 2022, an exam with 15 questions was given. These were mainly knowledge questions, one question requires transfer competence. All knowledge questions were answered very well by the large language model, similar to good students. No big difference between the performance of ChatGPT and the students can be seen.

The worst performance showed ChatGPT solving the transfer question. This task was to discuss the parameters of a point-to-point fiber optic connection. For students, this task was very easy and all students got full points. ChatGPT only got 30 % of the points; only very general statements were produced that can be found on the internet and no real discussion about the topic was given.

An interesting observation is that ChatGPT seemed to randomly output false statements mixed into correct responses. This is done because ChatGPT used the most probable words for the answer without having deep knowledge about the topic. In sum, ChatGPT performed “good” in this exam similar to an average student.

## 5 Analysis of Exam Examples

Even as the exams have been very different, common observations have been made for all exams. Grading of the ChatGPT output required a high effort due to the following reasons: All answers from ChatGPT have been very long, often containing redundant information compared to answers from humans. ChatGPT gave very much general information that is not closely related to the question. Most of the answers were partly correct and partly wrong; correct and false sentences were seemingly mixed in a random way giving the impression that ChatGPT mixed the answer from different internet sources with different lines of reasoning. Sometimes, single wrong words appeared within correct answers. Despite this, the answers were always syntactically correct and contained the correct wording that is typically expected from a high-quality answer. This provided the risk that the grader is lured into reconstructing meaning that was never there by connecting the presented expected technical terms with the required reasoning:

> Man usually believes, if only words he hears,
> That also with them goes material for thinking! Goethe, Faust I [21]

It can be seen that ChatGPT has a different answering "strategy" compared to humans. It has much larger knowledge than any human due to the intense training with billions of words. While humans with their limited knowledge try to solve complex knowledge questions using a transfer solution, ChatGPT can solve these by making use of this immense knowledge. To some extent, the language model simulates competence by knowledge while humans simulate (or recreate) knowledge by logical thinking and reasoning. Also, some questions that are easy to answer for a human with some knowledge in a very short way have generated an elongated response by ChatGPT often filled up with wrong or inconsistent statements. In several cases, at first sight, the answer looked interesting but after a thorough analysis, it became clear that the generated argumentation was completely void. The same behavior is seen elsewhere as already Marcus and Davis [14] stated that GPT can write perfect sentences but has no real understanding of the world. The two correctors normally agree about the student's performance when correcting exams together. When correcting ChatGPT, the number of points differed considerably. Due to the contradictory information from ChatGPT within one answer, the correctors evaluated these parts with different weights and gave a different number of points. After personal discussions, they agreed on a common number of points.

From the exam answers it can be seen that ChatGPT uses its artificial neural network to write answers that consist of the most probable sequence of words, as already Lee and Soylu [11] writes. Knowledge questions are answered well by ChatGPT when the training data set contained this knowledge, while transfer questions are answered by ChatGPT with general statements that are not solving the task.

## 6 Suggestions for Further Exam Development

It is seen from the research about the demands for future employees that the importance of pure knowledge will decrease with an increasing demand for different competencies. Therefore, future lectures and exams should prepare students for this new situation. If there are courses that are concentrated on knowledge transfer, it is not avoidable to also test this knowledge in exams. Generative AI chatbots with a powerful knowledge base can answer such questions similarly or even better compared to students. It is questionable if these courses educate students with competencies that are necessary for future working life. Therefore the authors are convinced that future lectures and exams should teach and measure competencies. It is seen from the exemplary exams that ChatGPT is not able to answer questions where problem-solving competencies are necessary. ChatGPT is therefore a good tool to test exam questions if competencies are required to solve these. If this is the case, ChatGPT answers these questions with very general statements that do not lead to the solution.

## 7 Conclusion and Outlook

The authors show in this investigation that ChatGPT is capable of answering knowledge exams with a similar performance or better than an average student. ChatGPT can not answer tasks that require using knowledge to create new and innovative solutions, which test problem-solving competencies. From this survey, the authors conclude that at first glimpse, it looks like ChatGPT passes the Turing Test [20] when answering the exam questions, but with more detailed analysis, ChatGPT behaves like a Chinese Room [19] where an immense amount of knowledge is replicated without deeper understanding. If exam tasks perceived as competence oriented are answered by ChatGPT, then these tasks do not need real competencies but can be solved with the large amount of knowledge that was used for training the artificial neural network. As it is well known, the competence orientation of an exam is spoiled if the exam questions are leaked beforehand. The challenge with language models is that they are extending the definition of "leaked" in this context to limits hitherto unknown. A truly competence-oriented task needs to be new and unique in such a sense that no solution to it can be found on the internet. The results of this article reflect the current status of AI capabilities. It is obvious that these will be increasing in the next years. The authors expect that the results of this article will remain valid as long as no strong AI is available that is able to replicate competence-oriented behavior. The authors plan to optimize their future exam design regarding competence orientation with the knowledge they gained from using ChatGPT for exam solving. Future exams will consist of fewer knowledge questions and more competence-oriented tasks to transfer knowledge and apply it to new solutions. So, the exams should be in line with the competence-oriented lecture for a future-proof education toward a new way how to work together with AI.

## References


1. Barbosa, C.E., de Lima, Y.O., Costa, L.F.C., dos Santos, H.S., Lyra, A., Argôlo, M., da Silva, J.A., de Souza, J.M.: Future of work in 2050: thinking beyond the Covid-19 pandemic. Eur. J. Fut. Res. **10**(1), 1–19 (2022)
2. Bloom, B.S., Engelhart, M.D., Furst, E.J., Hill, W.H., Krathwohl, D.R.: Taxonomy of educational objectives: the classification of educational goals. In: Handbook I: Cognitive domain. David McKay Company (1956)
3. Bordt, S., von Luxburg, U.: Chatgpt participates in a computer science exam (2023). arXiv preprint arXiv:2303.09461
4. Choi, J.H., Hickman, K.E., Monahan, A., Schwarcz, D.: Chatgpt goes to law school. Available at SSRN (2023)
5. de Winter, J.C.F.: Can ChatGPT pass high school exams on English language comprehension? (2023)
6. Dondi, M., Klier, J., Panier, F., Schubert, J.: Defining the Skills Citizens Will Need in the Future World of Work, June 25. McKinsey & Company (2021)
7. Gilson, A., Safranek, C.W., Huang, T., Socrates, V., Chi, L., Taylor, R.A., Chartash, D., et al.: How does ChatGPT perform on the United States medical licensing examination? The implications of large language models for medical education and knowledge assessment. JMIR Med. Educ. **9**(1), e45312 (2023)
8. Heaven, W.D.: ChatGPT is going to change education, not destroy it. MIT Technol. Rev. (2023)
9. Jalil, S., Rafi, S., LaToza, T.D., Moran, K., Lam, W.: ChatGPT and software testing education: promises & perils (2023). arXiv preprint arXiv:2302.03287
10. King, M.R.: ChatGPT: a conversation on artificial intelligence, chatbots, and plagiarism in higher education. Cell. Mol. Bioeng. **16**(1), 1–2 (2023)
11. Lee, J., Soylu, M.Y.: ChatGPT and assessment in higher education (2023)
12. Malik, A., Khan, M.L., Hussain, K.: How is ChatGPT transforming academia? Examining its impact on teaching, research, assessment, and learning, April 9, 2023
13. Malinka, K., Perešíni, M., Firc, A., Hujňák, O., Januš, F.: On the educational impact of ChatGPT: Is artificial intelligence ready to obtain a university degree? (2023). arXiv preprint arXiv:2303.11146
14. Marcus, G., Davis, E.: Gpt-3, bloviator: Openai's language generator has no idea what it's talking about. MIT Technol. Rev. (2020)
15. Pursnani, V., Sermet, Y., Demir, I.: Performance of ChatGPT on the us fundamentals of engineering exam: comprehensive assessment of proficiency and potential implications for professional environmental engineering practice (2023). arXiv preprint arXiv:2304.12198
16. Qureshi, B.: Exploring the use of ChatGPT as a tool for learning and assessment in undergraduate computer science curriculum: opportunities and challenges (2023). arXiv preprint arXiv:2304.11214
17. Rudolph, J., Tan, S., Tan, S.: ChatGPT: Bullshit Spewer or the end of traditional assessments in higher education? J. Appl. Learn. Teach. **6**(1) (2023)
18. Savelka, J., Agarwal, A., Bogart, C., Song, Y., Sakr, M.: Can generative pre-trained transformers (GPT) pass assessments in higher education programming courses? (2023) arXiv preprint arXiv:2303.09325

19. Searle, J.R.: Minds, brains, and programs. Behav. Brain Sci. **3**(3), 417–424 (1980)
20. Turing, A.M.: Computing machinery and intelligence. Mind LIX (236), 433–460 (1950)
21. von Goethe, J.W.: Faust I. Translated Into English in the original metres. Project Gutenberg (1808/2005). www.gutenberg.org/ebooks/14591

# Exploring Faculty Perceptions and Implementation of Learning Analytics in Higher Education

Anne Uukkivi(✉), Oksana Labanova, Karin Lellep, and Natalja Maksimova

TTK University of Applied Sciences, Pärnu mnt 62, 10315 Tallinn, Estonia
anne.uukkivi@tktk.ee

**Abstract.** This study aims to provide an overview of the Learning Analytics (LA) implementation process and to explore faculty members' perceptions of the use of Moodle LA tools in their teaching practices. It seeks to identify the benefits and challenges of implementing LA and provide insights into how faculty members perceive the use of Moodle LA tools in their teaching. Understanding these perceptions and challenges will assist in developing effective strategies to promote the use of LA and ultimately improve student learning outcomes. The paper reviews existing literature on LA implementation and faculty member perceptions and analyzes data from a survey. The study finds that developing a comprehensive strategy for implementing LA is crucial, as faculty members face various challenges such as lack of understanding, time, and motivation. Also, prerequisites like the need for constant Moodle course activities for using LA tools can be difficult. The study also suggests the 4 steps LA implementation strategy and the need for regular communication and training to enhance faculty members' skills and awareness of LA benefits.

**Keywords:** Learning analytics implementation process · Teacher's perceptions · Moodle learning analytics tools

## 1 Introduction

The concept of Learning Analytics was established in 2011 and is widely recognized as the assessment, collection, analysis, and disclosure of information regarding learners and their learning environments, to understand and improve learning outcomes and their associated contexts [1]. LA is employed in various contexts, including the identification of at-risk students, tracking online activities, providing automated feedback, enhancing learning strategies, and optimizing collaborative teamwork [2]. The implementation of LA has proven effective in enhancing student achievements in higher education [3]. Consequently, higher educational institutions strive to integrate LA into their educational practices. The implementation of LA requires a multidisciplinary approach [4, 5]. This complexity presents challenges in effectively implementing LA.

This article aims to provide an overview of the LA implementation process and explore faculty members' perceptions of Moodle LA tools in their teaching practices at TTK University of Applied Sciences (TTK UAS).


M. E. Auer et al. (Eds.): ICL 2023, LNNS 899, pp. 412–419, 2024.
https://doi.org/10.1007/978-3-031-51979-6_43

## 2 Insights into LA Implementation and Teachers' Perceptions

Updating something within an organization is not an easy task because people must agree with the updates, and adapting to them takes time [6]. That is why LA requires careful and responsible design and implementation processes. However, relatively little research has been done on it [7]. One of the challenges of implementing LA in higher education is getting teachers to use it. Teachers mostly face barriers like a lack of awareness, skills, motivation, and support in adopting LA [7–9].

Several studies have proposed strategies or frameworks for facilitating LA adoption by teachers in higher education. Arthars and Liu [7] suggest that when implementing learning analytics, institutions should adopt a bottom-up approach, support the organic growth of champions, provide flexible and customizable tools, foster a culture of innovation, establish a central support unit, develop a clear communication strategy, create a community of practice, and evaluate the impact and value of LA.

Greller et al. [10] have proposed a framework that consists of six dimensions: stakeholders, objectives, data, instruments, external constraints, and internal limitations. They suggested that teachers need to be involved in defining the purpose and scope of LA, selecting the data sources and instruments, addressing the ethical and legal issues, and overcoming the technical and organizational challenges. These recommendations were considered during the LA implementation process, which will be discussed in more detail in the next chapter.

Teachers' perceptions of using LA vary. However, according to Costa et al. [11], most teachers appreciate the benefits of LA and believe that LA is valuable in enhancing their teaching, as any data about learners could be useful for tailoring teaching materials to individual needs. Teachers find that LA solutions are crucial for providing feedback on students' learning experiences. Teachers argue that that information enables educators to compare students' current progress with the originally planned didactic-pedagogical actions. This helps them recognize the need for adjustments in their academic planning [12].

However, some skeptical viewpoints also arise. For example, teachers may encounter challenges in connecting the information from the LA reports to concrete interventions, which is partially influenced by the teacher's level of experience. Additionally, teachers may need additional types of information that are currently absent in LA reports [13]. Both pros and cons need to be considered when implementing LA to get better results.

## 3 LA Implementation Process

### 3.1 Developing a Moodle Course to Introduce Moodle LA Possibilities

Moodle has been used as an e-learning platform by the faculty of TTK UAS for quite a long time, but to date, its LA has not been widely used. To have a better understanding of how these tools could be used, research was done on how several other universities use them and how they rate their effectiveness. Although most of them stated that they are using also external LA tools in addition to Moodle, it was decided at this stage to proceed only with the possibilities that Moodle offers.

After having mapped LA practices in other universities, the next step was to map the possibilities in the current Moodle version used by the faculty. As Moodle has a wide variety of activities and resources that are located in different sections of the interface, it was necessary to gather them into one place. For that, a Moodle course named "Learning Analytics Possibilities in Moodle" was created.

This Moodle course consisted of an overview of 25 different LA tools available in Moodle. The main goal was to provide an overview of each of them – where they can be found and how to set them up; and to provide examples of where and how they can be used. They were at first grouped according to their placement in the Moodle interface – "Reports" and "Usage statistics" are the block "Navigation", "Reports" come from a participant's profile, and other possibilities are located in various parts of a course. However, as it became clear that quite often a lecturer will need the instructions of a certain LA tool, a secondary section was created where they were ordered alphabetically.

### 3.2 Testing Moodle LA Tools in a Small Group

After having completed the educational material showing how and where Moodle LA can be used, a small group was formed to test it. The test group consisted of lecturers from many different fields – mathematics, civil engineering, logistics, fashion engineering, service economy, etc. The testing period lasted for one semester.

The first step in the testing process was to share the Moodle course. After all the group members had a chance to get familiar with it, a session was organized to explain the broader goal and discuss opportunities. By the end of this first session, the purpose and idea were much clearer, and all members were able to start planning and setting up their upcoming courses (because most LA tools have specific requirements, that may not have been relevant before). Prerequisites like the need for constant activities on Moodle courses for using LA tools are needed. For a better overview of who would use which tool, a table was created, where each test group member marked down which of the 25 LA tools they planned to try. Before the start of the semester, another session was organized to discuss any questions that had arisen and to introduce what would be expected from them by the end of the semester.

The test groups' members' assignment was to use any LA tool they found to be useful as regularly as possible. No requirements or restrictions were set for them. A feedback form was created at the very beginning so that the group members could continuously write down their observations and comments as the semester progressed. At first, the goal was to get at least one answer for each LA tool, but halfway through the semester this idea was abandoned because the main goal was not to use everything, but to find the ones worth using.

Halfway into the semester, a quick overview session was organized to keep all test group members on track and help with any issues that had arisen. The final session was organized after the end of the 16-week semester. At this time, the test group members had to fill in the feedback form for each LA tool they had used, and also give a general conclusion regarding Moodle LA possibilities, e.g. if they would use LA tools in their future courses and if so which.

Based on the experiences and feedback of both the development team and the test group, a set of instructions will be compiled grouping different LA tools based on their usage, e.g. tools that can be used when a lecturer wishes to check a group's progress or tools which can help in the development of a course. These instructions will aim to provide lecturers with a quick guide for using LA without having to search through the several dozen tools available. The final stage of the process involves training sessions after which the lecturers will be supported by instructional designers and e-learning support specialists.

## 4 Methods

The study used a mixed-methods approach, including a survey and focus group interviews with students and faculty members. This paper presents only the survey results and faculty members' oral feedback collected during the overview sessions when the survey was discussed. The survey was compiled using a Google Form and conducted in the spring semester of 2023. The survey was distributed to the test group in January 2023. It consisted of three parts: general questions, an evaluation of LA tools, and final questions (feedback).

The statistical program R was used for data analysis. Structured questions were measured on a Likert scale from 0–5. Quantitative statistical analysis was used to analyze this data. The descriptive statistics and the response rate were used to present the results. Simple frequency analysis was used to extract keywords from text, which were visualized using word clouds and network analysis. The word cloud visualizes the importance and frequency of words, giving a quick overview of the content of the text or the focus of the topics. Visualizing text networks helps researchers to see the relationships between words.

The survey involved 18 lecturers (the test group) who have extensive experience in developing online courses and in using the Moodle platform in their teaching. They were asked to provide feedback on six questions, three of which were structured questions and three of which allowed them to express their opinions freely in a written format.

During the testing process, two meetings were held with the testers. The first meeting took place in a question-and-answer format. The recording of the meeting was transcribed, and subsequently, text analysis was conducted. In the second meeting, the LA tools were distributed according to the objectives.

## 5 Teachers' Perceptions of Using LA Tools

### 5.1 Overview of General Questions

The testers used their Moodle courses, which were partially held online and had varying numbers of students (11–158). The testers had varying levels of experience with learning analytics tools and were free to choose the ones that suited them.

During the evaluation of the tools, all 25 were offered for testing. The descriptive statistics and response rate are presented in Table 1. The evaluation scale ranges from 1 to 5, with point 3 representing a neutral state. A total of 23 tools were evaluated, as two of them (Participants and Logs) had no selections. Each tool was evaluated by an average of two individuals.

**Table 1.** Answers to the general questions.

| General questions | Completion rate | Mean | SD | Min | p25 | p50 | p75 | Max | Histogram |
|---|---|---|---|---|---|---|---|---|---|
| Number of testers | 0.96 | 2.04 | 1.62 | 0 | 1 | 2 | 2 | 7 |  |
| Efficiency | 0.84 | 4.21 | 0.78 | 2.5 | 4 | 4.5 | 5 | 5 |  |
| Usability | 0.80 | 4.39 | 0.50 | 3.5 | 4 | 4.5 | 5 | 5 |  |
| Time efficiency | 0.76 | 2.91 | 1.39 | 1 | 1.25 | 3.33 | 4 | 5 |  |

The efficiency and usability of LA tools were rated higher on average. Regarding time efficiency, a double-peaked distribution indicated that LA tools can be divided into two categories: those that require less time, and those that are time-consuming.

The tool with the lowest efficiency rating was Quizzes and Assignments, and for four tools (Hits distribution, Number of active students, Live logs, and Grades overview), the level of usefulness was not determined (efficiency rating of 3–3.5). Testers highly rated the tools Outline report, All logs, Activity report, Activity completion, and Complete report. The regression coefficient (r) was used to confirm the linear dependence of two numerical characteristics. Linear dependence is considered absent if $|r| \leq 0.2$ and weak if $|r| \leq 0.4$. The ratings for usability are weakly dependent on efficiency ($r = 0.35$), but there is no linear relationship between usability and time efficiency ($r = -0.05$).

## 5.2 Evaluation of LA Tools

An overview of the answers to the question "What was this tool used for?" shows (see Fig. 1) that the most frequently used words are Student, Overview, Activity, Monitor, and Material. Based on the responses regarding the reasons for using the tools, the following patterns of relationships were identified: monitoring assignment submissions, tracking and monitoring the use and views of learning materials, monitoring activity and participation, analyzing test performance and results, monitoring individual student activities and progress, an overview of student activity and behavior, and overall course monitoring and trend analysis.

When discussing the advantages of the tools (see Fig. 1), we can highlight the "Provide Overview" node. By examining the word relationships, we can observe that the nodes "Provide," "Overview," "Can," "See," and "Results" have more connections compared to others. The closer the nodes/words appear together and the more frequently they co-occur, the stronger the association between them, such as "can see" and "information overview." This indicates that one of the notable advantages is the ability to obtain an informative overview or see results through the tools.

These tools simplify course assessment for teachers and provide a convenient and comprehensive way to analyze learning. In summary, such tools offer timely submission

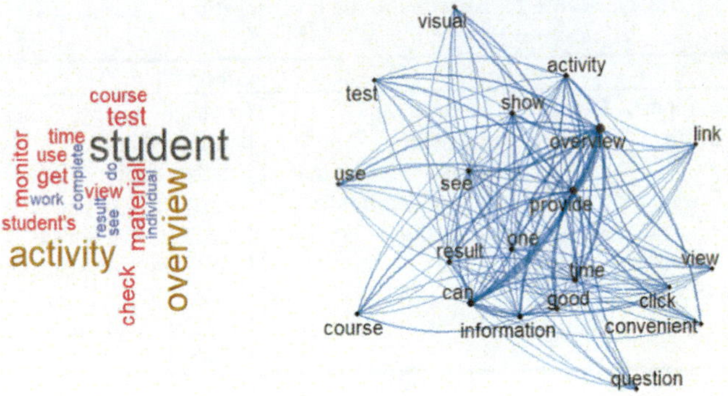


**Fig. 1.** Word clouds and relationships between words that describe the advantages of LA tools.

monitoring, activity and task tracking, test results and usage monitoring, individual and group-level analysis, and an overall overview.

The mentioned drawbacks were:

- Course setup difficulties: setting up the course requires a significant amount of time and must be done before the course starts. Subsequent changes may result in incorrect outcomes.
- Usability challenges: some users experienced difficulties using LA tools or found them hard to understand.
- Lack of clarity in group settings: when there are multiple groups with different assignment deadlines within an online course, the results are messy.
- Insufficient feedback on video completion: clicking on a URL does not indicate whether a learner has watched the video until the end.
- Lack of visual presentation: charts are not easily comprehensible and may sometimes be illegible.

### 5.3 Overview of Final Questions

In summary, it can be said that the testers broke the LA tools in Moodle into three groups. Group I consist of the tools they were willing to use in future courses, group II is a group of tools where the testers had differing opinions, and group III consists of the tools that teachers do not intend to use for various reasons.

Table 2 presents these groups where the usefulness of the tools is evaluated based on different factors: green – very useful ($t = 5$), black – somewhat useful ($t = (5;4)$), blue – $t = 4$, and if the tester rated the usefulness below 3, then brown.

It cannot be confirmed that willingness to use a certain LA tool in the future depends on efficiency ($r = 0.23$). Some highly rated LA tools were among those that will not be used in the future because several LA tools duplicate the results and each teacher chose the most suitable one among them.

**Table 2.** Answers to the final questions.

| Group I | Group II | Group III |
|---|---|---|
| Complete report, Graph of test results, Activity completion, Activity Report, Logs, All Logs, Hits distribution, Grades chart, Most used activities (9.40%) | Heatmap, Assignment submissions, Content accesses, Complete progress (4.17%) | Grades overview, Course participation, Outline report, Number of active students, Quizzes & Assignments, Live logs, Statistics (Course level), Statistics (Student level), Today's logs, Learning Analytics Dashboard (10.43%) |

### 5.4 Feedback from Overview Sessions

The feedback of testers revealed that LA was mostly used before classes to identify learners who hadn't familiarized themselves with the materials, or before assessments to identify learners who hadn't submitted certain assignments yet.

Difficulties in using LA tools were also addressed. Testers claimed that learning to use LA tools is initially time-consuming but ultimately saves time. They also mentioned that it takes more time to find the right tool than to use it and that Moodle's logic is not intuitive. Teachers tend to forget how to use a tool if they don't use it regularly. All of these responses are related to time efficiency and explain the data in Table 1.

In the second meeting, it had been determined beforehand that finding the right tool was time-consuming. A new section was added to the Moodle LA course where the LA tools are presented based on their intended use:

- To evaluate the quality and usefulness of learning materials: Activity Report, Content Accesses, Grades Chart, Graph of Test Results, Heatmap, Most Used Activities, Quizzes and Assignments.
- To monitor the progress of all the learners: Activity Completion, Assignment Submissions, Complete Progress, Content Accesses, Grades Overview, and Hits Distribution.
- To monitor the progress of a single learner: All Logs, Complete Progress, Complete Report, Course participation, Hits Distribution, Learning Analytics Dashboard, Live logs, Logs, Today's logs, Outline Report, and Statistics (Student level).
- To track statistics from the university's perspective: Number of active students, Outline Report, and Statistics (Course level).

## 6 Conclusion

The study concludes that developing a comprehensive strategy for implementing LA in HEIs is crucial. Successful implementation requires a multidisciplinary approach, involving teachers as key stakeholders and addressing teachers' barriers like lack of awareness, time, and support. The study's recommendations suggest the need for training faculty members to enable teachers to effectively utilize LA tools for personalized feedback and improved learning outcomes. The implementation strategy would have as follows:

1. Develop guidance to introduce how the LA tools can be used for various purposes.
2. Form a small test group of teachers to test the LA tools.
3. Provide support and guidance to the teachers throughout the testing process.
4. Collect and analyze the feedback from the teachers and use the results to improve the testing and constant use process of all the teachers.

The success of the implementation strategy could be verified by using the following indicators: the number of teachers who decide to continue using LA tools; the frequency and variety of LA tools which lecturers will continue to use; the impact of using LA tools on student activity, participation, performance, and outcomes.

## References


1. Long, P., Siemens, G.: Penetrating the fog: analytics in learning and education. Educause Rev. **46**(5), 31–40 (2011)
2. Kaliisa, R., Rienties, B., Mørch, A.I., Kluge, A.: Social learning analytics in computer-supported collaborative learning environments: a systematic review of empirical studies. Comput. Educ. Open **3**, 100073 (2022)
3. Foster, C., Francis, P.: A systematic review on the deployment and effectiveness of data analytics in higher education to improve student outcomes. Assess. Eval. High. Educ. **45**, 822–841 (2019)
4. Gasevic, D., Tsai, Y.-S., Dawson, S., Pardo, A.: How do we start? An approach to learning analytics adoption in higher education. Int. J. Inf. Learn. Technol. **36**, 342–353 (2019)
5. Hilliger, I., et al.: Identifying needs for learning analytics adoption in Latin American universities: a mixed-methods approach. Internet High. Educ. **45**, 100726 (2020)
6. Motivaator.ee: Muudatuste juhtimine ehk kuidas muudatusi meeskondades edukalt läbi viia? Meeskonnakoolitused, motivatsioonikoolitused ja juhtimiskoolitused, Nov. 28, 2019. https://motivaator.ee/kuidas-muutusi-edukalt-meeskondades-labi-viia/. Accessed 26 May 2023
7. Arthars, N., Liu, D.: How and Why Faculty Adopt Learning Analytics: Wide-Scale Learning Analytics Adoption Through a 'Diffusion of Innovation' Lens, pp. 201–220 (2020). https://doi.org/10.1007/978-3-030-47392-1_11
8. Herodotou, C., Rienties, B., Hlosta, M., Boroowa, A., Mangafa, C., Zdrahal, Z.: The scalable implementation of predictive learning analytics at a distance learning university: insights from a longitudinal case study. Internet High. Educ. **45** (2020). https://doi.org/10.1016/j.iheduc.2020.100725
9. Rehrey, G., Shepard, L., Hostetter, C., Reynolds, A.M., Groth, D.: Engaging faculty in learning analytics: agents of institutional culture change. Learn. Anal. **6**(2) (2019). https://doi.org/10.18608/jla.2019.62.6
10. Greller, W., Drachsler, H.: Translating learning into numbers: a generic framework for learning analytics. J. Educ. Technol. Soc. **15**(3), 42–57 (2012)
11. Bienkowski, M.-A., Means, B., Feng, M.: Enhancing Teaching and Learning Through Educational Data Mining and Learning Analytics: An Issue Brief. U.S. Department of Education (2014)
12. Costa, L.-A., Silveira, A., Souza, M., Salvador, L.: Investigating student and teacher perceptions in e-learning with learning analytics and ontologies. Int. J. Emerg. Technol. Learn. **18**(8) (2023)
13. Leeuwen, A.: Teachers' perceptions of the usability of learning analytics reports in a flipped university course: when and how does information become actionable knowledge? Educ. Technol. Res. Dev. **67**, 1043–1064 (2019)

# Deep Learning Based Audio-Visual Emotion Recognition in a Smart Learning Environment

Natalja Ivleva, Avar Pentel, Olga Dunajeva(✉), and Valeria Juštšenko

Virumaa College of Tallinn University of Technology, Järveküla tee 75, 30322 Kohtla-Järve, Estonia

{natalja.ivleva,avar.pentel,olga.dunajeva, valeria.justsenko}@taltech.ee

**Abstract.** The main goal of this work is to develop a method to monitor the emotional state of students and teachers during the study process based on voice and facial expressions, that can be applied in a smart learning environment. In this paper we create a multimodal emotion detection model based on voice and facial expression features using convolutional neural network (CNN) models. We describe the implementation of the created emotion detection model into the learning process as a web application to monitor the emotional state of students and teachers in a smart learning environment. In this work we compare three types of emotion detection models: models based on audio and facial features separately and for both features taken together and test the performance of these models in the simulation of study process. To evaluate and analyze the models' performances k-fold cross-validation is applied and classification accuracy, weighted F1 score, and confusion matrix are computed. The application developed in this study allows to identify the overall emotional background of the learning environment, determine the emotional state of students and academic staff during the learning process in near real-time.

**Keywords:** Emotion recognition · Deep learning · Smart learning environment · Audio and visual data · Real-time application

## 1 Introduction

The quality of teaching and the interest and involvement of learners in a classroom or e-learning environment can be assessed not only through exams and tests, but also through students' emotional state. The emotional state of students in the learning process can have a positive or negative impact on their learning outcomes. An emotionally positive background in the learning environment enhances students' ability to work, learning efficiency, academic performance, and the quality of material acquisition, as well as the students' motivation to learn. Emotion recognition in the learning process allows to determine the emotional state of students and provide real-time feedback to the teacher. By identifying students' emotions, teachers can make timely course adjustments and adapt their teaching methods and learning materials to facilitate students' learning, activate the classroom atmosphere and improve overall learning efficiency.


M. E. Auer et al. (Eds.): ICL 2023, LNNS 899, pp. 420–431, 2024.
https://doi.org/10.1007/978-3-031-51979-6_44

The main goal of this work is to develop a method to monitor the emotional state of students and teachers during the study process based on voice and facial expressions, that can be applied in a smart learning environment. In this paper we create a multimodal emotion detection model based on voice and facial expression features using deep learning techniques. In this work we compare three types of emotion detection models: models based on audio and facial features separately and for both features taken together and test the performance of these models in the simulation of study process. The system will classify the facial expressions in two ways – firstly, into one of seven expressions: anger, happiness, sadness, surprise, fear, neutral, disgust, and secondly in activation-valence scale: positive versus negative, active versus passive.

We describe the implementation of the created emotion detection model into the learning process as a web application to monitor the emotional state of students and teachers in a smart learning environment in near real-time. The application developed in this study allows to identify the overall emotional background of the learning environment and provide real-time feedback to the teacher. By analyzing this feedback, lecturers can adjust their teaching methods to facilitate students' learning, activate the classroom atmosphere and improve the quality of teaching.

## 2 Background and Related Work

Emotions play a crucial role in human communication and interaction, enabling individuals to express themselves beyond the confines of language. According to the emotion researcher Paul Ekman [1], there is a range of emotions that differ qualitatively in terms of the events that give rise to them, their appraisals, behavioral responses, and physical experience. He proved that different peoples of the world, irrespective of race, language, and cultural development, recognize at least six basic emotions (anger, fear, disgust, sadness, surprise, and joy), although there may be cultural differences in the way they are recognized.

Ekman [2] also proposed a pleasant-unpleasant and active-passive scale as sufficient to capture the difference among emotions. Russell [3] developed the circumplex model and proposed that all emotions can be arranged in a circle controlled by two independent dimensions: valence (pleasant-unpleasant) and arousal (active-passive). Russell's model was updated by Scherer [4] to represent a greater variety of emotions. Based on Russell's and Scherer's research Haq and Jackson [5] summarized the emotion distribution in two dimensions as shown in Fig. 1.

Emotion recognition has gained significant attention in the field of human-computer interaction and affective computing. Emotion recognition plays a crucial role in various domains, including human-computer interaction, mental health monitoring, social robotics, and also in the analytics of educational processes.

Video and audio data present rich sources of information that can be leveraged to infer and understand human emotions.

Facial expression recognition involves extracting facial features and analyzing them to infer emotions. Various methods such as geometric-based approaches, appearance-based approaches, and deep learning techniques have been explored. Studies by Ekman et al. [6] and Lyons et al. [7] introduced landmark-based facial feature extraction, while

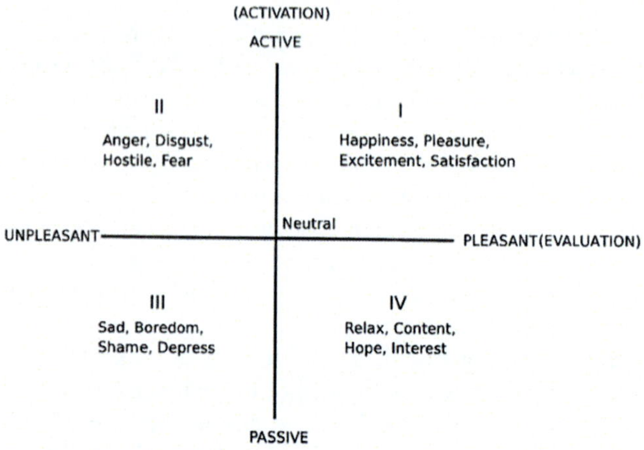


**Fig. 1.** Distribution of emotion in 2D space based on Russell and Scherer research [5].

more recent works by Goodfellow et al. [8] and Liu et al. [9] have utilized convolutional neural networks (CNNs) for accurate emotion recognition.

The Facial Action Coding System (FACS) introduced by Ekman and Friesen [10] provides a comprehensive framework for analyzing facial expressions. FACS has been widely adopted in emotion recognition research, facilitating the development of facial expression databases such as CK+ [11] and MMI [12] Additionally, Yang et al. [13] proposed a real-time emotion recognition system based on FACS that achieved high accuracy.

Speech Emotion Recognition, i.e., tonal analysis involves extracting features from audio signals, such as pitch, intensity, and spectral characteristics, to recognize emotions conveyed through speech. Studies by Scherer [14] and Schuller et al. [15] have explored acoustic-based feature extraction methods, while more recent works by Eyben et al. [16] and Han et al. [17] have utilized deep learning techniques to achieve improved accuracy in speech emotion recognition.

Prosody and paralinguistic analysis focus on the study of non-verbal vocal cues, including intonation, rhythm, and speech rate. Researchers such as Banse and Scherer [18] and Bänziger and Scherer [19] have highlighted the importance of prosody in emotion recognition. Notably, Juslin and Laukka [20] proposed the Geneva Emotional Music Scales (GEMS) to assess emotional responses in music.

Several studies have explored the integration of facial and tonal analysis for multimodal emotion recognition. Martinez et al. [21] combined facial expressions and speech signals to achieve improved emotion recognition accuracy. Furthermore, Liu et al. [22] proposed a multimodal emotion recognition system that integrated facial features, speech features, and textual features.

Despite significant advancements, emotion recognition in video and audio still faces challenges such as individual differences, data variability, and the need for large, labeled datasets. Future research should focus on addressing these challenges and exploring novel approaches such as transfer learning, attention mechanisms, and multimodal fusion techniques.

This work is the continuation of our previous project [23]. The main objective of this paper is to improve the emotion identification models based on facial expressions that were developed in our previous study [23]. The goal of this project is to create a multimodal system that can classify emotions by utilizing facial expressions and voice. The final model aims to determine the expressed emotion of individuals in an audio-video stream. The system incorporates cropped face images extracted from a video stream, and audio features extracted from an audio stream as inputs for two separate Convolutional Neural Networks. The scores from these networks are fused to generate the predicted emotion label.

## 3 Methods

### 3.1 Datasets and Data Preprocessing

In this work, the models for facial expression recognition are trained using the FER2013 dataset and face images extracted from video files of the RAVDESS dataset. For voice emotion recognition, several audio recording datasets, namely TESS, SAVEE, RAVDESS, CREMA, and EMO, are utilized. The selection of the FER2013 dataset is motivated by the goal of comparing the results with our previous work. All the utilized datasets were appropriately labeled for three types of emotion classification:

- classification into 7 emotion classes: surprise, angry, fear, sad, neutral, happy, disgust;
- classification into 2 classes: positive (happy, calm) and negative (angry, fear, sad, disgust) emotions;
- classification into 2 classes: active (angry, disgust, fear, happy, surprise) and passive (sad, calm) emotions.

In this study all data processing and analysis was implemented with Python v.3.9 [24].

**FER2013 Dataset**. Facial Expression Recognition 2013 (FER2013) dataset [25] contains images similar to the real-life situations such as age, gender, ethnicity, head poses, lighting conditions, and viewed from different angles. The database was created using the Google Image Search API and the facial regions were automatically detected, centered , resized and cropped so that the face is more or less in the middle and takes up about the same amount of space in each image. FER2013 consists of 35,887 labelled 48 × 48 pixels' grayscale face images with seven different emotions: 0 = angry (13.80%), 1 = disgusted (1.52%), 2 = fear (14.27%), 3 = happy (25.05%), 4 = sad (16.93%), 5 = surprised (11.15%), 6 = neutral (17.27%). The training set consists of 28,709 samples and the test set of 7,178 samples.

The FER2013 dataset initially included images with various issues, such as photos without faces, sleepy faces, text images, and incorrectly labeled images. To address these issues, the face_recognition library [26] was employed to verify and filter out non-face images from the dataset. Additionally, to remove incorrectly labeled images and to balance the dataset, the Deepface framework [27] was utilized in this study. The framework facilitated the selection of the top 1000 images with the highest emotion recognition rate from the training set. Similarly, from the test set, images with a recognition rate exceeding 90% were chosen for further research.

In this study, the FER2013 dataset was extended by incorporating face images extracted from video files of the RAVDESS dataset. The MTCNN [28] was used to crop the facial region in the image and the Deepface framework was employed to select images with an emotion recognition rate of over 90%.

Subsequently, the photos were converted into arrays of numbers, which were then used as inputs to feed the CNN (Convolutional Neural Network) model.

The final distribution of emotions in the dataset is presented in Table 1.

**Table 1.** Summary of the final dataset.

| Emotions | Angry | Disgusted | Fear | Happy | Sad | Surprised | Neutral | Total (%) |
|---|---|---|---|---|---|---|---|---|
| Negative | + | + | + | – | + | – | – | 3798 (74%) |
| Positive | – | – | – | + | – | – | – | 1300 (26%) |
| Active | + | + | + | + | – | + | – | 5358 (83%) |
| Passive | – | – | – | – | + | – | – | 1093 (17%) |
| Total | 1189 | 370 | 1146 | 1300 | 1093 | 1353 | 1324 | 7775 |
| % | 15 | 5 | 15 | 17 | 14 | 17 | 17 | 100 |

**Audio Recording Datasets.** To improve the efficiency of the developed system in this work, five datasets: TESS, SAVEE, RAVDESS, CREMA, and EMO, were combined into one.

*TESS Dataset.* The TESS dataset [29] contains 2800 WAV audio clips. The dataset consists of a set of 200 target words spoken in the phrase "Say the word _____." Two female voices were used to record this dataset, and it includes 7 considered emotions.

*SAVEE Dataset.* The SAVEE dataset [30] is recorded by four male actors and is labeled with 7 considered emotions. The dataset contains 480 WAV audio clips, with 15 sentences for each of the 7 emotion categories.

*RAVDESS Dataset.* The RAVDESS dataset [31] contains audio and video recordings from 24 professional actors, with an equal split of 12 male and 12 female actors. Each actor performed two statements ("Kids are talking by the door" and "Dogs are sitting by the door") in a neutral North American accent. The speech includes the following emotions: calm, happy, sad, angry, fearful, surprised, disgusted and neutral. Each expression is created at two levels of emotional intensity (normal, strong). The audio is available in WAV format. The dataset includes 1440 files, with 60 recordings per actor. For further research, the "calm" emotion in the RAVDESS dataset was merged with the "neutral" emotion in case of the classification with 7 emotion classes. For the purpose of classifying into 2 classes (positive/negative), the 'calm' emotion was added to the positive class. For classifying into 2 classes (active/passive), the 'calm' emotion was added to the passive class. The same data preprocessing steps for the 'calm' emotion were applied to face images extracted from video files of the RAVDESS dataset.

*CREMA Dataset.* The CREMA-D dataset [32] includes 7442 clips from 91 actors of diverse ethnic backgrounds (48 males and 43 females) recorded by professional theater directors. The actors were instructed to convey specific emotions by speaking 12 specific sentences with different intonations to evoke the target emotion. The sentences were presented in six different emotions: anger, disgust, fear, happiness, neutral, and sadness, with four levels of emotional intensity. The audio is available in WAV format.

*EMO Dataset.* The Berlin Emotional Speech Dataset [33] contains approximately 500 audio recordings of actors expressing 6 emotions: anger, disgust, fear, happiness, neutral, and sadness. The statements were recorded by 10 different actors using 10 different texts. The audio is available in WAV format.

*Dataset Merging and Feature Extraction.* After merging the 5 audio recording datasets, the final audio recording dataset consisted of 12,616 WAV audio clips, including both male and female voices, and labeled with 7 emotions. The distribution of emotions in the final dataset is presented in Fig. 2.

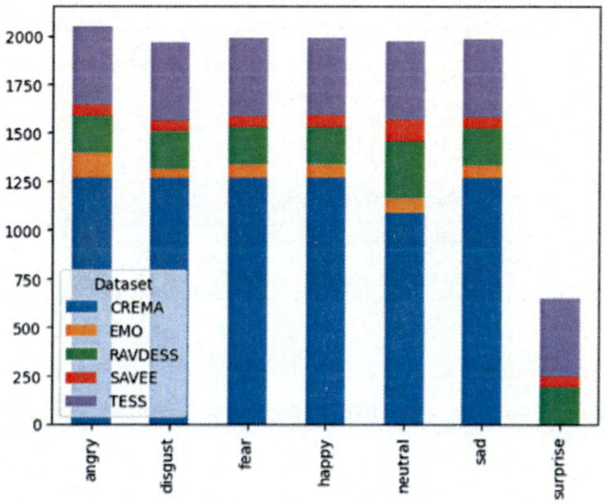


**Fig. 2.** The distribution of emotions in the final dataset.

To augment the audio data, noise injection was applied by adding random values to the data, and pitch shifting was performed using the pitch_shift function from the librosa library [34]. For emotion detection from the voice, the following audio features were extracted using the librosa library: Root Mean Square Energy, Zero-Crossing Rate and Mel Frequency Cepstral Coefficients [35]. The library function takes the file path and loads the audio file for feature extraction, which is then concatenated and returned as a NumPy array. The number of features obtained was 2376.

Table 1 and Fig. 2 show that our dataset has unbalanced class distributions. In this case the comparison between models was made based on the 10-fold cross-validation, confusion matrix and mean of F1 score.

### 3.2 Experiments

In this study, the CNN algorithms were implemented in Keras running on the TensorFlow [36, 37] backend. All experiments were performed on Google Colab TPU runtime [38].

In this study, face and voice features were combined by employing two independent models connected through a late fusion strategy. The proposed architecture includes two CNN systems: the speech emotion recognizer and the facial emotion recognizer. The scores from these models are fused together to obtain the predicted emotion label by averaging the probabilities from the softmax functions for each class (Fig. 3).

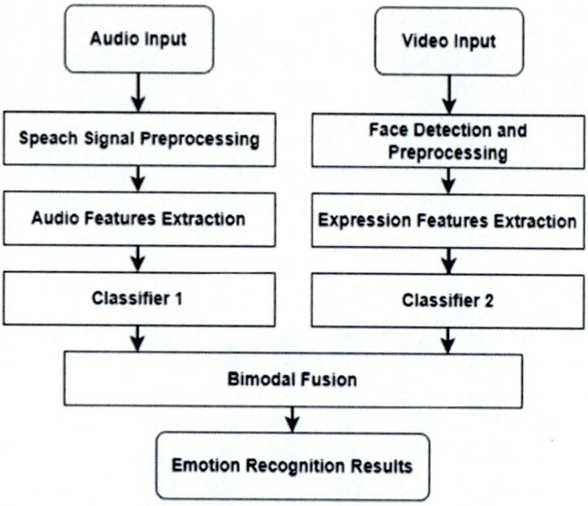


**Fig. 3.** Bimodal emotion detection system.

**Facial Expression Recognition.** The architecture of our CNNs for facial expression recognition consists of 3 consecutive 2D convolution blocks which are made of a 2D convolutional layer with "ReLU" activation function, a batch normalization layer, a maximum pooling aggregation layer with a 2 by 2 filter and a dropout regularization layer with a rate of 0.5, and of 1 consecutive dense layer with batch normalization and dropout. The last dense layer contains neurons according to the number of labels in our data with "Softmax" activation.

The Python Keras Tuner library [39] was used to define the best hyperparameters for building our CNNs, optimization results used in our CNNs are shown in Table 2.

**Voice Emotion Recognition.** The architecture of our CNNs for voice emotion recognition consists of 4 consecutive 1D convolution blocks which are made of a 1D convolutional layer with "ReLU" activation function, a batch normalization layer, a maximum pooling aggregation layer with pool size = 5 and stride = 2, and a dropout regularization layer with a rate of 0.25, and of 1 consecutive dense layer with batch normalization and dropout. The final dense layer size is determined by the number of target classes in the model, and it uses a "Softmax" activation function to generate a value between 0 and 1

**Table 2.** Values of hyperparameters in facial expression recognition CNNs.

| Hyperparameter | All 7 emotions | Active/passive | Positive/negative |
|---|---|---|---|
| conv_1_filter | 64 | 112 | 112 |
| conv_1_kernel | 3 | 3 | 3 |
| conv_2_filter | 32 | 64 | 128 |
| conv_2_kernel | 3 | 3 | 3 |
| conv_3_filter | 112 | 224 | 64 |
| conv_3_kernel | 3 | 6 | 3 |
| dense_1 | 16 | 96 | 48 |
| learning_rate | 0.01 | 0.001 | 0.001 |
| decay | 1e−06 | 1e−06 | 1e−06 |

**Table 3.** Values of hyperparameters in voice emotion recognition CNNs.

| Hyperparameter | Value |
|---|---|
| conv_1_filter | 256 |
| conv_1_kernel | 5 |
| conv_2_filter | 256 |
| conv_2_kernel | 5 |
| conv_3_filter | 128 |
| conv_3_kernel | 5 |
| conv_4_filter | 64 |
| conv_4_kernel | 5 |
| dense_1 | 32 |
| learning_rate | 0.0001 |

for each class. The hyperparameters for building voice emotion recognition CNNs are shown in Table 3.

The Adam optimizer was used to train all neural networks.

## 4 Results

To evaluate the classification models' performances 10-fold cross-validation was applied and the average of the values of classification accuracy, weighted F1 score and confusion matrix were computed. The models were tested on a subset of the RAVDESS dataset, specifically on the video recordings of 4 out of the total 24 actors, that were not utilized during the training of the models. The testing results for three types of emotion detection models: models based on audio and facial features separately, as well as models combining both features, are presented in Table 4.

**Table 4.** Comparison of models.

| Emotion | CNN (audio-only) | | CNN (video-only) | | CNN (audio-video) | |
|---|---|---|---|---|---|---|
| | F1 | Accuracy | F1 | Accuracy | F1 | Accuracy |
| All 7 emotions | 0.61 | 0.59 | 0.69 | 0.70 | 0.71 | 0.71 |
| Positive/negative | 0.76 | 0.76 | 0.82 | 0.82 | 0.89 | 0.91 |
| Active/passive | 0.78 | 0.79 | 0.79 | 0.80 | 0.83 | 0.87 |

As shown in Table 4, the best performing algorithm was the CNN trained with audio and facial features, with the largest average cross-validation accuracy 91% for classification into 2 classes: positive and negative emotions.

All CNN classification models were implemented for testing in a web application. The purpose of this web application is to analyze and determine the emotional state of students and teachers in a digital learning environment in near real-time. The PostgreSQL database and the Python Streamlit library [40] were used to develop the web application. The emotion detection process is shown in Fig. 4.

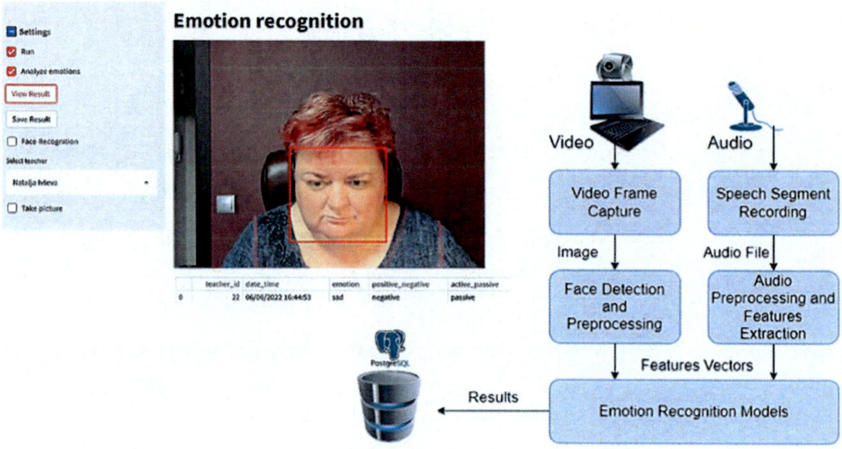


**Fig. 4.** Emotion recognition web application prototype.

## 5 Conclusion

The goal of this research was to create an emotion recognition model and the prototype application that could be integrated into online learning systems. The emotional state of students in the learning process can have a positive or negative impact on their learning outcomes. An emotionally positive background in the learning environment increases the efficiency of students, the effectiveness of learning, as well as the learning motivation of students. Emotion recognition in the learning process makes it possible to determine the emotional state of students and provide feedback to the teacher in real time.

We used two-dimensional model of emotions, where emotions are measured on two scales – as positive versus negative and as active and passive. We used labeled facial and audio data to train separate models - with facial image data, audio data and combined facial and audio data. The model trained with facial and audio attributes led to the best prediction accuracy classifying positive and negative emotions over 90% of accuracy.

Additionally, we also created a prototype application that detects emotions based on real-time audiovisual data, extracting facial images and audio features from the video stream and feeding them to the model which outputs prediction.

In future research, it is planned to work with the created prototype application and to use real-time audiovisual data for evaluation our model. We plan to use various sensors such as galvanic skin response or EEG sensors to determine the ground truth about real emotions of observed people.

## References


1. Ekman, P.: Basic emotions. In: Dalgleish, T., Power, M.J. (eds.) Handbook of Cognition and Emotion, pp. 45–60. Wiley (1999). https://doi.org/10.1002/0470013494.ch3
2. Ekman, P.: A methodological discussion of nonverbal behavior. J. Psychol. **43**(1), 141–149 (1957). https://doi.org/10.1080/00223980.1957.9713059
3. Russell, J.: A circumplex model of affect. J. Pers. Soc. Psychol. **39**(6), 1161–1178 (1980). https://doi.org/10.1037/h0077714
4. Scherer, K.: What are emotions? And how can they be measured? Soc. Sci. Inf. **44**(4), 695–729 (2005). https://doi.org/10.1177/0539018405058216
5. Haq, S., Jackson, P.: Multimodal emotion recognition. In: Wang, W. (ed.) Machine Audition: Principles, Algorithms and Systems, pp. 398–423. IGI Global (2011). https://doi.org/10.4018/978-1-61520-919-4.ch017
6. Ekman, P., Friesen, W.V., Ellsworth, P.: Emotion in the Human Face: Guidelines for Research and an Integration of Findings. Pergamon Press (1971)
7. Lyons, M., Akamatsu, S., Kamachi, M., Gyoba, J.: Coding facial expressions with Gabor wavelets. In: Proceedings of the IEEE International Conference on Automatic Face and Gesture Recognition, pp. 200–205 (1998)
8. Goodfellow, I., Bengio, Y., Courville, A.: Deep Learning. MIT Press (2013)
9. Liu, M., Zhang, D., Zhang, D.: Deep convolutional neural network-based facial expression recognition on small-scale datasets. IEEE Trans. Multimed. **19**(6), 1276–1285 (2017)
10. Ekman, P., Friesen, W.V.: Facial Action Coding System. Consulting Psychologists Press (1978)
11. Kanade, T., Cohn, J.F., Tian, Y.: Comprehensive database for facial expression analysis. In: Proceedings of the IEEE International Conference on Automatic Face and Gesture Recognition, pp. 46–53 (2000)
12. Pantic, M., Rothkrantz, L.J., Pelachaud, C.: Toward an affect-sensitive multimodal human-computer interaction. Proc. IEEE **93**(9), 1693–1712 (2005)
13. Yang, H., Zhao, G., Zhang, L., Zhu, N., He, Y., Zhao, C.: Real-time emotion recognition framework based on convolution neural network. In: Pan, J.S., Li, J., Tsai, PW., Jain, L. (eds.) Advances in Intelligent Information Hiding and Multimedia Signal Processing. Smart Innovation, Systems and Technologies, vol. 157. Springer, Singapore (2020)

14. Scherer, K.R., Johnstone, T., Klasmeyer, G.: Vocal expression of emotion. In: Davidson, R.J., Scherer, K.R., Goldsmith, H.H. (eds.) Handbook of Affective Sciences, pp. 433–456. Oxford University Press (2003)
15. Schuller, B., Batliner, A., Steidl, S., Seppi, D.: Recognising realistic emotions and affect in speech: state of the art and lessons learnt from the first challenge. Speech Commun. (North-Holland) **53**(9–10), pp. 1062–1087 (2011)
16. Eyben, F., Weninger, F., Gross, F., Schuller, B.: Recent developments in openSMILE, the Munich open-source multimedia feature extractor. In: Proceedings of the ACM International Conference on Multimedia, pp. 835–838 (2016)
17. Han, K., Yu, C., Vasconcelos, N.: Speech emotion recognition using deep convolutional neural network and discriminant temporal pyramid matching. IEEE Trans. Affect. Comput. **10**(4), 607–619 (2019)
18. Banse, R., Scherer, K.R.: Acoustic profiles in vocal emotion expression. J. Pers. Soc. Psychol. **70**(3), 614–636 (1996)
19. Bänziger, T., Scherer, K.R.: The role of intonation in emotional expressions. Speech Commun. **46**(3–4), 252–267 (2005)
20. Juslin, P.N., Laukka, P.: Impact of intended emotion intensity on cue utilization and decoding accuracy in vocal expression of emotion. Emotion **1**(4), 381–412 (2001)
21. Martinez, B., Valstar, M.F., Jiang, B., Pantic, M., Valstar, M.: Automatic analysis of facial actions: a survey. IEEE Trans. Affect. Comput. **9**(3), 378–390 (2017)
22. Liu, W., Fu, Z., Huang, B.: Multimodal emotion recognition based on deep neural networks. IEEE Trans. Affect. Comput. **10**(4), 542–555 (2019)
23. Ivleva, N., Pentel, A., Dunajeva, O., Juštšenko, V.: Machine learning based emotion recognition in a digital learning environment. In: Auer, M.E., Pachatz, W., Rüütmann, T. (eds.) Learning in the Age of Digital and Green Transition. ICL 2022. Lecture Notes in Networks and Systems, vol. 633, pp. 405–412. Springer, Cham. https://doi.org/10.1007/978-3-031-26876-2_38
24. Python Software Foundation. https://www.python.org/. Last accessed 29 May 2023
25. FER-2013. https://www.kaggle.com/datasets/msambare/fer2013. Last accessed 24 March 2023
26. face_recognition. https://github.com/ageitgey/face_recognition. Last accessed 29 May 2023
27. Serengil, S.I., Ozpinar, A.: LightFace: a hybrid deep face recognition framework. In: 2020 Innovations in Intelligent Systems and Applications Conference (ASYU), pp. 23–27 (2020). https://doi.org/10.1109/ASYU50717.2020.9259802
28. Zhang, K., Zhang, Z., Li, Z., Qiao, Y.: Joint face detection and alignment using multitask cascaded convolutional networks. IEEE Signal Process. Lett. **23**(10), 1499–1503 (2016). https://doi.org/10.1109/LSP.2016.2603342
29. Pichora-Fuller, M.K., Dupuis, K.: Toronto emotional speech set (TESS). Borealis **V1**. https://doi.org/10.5683/SP2/E8H2MF
30. Surrey Audio-Visual Expressed Emotion (SAVEE) Database. http://kahlan.eps.surrey.ac.uk/savee/Database.html. Last accessed 29 May 2023
31. Livingstone, S.R., Russo, F.A.: The Ryerson Audio-Visual Database of Emotional Speech and Song (RAVDESS): a dynamic, multimodal set of facial and vocal expressions in North American English. PLoS ONE **13**(5), e0196391 (2018). https://doi.org/10.1371/journal.pone.0196391
32. Cao, H., Cooper, D.G., Keutmann, M.K., Gur, R.C., Nenkova, A., Verma, R.: CREMA-D: crowd-sourced emotional multimodal actors dataset. IEEE Trans. Affect. Comput. **5**(4), 377–390 (2014). https://doi.org/10.1109/TAFFC.2014.2336244
33. Berlin Database of Emotional Speech. http://emodb.bilderbar.info/docu/. Last accessed 29 May 2023

34. librosa/librosa: 0.10.0.post2. https://doi.org/10.5281/zenodo.7746972. Last accessed 29 May 2023
35. Er, M.B.: A novel approach for classification of speech emotions based on deep and acoustic features. IEEE Access **8**, 221640–221653 (2020). https://doi.org/10.1109/ACCESS.2020.3043201
36. TensorFlow. https://www.tensorflow.org/. Last accessed 29 May 2023
37. Keras. https://keras.io/. Last accessed 29 May 2023
38. Google Colaboratory. https://colab.research.google.com/. Last accessed 29 May 2023
39. O'Malley, T., Bursztein, E., Long, J., Chollet, F., Jin, H., Invernizzi, L.: Keras Tuner (2019). https://github.com/keras-team/keras-tuner
40. Streamlit. https://streamlit.io/. Last accessed 29 May 2023

# Implementation and Evaluation of an Automatic Scoring System for Experimental Reports Based on ChatGPT

Xingwei Zhou[1], Wenshan Hu[1], Zhongcheng Lei[1](✉), and Guo-Ping Liu[2]

[1] Wuhan University, Wuhan Luojia Hill, Wuhan, China
zhongcheng.lei@whu.edu.cn
[2] Southern University of Science and Technology, Shenzhen No. 1088, Xueyuan Avenue, Shenzhen, China

**Abstract.** This paper introduces an automatic scoring system based on ChatGPT for teachers in online laboratories. Students used to be requested to complete experimental reports after the experiments in whether conventional laboratories or online laboratories. However, it is repetitive and tiring work for teachers to mark dozens or even hundreds experimental reports. Moreover, evaluations of students experiment performance are based on teachers' teaching experience and sometimes are influenced by more subjective factors, so grades marked maybe not objective and fair. Take disadvantages of manual evaluations above into consideration, this paper proposes an automatic scoring system based on ChatGPT where each student's score is assessed automatically based on his experiment steps and answers to questions in online laboratories. This system is embedded into the Networked Control System Laboratory (NCSLab) and applied into grade evaluations of the *Automatic Control Theory* experiment course in Wuhan University. Through testing and validation,the system performs the high exactitude and reliability.

**Keywords:** ChatGPT · Automatic scoring system · Remote laboratories · Engineering education · NCSLab · Experiment courses

## 1 Introduction

With the development of Internet technologies, a huge number of remote laboratories, virtual laboratories and hybrid laboratories are implemented in recent decades to provide experiment services online. Any one who can access to the Internet can visit these online laboratories to obtain experimental resources online, which greatly improves the efficiency and reduces the maintenance cost of experimental resources.

Experimentation is an important part of engineering education, which builds a bridge between theories and practice application [1,2]. Putting the blame on


M. E. Auer et al. (Eds.): ICL 2023, LNNS 899, pp. 432–441, 2024.
https://doi.org/10.1007/978-3-031-51979-6_45

the outbreak of COVID-19, people are under restricted movements, in which case, students are unable to assemble in conventional laboratories to conduct experiments. Online laboratories, the substitute for conventional laboratories, play an important role in opening experimental resources to the world under the condition of restricted people movements.

While, despite online laboratories' rapid development, most online laboratories focus more on the implementation of remote or virtual experiments, rather than the supervision of students' experimental processes and the automatic scoring of experimental reports. Due to processes of experiments are fulfilled online, teachers hardly supervise students' actions, which would, in turn, make evaluation difficult in some degree. In addition, students are requested to submit experimental reports in experiment courses according to the usual practice, which is a repetitive and tiring work for teachers to mark dozens or even hundreds experiment reports. Under such circumstance, a system that can capture students' experiment actions and grade automatically for students' experiment actions and experimental reports is desired for teachers.

Several online laboratories have developed teacher supervision systems or even automatic scoring systems at present. For instance, the virtual laboratory proposed by [3] can record processes of students' experiments and allow teachers to analyze and master students' experimental behaviors. Beijing Institute of Technology proposed an automatic evaluation system for experimental reports in the field of university computer virtual experiment [4]. Networked Control System Laboratory (NCSLab) [5–9] has developed a face recognition system with *Faceapi.js, ch45ZhouFace* to supervise students during experiments. Moreover, NCSLab also has proposed an assessment method which analyzes and grades students' activities and behaviors [11,12]. Universities such as [13,14] and some other institutions also take the students' online experimental supervision into consideration and have done some practical work towards it.

Recently, natural language processing (NLP) develops rapidly and achieves remarkable success, especially those who based on the large language models [15]. These large language models can learn from large amounts of data to make the output information more accurate [16]. As one of the NLP models, ChatGPT has made an amazing performance in NLP. ChatGPT was developed by OpenAI in 2022 for open-ended conversations, which is based on GPT-3.5, the third-generation language processing model [16] having 175 billion parameters by 2020 [17]. And the forth-generation newest GPT-4 was released in Mar 15, 2023, greatly improved on GPT-3. Comparing with GPT-3, GPT-4 is able to support multimodal input, including texts, images, videos, etc., which means that it can understand reports in the PDF or DOC form.

In order to supervise students' behavior during experiments, relieve teachers of the burden of marking experimental reports and improve the fairness of grades, an automatic scoring system which can grade both students' experiment processes and experimental reports is in demand. With the excellent performance of ChatGPT, GPT-3 is used for semantic analysis of experimental reports to grade them comparing with standard answers and detect whether there is plagiarism. This work is applied in NCSLab and plays an important role.

In this paper, sections are organized as follows. Section 2 introduces the previous works about NCSLab and experimental behaviors' evaluation based on NCSLab. Section 3 explains the implementation and evaluates the accuracy of the automatic scoring system for experimental reports based on ChatGPT. Section 4 summarizes the whole paper. Section 5 presents the disadvantages and the future work of the automatic scoring system.

## 2 Previous Works

This section mainly introduces the current overall structure of NCSLab and the application of experimental behaviors' evaluations in NCSLab.

### 2.1 Overall Structure of NCSLab

NCSLab is an online laboratory which is established in 2006 and has been achieved fast development in Wuhan University recent years [18,19]. The most recent overall structure in Wuhan University is presented in Fig. 1. Three layers clearly divide the whole system, which are the client layer, server layer and device layer. The client layer provides web-browser access for users with different terminals such as personal computers (PCs) and mobile devices. The server layer deals with requests sent from the client layer and sends corresponding responds back. The device layer takes responsibilities of executing algorithms and commands issued from upper layers and uploading signal data back in real time. The detailed description about the over structure of NCSLab can be found in [5–9].

### 2.2 Assessment Based on Experimental Behaviors

In the user interface of NCSLab, students can choose task lists which are issued by system administrators who committed teaching for corresponding experiment courses. And users should follow experimental steps requested in task lists to complete experiments. Lab work assessments are completed through recording users' experimental steps and results during experiments in NCSLab based on the fuzzy inference system [11,12]. The implementation of assessment based on experimental behaviors in NCSLab is as Fig. 2 illustrated.

These works have made users can conduct experiments online and get fair and objective scores according experimental behaviors in some degree. However, experimental reports are still cost teachers a huge number of time to read over and give remarks. Based on the idea of completely releasing the burden of teachers, the automatic scoring system for experimental reports is proposed to complement the NCSLab's functions.

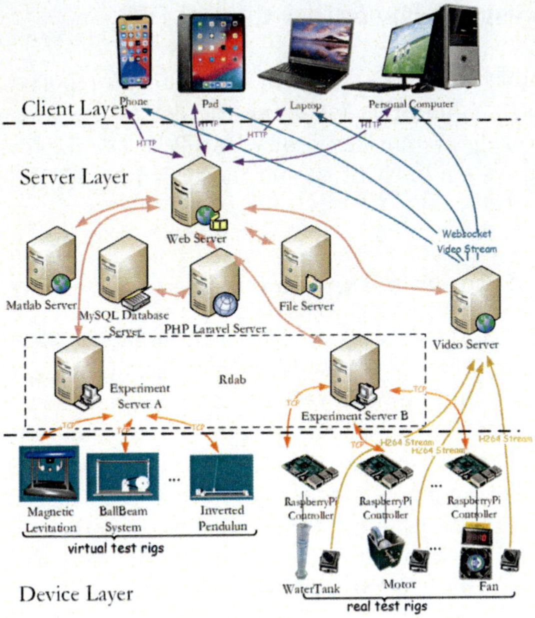


**Fig. 1.** The overall structure of NCSLab.

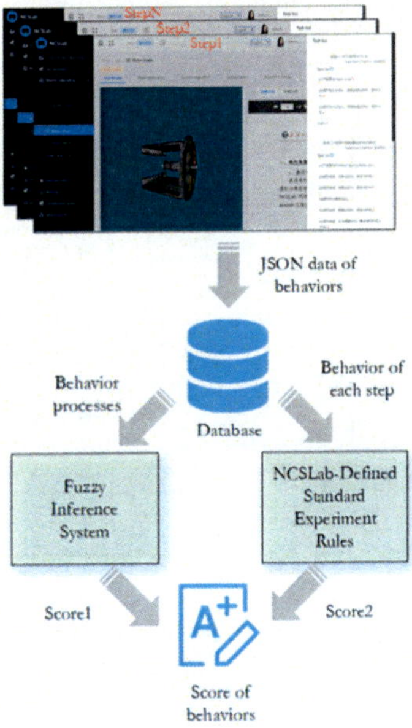


**Fig. 2.** The implementation of assessment based on experimental behaviors in NCSLab.

## 3 Implementation and Evaluation Towards Automatic Scoring System Based on ChatGPT

This Sect. 1 introduces the application of experimental reports' automatic scoring system based on ChatGPT, including the design of the automatic experimental reports scoring system based on ChatGPT, the implementation of the automatic scoring system in NCSLab and the evaluation for the automatic scoring system.

### 3.1 Design of Automatic Experiment Reports Scoring System

In NCSLab, users are required to upload their electronic experimental reports online, which are then stored on the file server. Additionally, GPT-3, which is adopted in NCSLab, only supports text input. Therefore, document preprocessing is necessary. Experiment reports are typically in standardized formats, in which case, it is easy for text extraction and content partitioning. After text extraction, different contents (e.g. identify information, experimental steps, discussion & practice questions, etc.) in experimental reports should be automatically composed in JSON format data. In order to uncover the highly approximate and the same experimental reports, the cheat detection system is also in need. The cheat detection system focus more on answers of practice questions. Therefore, the execution steps of automatic experimental reports scoring system are document preprocessing, each part score's generation with ChatGPT, cheat detection and total scores generation, respectively, which is shown in Fig. 3.

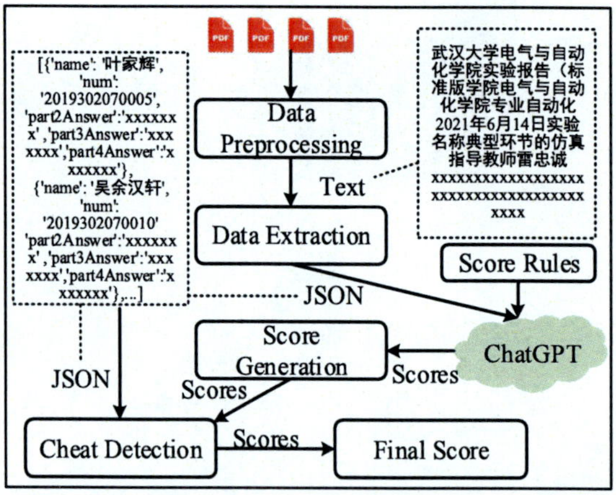


**Fig. 3.** Automatic scoring steps for experimental reports.

1. Document preprocessing. Experiment reports are normally in the form of '.pdf', while GPT-3 can only deal with text, therefore, text should be

extracted from the experimental reports. Besides, text directly extracted must be chaotic, in which case, the useful information such as student name, student number, questions and answers in each section, etc. have to be extracted in JSON form. All students' experimental reports constitute an array, which is a convenience for information finding and management.
2. Partly scores generation with ChatGPT. Each experimental report is split into several sections, such as experimental steps, experimental results, discussions, etc. Each section's content, together with corresponding score rule(or prompt), are sent to ChatGPT. ChatGPT will return its scoring results and reasons back with its powerful NLP abilities. Similarly, the scores of all sections are extracted and provisionally stored in the form of JSON.
3. Cheat detection. It is very easy for students to copy answers from others, especially when required to submit electronic experimental reports. Therefore, cheat detection is a necessary step in the automatic scoring system. Due to all students' experimental reports information is organized in the form of JSON, the detection is very simple. Besides, the result of cheat detection is very accurate. The cheat detection step will record the maximum text similarity and its corresponding student. When the similarity is higher than the set threshold, the score of the corresponding section will be set to zero directly.
4. Total scores generation. The total scores are generated at last, which is made up of scores of all the sections. And these scores together with students' names, numbers, plagiarism, etc. are stored into the database of NCSLab and will presented in the interface of teachers.

The above processes are the principle and implementation steps of automatic scoring system.

### 3.2 Implementation of Automatic Experiment Reports Scoring System in NCSLab

The automatic scoring system is essentially a python script, which can be executed locally by teachers. However, it is more convenient for teachers to use this system in NCSLab, because it is more easy to query and export students' scores. Thus, a scoring system for teachers in NCSLab should be implemented. Students upload their experimental reports to corresponding classes in NCSLab, so that teachers can keep track of the students' experimental reports submission in NCSLab. Teachers input scoring rules (or prompts) in NCSLab and the scores of experimental reports will be obtained after teachers submit their rules. The user interfaces of students' submission of experimental reports and teachers' scoring management are shown in Fig. 4.

As can be seen in Fig. 4(b), students submit their experimental reports to their experiment classes. In the user interface of teachers in Fig. 4(a), they are required to input scoring rules of different sections and experiments. The scores will be automatically obtained based on ChatGPT and displayed in the teachers' interfaces, where detail information such as scores, cheat results, etc. are presented.

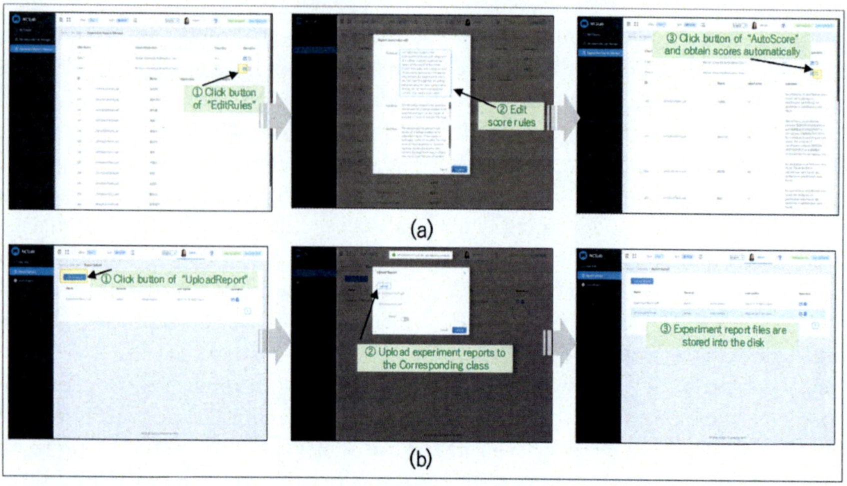


**Fig. 4.** User interfaces of NCSLab. (a) The automatic scoring system in the teacher management interface. (b) The student interface of uploading experimental reports

### 3.3 Evaluation for the Automatic Scoring System

As a language model, ChatGPT can certainly assist in scoring experimental reports in engineering education, but the accuracy of the scores will depend on various factors such as the quality of the training data, the quality of the model, and the complexity of the task at hand. To judge the accuracy of the scores generated by ChatGPT, this paper compares the scores generated by ChatGPT with scores generated by human graders to assess the agreement between them. In this paper, 12 experimental reports were graded by teachers first and then graded by the automatic scoring system to judge the accuracy of ChatGPT. The result is presented in Fig. 5. The scores of the second student and the twelfth one generated by ChatGPT vary from those generated by teachers, that is because the two reports are highly similar, with a 99.99% similarity, in which case, ChatGPT gave a score of zero in plagiarized sections. In addition, the scores of the third student, the seventh one and the tenth one generated by ChatGPT vary from those generated by teachers, which is caused by lots of pictures instead of words in the experimental reports. While, these pictures did not been used by ChatGPT. Therefore, the scores generated by ChatGPT were lower than that generated by teachers. Other than that, the scores generated by ChatGPT and teachers were pretty much the same.

It's important to note that ChatGPT may not always produce scores that perfectly align with human grading, and it may make mistakes or misunderstand the meaning of certain phrases or sentences. Besides, the scoring system is based on GPT-3, which can only deal with text, therefore, the images in the experimental reports have not been used yet. That greatly affects the overall score of the experimental reports. When a student used a lot of pictures instead of words

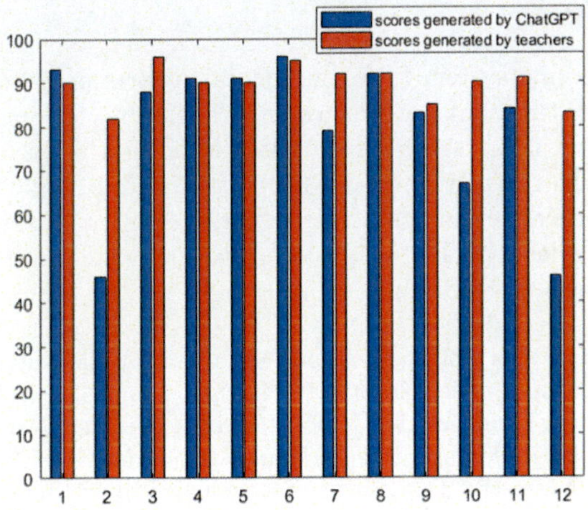


**Fig. 5.** A comparison of the score generated by ChatGPT and the score generated by teachers

in the experimental report, his score was usually graded lower by ChatGPT. In summary, ChatGPT is capable of producing relatively accurate scores for experimental reports in engineering education, provided that it has been properly trained on a high-quality dataset and tuned for the specific task at hand. However, the accuracy of the scores generated by ChatGPT may not always be perfect and can vary depending on the complexity of the task, the quality of the input data, and the accuracy of the scoring rubric.

The cheating detection in the scoring system is very accurate and reliable. It precisely reports to the teachers who copied whose experimental report and the similarity between the copied experimental reports.

## 4 Conclusion

In order to deal with dozens of experimental reports, this paper proposes an automatic scoring system based on ChatGPT. The design, implementation and application have been fully presented in Sect. 3 of this paper. After testing, the automatic scoring system can grade experimental reports and with little difference from the teachers' grade. Besides, the cheat detection can point out whether students plagiarize and who do they copy. In the whole, the scoring system for experimental reports proposed in this paper saves teachers from repeating and tiring grading work. Two scoring systems, the scoring system for user behaviors and the scoring system for experimental reports, work together to grade students' learning level of online experimental courses together, greatly reducing the burden of teachers.

## 5 Future Work

The automatic scoring system for experimental reports proposed in this paper can only deal with text, but not images, etc. Therefore, the image information in the experimental reports cannot be used, which is the disadvantage of the scoring system. The recently opened GPT-4 is capable of handling multimodal inputs, directly processing images, documents, etc. In the future work, GPT-4 will be used to replace GPT-3 as a scoring tool for the automatic scoring of experimental reports in NCSLab.

**Acknowledgments.** This work was supported in part by the National Natural Science Foundation of China under Grant 62073247 and Grant 62103308, in part by the Fundamental Research Funds for the Central Universities under Grant 2042023kf0095, in part by the China Postdoctoral Science Foundation under Grant 2022T150496, and in part by Wuhan University Experiment Technology Project Funding under Grant WHU-2022-SYJS-10.

## References

1. Achumba, I.E., Azzi, D., Dunn, V.L., Chukwudebe, G.A.: Intelligent performance assessment of students' laboratory work in a virtual electronic laboratory environment. IEEE Trans. Learn. Technol. **6**(2), 103–116 (2013). https://doi.org/10.1109/TLT.2013.1
2. Fraire, J.A., Durán, J.E.: Revising computer science networking hands-on courses in the context of the future internet. IEEE Trans. Educ. **64**(2), 133–138 (2021). https://doi.org/10.1109/TE.2020.3015673
3. Guan, S., Zhao, H., Tan, S., Xu, L.: Conducting experiments with real-time cloud computing models: a virtual laboratory for massive open online courses on computer control systems [focus on education]. IEEE Control Syst. Mag. **41**(6), 97–107 (2021). https://doi.org/10.1109/MCS.2021.3107640
4. Chen, Y., Liu, X., Huo, P., Li, L., Li, F.: The design and implementation for automatic evaluation system of virtual experiment report. In: 2017 12th International Conference on Computer Science and Education (ICCSE), pp. 717–721 (2017). https://doi.org/10.1109/ICCSE.2017.8085587
5. Hu, W., Lei, Z., Zhou, H., Liu, G.P., Deng, Q., Zhou, D., Liu, Z.W.: Plug-in free web-based 3-d interactive laboratory for control engineering education. IEEE Trans. Industr. Electron. **64**(5), 3808–3818 (2017). https://doi.org/10.1109/TIE.2016.2645141
6. Lei, Z., Zhou, H., Hu, W., Deng, Q., Zhou, D., Liu, Z.W.: Html5-based 3-d online control laboratory with virtual interactive wiring practice. IEEE Trans. Industr. Inf. **14**(6), 2473–2483 (2018). https://doi.org/10.1109/TII.2017.2769883
7. Lei, Z., Zhou, H., Hu, W., Liu, G.P., Deng, Q., Zhou, D., Liu, Z.W., Gao, X.: Unified 3-d interactive human-centered system for online experimentation: current deployment and future perspectives. IEEE Trans. Industr. Inf. **17**(7), 4777–4787 (2021). https://doi.org/10.1109/TII.2020.3019238
8. Lei, Z., Zhou, H., Hu, W., Deng, Q., Zhou, D., Liu, Z.W., Gao, X.: 3-d interactive control laboratory for classroom demonstration and online experimentation in engineering education. IEEE Trans. Educ. **64**(3), 276–282 (2021). https://doi.org/10.1109/TE.2020.3041070

9. Lei, Z., Zhou, H., Hu, W., Liu, G.P.: Unified and flexible online experimental framework for control engineering education. IEEE Trans. Industr. Electron. **69**(1), 835–844 (2022). https://doi.org/10.1109/TIE.2021.3053903
10. Zhou, X., Hu, W., Liu, G.P., Pang, Z.: Face recognition system based on ncslab for online experimentation in engineering education. In: 2022 41st Chinese Control Conference (CCC), pp. 4390–4394 (2022). https://doi.org/10.23919/CCC55666.2022.9902741
11. Xiao, H., Hu, W., Liu, G.P.: Big data assessment methods for students' online laboratory work based on NCSLab. In: 2021 40th Chinese Control Conference (CCC), pp. 8709–8714 (2021). https://doi.org/10.23919/CCC52363.2021.9549837
12. Xiao, H., Hu, W., Liu, G.P.: Students' online laboratory work assessment based on learning task lists and behavior data. IEEE Trans. Learn. Technol. 1–12 (2022). https://doi.org/10.1109/TLT.2022.3213751
13. Considine, H., Nedic, Z., Nafalski, A.: Automation of basic supervision tasks in a remote laboratory—case study Netlab. In: 2019 5th Experiment International Conference (exp.at'19), pp 189–192 (2019). https://doi.org/10.1109/EXPAT.2019.8876508
14. Banky, G.P., Blicblau, A.S.: Doing it with a glass onion: investigating affordances provided by a head-mounted augmented reality immersive device for the real-time online supervision of experimental learning. In: 2017 IEEE 3rd International Conference on Engineering Technologies and Social Sciences (ICETSS), pp 1–6 (2017). https://doi.org/10.1109/ICETSS.2017.8324168
15. Du, H., Teng, S., Chen, H., Ma, J., Wang, X., Gou, C., Li, B., Ma, S., Miao, Q., Na, X., Ye, P., Zhang, H., Luo, G., Wang, F.Y.: Chat with ChatGPT on intelligent vehicles: An IEEE TIV perspective. IEEE Trans. Intell. Veh. 1–7 (2023). https://doi.org/10.1109/TIV.2023.3253281
16. Abdullah, M., Madain, A., Jararweh, Y.: ChatGPT: fundamentals, applications and social impacts. In: 2022 Ninth International Conference on Social Networks Analysis, Management and Security (SNAMS), pp. 1–8 (2022). https://doi.org/10.1109/SNAMS58071.2022.10062688
17. Wang, F.Y., Miao, Q., Li, X., Wang, X., Lin, Y.: What does ChatGPT say: the DAO from algorithmic intelligence to linguistic intelligence. IEEE/CAA J. Automatica Sinica **10**(3), 575–579 (2023). https://doi.org/10.1109/JAS.2023.123486
18. Zhou, X., Hu, W., Liu, G.P.: React-native based mobile app for online experimentation. In: 2020 39th Chinese Control Conference (CCC), pp. 4400–4405 (2020). https://doi.org/10.23919/CCC50068.2020.9189636
19. Qiao, Y., Liu, G.P., Zheng, G., Hu, W.: Ncslab: a web-based global-scale control laboratory with rich interactive features. IEEE Trans. Industr. Electron. **57**(10), 3253–3265 (2010). https://doi.org/10.1109/TIE.2009.2027924

# TRACE: A Conceptual Model to Guide the Design of Educational Chatbots

Juan Carlos Farah[1,2(✉)], Basile Spaenlehauer[1], Sandy Ingram[2], Fanny Kim-Lan Lasne[1], María Jesús Rodríguez-Triana[3], Adrian Holzer[4], and Denis Gillet[1]

[1] École Polytechnique Fédérale de Lausanne (EPFL), Lausanne, Switzerland
juancarlos.farah@epfl.ch, juancarlos.farah@hefr.ch, basile.spaenlehauer@epfl.ch, fanny.lasne@epfl.ch, denis.gillet@epfl.ch
[2] University of Applied Sciences (HES-SO), Fribourg, Switzerland
sandy.ingram@hefr.ch
[3] Tallinn University, Tallinn, Estonia
mjrt@tlu.ee
[4] University of Neuchâtel, Neuchâtel, Switzerland
adrian.holzer@unine.ch

**Abstract.** Driven by the rising popularity of chatbots such as ChatGPT, there is a budding line of research proposing guidelines for chatbot design, both in general and specifically for digital education. Nevertheless, few researchers have focused on providing conceptual tools to frame the chatbot design process itself. In this paper, we present a model to guide the design of educational chatbots. Our model aims to structure participatory design sessions in which different stakeholders (educators, developers, and learners) collaborate in the ideation of educational chatbots. To validate our model, we conducted an illustrative study in which 25 software design students took part in a simulated participatory design session. Students were divided into eight groups, assigned the role of one of the different stakeholders, and instructed to use our model. The results of our qualitative analysis suggest that our model helped structure the design process and align the contributions of the various stakeholders.

**Keywords:** Educational chatbots · Participatory design · Conceptual model

## 1 Introduction

Over the past decade, an increasing interest in integrating conversational agents into educational contexts has motivated the design, deployment, and evaluation of *pedagogical* conversational agents, also referred to as *educational chatbots*. Indeed, there is a wide variety of educational chatbot designs and architectures, with one review of the literature noting that "there exists as much technology used in the development of chatbots as there are educational chatbots" [16]. In


M. E. Auer et al. (Eds.): ICL 2023, LNNS 899, pp. 442–454, 2024.
https://doi.org/10.1007/978-3-031-51979-6_46

light of this, a recent survey of the principles grounding the design of educational chatbots emphasized that "researchers should explore devising frameworks for designing and developing educational chatbots to guide educators to build usable and effective chatbots" [13]. Our work aims to address this gap by proposing a conceptual model that can guide the participatory design of educational chatbots. By providing the conceptual tools necessary to define the context in which an educational chatbot is deployed as well as the interaction it has with learners, our model could serve to structure participatory design sessions involving educators, developers, and learners.

## 2 Background and Related Work

Concerning general-purpose chatbots, Ferman Guerra identified 18 chatbot best practices, spanning user-chatbot communication, chatbot features, and human factor concerns [9]. More recently, Feine et al. designed a chatbot social cue configuration system "to support chatbot engineers in making justified chatbot social cue design decisions" [6]. They evaluated this configuration system with a focus group and two practitioner symposia, receiving positive feedback. In a subsequent publication, Feine et al. also proposed an interactive chatbot development system designed to encourage collaboration between domain experts and chatbot developers [7]. Their evaluation of an implementation of this development system through an online experiment in the context of customer service chatbots showed that this system improved subjective and objective engagement.

In education, Griol and Callejas proposed a modular architecture to integrate chatbots into multimodal applications for education, featuring the ability to easily adapt technical and pedagogical content [10]. More generally, Farah et al. proposed a technical blueprint for integrating task-oriented agents in education along with a proof-of-concept implementation of this blueprint [5], while Jung et al. proposed a set of chatbot design principles derived from a literature review of empirical studies [12].

Nevertheless, few researchers have focused on providing conceptual tools to guide the process of designing educational chatbots, including the participatory design sessions aimed at ideating these chatbots. Bahja et al. proposed an iterative step-by-step user-centric methodology for educational chatbots whereby learners and teachers actively collaborate during the requirements analysis, design, and validation phases [1]. Furthermore, Durall Gazulla et al. adopted a collaborative approach, involving students in the design of chatbots for reflection and self-regulated learning in higher education, and focusing their study on challenges encountered during the co-design process [3]. While these studies have incorporated participatory design sessions for the co-creation of educational chatbots, to the best of our knowledge, no study has proposed chatbot-oriented conceptual tools to structure these sessions. The model we propose in this paper aims to serve this purpose.

# 3 Design

Our model was conceived to guide educators, developers, and learners in designing task-oriented chatbots for education. The aim is to provide a conceptual tool that can serve to structure participatory design sessions in which stakeholders collaborate to design these chatbots. The model provides two guidelines. First, it proposes five components considered essential in defining a chatbot's integration into a learning activity (i.e., *Tasks*, *Resources*, *Applications*, *Cues*, and *Exchanges*). Second, it outlines a procedure for how the design process should unfold, assigning specific tasks to each of the stakeholders involved (i.e., *Educators*, *Developers*, and *Learners*).

## 3.1 Backdrop

The backdrop for our model's components and the process in which these components are defined is a *learning activity*. Depending on the scope of the design session, this activity can be selected in advance, chosen by all participants, by each group, or ultimately by each educator present in the session. This activity can be as general (e.g., learning to draw) or as specific (e.g., learning to draw seagulls perched on a boat) as necessary. For illustrative purposes, we will use *Learning the Python Code Style Standards* as an example of a learning activity.

## 3.2 Stakeholders

A wide range of stakeholders, from administrators to parents, are involved in the design and implementation of educational technologies. Nevertheless, three key actors stand out given their roles in how these technologies are developed, integrated into the classroom, and exploited. These actors are (i) the *educators* who choose which technologies to integrate into their teaching practices, (ii) the *developers* of these technologies, and (iii) the *learners* who will eventually use these technologies. In this section, we outline their role in the participatory design process structured by our model.

**Educators** are often the ones who select the material and technology to be used in their practices [18]. In our model, educators initiate the design process by selecting the *tasks* and *resources* that will serve to scaffold the pedagogical scenario. The selection of these two elements serves to frame the chatbot integration with learning scenarios that are relevant to an educator's practice.

**Developers** in education have the important role of building the technology used in practice, with research suggesting that developers benefit greatly from collaborations with educators when building educational technologies [4]. In our model, developers bridge the scenario and interaction aspects of a chatbot's integration. That is, developers are the ones in charge of designing the *applications* that will both feature the resources selected by the educator and host the chatbot that will interact with the learner.

**Learners** are the ones who will interact with the chatbot and, therefore, the end users of this technology. Of particular importance is that learners consider the interactions timely and engaging, all the while being relevant to the learning activity at hand. In our model, learners are the ones that will lead the process of defining the *cues* used to provide ways to trigger the chatbot interaction, as well as the content of these *exchanges* themselves.

### 3.3 Components

Our model—summarized in Fig. 1—focuses on five components to guide the integration of chatbots into educational contexts. Indeed, the model takes its acronym (*TRACE*) from the components it defines. The first two components (*Tasks* and *Resources*) focus on the scenario in which the interaction will take place, while the last two (*Cues* and *Exchanges*) focus on the interaction itself. The central component (*Applications*) serves as a bridge between the scenario and the interaction and is responsible for hosting the chatbot. The design decisions made for each component are led by one of the stakeholders involved in the participatory design session. In this section, we present each component, describe it, and provide illustrative examples.

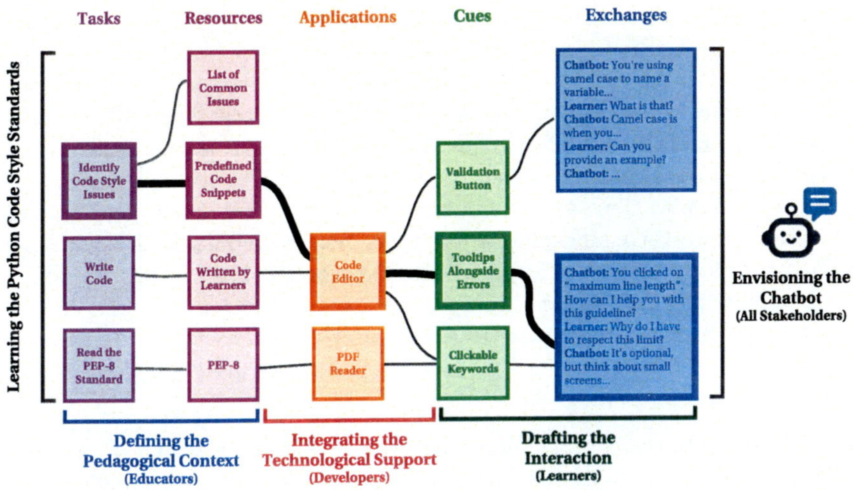


**Fig. 1.** Overview of the TRACE model including its components and the tasks associated with each stakeholder. The thick line represents how the stakeholders can then *trace* a line through the components they select for their design [19].

**Tasks.** A task in our model can be conceived as a structured step that is aligned with the objectives of the learning activity. This is based on the idea that learning tasks are "an interface between the learners and the information offered in the learning environment" and "serve to activate and control learning processes in order to facilitate successful learning" [17]. Educators take the lead in defining the tasks that are best aligned with the learning activity. *Examples: (i) Learners have to identify code style issues in snippets of code. (ii) Learners have to write a snippet of code following the official style guide for Python code (PEP-8).*

**Resources**. A resource follows from the definition of a learning object as "any digital resource that can be reused to support learning" [20], but specifically applied to the learning tasks chosen for a given learning activity. Furthermore, a resource should be able to be featured in (e.g., embedded in, interfaced through) a software application that can make it accessible to both the learner and the chatbot. In that sense, resources in our model can be any of the examples provided by Wiley ("digital images or photos, live data feeds..., live or prerecorded video or audio snippets, small bits of text, animations, and smaller Web-delivered applications", as well as "entire Web pages that combine text, images, and other media or applications") [20], as long as they can be integrated into a software application compatible with an educational chatbot. *Examples: (i) Snippets of code provided by the educator. (ii) Code written by learners.*

**Applications**. Given that chatbots are a type of digital education technology, it is imperative that the bridge between the pedagogical scenario and the interaction with the educational chatbot be mediated by a software application. Applications can be contextualized to provide different interfaces to different stakeholders. That is, educators can access a dedicated interface where they can configure the application, select the resources that it will feature, and connect the educational chatbot that will be hosted therein. Learners can then use a different interface to access the resources selected by the educator. This learner interface also provides the affordances through which the learner will interact with the educational chatbot. Given the prevalence of web technologies in digital education, applications are often built to run in web browsers and be compatible with digital learning platforms. Nevertheless, applications can also be standalone desktop or mobile applications. Developers take the lead in designing the application that is most appropriate for the selected resource, as they have the required technical expertise. *Examples: (i) An application where learners can write, annotate, and execute code. (ii) An interactive reader featuring the Python documentation.*

**Cues**. To integrate the chatbot interaction into the learning activity, we rely on the notion of an *interaction cue*, often employed in educational technology [11]. Interaction cues serve to inform users of the actions they can take and to guide them toward a particular action [2]. The function of a cue in our model is to trigger an exchange between the learner and the chatbot. Cues are digital affordances that are an essential part of the application's learner interface. Cues can be graphical or textual and can be linked to actions that the learner takes, featured in elements within the resource embedded in the application, or exposed permanently in the interface. As is often the case, these cues could be paired with visual affordances (e.g., buttons, tooltips) that the learner can interact with. These cues should be featured in the resource and accessible via the application.

Learners take the lead in defining the cues, as they will be the ones triggering and following these cues. *Examples: (i) A tooltip appears next to an error as the learner types. (ii) A button that the learner can click on to have the chatbot inspect the code.*

**Exchanges**. The final component serves to illustrate what a conversational exchange between a learner and a chatbot would look like. By providing examples of useful interactions, learners can both highlight their expectations of the interaction and indicate the design features, conversational capabilities, and social characteristics they expect the chatbot to have. This is an important aspect of the design of interactive agents and is guided by the Computers Are Social Actors (CASA) framework [14]. As such, understanding—and possibly curbing—learners' expectations of these interactions can serve to design a chatbot that is better adapted to the learning activity. Furthermore, these exchanges could be used to create longer examples that could eventually serve to prompt the language models used by the chatbots, as some of these models can improve their performance based on a few samples. Once again, learners take the lead in defining the exchanges, since they will be the ones interacting with the chatbot.

*Example*:

- ***Chatbot***: *Hey! You're using* `camelCase` *to name a variable. Did you know that you have to use* `snake_case` *for variable names?*
- ***Learner***: *No. Why* `snake_case` *and not* `camelCase`?
- ***Chatbot***: *Well,* `snake_case` *is the default for variables in Python...*

### 3.4 Process

The following process is one way in which the components proposed by our model can be defined by the stakeholders involved in the participatory design session. This process consists of four phases and can be integrated as the central activity of a workshop, following initial icebreakers and the selection of the learning activities that will serve as a backdrop. In this section, we describe these phases, providing sample ***questions*** that can guide the discussion among stakeholders, as well as a list of the ***outcomes*** that should be produced in each phase.

**Defining the Pedagogical Scenario**. In this first phase, educators take the design lead. The educator will start by proposing tasks that learners could do in relation to the selected learning activity and then listing the resources that could support these tasks. There is a many-to-many mapping between resources and tasks, as the same resource can support multiple tasks and one task can be supported by many resources. Once the tasks and resources have been mapped, the educator can choose the resources that they think will be most relevant to their practice. ***Questions***: *(i) What tasks can support this learning activity? (ii) What resources are traditionally used for these tasks? (iii) Can these resources be delivered digitally?* ***Outcomes***: *(i) A description of tasks that the learner could engage in. (ii) A description of resources that can be used to support these tasks. (iii) A mapping between the tasks and the resources.*

**Integrating the Technological Support.** The second phase concerns the technological scaffolding that will serve as a bridge between the pedagogical scenario and the interaction with the chatbot. The developer leads this phase and needs to sketch out an application that could feature the resources selected by the educator. If a resource cannot be embedded in or handled by an application, then a different resource needs to be selected. Once one or more applications have been sketched out, these applications can be mapped back to all the other resources in the list that they can potentially support. Once again, this is a many-to-many mapping. ***Questions***: *(i) How can the selected resources be embedded into an application? (ii) What devices would this application run on? (iii) What learning platforms is this application compatible with?* ***Outcomes***: *(i) Descriptions of applications that can feature the resources selected in the previous phase. (ii) Sketches or mockups of how the applications will feature the respective resources. (iii) A mapping between the resources and the applications.*

**Drafting the Interaction.** The third phase is led by the learner, who will first identify what cues within the application (or the resource embedded therein) should prompt the chatbot to start an interaction with the learner. These cues can also define where the interaction takes place within the interface. The developer needs to ensure that the cues are compatible with the application and can be displayed to learners. Once the cues are defined, the learner can choose one or more cues to construct sample exchanges between a learner and the chatbot. These exchanges do not need to be long or detailed but should contain enough information so as to envision what a dialog would look like. ***Questions***: *(i) What elements or affordances in the resource or application can serve to cue the chatbot interaction? (ii) Are these cues specific to one (type of) resource or are they applicable to other (types of) resources? (iii) Are the exchanges aligned with the goals of the learning activity?* ***Outcomes***: *(i) A description of one or more cues that could be present in the selected applications. (ii) One or more sample dialogs illustrating an exchange between the chatbot and the learner. (iii) A mapping between the cues and the exchanges.*

**Envisioning the Chatbot.** At this point in the process, stakeholders will have produced a complete mapping of all the components that could serve to support the integration of chatbots into the learning activity. In the final phase, participants use the five components defined in the previous phases to envision the chatbot. Stakeholders can then highlight the examples of each component that are most appropriate for the chatbot integration and *trace* a line from learning activity to chatbot, as shown in Fig. 1. This line serves to visualize the interactions that learners will have with the chatbot, linking particular examples of exchanges with learning tasks, via the corresponding cues, applications, and resources. Once this link is established, stakeholders can define the chatbot's identity, what social cues it will be equipped with, what it will look like, and what strategies it will use to support these interactions. All stakeholders are invited to be equally active in this phase, as the chatbot's identity serves to summarize various aspects defined in the previous phases. The final outcome of this phase could be the starting point for future participatory design sessions, an initial prototype, or other iterations of this exercise. ***Questions***: *(i) What is the chatbot's name and what does it look like? (ii) What social cues can the chatbot harness in its interactions with users? (iii) What technologies power the chatbot so that it can support the sample exchanges?* **Outcomes**: *(i) A description of the chatbot's identity. (ii) A list of the technologies needed to support how the chatbot interacts with learners (e.g., rule-based scripts, AI-based solutions). (iii) A sketch or mockup of the chatbot embedded in the application.*

## 4 Methodology

To validate our model, we first conducted two pilot studies comprising a workshop with eight researchers and developers in education and a case study with an undergraduate student completing a semester project on designing educational chatbots. We then conducted an illustrative within-subject study in which 25 students—all enrolled in a course on software design—took part in a simulated participatory design session. The purpose of these studies was to address one main research question: *Does the TRACE model help guide educators, developers, and learners in collaboratively designing educational chatbots?* Our analysis focused on two aspects of this research question: (i) alignment between stakeholders and (ii) feedback provided about our model. In this section, we present the methodology followed for our illustrative study.

### 4.1 Participants and Procedure

The purpose of our main study was to demonstrate the feasibility of our proposal and was conducted as an hour-long role-play activity as part of a software design course at the University of Neuchâtel, Switzerland. A total of 25 students (7 female, 18 male) were recruited as participants for the session, which took place in December 2022. At the beginning of the session, participants were divided into eight groups of three or four. These groups corresponded to groups in which the students had been working on for one of the course assignments. As such, students within each group were well acquainted with each other.

To motivate the session, after a short introduction, the groups were first asked to come up with a learning activity or choose one from a list (e.g., *how to cook pasta*, *calculating the area of a circle*). Participants were then asked to interact with a chatbot powered by the GPT-3 language model for a few minutes. The topic of this interaction was supposed to be the topic they had come up with or selected from the list. Within each group, participants were then assigned the role of *educator*, *developer*, or *learner*. They then completed two exercises. In the first exercise, participants were asked to provide a short answer describing—from the point of view of their role—how they would integrate a chatbot into the learning activity chosen by their group. After this exercise, they were introduced to our model through a short presentation. This presentation constituted the intervention in our within-subject setup.

The short presentation consisted of five slides in which the different components of the model were outlined and examples of each component were proposed. These examples specifically covered the PEP-8 use case. After the intervention, participants completed a second exercise. In this second exercise, participants were instructed to use our model to collaborate on the design of the chatbot integration. At the end of the exercise, the groups provided short descriptions of each of our model's components in the context of their respective learning activity, as well as an optional mockup of the chatbot integration. Finally, qualitative feedback was captured through an open-ended question that asked participants if they found the model useful.

### 4.2 Instruments and Data Analysis

We captured qualitative responses through short answers to a series of open-ended questions. In the first exercise—before the presentation of our model—participants were asked to specify how they would integrate the chatbot in a few phrases. Although they worked as a group, each student was asked to provide an answer from the point of view of the role they had been assigned. In the second exercise, participants had to specify each component of our model in a separate input box. Finally, groups were asked to provide feedback about the model.

All student responses were analyzed using line-by-line data coding. Furthermore, responses concerning the alignment aspect were tagged as either *aligned* or *misaligned* depending on whether they were compatible with other responses from the same group. Component descriptions were also tagged as *valid* or *invalid* depending on whether they were applicable to the respective component.

## 5 Results

In the first exercise, seven groups provided more than one answer on how they would integrate the chatbot into the chosen learning activity. Only in two of these groups were the answers provided aligned. In the second exercise—after being introduced to the TRACE model—all groups provided descriptions of each of the components of the model. For all groups, the descriptions of all components were aligned within each group. Furthermore, four groups provided valid descriptions for all five components, while two groups did so for four components, and the other two groups for only three components. To illustrate the answers provided, Fig. 2 presents word clouds corresponding to each component.

Finally, four groups provided feedback on the usefulness of the model. Three groups responded positively, while one group described the explanation of the model as *complicated*. This last group provided the following feedback: "*It was complicated to understand what was asked and the explanation of [TRACE] was really fast so we didn't have time to understand.*" Nonetheless, two groups specifically referred to the model's ability to *structure* the design, while one group appreciated how it was *inclusive* in the sense that it took input from multiple stakeholders into account: "*Creates some structure. Allows [one] to think about*

**Fig. 2.** How groups defined the different components of the TRACE model can be illustrated with word clouds to highlight important keywords.

*all points of view and not [miss] one... This framework allows therefore to take on every stakeholder.*"

## 6 Discussion

Our current findings are promising. While preliminary, results from our illustrative study suggest that our model helped students collaborate more efficiently by aligning the contributions of the different stakeholders and providing a structure with which to reason about the educational chatbot's implementation. Aligning the expectations of different holders was one of the challenges highlighted by Durall Gazulla et al., who specifically noted that one obstacle they faced in the design of their chatbot was the "challenge [of] addressing diverse needs, while ensuring the relevance of the solutions envisioned" [3]. By assigning specific responsibilities to different stakeholders, but inviting them to participate in the full design process in order to align their different needs, the TRACE model could help address this challenge.

Furthermore, the fact that the model was reported to provide *structure* for the design process addresses another challenge highlighted by Durall Gazulla et al. [3]. Namely, the challenge of translating research into practice. One outcome of a participatory design workshop structured with the TRACE model is a diagram that can serve as a blueprint to further design and develop the educational chatbot in question. While it is not a functioning chatbot, this outcome can be shaped into a sketch, a mockup, or even a prototype. In essence, it serves as a way to translate the ideas emerging from the participatory design session into actionable tasks for the stakeholders who will implement this chatbot in practice.

However, it is important to note that one group described the model as *complicated* and that four groups did not provide valid descriptions for all components. Although misunderstandings of the TRACE model could be mitigated by allotting more time to present the model, participatory design workshops are also limited by time constraints and include stakeholders with different backgrounds and technical aptitudes [3]. Hence, long presentations featuring abstruse terminology and complex definitions should be best avoided. Instead, ensuring that component definitions are clear and accessible to a wide variety of stakeholders could be crucial to maximizing adoption in participatory design practices.

Our model could also help educators adapt their teaching practices in light of the impact large language models (e.g., ChatGPT [15]) are having on education. It has recently been highlighted that "occupations in the field of education are likely to be relatively more impacted by advances in language modeling than other occupations" [8]. The TRACE model could serve as a canvas for educators to collaborate (i) with developers to better understand the opportunities and limitations of large language models and (ii) with learners to better understand how learners envision using chatbots to support their studies.

## 7 Conclusion, Limitations, and Future Work

The model proposed in this paper addresses the lack of bespoke conceptual tools for structuring the participatory design of chatbots. By outlining five different components that can be defined over a four-step process, TRACE breaks down the task of specifying how chatbots can support a given learning activity, which could be interpreted differently by different stakeholders. To maximize relevance in practice, the model places educators first, allowing them to define the pedagogical scenario that will guide the design process. Developers, due to their technical expertise, are also central in our model, ensuring that the pedagogical scenario can be supported by the technology in which the chatbot will be embedded, and bridging the scenario with the interactions that will occur between the chatbot and the learner. Finally, learners are tasked with identifying both the timing and content of these interactions, ensuring that the chatbot adds value to the learning experience, rather than being an element of distraction. The result is a model that can be used to produce a blueprint of how an educational chatbot could be integrated into a learning activity. This blueprint could be the input to future participatory design sessions, guide educators in adapting their lesson plans to make room for learner-chatbot interactions, or serve as a starting point for a technical specifications document or prototype.

Nevertheless, our study has limitations worth addressing. First, while we explicitly chose students from a software design course to maximize the number of participants that could play the role of the different stakeholders, our role-playing study is not indicative of how professional educators, developers, and learners might judge our conceptual model. Conducting a formal participatory design workshop with actual stakeholders could help improve the ecological validity of our proposed tool. Second, the limited time that participants had to interact might have affected their ability to efficiently understand and harness the model. Extending the workshop to a two or three-hour session could give participants time to assimilate the concepts presented in TRACE. We aim to address these limitations in future work.

## References

1. Bahja, M., Hammad, R., Butt, G.: A user-centric framework for educational Chatbots design and development. In: Stephanidis, C., Kurosu, M., Degen, H.,

Reinerman-Jones, L. (eds.) HCI International 2020—Late Breaking Papers: Multimodality and Intelligence, vol. 12424, pp. 32–43. Springer, Cham, Switzerland (2020). https://doi.org/10.1007/978-3-030-60117-1_3

2. Dillman, K.R., Mok, T.T.H., Tang, A., Oehlberg, L., Mitchell, A.: A visual interaction cue framework from video game environments for augmented reality. In: Proceedings of the 2018 CHI Conference on Human Factors in Computing Systems, pp. 1–12. ACM, Montreal, QC, Canada (2018). https://doi.org/10.1145/3173574.3173714
3. Durall Gazulla, E., Martins, L., Fernández-Ferrer, M.: Designing learning technology collaboratively: analysis of a chatbot co-design. Educ. Inf. Technol. **28**(1), 109–134 (2023). https://doi.org/10.1007/s10639-022-11162-w
4. Farah, J.C., Ingram, S., Gillet, D.: Supporting developers in creating web apps for education via an app development framework. In: HEAd'22 Conference Proceedings, pp. 883–890. Editorial Universitat Politècnica de València, Valencia, Spain (2022). https://doi.org/10.4995/HEAd22.2022.14587
5. Farah, J.C., Spaenlehauer, B., Ingram, S., Gillet, D.: A blueprint for integrating task-oriented conversational agents in education. In: Proceedings of the 4th Conference on Conversational User Interfaces, CUI'22. ACM, Glasgow, United Kingdom (2022). https://doi.org/10.1145/3543829.3544525
6. Feine, J., Morana, S., Maedche, A.: Designing a Chatbot social cue configuration system. In: Proceedings of the 40th International Conference on Information Systems (ICIS) (2019)
7. Feine, J., Morana, S., Maedche, A.: Designing interactive Chatbot development systems. In: Proceedings of the 41st International Conference on Information Systems (ICIS) (2020)
8. Felten, E., Raj, M., Seamans, R.: How will language modelers like ChatGPT affect occupations and industries? (2023)
9. Ferman Guerra, M.A.: Towards best practices for Chatbots. Master's thesis, University of Victoria, Victoria, BC, Canada (2018)
10. Griol, D., Callejas, Z.: An architecture to develop multimodal educative applications with Chatbots. Int. J. Adv. Robot. Syst. **10**(3) (2013). https://doi.org/10.5772/55791
11. Hu, X., Moore, A., Coleman Eubanks, J., Aiyaz, A., McMahan, R.P.: Evaluating interaction cue purpose and timing for learning and retaining virtual reality training. In: Proceedings of the 2020 ACM Symposium on Spatial User Interaction, SUI '20. ACM, Virtual Event, Canada (2020). https://doi.org/10.1145/3385959.3418448
12. Jung, H., Lee, J., Park, C.: Deriving design principles for educational chatbots from empirical studies on human-chatbot interaction. J. Digi. Contents Soc. **21**(3), 487–493 (2020). https://doi.org/10.9728/dcs.2020.21.3.487
13. Kuhail, M.A., Alturki, N., Alramlawi, S., Alhejori, K.: Interacting with educational chatbots: a systematic review. Educ. Inf. Technol. **28**(1), 973–1018 (2023). https://doi.org/10.1007/s10639-022-11177-3
14. Nass, C., Steuer, J., Tauber, E.R.: Computers are social actors. In: Proceedings of the SIGCHI Conference on Human Factors in Computing Systems, CHI '94, pp. 72–78. ACM, Boston, MA, USA (1994). https://doi.org/10.1145/191666.191703
15. OpenAI: Introducing ChatGPT (2022). https://openai.com/blog/chatgpt
16. Quiroga Pérez, J., Daradoumis, T., Marquès Puig, J.M.: Rediscovering the use of chatbots in education: a systematic literature review. Comput. Appl. Eng. Educ. **28**(6), 1549–1565 (2020). https://doi.org/10.1002/cae.22326

17. Richter, S.: Learning tasks. In: Seel, N.M. (ed.) Encyclopedia of the Sciences of Learning, pp. 1975–1979. Springer, Boston, MA, USA (2012). https://doi.org/10.1007/978-1-4419-1428-6_342
18. Rodríguez-Triana, M.J., Martínez-Monés, A., Villagrá-Sobrino, S.: Learning analytics in small-scale teacher-led innovations: ethical and data privacy issues. J. Learn. Anal. **3**(1) (2016). https://doi.org/10.18608/jla.2016.31.4
19. Vector Stall: Assistant free icons. https://www.flaticon.com/free-icon/assistant/_4818971
20. Wiley, D.A.: Connecting learning objects to instructional design theory: a definition, a metaphor, and a taxonomy. In: The Instructional Use of Learning Objects, pp. 3–23. AIT/AECT (2002)

# Harnessing Rule-Based Chatbots to Support Teaching Python Programming Best Practices

Juan Carlos Farah[1,2(✉)], Basile Spaenlehauer[1], Sandy Ingram[2], Aditya K. Purohit[3], Adrian Holzer[4], and Denis Gillet[1]

[1] École Polytechnique Fédérale de Lausanne (EPFL), Lausanne, Switzerland
{juancarlos.farah,basile.spaenlehauer,denis.gillet}@epfl.ch
[2] University of Applied Sciences (HES-SO), Fribourg, Switzerland
{juancarlos.farah,sandy.ingram}@hefr.ch
[3] Radboud University, Nijmegen, The Netherlands
aditya.purohit@ru.nl
[4] University of Neuchâtel, Neuchâtel, Switzerland
adrian.holzer@unine.ch

**Abstract.** In recent years, the use of chatbots in education has been driven by advances in natural language processing and the increasing availability of digital education platforms. Although the added value of educational chatbots appears promising, researchers have noted that there is a need for empirical studies that explore the effects of incorporating chatbots into different learning scenarios. In this paper, we report on the integration of a rule-based chatbot into an information technology course. We conducted a controlled experiment in which half of the students were able to interact with the chatbot during Python lab sessions while the other half completed the sessions without the chatbot. Our results suggest that educational chatbots powered by short, simple, interactive scripts could have a positive impact on the user experience offered by learning technologies and could be pertinent to educators looking to integrate chatbots into their practice.

**Keywords:** Educational chatbots · Digital education · Python · Empirical study · Programming best practices

## 1 Introduction

Chatbots have been poised to disrupt educational technologies for well over a decade [30]. Nevertheless, researchers have found that there is still a lack of empirical studies on the use of chatbots in education [15]. In this study, we present results from a controlled experiment conducted within a university-level information technology course. Our study aimed to shed light on how rule-based chatbots could be harnessed to support students learning Python programming best practices as defined by the PEP-8 standard [24]. To that end, we configured a


M. E. Auer et al. (Eds.): ICL 2023, LNNS 899, pp. 455–466, 2024.
https://doi.org/10.1007/978-3-031-51979-6_47

chatbot to follow a predefined script designed to show students that a particular code style guideline was useful in practice. The results from our case study contribute to the growing body of research on educational chatbots and may be relevant to practitioners looking to integrate these chatbots into their courses.

## 2 Background and Related Work

Recent reviews of the literature have highlighted that chatbots are often used to teach computer-related topics [18,31]. In software engineering education, educational chatbots have been specifically harnessed in database, programming, computer networks, and compiler courses [18]. For example, Coronado et al. proposed a personal agent to support students in learning the Java programming language [5]. Their empirical evaluation, which relied on both objective and subjective metrics, showed that incorporating social dialog through question-answering agents increased user satisfaction and engagement with the system in which the agents were deployed. Mad Daud et al. also proposed a chatbot for learning Java [22]. By generating different code control structures, the proposed *e-Java* chatbot helped students in learning different ways of coding solutions for the same problem. Custom surveys were used as evaluation metrics to assess the perceived usefulness of the proposed chatbot. Going beyond self-reported metrics, an empirical study conducted by Winkler et al. evaluated information retention and transfer ability using different types of conversational agents (e.g., scaffolding vs. non-scaffolding, text vs. voice-based) [29]. Their evaluation using introductory Python programming tutorials achieved positive results.

Our work adds to this growing body of research by addressing the use of chatbots to support conducting code reviews in educational contexts. Conducting peer reviews for software verification and validation is one of the topics that the IEEE Computer Society and the Association for Computing Machinery recommend for computer science curricula [16]. Indeed, code reviews are an integral part of the software development process [28] and while there are several tools to support code reviews [2, 13], most collaborative software development involving the code review process currently takes place on social coding platforms. These platforms feature interfaces for reviewing and annotating code, providing a backdrop for discussions between developers. Chatbots can help with this process and have thus become a common feature of the social software development experience [20], helping to reduce manual labor, improve code quality, and increase productivity [26,27]. However, very few studies have focused on the use of chatbots to support the code review process in formal education settings. We build on a previous Wizard of Oz [6] experiment [10] and a pilot study [8] to explore the impact that rule-based chatbots supporting code review exercises could have on the learning experience. In the following section, we present our guiding research question and the methodology we followed for our evaluation.

# 3 Methodology

Our evaluation aimed to address the following research question: *What are the effects of a rule-based chatbot designed to support Python programming lessons on students' learning experiences?* We addressed this research question by conducting a between-subjects controlled experiment comprising one control and one treatment. In both conditions, students were presented with the same lesson. The conditions differed only in the way we explained the code style issues illustrated by the example code snippets. To frame our evaluation, we focused specifically on four aspects of the learning experience: (i) learning gains achieved, (ii) perceived usefulness of the material, (iii) user experience of the lesson, and (iv) feedback. In this section, we explain the methodology of our evaluation in detail.

## 3.1 Pedagogical Scenario

The main evaluation took place within the five-week Python programming component of a 14-week information technology course tailored for students completing a bachelor's degree in economics and business at the University of Neuchâtel, Switzerland. A total of 97 students were enrolled in the course. Before the beginning of the course, students were randomly assigned to the treatment (chatbot) or control (no chatbot) group. As outlined in Table 1, each week, students took part in an in-class lecture. The same week, they were able to attend a lab session in which the topics covered in the lecture were reviewed and a code style exercise was presented. Participation, however, was not mandatory. The course was conducted in French and all the material—including the chatbot scripts—was presented in French. For convenience, scripts, screenshots, student responses, and other material presented herein have been translated into English by the authors.

**Table 1.** Weekly lecture topics included in the Python programming component of the course alongside their corresponding lab session and code style exercises.

| Week | Lecture | Lab | Code style exercise |
|---|---|---|---|
| 1 | Conditions | Conditions | Pre-test |
| 2 | Loops | Loops | Indentation, whitespace, constants |
| 3 | Lists | Lists | Comparisons, negating, comparing booleans |
| 4 | Functions | Functions | Max length, function names, descriptive names |
| 5 | – | Review | Post-test |

Each lab session was conducted in French and included an exercise on Python code style guidelines based on the PEP-8 standard. These exercises were structured as code review notebooks [9] and followed the *Fixer Upper* pedagogical pattern both for explanation and evaluation [3]. That is, students were presented with code snippets that included code styling violations and were shown how to correct them (explanation) or asked to identify the issues present (evaluation). The first lab introduced students to the exercises and included a short activity that served as a pre-test. The second, third, and fourth labs each included explanations covering code layout, coding standards, and naming standards, respectively. In the final session, students completed a second activity that served as a post-test. Sessions were supported by the *Graasp* learning experience platform [11] and *Code Review*, an application that allows students to annotate code and supports dialogs with chatbots [9].

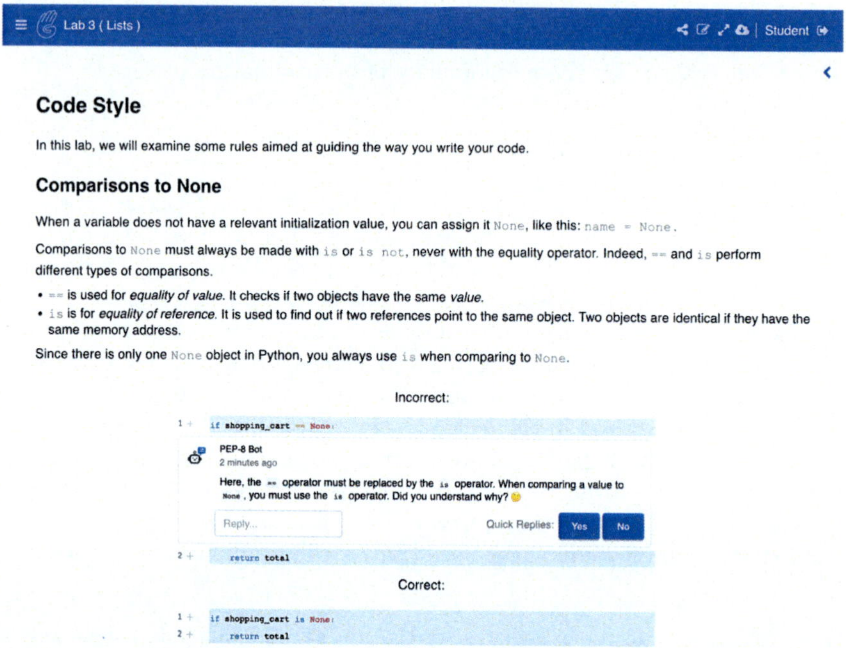


**Fig. 1.** Our chatbot was integrated into an application that was embedded in a code review notebook aimed at teaching Python programming.

**Chatbot** For this case study, we equipped Code Review with *PEP-8 Bot*. This chatbot was used to annotate the lines of code that contained potential code style issues within a series of Python code snippets used in Labs 2–5. The chatbot was configured to engage students by asking them if they understood and agreed with the logic behind the code style issue at hand, providing explanations of Python programming best practices, and motivating the reasons behind those best practices. An example of the chatbot embedded in the code review notebook is shown in Fig. 1. Furthermore, we included interactions featuring emoji and animated gifs, taking advantage of the Markdown [12] support provided by Code Review. This allowed our chatbot to express—among other emotions—humor and confusion, as shown in Fig. 2.

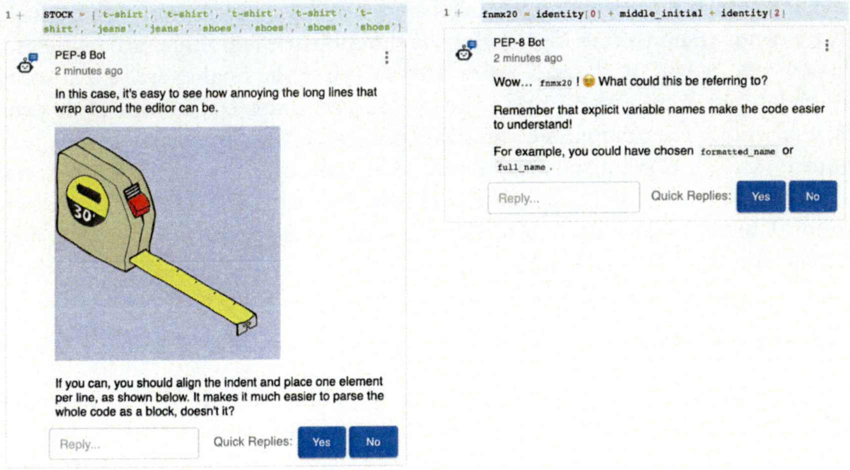


**Fig. 2.** Some of the chatbot's comments featured animated gifs (left) and emojis (right) to elicit humor and express confusion, respectively.

## 3.2 Procedure

At the beginning of each lab session, students were asked to complete a short code style exercise. Each exercise was meant to take approximately five minutes to complete. Lab 1 included the pre-test and was identical for all students. In Labs 2–4, each code style exercise covered three guidelines, spanning a total of nine guidelines (see Table 1). Finally, in Lab 5, students completed the post-test, after which they were shown the solutions to the test. For each guideline—and for the solutions to the post-test—students were presented with a short explanation followed by a code snippet containing an issue (incorrect snippet) and a code snippet with a correction of the issue (correct snippet). In the incorrect snippet, on the line containing the issue, students in the treatment group were also shown a comment from PEP-8 Bot that provided a further explanation and prompted the student to start a dialog about the usefulness of the guideline. Students in the control group did not see this comment.

## 3.3 Participants

There were 97 students enrolled in the course (44 female, 53 male). A total of 89 students accessed at least one of the exercise sessions and 65 students accessed all the sessions.

## 3.4 Instruments

We operationalized the four aspects of the learning experience as follows. Learning gains were calculated by taking the difference between students' scores on the pre-test and the post-test. This yielded a learning gain that could range from −100% to 100%. Usefulness was measured using a seven-point Likert scale. Students were asked to rate how useful they found each individual guideline on a scale of 1 (*not useful at all*) to 7 (*very useful*). After the final exercise, using the same scale, they were asked to rate the overall usefulness of the code style exercises as a whole. User experience was captured with the *User Experience Questionnaire* (UEQ), a standard instrument that measures user experience across six dimensions [19]. Finally, feedback was assessed using the following question: "*Do you have any suggestions or comments on the parts of the labs that dealt with code style guidelines?*"

### 3.5 Data Analysis

We analyzed quantitative data using descriptive and inferential statistics, reporting sample means ($\bar{x}$), medians ($\tilde{x}$), and standard deviations ($s_x$), as well as results from two-sample $t$-tests, where applicable. Qualitative feedback was analyzed using line-by-line data coding [4].

## 4 Results

In this section, we highlight our results with respect to each aspect considered.

### 4.1 Learning Gains

A total of 25 students (10 control, 15 treatment) completed the post-test required to calculate learning gains. In both groups, as shown in Fig. 3, learning gains were positive. The students in the control group achieved a mean learning gain of 36.0% ($s_x$ = 31.3%), while those in the treatment group achieved a mean learning gain of 36.7% ($s_x$ = 30.6%). As expected, the results of a two-sample $t$-test did not yield significant results ($p = 0.958$).

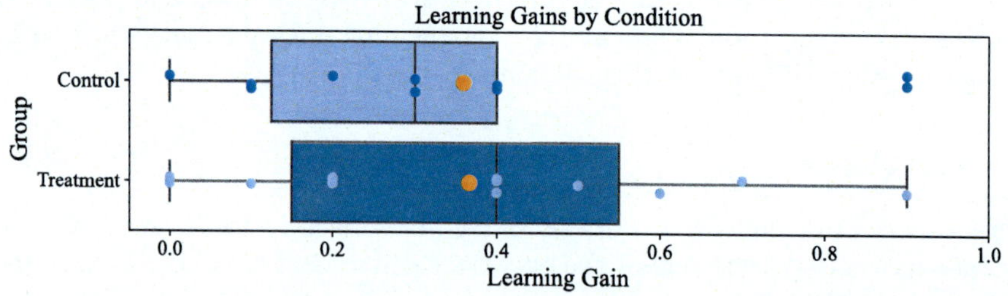


**Fig. 3.** Learning gains were positive for both groups, but there were no significant differences across conditions.

### 4.2 Perceived Usefulness

Regarding perceived usefulness, 46 students (26 control, 20 treatment) provided a total of 232 ratings distributed across the nine different guidelines, while 14 students (5 control, 9 treatment) rated the overall usefulness of the code style exercises. As shown in Fig. 4, ratings were on average positive (above the median rating of 4), both overall and across individual guidelines. Nevertheless, two-sample $t$-tests did not yield significant differences across the groups. Furthermore, the number of students who provided the ratings decreased over time. While 34 students rated the first guideline, only 17 students rated the last one. It is important to note, however, that there were no significant differences in how this diminishing trend manifested itself in each of the two conditions.

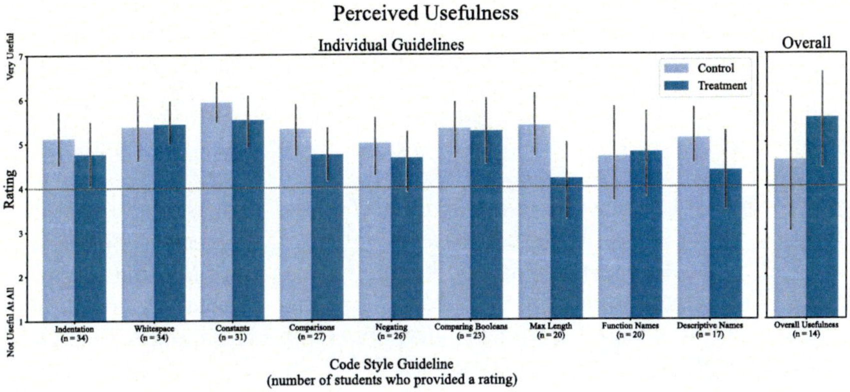


**Fig. 4.** On average, students rated how useful they found each guideline and the overall usefulness of the code style exercises positively (above the median rating of 4), but there were no significant differences across groups.

### 4.3 User Experience

A total of 27 students (14 control, 13 treatment) completed the UEQ. On average, students in the control group provided negative ratings for four of the six dimensions, while students in the treatment group provided positive ratings across all dimensions (see Fig. 5). This difference was more pronounced in the efficiency ($p = 0.0127$), dependability ($p = 0.0565$), and perspicuity ($p = 0.0807$) dimensions.

### 4.4 Feedback

A total of 10 students (5 control, 5 treatment) provided qualitative feedback in the form of short open-ended responses. Four themes emerged in the responses:

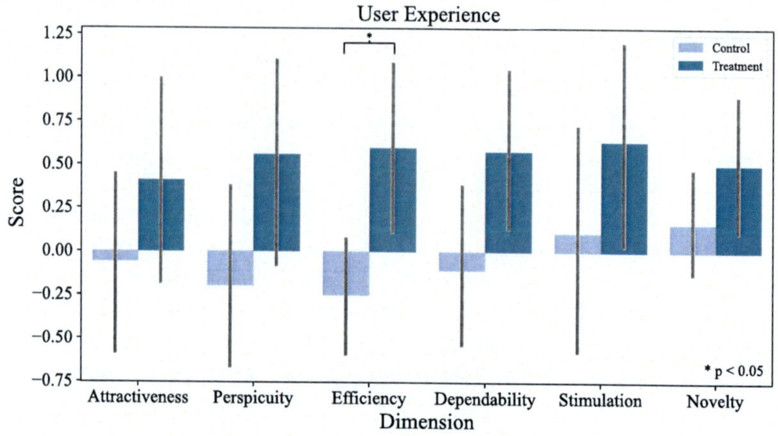


**Fig. 5.** User Experience Questionnaire (UEQ) ratings were consistently higher for the treatment group, especially for the efficiency ($p = 0.0127$), dependability ($p = 0.0565$), and perspicuity ($p = 0.0807$) dimensions.

(i) *more support*, (ii) *more exercises*, (iii) *useful*, and (iv) *useless*. First, *more support* was requested by four students (one control, three treatment), who suggested that instructors help clarify doubts and provide more information on the exercises. These requests are exemplified by a comment from a student in the treatment condition, who underlined that a "human touch would be nice" when referring to the lesson. Second, two students in the control group requested *more exercises* as a way to better assimilate the material. Third, two students in the treatment condition expressed that the exercises were *useful*, with one student providing the following comment:

> Guidelines that addressed code style that made code more readable (layout, balance between upper and lower case, convention in writing style) were very helpful. The little touches of humor in [Lab 4] made the [lab] much more interesting. In general, the practical guidelines of the code allowed the course to be more complete.

Finally, two students in the control group referred to the exercise as *useless*, with one explicitly saying that they were "quite useless and boring compared to normal lab exercises".

## 5 Discussion

The results of our evaluation show that although our chatbot integration did not affect learning gains or the perceived usefulness of the material, it did have an impact on the user experience of the lesson. On the one hand, results regarding learning gains could be explained by the fact that the chatbot interaction was primarily designed to reiterate the explanation that was already present in the text and—if needed—persuade the student that the guideline in question was useful in practice. On the other hand, for the same design reason, we would have expected students who were exposed to the chatbot to perceive the code style guidelines as more useful than students who did not hold those interactions, which was not the case. However, as evidenced by the decreasing number of students who provided ratings as the course progressed, this result could have been influenced by a diminishing novelty effect that curbed student interest in the code style exercises.

The lack of significant differences in learning gains between conditions could be interpreted positively. In line with a suggestion made by Hobert [14], this result bolsters the idea that educational chatbots could support learners when teaching staff are not available or in scenarios with large numbers of students. Given that learning is not impacted, educational chatbots could serve as an additional layer of interaction for learners seeking more information.

The differences in user experience, however, were evident—albeit not always significant—across all dimensions of the UEQ, and are very promising. Results from the UEQ suggest that incorporating the chatbot into the interface improved student perception of the user experience of the lesson, especially in terms of efficiency, dependability, and perspicuity. This improvement could have been mediated by the fact that including the chatbot added interactivity to a lesson that was otherwise primarily explanatory.

Improvements in user experience that are not accompanied by improvements in learning gains have been observed in the literature. As discussed by Davids et al., this lack of correlation could be due to the type of learners that are participating in the learning activity or the interface that is being optimized [7]. In the case of Davids et al., their study was conducted with practicing clinicians who the authors describe as possibly being highly motivated and therefore less affected by the user experience improvements the authors were testing. In our case, the fact that the exercises were not mandatory might have led to a selection bias, where the most motivated students took part in code style exercises included as part of their lab work, possibly leading to a similar effect as the one observed by Davids et al. [7]. Nevertheless, an improvement in perceived user experience can have an impact on other dimensions, such as task completion rate [17], self-regulation [21], and motivation [32].

Further insights were provided by the qualitative feedback. Positive comments regarding the *usefulness* of the exercises were present exclusively in the treatment condition, while negative comments about the *uselessness* of the exercise were present exclusively in the control condition. This contrast suggests that including the educational chatbot made the exercises more meaningful, possibly shedding some light on why perceived user experience was higher in the treatment condition. Nonetheless, the fact that more support was requested by four students, including three in the treatment condition, indicates that even with the inclusion of a chatbot, students still require a "human touch" in the learning process.

## 6 Conclusion, Limitations, and Future Work

In this paper, we presented results from an empirical case study assessing the effects of integrating educational chatbots into blended learning scenarios aimed at teaching Python programming best practices. The findings of our controlled experiment show that there were no significant differences in the learning gains achieved and the perceived usefulness of the lessons between students who completed the exercises alone and those who completed them with support from the chatbot. However, students who had access to the chatbot rated the user experience of the lesson more positively, particularly in the efficiency, dependability, and perspicuity dimensions of the UEQ. This improved user experience could motivate the integration of chatbots into the type of lessons used in this study.

It is important to note that this study has some limitations worth considering. First, given that the exercises were not mandatory, there could have been some selection bias in our sample, as possibly only the most motivated students interacted with the lesson. Second, while the rule-based scripts ensured that student interaction with the chatbot was on topic and pertinent to the lessons, the limited scope of these exchanges could have diminished how natural they appeared to the student and, therefore, discouraged students from interacting with the chatbot. These limitations could be addressed by (i) ensuring that all students are exposed to the exercises and (ii) equipping the chatbot with a generative language model. We will address these limitations and explore possible improvements through the use of large language models—such as those powering ChatGPT [23]—in future work.

**Acknowledgments.** Images used in this study include icons made by Vector Stall [25] and Bad Arithmetic [1].

## References

1. Bad Arithmetic: Measuring tape measure up (2017). https://giphy.com/embed/3og0IQttlo3NfcsIiQ
2. Baum, T., Schneider, K.: On the need for a new generation of code review tools. In: Abrahamsson, P., Jedlitschka, A., Nguyen Duc, A., Felderer, M., Amasaki, S., Mikkonen, T. (eds.) Product-Focused Software Process Improvement, vol. 10027, pp. 301–308. Springer, Cham, Switzerland (2016). https://doi.org/10.1007/978-3-319-49094-6_19
3. Bergin, J.: Fourteen pedagogical patterns. In: Devos, M., Rüping, A. (eds.) Proceedings of the 5th European Conference on Pattern Languages of Programs (EuroPLoP 2000). Universitaetsverlag Konstanz, Irsee, Germany (2000)
4. Charmaz, K.: Constructing Grounded Theory: A Practical Guide through Qualitative Analysis. Sage, London, UK (2006)
5. Coronado, M., Iglesias, C.A., Carrera, Á., Mardomingo, A.: A cognitive assistant for learning java featuring social dialogue. Int. J. Hum Comput Stud. **117**, 55–67 (2018). https://doi.org/10.1016/j.ijhcs.2018.02.004
6. Dahlbäck, N., Jönsson, A., Ahrenberg, L.: Wizard of Oz studies—Why and how. Knowl.-Based Syst. **6**(4), 258–266 (1993)

7. Davids, M.R., Chikte, U.M.E., Halperin, M.L.: Effect of improving the usability of an e-learning resource: a randomized trial. Adv. Physiol. Educ. **38**(2), 155–160 (2014). https://doi.org/10.1152/advan.00119.2013
8. Farah, J.C., Spaenlehauer, B., Bergram, K., Holzer, A., Gillet, D.: Challenges and opportunities in integrating interactive Chatbots into code review exercises: a pilot case study. In: EDULEARN22 Proceedings, pp. 3816–3825. IATED, Valencia, Spain (2022). https://doi.org/10.21125/edulearn.2022.0932
9. Farah, J.C., Spaenlehauer, B., Rodríguez-Triana, M.J., Ingram, S., Gillet, D.: Toward code review notebooks. In: 2022 International Conference on Advanced Learning Technologies (ICALT), pp. 209–211. IEEE, New York, NY, USA (2022). https://doi.org/10.1109/ICALT55010.2022.00068
10. Farah, J.C., Spaenlehauer, B., Sharma, V., Rodríguez-Triana, M.J., Ingram, S., Gillet, D.: Impersonating Chatbots in a code review exercise to teach software engineering best practices. In: 2022 IEEE Global Engineering Education Conference (EDUCON), pp. 1634–1642. IEEE, New York, NY, USA (2022). https://doi.org/10.1109/EDUCON52537.2022.9766793
11. Gillet, D., Vonèche-Cardia, I., Farah, J.C., Phan Hoang, K.L., Rodríguez-Triana, M.J.: Integrated model for comprehensive digital education platforms. In: 2022 IEEE Global Engineering Education Conference (EDUCON), IEEE Global Engineering Education Conference, pp. 1586–1592. IEEE, New York, NY, USA (2022). https://doi.org/10.1109/EDUCON52537.2022.9766795
12. Gruber, J., Swartz, A.: Markdown (2004). https://daringfireball.net/projects/markdown/
13. Hedberg, H.: Introducing the next generation of software inspection tools. In: Kanade, T., Kittler, J., Kleinberg, J.M., Mattern, F., Mitchell, J.C., Nierstrasz, O., Pandu Rangan, C., Steffen, B., Terzopoulos, D., Tygar, D., Vardi, M.Y., Bomarius, F., Iida, H. (eds.) Product Focused Software Process Improvement, vol. 3009, pp. 234–247. Springer, Berlin, Germany (2004). https://doi.org/10.1007/978-3-540-24659-6_17
14. Hobert, S.: Say hello to 'Coding Tutor'! Design and evaluation of a Chatbot-based learning system supporting students to learn to program. In: 40th International Conference on Information Systems (ICIS 2019), vol. 3, pp. 1776–1792. Curran Associates, Inc., Red Hook, NY, USA (2020)
15. Hwang, G.J., Chang, C.Y.: A review of opportunities and challenges of Chatbots in education. Interact. Learn. Environ. (2021). https://doi.org/10.1080/10494820.2021.1952615
16. Joint Task Force on Computing Curricula: Software Engineering 2014: Curriculum Guidelines for Undergraduate Degree Programs in Software Engineering. Technical reports. IEEE & ACM (2015)
17. Kanuka, H., Szabo, M.: Conducting research on visual design and learning: pitfalls and promises. Canadian J. Learn. Technol./La revue canadienne de l'apprentissage et de la technologie **27**(2) (1999). https://doi.org/10.21432/T2SW37
18. Kuhail, M.A., Alturki, N., Alramlawi, S., Alhejori, K.: Interacting with Educational Chatbots: a Systematic Review. Educ. Inf. Technol. **28**(1), 973–1018 (2023). https://doi.org/10.1007/s10639-022-11177-3
19. Laugwitz, B., Held, T., Schrepp, M.: Construction and evaluation of a user experience questionnaire. In: Holzinger, A. (ed.) HCI and Usability for Education and Work. Lecture Notes in Computer Science, vol. 5298, pp. 63–76. Springer, Berlin, Germany (2008). https://doi.org/10.1007/978-3-540-89350-9_6
20. Lebeuf, C., Storey, M.A., Zagalsky, A.: Software bots. IEEE Softw. **35**(1), 18–23 (2018). https://doi.org/10.1109/MS.2017.4541027

21. Liaw, S.S., Huang, H.M.: Perceived satisfaction, perceived usefulness and interactive learning environments as predictors to self-regulation in e-learning environments. Compu. Educ. **60**(1), 14–24 (2013). https://doi.org/10.1016/j.compedu.2012.07.015
22. Mad Daud, S.H., Ibrahim Teo, N.H., Mat Zain, N.H.: E-JAVA Chatbot for learning programming language: a post-pandemic alternative virtual tutor. Int. J. Emerg. Trends Eng. Res. **8**(7), 3290–3298 (2020). https://doi.org/10.30534/ijeter/2020/67872020
23. OpenAI: Introducing ChatGPT (2022). https://openai.com/blog/chatgpt
24. van Rossum, G., Warsaw, B., Coghlan, N.: Style guide for Python code. PEP 8, Python Software Foundation (2001). https://www.python.org/dev/peps/pep-0008/
25. Vector Stall: Assistant free icons. https://www.flaticon.com/free-icon/assistant_4818971
26. Wessel, M., Serebrenik, A., Wiese, I., Steinmacher, I., Gerosa, M.A.: Effects of adopting code review bots on pull requests to OSS projects. In: Proceedings of the 2020 IEEE International Conference on Software Maintenance and Evolution (ICSME). IEEE, Adelaide, Australia (2020). https://doi.org/10.1109/ICSME46990.2020.00011
27. Wessel, M., Serebrenik, A., Wiese, I., Steinmacher, I., Gerosa, M.A.: What to expect from code review bots on GitHub?: A survey with OSS maintainers. In: Proceedings of the 34th Brazilian Symposium on Software Engineering, pp. 457–462. ACM, Natal, Brazil (2020). https://doi.org/10.1145/3422392.3422459
28. Wiegers, K.E.: Peer Reviews in Software: A Practical Guide. Addison-Wesley, Boston, MA, USA (2002)
29. Winkler, R., Hobert, S., Salovaara, A., Söllner, M., Leimeister, J.M.: Sara, the Lecturer: Improving learning in online education with a scaffolding-based conversational agent. In: Proceedings of the 2020 CHI Conference on Human Factors in Computing Systems. ACM, Honolulu, HI, USA (2020). https://doi.org/10.1145/3313831.3376781
30. Winkler, R., Soellner, M.: Unleashing the potential of chatbots in education: a state-of-the-art analysis. Acad. Manag. Proc. **2018**(1), 15903 (2018). https://doi.org/10.5465/AMBPP.2018.15903abstract
31. Wollny, S., Schneider, J., Di Mitri, D., Weidlich, J., Rittberger, M., Drachsler, H.: Are we there yet?—a systematic literature review on chatbots in education. Front. Artif. Intell. **4**, 654924 (2021). https://doi.org/10.3389/frai.2021.654924
32. Zaharias, P., Poylymenakou, A.: Developing a usability evaluation method for e-learning applications: beyond functional usability. International Journal of Human-Computer Interaction **25**(1), 75–98 (2009). https://doi.org/10.1080/10447310802546716

# Providing a Natural Language Processing App for Language Teachers

Alexandra Posekany[1,2](✉) and Dominik Dolezal[2,3]

[1] TU Vienna University of Technology, Vienna, Austria
aposekany@tgm.ac.at
[2] TGM—Vienna Institute of Technology, Vienna, Austria
[3] University of Vienna, Vienna, Austria

**Abstract.** Natural Language Processing (NLP) is a common application for Artificial Intelligence. The goal is to provide language teachers with a simple to apply tool for topic model analyses to integrate into their classroom. The project also involves project based learning for students programming the actual AI web application. The original notion is to provide language teacher with AI methodology without requiring any technical knowledge in AI or any programming skills. Natural Language Processing provides various tools for word frequencies, but also topic modelling, allowing to track the relevance of topics over time in the media or in the literature. In collaboration with University of Technology linguistics, we intend to provide a corpus of classical English and German literature, as well as the option of uploading your own corpus which can be obtained from webscraping or other sources. A team of students of the vocational high school TGM Wien specialised in IT and Software Development is working on the design of the interactive GUI for this NLP application, learning in this way the methods of Natural Language Processing and Artrificial Intelligence in a project based setting. For this the statistical programming language R is utilized which already provides packages with implementation for Natural Language Processing and in addition the shiny package which allows to develop interactie web apps without additional web and app programming. A team of teachers supervises and supports the students during the development process, providing expertise in AI and NLP, in web and app programming, as well as server management. Two intended outcomes exist. Ont the one hand, we want our students to learn Natural Language Processing first hand through development of this application. On the other hand, we intend to obtain an interactive AI tool which can assist language teachers and their students on the long term in the classroom. In times of GPT3 and GPT4 dominating the media and perception of Artificial Intelligence, we intend to teach students about the methodology involved first hand by having them participate in the development of such an interactive application for Natural Language Processing. Through a project based learning approach, we involve them in all steps of the development over a times of one and a half years, two school years. The final product of this project is an interactive web application for topic modelling and visualising word frequency counts in various ways.


M. E. Auer et al. (Eds.): ICL 2023, LNNS 899, pp. 467–473, 2024.
https://doi.org/10.1007/978-3-031-51979-6_48

**Keywords:** Artificial intelligence · Natural language processing · Interactive web application · Project based learning · Software development

## 1 Introduction

Cohen and Demner-Fushman define the term Natural Language Processing (NLP) as "a subfield of linguistics and computer science that deals with computer applications whose input is natural language [...] [and] with building applications that process language at some given level of linguistic structure" [1, p. 1]. In the field of Natural language processing (NLP) several disciplines meet and participate, such as Artificial Intelligence, Statistics, Linguistics. Starting its life circle in the 1950s with symbolic NLP [2] through programs such as ELIZA [3] which stayed state of the art until the 1980s. In the 1990s, the shift toward statistical NLP based on corpora, embeddings and probabilistic models happened which include Latent Dirichlet Allocation and topic models, cf. [4]. In recent years, the field of NLP has exploded [5] and its research has shifted "to robust learning and processing systems applied to large corpora" [5, p. 1]. Despite their different meanings, NLP is often used synonymously with the terms "computational linguistics" (development of human language models) and "data mining" (applications performing specific tasks) [1]. Data science "is the systematic analysis of data within a scientific framework"[6, p. 1] and belongs to the fastest growing fields in the world [6].

Especially considering recent developments in Artificial Intelligence, which "is a computer system that can do tasks that humans need intelligence to do" [7, p. 1], the relevance of NLP and data science is indisputable. Hassani and Silva investigated the use of ChatGPT for data science applications [8]. They found that ChatGPT may assist in various ways, including data handling, model training, interpretation of results, analyzing unstructured data, synthetical generation of data, and natural language-related work, such as translations, sentiment analysis, and classifications. Hassani and Silva conclude that ChatGPT has the potential to increase productivity and precision in this field through automation and revolutionize the data science field as a whole.

While the relevance of NLP and data science research in general is increasing, the skills required to be competent in the modern digital world are experiencing an ongoing shift. It is no longer sufficient to possess subject-related knowledge and skills; instead, so-called "future skills", also known as "21st century skills", "transversal skills", "soft skills", "non-cognitive skills" [9], and "transformative competencies" [10], have become more and more important over the last decades.

According to Trilling and Fadel, 21st century skills include "digital literacy" and "information literacy skills" [11], which encompass the reflected usage and handling of data and information. Furthermore, modern frameworks for digital competence – which is tightly interwoven with 21st century skills [12] – also assign value to reflectively working with data: The European DigComp Framework, which describes the essential digital skills European citizens are expected to possess, comprises five areas in its current version 2.2, including "information and data literacy" [13].

Considering that 21st century skills should be taught together with subject-related knowledge and skills (see, e.g., [14–16]), most frameworks recommend an interdisciplinary integration of 21st century sills in school subjects [17]. Therefore, our approach of the present work is to include the acquisition of information and data literacy skills into language subjects.

This paper deals with the following research questions:

**RQ1:** Can we teach vocational high school students the basics of machine learning, NLP and app development through project based learning?

**RQ2:** How can a web-based application developed to visualize and interactively experience German and English bodies using R Shiny be useful for teaching and learning languages and literature?

## 2 Methods

Given the practical nature of this work which is still work in progress, we employed a two-stage strategy of student-focussed learning. In the first stage, University students from TU Vienna focussed on the literature review of NLP and existing packages in R as part of a bachelor thesis project at TU University of Technology Vienna. Since R provides many packages which support computational linguists in conducting analysis of written and oral language on a plethora of levels - setting focus on syntax, semantics, words as well as pragmatics, we intended to apply R to develop a natural language processing software for German and English corpora. The quanteda package provides a suite of tools for text analysis, including tokenization, stemming, stopword removal, n-grams, and more. The openNLP package provides an interface to the Apache OpenNLP library for natural language processing. The additional challenge for the university level students was to come up with implementations for visualising the outputs and text corpora appropriately. As a example data, we use the publicly available data set of speeches of U.S. presidents which provide a reasonably small scenario for development and intuitively interpretable visual results.

The second stage of our project involves vocational high school students to learn about the concepts of Data Science, machine learning and natural language processing through a project-based work which is accompanied by lectures and practical exercises teaching concepts of Data Science and machine learning. For this, we have chosen R shiny [18] as a reasonable software package that allows developers to build interactive web applications using R which can be used by people who do not know how to code. The golem package as described by [8] is a framework for building production-grade shiny applications which has been developed to abstract away the most common engineering tasks (for example, module creation, addition and linking of an external CSS or JavaScript file, etc.), so that the developer can focus on what matters: building the application, also guiding through testing and bringing tools for deploying to common platforms.

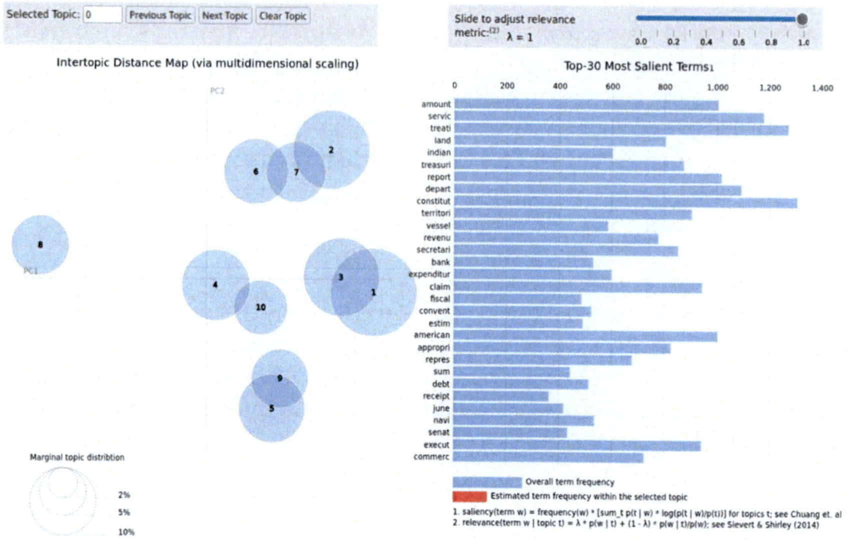


**Fig. 1.** Example of Latent Dirichlet Allocation (LDA) outcome visualised by salience and relevance based on term frequencies researched during her bachelor thesis.

This feature makes it particularly useful for vocational high school students specialised on IT who learn their footing in their first software development project. Within the students' class schedule, they are provided with 4 h per week to dedicate to the project in addition to 2 h of theoretical and 2 h of practical lessons in Data Science and machine learning.

The third stage, has been to conduct personal interviews with 17 language teachers of different schools showing them the app prototype and asking for an assessment of its usefulness for teaching languages and well as its usability, how easily and intuitively the app can be used by its target user group.

## 3 Results and Discussion

Since all showed work is still work in progress, we can only provide preliminary results based on the current stage of theses and projects. Figure 1 shows the visualisation of LDA, a type of topic model, which shows the most frequent terms co-occurring in common topics automatically recognised by the algorithmm as adapted by the University stduents during their bachelor thesis. Figure 2 shows the additional options for visualisation provided by the interactive web app prototype which the vocational high school students have developed during their project. The app does not only allow interaction through the the sliders, check boxes and input fields, but allows to click the topic circles of Fig. 1 and the single colored terms of Fig. 2.

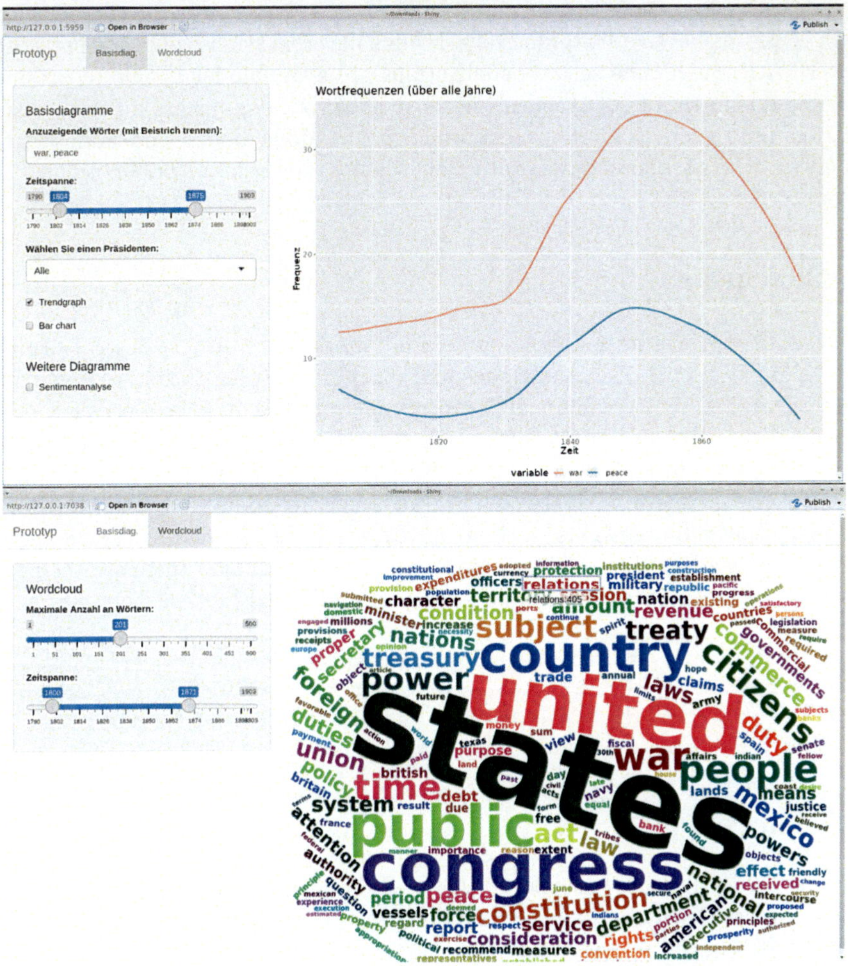


**Fig. 2.** Examples of visualisation of time series of terms occurring in example data (top) and topic model outcome visualised through word cloud in interactive app designed by our vocational highschool students.

Regarding the project based approach to learning about NLP, the group of five vocational high school students have not only been highly motivated to learn about methods beyond the covered scope of their classes, but also volunteered to continue working on the project during their final year in school and dedicating their final thesis to the work on this project. In addition to the machine learning contents which have captivated their interest, they provided very positive feedback to the clear app development structure of R golem.

Regarding our second research question which builds on the outcome of the first, interviews with fellow teachers teaching languages (English, German) have unisonously shown that providing intuitive visualisations will allow users teaching or learning languages to gain a better idea of how contents and topics develop which is impossible to do interactively through classical studies of literature word by word. Although time series visualisations are available for google terms and available documents from googles corpus to show trends, uniting different visualisations and models in the background is an added bonus and addition to dispersing worries about data protection which is an issue in public schools in Europe.

## 4 Conclusion

To conclude we have been able to successfully implement project based learning for teaching concepts of Data Science and machine learning through implementation of an Interactive shiny app prototype with the R golem package in a vocational high school setting after transferring knowledge from university to high school level. In addition, our intended user group of language teachers have provided very positive feedback and willingness to participate in the project a testers and providing insights to the students which also helps them learn about their project's advantages and disadvantages as well as makes them see the benefit their work provides to others provding an additional motivation.

**Acknowledgments.** We thank foremost our students for their active and highly motivated work on the project. Additionally, we thank our fellow teachers who have supported this project as well as the Austrian Ministry of Education for supporting our work though an "Innovationsprojekt des BMBWF" during a time frame of two years.

## References

1. Cohen, K.B., Demner-Fushman, D. (eds.): Biomedical Natural Language Processing, Natural Language Processing, vol. 11. John Benjamins Publishing Company, Amsterdam and Philadelphia (2014)
2. Hutchins, J. (2005). http://www.hutchinsweb.me.uk/Nutshell-2005.pdf
3. Weizenbaum, J.: Eliza—a computer program for the study of natural language communication between man and machine. Commun. ACM **9**, 35–36 (1966)
4. How the statistical revolution changes (computational) linguistics. In: Proceedings of the EACL 2009 Workshop on the Interaction between Linguistics and Computational Linguistics: Virtuous, Vicious or Vacuous? Association for Computational Linguistics (2009)
5. Clark, A., Fox, C., Lappin, S. (eds.): The Handbook of Computational Linguistics and Natural Language Processing. Blackwell Handbooks in Linguistics. Wiley-Blackwell, West Sussex, England (2010)
6. Larose, C.D., Larose, D.T.: Data Science using Python and R. Wiley, Hoboken, NJ (2019)
7. Healey, J. (ed.): Artificial Intelligence, Issues in Society, vol. 450. Spinney Press, Thirroul, NSW (2020)
8. Hassani, H., Silva, E.S.: The role of ChatGPT in data science: how AI-assisted conversational interfaces are revolutionizing the field. Big Data Cogn. Comput. **7**(2), 62 (2023)

9. Global Partnership for Education: 21st century skills: What potential role for the Global Partnership for Education? (2020). https://www.globalpartnership.org/content/21st-century-skills-what-potential-role-global-partnership-education
10. OECD: Transformative Competencies for 2030 (2020). https://www.oecd.org/education/2030-project/teaching-and-learning/learning/transformative-competencies/in_brief_Transformative_Competencies.pdf
11. Trilling, B., Fadel, C.: 21st Century Skills: Learning for Life in our Times. Jossey-Bass, San Francisco (2012)
12. van Laar, E., van Deursen, A.J., van Dijk, J.A., de Haan, J.: The relation between 21st-century skills and digital skills: a systematic literature review. Comput. Hum. Behav. **72**, 577–588 (2017)
13. Vuorikari, R., Kluzer, S., Punie, Y.: DigComp 2.2, The Digital Competence Framework for Citizens: With New Examples of Knowledge, Skills and Attitudes, EUR, vol. 31006. Publications Office of the European Union, Luxembourg (2022)
14. Voogt, J., Erstad, O., Dede, C., Mishra, P.: Challenges to learning and schooling in the digital networked world of the 21st century. J. Comput. Assist. Learn. **29**(5), 403–413 (2013)
15. Valtonen, T., Hoang, N., Sointu, E., Näykki, P., Virtanen, A., Pöysä-Tarhonen, J., Häkkinen, P., Järvelä, S., Mäkitalo, K., Kukkonen, J.: How pre-service teachers perceive their 21st-century skills and dispositions: a longitudinal perspective. Comput. Hum. Behav. **116**, 106643 (2021). https://www.sciencedirect.com/science/article/pii/S0747563220303903
16. Silva, E.: Measuring skills for 21st-century learning. Phi Delta Kappan **90**(9), 630–634 (2009)
17. Voogt, J., Roblin, N.P.: A comparative analysis of international frameworks for 21st century competences: Implications for national curriculum policies. J. Curric. Stud. **44**(3), 299–321 (2012)
18. Wickham, H.: https://shiny.rstudio.com/tutorial/

# Tackling Learning Obstacles in Learning Videos by Thematic Ad-Hoc Recommendations

Alexander Lehmann(✉) and Dieter Landes

University of Applied Sciences and Arts, 96450 Coburg, Germany
{alexander.lehmann,dieter.landes}@hs-coburg.de

**Abstract.** Learning videos enjoy great popularity in a digitalized world, especially since their use is usually possible regardless of time and location. Learners use this advantage mainly in self-study. Supervision, as for example in classroom teaching, is rather difficult and learners are usually left to their own devices when learning obstacles arise. However, not treating learning obstacles can have serious consequences, ranging from a gradual loss of the learners' motivation to the termination of the learning project. Consequently, learning obstacles must be identified and treated in order to support an efficient learning process. Fortunately, a digital learning environment opens up many opportunities to support learners automatically. This paper explains an approach to identify potential learning obstacles in video learning based on indirect feedback. The first part of the approach to removing learning obstacles in learning videos is based on an analysis of learners' click interaction within a video to identify potential problem areas. Building on this, the second part provides first thematically relevant ad hoc video recommendations for the potentially identified learning obstacle. In order to verify whether there is actually a learning obstacle, the third part explicitly induces learners to give indirect or direct feedback on whether the recommendations have helped them and, consequently, whether they have removed an actual learning obstacle.

**Keywords:** Adaptive learning · Video learning · Educational recommender system

## 1 Introduction and Motivation

Video Learning enjoys great popularity and is often used in both pure e-learning and blended learning [1]. Since the use of learning videos is usually independent of time and location, learning videos are usually used in self-study and consequently viewed alone. Hence, learners are often left to their own devices since physical supervision as in face-to-face teaching is hardly possible. Thus, learning obstacles can only be removed by the learner, yet with considerable additional effort. Although learning video environments offer almost by default the possibility to comment on videos in order to discuss problems, these possibilities are rarely used by learners, as they often do not want to comment on their problems in order not to be exposed. Consequently, learning obstacles are very rarely detected and removed. This in turn can lead to a loss of motivation of the learners or to the termination of the learning project [2].


M. E. Auer et al. (Eds.): ICL 2023, LNNS 899, pp. 474–481, 2024.
https://doi.org/10.1007/978-3-031-51979-6_49

This paper presents an approach to identify possible problem areas by means of user interactions within a learning video, in particular by analyzing click positions and repeatedly viewed sequences. Using a light-weighted ontology, which contains the individual topics covered in a video course and is linked to the video sequences, it is possible to infer individual topics that may not have been understood correctly. As a basis, we use a proprietary video platform that allows the analysis of user interactivity in learning videos. The videos are provided and integrated by another platform, the Panopto video platform. Panopto provides a robust video search function which allows not only to search for individual videos but also to search for relevant positions within the videos, so recommendations could be more specific. If a problematic position is identified, Panopto helps us to offer user topic-specific ad hoc video recommendations. In the last step, the learners are prompted to evaluate the recommendations' relevance in order to be able to conclude whether a learning obstacle has actually occurred and finally been removed.

## 2 Research Questions

This paper presents an experimental approach for ad-hoc recommendations to overcome learning obstacles when watching learning videos. The underlying research questions are as follows:

- Can learning obstacles be identified based on user interactions (click data) in learning videos?
- Can ad-hoc recommendations support learners quickly and appropriately in case of learning obstacles?

## 3 Related Work

The identification of problem areas (comprehension problems, learning obstacles) was already discussed in a previous paper [3]. Here, learners could anonymously tag problematic places in videos the with so-called problem tags and, thus, give direct feedback when they encountered a learning obstacle. Surveys showed that learners would like to use problem tags, but widespread use could only be observed if learners were continuously encouraged to use these features. Although learning obstacles can in principle be identified through direct feedback, reality shows that an approach that only involves direct feedback tends to fail. Additional indirect feedback is therefore urgently needed (hybrid approach). Therefore, the presented approach builds on the previous paper and focuses on indirect feedback in the form of user interactivity within videos.

The terms "indirect" and "direct" feedback have the following meaning throughout this paper:

**Direct feedback** denotes that learners explicitly state in the video platform that they have a problem or explicitly respond to questions or the like. **Indirect feedback** designates normal user interaction in the learning platform (especially click data) which we use to conclude on certain needs and problems.

The analysis of click data in e-learning environments and especially in video environments is not new.

Belarbi et al. already presented an approach to cluster users' profiles based on user interactions in videos and recommend additional videos based on the interests analyzed [4].

In MOOCs (Massive Open Online Courses) similar studies have already been conducted to predict user behavior based on clickstream data in videos [5].

Recommendation systems, as known from entertainment media and online shops, are becoming increasingly popular in the learning domain. Usually, such so-called Educational Recommender Systems build on the preferences and interests of the respective users [6], similar to Belarbi et al. [4].

Our work is also inspired by Belarbi et al. [4], who demonstrate that click data can be used to infer certain behavior. Although we follow the approach of a recommendation system, we do not make recommendations based on interests [6] in the first place, but rather on problems that arise. Such approaches are currently not yet widely used.

## 4 Tackling Learning Obstacles

This work presents an approach how learning obstacles can be identified through indirect feedback in learning videos and how they can be removed through targeted video recommendations. For our approach we use a specially developed learning platform with the possibility to track user interactions in videos in combination with the video platform Panopto. Currently, we use a test environment that covers 3 courses in the domain of programming and software engineering with about 100 videos per course. Panopto videos are integrated into our platform for analysis and further recommendations.

Figure 1 explains our approach. In a first step we capture indirect feedback in the form of user interactions within a learning video. Relevant for us are primarily longer stops at one position and the increased viewing of individual sequences. In this respect, we follow a similar approach as Teusner et al. [7], who assumed that learner have a problem if they need more time than on average to complete programming tasks. In this case, automated intervention in the learning process took place.

Our interactions also serve as a kind of trigger. If a learner stops at one position for a longer period of time or watches sequences repeatedly, we assume that he has encountered a learning obstacle. If such an event occurs, we also intervene automatically. In addition, we follow a collaborative approach by matching a user's frequent click areas in videos with the ones of other learners to infer similar behavior.

Our intervention consists of recommending suitable learning videos for the problem area to the learners in order to overcome potential learning obstacles. To capture the topic at a particular video position, individual video sequences are linked to a lightweight ontology that characterizes the learning domain, such as programming or software engineering.

In a second step, our approach searches for appropriate video sequences in the video repository. Searching is delegated to the Panopto video platform for this purpose, which enables us to conduct an efficient video search. In the last step, we use targeted interventions to induce learners to evaluate the relevance of the video recommendations in order to be able to conclude whether there was actually a learning obstacle in the video and whether it has been removed.

From our current collection of data, we can conclude that intense click interactions in combination with back jumps in a video are indeed indicative of learning problems. This can be confirmed by the fact that parts of the learning community often set problem tags at these points and consequently draw attention directly to problems. The conclusion is that it is possible to identify problematic positions in videos based on interactions. For this reason, our approach relies on click-data for analyzing learning obstacles.

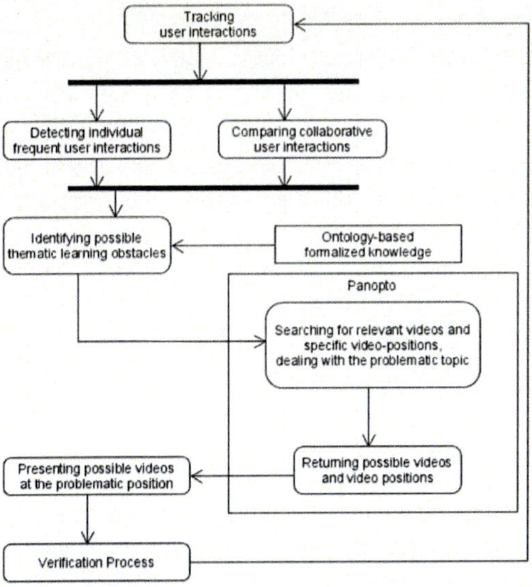


**Fig. 1.** Approach of identifying learning obstacles and targeted recommendations

### 4.1 Detecting Potential Problem Areas

In order to draw conclusions on possible problem areas, interactions of learners within the videos must first be tracked and analyzed. In particular, all click interactions are recorded as so-called "video actions". Currently, this is limited to four types of actions, namely Start, Stop, Pause, and Jump.

The analysis of click data consists of two parts; On the one hand, we are primarily interested in the positions with which a particular learner interacts via clicks, on the other hand we want to compare individual click data with those of other learners for figuring out if click areas are similar across learners. Such a collaborative approach can also help to identify problem areas more easily. We use two hypothetical methods to identify potential learning obstacles:

As described in Sect. 3, staying at one position for a longer time is presumably another indicator that a learner encountered a problem. Likewise, a learning obstacle may be expected if a learner watches individual sequences more often.

Increased click rates near the identified spots (compared to other learners) are indicators for problems that have occurred. Depending on the user behavior, the proper timing for an automated intervention in the learning process now needs to be determined, i.e. when a recommendation would make sense. We have defined experimental threshold values for this (see Table 1).

**Table 1.** Video action and trigger thresholds

| Video action | Trigger |
|---|---|
| Stopping at a specific position | Possible intervention after 10 s |
| Repeated viewing of a sequence | Possible intervention after 3rd repetition during a view |

Since these thresholds are only a tentative assessment of whether a learning obstacle have occurred, automated interventions should be carried out at random so that the learning process is not interrupted at the slightest sign. We could infer from initial experiments that interventions should take place in 25% of all cases. If, however, the problem is located within a frequent click area (in comparison to other learners), intervention should take place in 50% of all cases. By means of this process we try to identify potential learning obstacles and are now in a position to take reactive measures.

### 4.2 Formalized Knowledge

Section 4.1 outlines how potential learning obstacles in learning videos can be spotted. In order to be able to react to these problems appropriately, however, we need to know what is at stake at each point. There are two main approaches to accomplish this:

- identification of the topics dealt with at this point using text mining techniques;
- manually linking video sequences to topics.

Currently, we take the second approach. Yet, simple labeling of video sequences is not always sufficient since dependencies between different topics are not considered, though potentially relevant. For instance, a learner encounters a learning obstacle because he or she lacks various prerequisites for the topic being covered. For this reason, not only the actual topics in a video sequence are relevant, but possibly also its prerequisite topics, consequently we need to know these topics as well [8].

We tackle this issue by means of a light-weight ontology that describes the specific domain of learning [9]. In particular, we refine complex topics into (more) elementary sub-topics that are put in relation to each other. Currently, we only consider two types of relationships, namely **is-a** and **kind-of** relations.

Figure 2 presents an example from computer programming to explain the idea behind the domain ontology: A **loop** is a **control structure**, a **for-loop** is a **loop**. Whereas **constructors** and **inheritance** are kinds of **oop** (object-oriented programming) techniques.

Once such an ontology has been set up, individual sequences of a video can be linked to the corresponding nodes in the ontology to describe the contents of the respective sequence. For instance, if we tag a sequence with the topic while loop and in such a sequence a potential learning obstacle is detected, we can use our ontology to classify dependent topics as relevant and include them in the following recommendation process. Topics that have a is-a relationship with each other may be more relevant than kind-of relationships because the former have a closer relationship. Consequently, we rank topics with is-a relationships higher in the recommendation process.

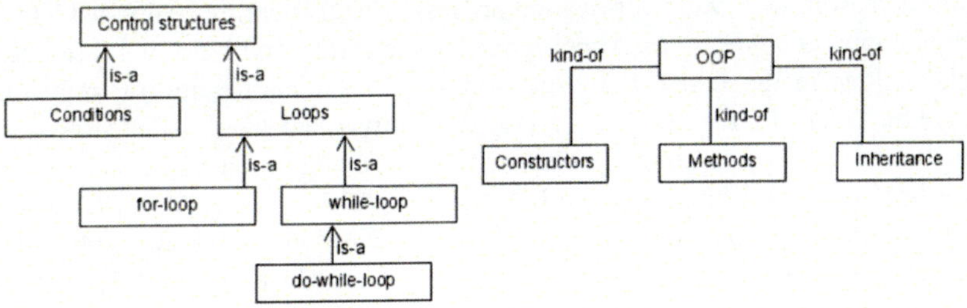


**Fig. 2.** Sample excerpt from the domain ontology for programming skills

### 4.3 Thematic Ad-Hoc Recommendations Using Panopto

After identifying relevant topics that could be considered for a recommendation, suitable learning videos and video sequences are searched for. This is where Panopto's Rest API comes into play, via which we outsource the retrieval process. Panopto has a powerful video search function which returns not only individual videos, but also relevant sequences. Thus, we may get much more relevant results than if only full videos were included in the recommendations. Learners can therefore jump directly to the relevant positions.

At this point we follow a top N strategy and return the first n video (sequences) as recommendations. As search terms we currently use the relevant topics as keywords. This means that we start a search process for each relevant topic and assemble the presumably most relevant results into a result list. This provides us with video recommendations that refer not only to the topic that has been explicitly dealt with at the problem location, but also to related topics in terms of the domain ontology.

Informal case studies indicate that learners should be provided with a certain choice of options. Even though they may not be able to fully grasp their problem, they can often decide what they need when making a selection [10].

In order not to overburden the learners with the selection, we limit ourselves to a maximum of five recommendations per problem area. It is necessary to make recommendations on an ad-hoc basis, i.e. when a learner may need support. The timing of an intervention (a recommendation) therefore depends on the instant when a potential learning obstacle is identified (see Sect. 4.1). Recommendations are immediately presented to the learner in the video with a reference to further information.

### 4.4 Verification Process

Identifying real learning obstacles is not easy. Only the learners themselves can give absolutely reliable confirmations whether they encountered a learning obstacle. Various criteria, like in our example the user interactions, can give at least a rough estimate.

Nevertheless, we have to ask the learner whether he or she has encountered a learning obstacle and whether the learning obstacle has been overcome.

For this reason, we need a verification process, which is currently divided into two parts in our approach. We try to explicitly induce the learner to give direct feedback.

This is done in the first part by intervening in the identification of problematic areas (see Sect. 4.1). In this case, learners are randomly asked whether they need further information on this topic. A simple yes/no answer is sufficient to conclude that a learning obstacle has occurred. This procedure helps us to confirm the existence of a learning obstacle. In the second part we need a confirmation whether the recommended videos were useful. For this purpose, our platform allows the evaluation of videos at the end of a video. The learners are explicitly asked 5 s before the video ends to rate whether the video helped them. In this way we can track whether the recommendations were useful. Unfortunately, this excludes learners who do not watch a recommended video to the very end and therefore do not rate it. Among other things, the issue how to deal with this type of learner is subject of future work.

## 5 Discussion and Outlook

Learning videos enjoy great popularity, especially since their use is usually possible regardless of time and location. Learners use this advantage mainly in self-study. Yet, in contrast to classroom teaching, supervision during self-study is rather difficult: learners are usually left to their own devices when learning obstacles arise. This paper presents an approach to identifying and removing learning obstacles in learning videos using click data from learners within the videos.

Our findings indicate that problem areas in learning videos are often associated with increased click interactions with particular positions in the video. Therefore, our approach to identify problem areas in videos is based on click data. Videos are linked to a domain ontology which describes the subject area of the videos. This enables us to deduce potentially problematic topics from positions with intense interactions. Suspected topics, in turn, can be used to derive thematic ad-hoc video recommendations that should support the learners in a way that is appropriate to the current situation. In order to substantiate the existence of particular learning obstacles and to confirm the relevance of a specific set of video recommendations, learners are induced to give direct feedback in the form of assessments in a last step of our approach.

Currently only preliminary insights are available, but our analyses clearly indicate that video sequences with frequent interactions are indeed partly problematic. This can be seen from the fact that these positions have a marked overlap with positions that bear problem tags, i.e. where learners have explicitly stated that they actually have a problem. The conclusion is that it is possible to identify problematic positions in videos based on interactions. Further evaluations are pending.

Our initial findings lay the ground for future work. Among others, the following questions need to be answered: Which types of click data in videos (e.g. start, stop, pause, jump) are really relevant for the recognition of learning obstacles? Is a lightweight ontology sufficient for the detection of relevant topics? Do we need to extend or refine the domain ontology to specify problematic topics more precisely (e.g. by means of more specific relations between topics)? Have the correct thresholds been set for the interventions and how would they need to be adjusted if necessary?

This should be investigated in further work.

## References


1. Dede, C.: The evolution of distance education: emerging technologies and distributed learning. Am. J. Distance Educ. (1996)
2. Giannakos, M.N., Aalberg, T., Divitini, M., Jaccheri, L., Mikalef, P., Pappas, I.O., Sindre, G.: Identifying dropout factors in information technology education: a case study. In: Proceedings of 2017 IEEE Global Engineering Education Conference (EDUCON), pp. 1187–1194. IEEE, Piscataway, NJ (2017)
3. Lehmann, A.: Problem tagging and solution-based video recommendations in learning video environments. In: Ashmawy, A.K., Schreiter, S. (eds.) Proceedings of 2019 IEEE Global Engineering Education Conference (EDUCON), pp. 365–373. IEEE, Piscataway, New Jersey (2019)
4. Belarbi, N., Chafiq, N., Talbi, M., Namir, A., Benlahmar, E.: User profiling in a SPOC: a method based on user video clickstream analysis. Int. J. Emerg. Technol. Learn. (2019)
5. Sinha, T., Jermann, P., Li, N., Dillenbourg, P.: Your click decides your fate: inferring information processing and attrition behavior from MOOC video clickstream interactions. In: Rose, C., Siemens, G. (eds.) Proceedings of the EMNLP 2014 Workshop on Analysis of Large Scale Social Interaction in MOOCs, pp. 3–14. Association for Computational Linguistics, Stroudsburg, PA, USA
6. Manouselis, N.: Recommender Systems for Learning. Springer Briefs in Electrical and Computer Engineering. Springer, New York (2013)
7. Teusner, R., Hille, T., Staubitz, T.: Effects of automated interventions in programming assignments. In: Luckin, R., Klemmer, S., Koedinger, K., Koedinger, K.R. (eds.) Proceedings of the Fifth Annual ACM Conference on Learning at Scale, pp. 1–10. ACM, New York, NY (2018)
8. Lehmann, A.: Adaptive starting points in video learning environments for new learners based on video and topic tree relations. In: Auer, M.E., Rüütmann, T. (eds.) Educating Engineers for Future Industrial Revolutions, vol. 1329. Advances in Intelligent Systems and Computing, pp. 808–818. Springer International Publishing, Cham (2021)
9. Stuckenschmidt, H.: Ontologien. Konzepte, Technologien und Anwendungen. Informatik im Fokus. Springer, Berlin Heidelberg, Berlin, Heidelberg (2009)
10. Middleton, S.E., de Roure, D., Shadbolt, N.R.: Ontology-based recommender systems. In: Staab, S., Studer, R. (eds.) Handbook on Ontologies. International Handbooks on Information Systems, pp. 477–498. Springer Berlin Heidelberg, Berlin, Heidelberg (2004)

# Internationalization at Home by Bringing English into the Lecture Hall

Iris Gross(✉), Doerthe Vieten, and Alexandra Reher

Bonn Rhein Sieg University of Applied Science, 53757 Sankt Augustin, Germany
iris.gross@h-brs.de

**Abstract.** The engineering department of the Hochschule Bonn Rhein Sieg (University of Applied Sciences) wished to significantly increase the number of English-language courses offered for incoming students, since this would be the basis for participating in international student exchange and build up partnerships with universities abroad. But for lecturers it is a high effort besides their daily work, to shift the scripts and presentation to English; using the correct technical terms. Hence, skilled student assistants were recruited to translate the learning material for more than 20 courses with the help of the artificial intelligence-based translation tool "deep L", to support the lecturers. The aim was to either switch a full course into English or to offer a mixed language teaching course, where international students have access to the complete material in English and join in exercises in English, while the lecture itself will be held in German. Besides enabling international exchange, switching the language to English in a greater number of modules shall help to promote technical language competencies of the German students. The resulting teaching materials already have been used in international cooperations.

**Keywords:** Project based learning · AI based translation · Social engagement in university

## 1 Introduction

Internationalization has become a natural part of everyday working life and intercultural competences are taken for granted when entering a company as a graduate. The Hochschule Bonn Rhein Sieg (University of Applied Sciences) addresses these developments through corresponding goals and measures for Internationalization, as stated in the actual University Development Plan 3 [1]. Measure for the implementation of these plans are taken in specialized central organizations such as the Language Centre and the International Office, but also in the departments. The aim is to make international experiences tangible for as many students as possible. One successful approach pursued by the university is digital internationalization [2]. Nevertheless, personal exchange, such as stay abroad or exchange of international guest students is still an indispensable element. Through international partnerships, an increasing number of students shall be given the opportunity to spend a semester abroad to gain experience in an international environment and expand their language and intercultural skills. In return, incoming students are


M. E. Auer et al. (Eds.): ICL 2023, LNNS 899, pp. 482–489, 2024.
https://doi.org/10.1007/978-3-031-51979-6_50

welcomed, which gives the students staying at home also a chance to get in contact with other cultures and to develop English communication and intercultural competencies (as "Internationalization at home").

Up to now, the department of Electrical Engineering, Mechanical Engineering, and Technical Journalism at the Hochschule Bonn Rhein Sieg University of Applied Sciences was not very successful in taking part in the international exchange and developing partnerships for mutual student exchange, although the teaching contents regularly generated great interest among potential partners, since they are very relevant actual. The internationalization efforts were limited by the fact that all courses were held in German, so the department was not able to host students without sufficient language skills in German language. However, English is our lingua franca in teaching [3] and makes exchanges with international universities possible in the first place. Whatever attractive the teaching contents and teaching methods may be, for international cooperation and exchange programs, courses merely offered in other languages than English can hardly attract guest students for a semester-long visit and are not a suitable basis for mutual exchange programs.

## 2 Aim

To significantly increase the range of courses offered either completely in English or with English self study elements, the department has applied for financial funding within the framework of the University Development Plan 3. With that funding, lecturers should be enabled to transfer their course into English so that consequently international students have a chance to participate in the modules based on English-language teaching materials. This aim aligns with the recommendation by the HRK (the conference of higher education commissioners, which as a voluntary association of German higher education institutions represents them in political and public affairs). In their guide for legal and organizational general conditions for foreign and multilingual study programs they recommend taking further steps to implement foreign and multilingual study programs [4, p. 3].

The aim of the project depicted here was to create English teaching materials of high quality that increases the international visibility and attractiveness of the university for incomings. As a side effect, these teaching materials should also enable the participation in international capacity building cooperations.

In addition, the German-speaking students should also benefit from the English teaching materials. Technical language skills make it possible to better understand foreign-language technical literature and improve career opportunities in the global labor market. High-quality translated teaching materials enable students to familiarize themselves with technical terminology in a targeted manner. This can be supported by the fact that students also must prepare English-language papers as part of the module examinations for the relevant modules.

To fulfil the aims discussed above, a project for the translation of teaching materials of the mentioned department was launched. It will be examined in more depth in this contribution.

## 3 Approach

To broaden the opportunities for international exchange, as well as an offer for incoming students, the number of English-language courses offered should be enlarged within the mentioned framework. To increase the number of English-speaking courses the department asked the lectures to join the initiative and to consider if it was possible, to either completely converse to English or to offer mixed forms with, if necessary, self-learning components like e.g. videos. The decision should be taken depending on the suitability of the subject and on personal preferences. Lecturers should take into account, that in the first semesters the student might struggle with the study topic itself, even when taught in German, so here a complete shift to English might be too demanding for them. Mixed forms would be, for example, a lecture in German accompanied by English-language teaching videos and a bilingual exercise program. One lecturer for example explains his plans: "The materials are offered to English-speaking foreign students to accompany the lecture. The lecture format is still German. Students are allowed to ask questions in English." Then step by step the course materials were translated, as described in the next chapter.

## 4 Method

To fulfil this approach, a translation team of five students was assembled for translation of selected course material. At technical level, the translation was supported by an AI application, DeepL [5].

The lecturers were asked to enter their translation requests in a table, and in spring 2021, more than 30 courses, with a varying degree of need for translation, were listed (Fig. 1).

**Fig. 1.** Screenshot from Excel file to sort out the need of support and possible starting date

A translation team of 5 student assistants was selected, based on their results in the "Oxford Placement Test" and a consistently demonstration of very good English skills. For a final quality assessment, one person could be required for the team, who had native level language competence in English, (through living long time abroad and holding a bachelor's degree in English Studies, although now studying electrical engineering).

Before the start of the translation work, the student assistants were also trained in optimized slide design and use of Microsoft PowerPoint.

The translation of the documents was mostly done in two steps: First, the documents were translated into English using the translation program DeepL Translate. The full version of DeepL can translate a limited number of complete PDF, Word, and PowerPoint files with a file size of up to 20 MB. For this purpose, three licenses for the full version were available. In the second step, the translation, as created by Deep L, was then revised manually. Especially for technical terms the program often does not choose the correct translation, so that manual checking and reworking by the student assistants was still necessary. To guarantee the translation quality for these technical terms, glossaries were developed for the modules. DeepL is generally able to translate text elements that are embedded in PowerPoint elements. But since the English terms would not automatically fit in the provided spaces on a slide, there was need of formatting afterwards. Also, the slides often contained pictures including text in German (in pixel format), which had to be fixed by using overlaying text blocks. This also showed to be a time consuming work.

Afterwards, the translations were passed on to the proofreader, a student with language skills at native-speaker level, who revised the documents once again to ensure that they were free of linguistic errors and formulated at a high linguistic level.

To coordinate the cooperation, weekly meetings were held, especially during the start-up phase, in which the progress and challenges of the translations were discussed jointly between the translation team and the first author as project coordinator.

To provide more details for possible exchange, Table 1 shows the translated courses and self-study opportunities for bachelor's and master's Degree students at the department.

## 5 Challenges and Solutions

Various challenges arose during translation, some of which could be solved and some of which could be circumvented. The biggest challenge, as mentioned, was the quality of automated translation regarding technical language of mechanics, physics and mathematics. Further challenges included inconsistencies in technical terms, necessitating different translations for the same term in different contexts. To remedy, the translated texts were checked critically, and technical terms were always researched in English-language technical literature or the Internet. A very good source for this was Wikipedia in English, since it could be easily compared to the German page of the same topic. For the critical review of the texts and the localization of possible translation errors, the translators' language experience was an advantage, as was an affinity with the translated module, which professional translation agencies would not bring with them.

Another problem was the inconsistency of technical terms on the English side. In different contexts, different translations are often used for the same German term, which

**Table 1.** Translated courses and self-study opportunities for bachelor's and master's Degree students. * Bachelor's degree course. ** Master's degree course.

| Translated courses and self-study opportunities for bachelor's and master's degree students | | |
|---|---|---|
| | Winter semester 22/23 | Summer semester 2023 |
| Courses in English | Scientific programming with Python* | Social media content* |
| | Industrial robotics/automation* | Technology policy* |
| | Journalism and PR digital I* | FPGA vision open online course* |
| | Innovative materials technology** | Control of grid-connected inverter* |
| | Networked systems** | Power hardware in the loop* |
| | Sector coupling** | Cost- and production management F.s* |
| | Energy 4.0** | Vehicle development F. S* |
| Self-study (mixed) | Electronics* | Advanced design methods and tools* |
| | Engineering mechanics 1* | LCA and sustainability analysis* |
| | Object oriented design and modelling** | Machine dynamics* |
| | Life cycle assessment** | Physics** |
| | Rapid control prototyping** | Advanced mathematics** |

makes the selection of suitable words even more difficult. To solve this problem, student translators created a glossary for the specialist vocabulary and shared it with the lecturers for approval. Especially whenever more than one student translator worked on the same course material, to avoid inconsistencies in terminology, shared glossaries were used. The lecturers were mostly quite experienced in English in their field of specialization but in the contrary, for many very basic subjects had surprisingly few knowledge about the correct English terms and not yet dealt with English-language literature, as this is probably used more for in-depth study and specialization and so had a learning effect as well.

Furthermore, the German language is famous for its composites, which form a fixed term in German but often cannot be translated directly and instead have to be replaced with paraphrases. Also here, extensive research was conducted to ensure the correct translation.

The formatting of the files as well as minor revisions, optical corrections when graphics or texts inserted as image files had to be translated and reworked manually in image editing programs made up a considerable part of the translation team's workload.

The complexity and variety of the tasks to be completed, as well as the required understanding of the technical language and close coordination with the clients, was also the decisive reason why no translation agency was entrusted with this task - and the effort involved explains why lecturers tended not to be able to manage this alongside their actual duties.

## 6 Actual Outcomes

Within the framework of the project, materials from a total of 22 modules as well as the module handbook of a bachelor's degree program were translated. These included 18 lectures in the form of PowerPoint files and three lecture notes as text files. In addition, six exercise collections (task collections partly with solutions) and three practical courses were translated and English explanatory videos for the module *Engineering mechanics 1* were created as part of the project. The number of translated courses and modules can be seen in Table 2.

**Table 2.** Number of translated courses and modules.

| Number of translated courses and modules | |
|---|---|
| Lectures | 18 |
| Exercises | 6 |
| Practical courses | 3 |
| Scripts | 5 |
| Videos | 1 |
| *Total modules* | 22 |

The satisfaction of the lecturers with the translations was checked by means of a survey and proved to be very high on average.

Nine of the 10 lecturers whose materials were translated responded to the request for feedback. Seven of the lecturers rated the quality of the translation as very good, whereby the good process and communication were emphasized positively. The exchange about the translation of certain technical terms was also rated as very useful, as the translated materials could be used directly. Two of the lecturers rated the quality of the translations as good. They criticized the fact that some of the translations of technical terms required additional work on their part. Overall, the translated work can be seen as a full success which can be underlined by a citation of a lecturer involved: "The quality of the translations and the support were very good, thank you very much!". Additionally, one claims that "The translation process and quality were excellent.".

The lecturers were asked about their plans for using the materials. Most of them stated that they (want to) use the materials in a mixed format, for example by giving a lecture in German and providing the teaching materials in English. Students should be able to ask questions in English either directly in the lecture or in separate consultation hours.

The translated materials concerning "Circuit Technology" and the materials of the the topic "Advanced Design, Methods, and Tools", in short: ADMT (by the 1. Author herself), have also already been used for an international cooperation with a university in Jordan to present the concept of take-home labs and to provide accompanying teaching materials. In addition, the students in ADMT now have to prepare their final presentation in English and access the English teaching materials with the corresponding specialist vocabulary for this purpose.

Some of the translated materials will be incorporated into an EU research project called "GREATER". The aim of the project is to bring the contents of the Sustainable Engineering degree program closer to several Rwandan universities. The professor in charge again states that "Furthermore, parts of it should in turn make it possible to offer English-language courses for visiting students in the department."

Several lecturers also hold out the prospect of giving the lecture in English or recording it in English if students who do not speak German want to take part in the modules. After having presented these results, the department now starts an initiative for more internationalization and there are further translation requests in between. In some modules that deal with topics that require constant updating, rework will be permanently necessary in the future to keep the English-language documents up to date. Depending on the scope, this will be done by the lecturers themselves.

## 7 Conclusion

Overall, the project *Translation of teaching materials into English* can be rated as very successful. Thanks to the translations, international students can now be offered additionally more than 15 bachelor's modules and nine master's modules in the degree programs of the faculty.

The translations could be produced in high quality by students and with on-board resources, and the workflow worked out very well. The glossaries created can be used also for future translation projects.

Within the framework of the project, materials from a total of 22 modules as well as the module handbook of a bachelor's degree program were translated. As quality management, lecturers whose materials were translated were asked for feedback, resulting in nine out of 10 quality assessments. 7 lecturers rated the quality of the translation as very good. The exchange about translation of certain technical terms was also rated as very useful. Two of the lecturers rated the quality of the translations as good. Most of the lecturers stated they would use the materials in a hybrid format by teaching in German language and providing the teaching materials in English.

Now that universities are on their way to establish foreign and multilingual courses, the current legislation stands ready to support the rights and obligations of the actors involved [4, p. 6]. Thus, the project work done contributes not only to the aims discussed but also brings the important topic of multilingualism in higher education into the focus of universities and politics [4, p. 3]. The authors are willing to promote these developments also in the future and in the next step hope to report on multilingual experiences in teaching.

## References


1. Das Präsidium der Hochschule Bonn-Rhein-Sieg: Hochschulentwicklungsplan 2021 bis 2025 [Online]. https://www.h-brs.de/sites/default/files/related/h-brs_broschuere_hep3_bf.pdf. Accessed 20 May 2023
2. Fonseca, P., Julian, K., Hulme, W., Martins, M.D.L., Brautlacht, R.: The multi-disciplinary approach to an interdisciplinary virtual exchange. In: Satar, M. (ed.) Virtual Exchange: Towards Digital Equity in Internationalization, pp. 41–49. Research-publishing.net (2021). https://doi.org/10.14705/rpnet.2021.53.1288
3. Decke-Cornill, H.: We would have to invent the language we are supposed to teach': the issue of English as Lingua Franca in language education in Germany. Lang. Cult. Curric. **15**(3), 251–263 (2002). https://doi.org/10.1080/07908310208666649
4. HRK Advance. Governance und Prozesse der Internationalisierung optimieren. Rechtliche und organisatorische Rahmenbedingungen fremd- und mehrsprachiger Studiengänge. Handreichung [Online]. https://www.hrk.de/themen/internationales/internationalisierung-in-lehre-und-forschung/hrk-advance-governance-und-prozesse-der-internationalisierung-optimieren/. Accessed 20 May 2023
5. DeepL SE. https://www.deepl.com/. Accessed 20 May 2023

# The Performance Evaluation of E-learning During the Emergency Using Machine Learning

Hosam F. El-Sofany[1,2] and Samir A. El-Seoud[3(✉)]

[1] King Khalid University, Abha, Kingdom of Saudi Arabia
helsofany@kku.edu.sa

[2] Cairo Higher Institute for Engineering, Computer Science and Management, Cairo, Egypt

[3] Faculty of Informatics and Computer , British University in Egypt (BUE), Cairo, Egypt
samir.elseoud@bue.edu.eg

**Abstract.** E-learning is one of the educational alternatives available to students who need assistance during an emergency. (e.g., Covid-19 pandemic, bad climate, etc.). Most educational institutions are moving a significant portion of their curriculum toward an online learning paradigm to reduce the amount of face-to-face interaction between students and faculty members during times of emergency (e.g., in the case of Covid-19 pandemic). The success of E-learning is conditional on a wide range of aspects, such as students' and teachers' levels of self-efficacy, attitudes toward, and confidence in making use of the relevant technology; the instructional approaches that are utilized; the capacity to monitor and evaluate educational outcomes; and students' levels of motivation. The performance and circumstances of students who are engaged in e-learning are analyzed in this paper. The research investigates and evaluates predictions made by a model that is based on machine learning techniques. Predicting the degree to which students are delighted with the online mode of instruction by considering several parameters, including internet capability, and involvement in the online mode of instruction.

**Keywords:** E-learning · Education in an emergency · Learning outcomes · Machine learning (ML)

## 1 Introduction

Students have no option but to take E-learning courses during times of emergencies (i.e., bad climate, spread of pandemics, etc.). Students and teachers alike have had to adapt to new challenges brought about by the shift from traditional classrooms to online education. How did students feel about this unexpected use of e-learning? How did they find the time to study? In what ways did students feel that this sudden education met their needs? The purpose of this research is to discover how students feel about using E-learning in times of emergencies.

E-learning offers numerous benefits, the most notable of which are adaptability, portability, individualization, consistency, and scalability. Learning something through the internet can save you both time and money, and it also makes it possible to obtain educational information and materials whenever and wherever you want [1].


M. E. Auer et al. (Eds.): ICL 2023, LNNS 899, pp. 490–498, 2024.
https://doi.org/10.1007/978-3-031-51979-6_51

E-learning has been given the ability to achieve success on a variety of different platforms because of advances in technology. E-learning can be made more engaging and accessible thanks to the many features offered by these different platforms. Students can participate in group activities and demonstrate how they are feeling using platform-specific emoticons even when they are not in the same room as their teachers. Researchers from all over the world are becoming increasingly concerned about the impacts that learning through the Internet has on students' mental health. This is the case even though the E-learning education system brings with it convenience and adaptability. Professors and instructors have to deal with increased workloads and the expectation of delivering quality education without face-to-face interaction, while students are required to spend hours every day in front of platforms on the internet without the opportunity for social interaction or playtime [2]. Individuals who participate in lengthy online courses or video conference sessions may end up feeling exhausted, and information overload may also lead to exhaustion. This may cause nervousness and tension [3]. A person's mental health may also be negatively affected by variables such as a lack of group contact and social seclusion. The online educational approach also disrupts a person's daily schedule by placing a physical distance between them and their family and friends. Home distractions are a common reason why students don't get their work done on time and why they struggle to focus while studying. The general online learning approach may also affect other family members who are already stretched thin between work and helping their children with their schoolwork. E-learning may influence people's emotional health, but it also has the potential to improve the connections and relationships within families [4]. Students who, in traditional classroom settings, are more likely to be bullied can benefit greatly from engaging in e-learning activities. Productivity is increased when one feels secure in their own house. It is essential to analyze and evaluate the effect that E-learning has on most of the audience to keep in mind the benefits and drawbacks that come along with it [5].

When schools and universities are closed due to an emergency, e-learning is a highly efficient means of maintaining students' education. It is important to understand if E-learning will be effective in the long run, or if it can be completely adopted in the long term because the peak in the number of cases is not constant during the emergency and fluctuates across the year, which could necessitate conducting online classes more frequently. The purpose of conducting a psychometric study on a person is to obtain quantitative information about their psychological traits and characteristics. Perspectives that consider all aspects of a person's life may shed light on their personality and mental health. Students can learn a great deal about their skills and areas for improvement by taking these exams, and they can also be used to gain insight into an individual's mental health. To solve these problems, a quantitative study of the effect of E-learning on pupils' psychological health is required. The following are the major concepts that we analyze and integrate in this research; these are also the study's most significant accomplishments: (1) Evaluating the academic standing and health of students through the medium of E-learning, (2) Evaluating combination model predictions and proposing layered models, (3) To what extent students are happy with the e-learning format can be predicted based on variables such as the availability of the internet and their ability to communicate in an online setting.

The purpose of this paper is also to give an analysis of how E-learning compares to more traditional methods of education in times of emergency. (e.g., Covid-19 pandemic, bad climate, etc.). This research shows promise for the use of E-learning platforms to achieve learning goals through the utilization of ML techniques.

## 2 Methodology

This section provides a concise explanation of the methodology and stages involved in the process of predictive modeling applying ML techniques. As a result, a detailed description of the dataset, models, and the proposed system will be provided in the subsequent sections [6].

### 2.1 ML Classifies

- *XGBoost*: XGBoost (i.e., Extreme Gradient Boosting) is an ensemble machine learning technique that is based on decision trees (DT) and makes use of the gradient boosting framework. During the training phase, XGBoost makes use of multiple decision trees for the prediction of a variety of tasks, including regression and classification. The most important characteristic of XGBoost is that it has a regularization mechanism to prevent excessive fitting and uses a parallelized implementation to approach the sequential tree-building process to decrease the amount of computing required [7]. When applied to tabular data in several industries, such as retail banking, the turnover of customers prediction, and other areas, XGBoost is frequently utilized to provide comparative findings that are outstanding to those produced by other algorithms. The most important settings for XGBoost are the *n* estim*ators'* parameter, which specifies the number of trees that should be formed, and the *max_depth* parameter, which specifies the criterion for when tree pruning should cease [8]
- *Random Forest*: It is an algorithm based on DT that is both supervised and ensemble based. When Random Forest is trained, it generates many DTs, or a forest, on data samples. After that, it obtains the prediction from each of the trees individually, and to determine the final prediction, it applies a voting or bagging strategy on these individuals. Random Forest is a collective learning method [9]. When it is trained, it creates a forest. It accomplishes this by integrating several somewhat ineffective learning algorithms to reach better levels of prediction and accuracy. The primary distinction between DTs and random forests is that choice trees take into account each feature and row, whereas random forests choose characteristics and rows at random to construct numerous trees and then take the average of the resulting trees' predictions [10]. The variance is reduced, and more accurate predictions are produced by random forests since they include the construction of numerous trees. It is frequently used in the resolution of difficult regression and classification issues. The output of the model is mostly determined by the primary parameters, which are the number of trees and the depth of the trees [11].
- *Gradient Boosting*: It's a top method for developing precise prediction models. The CART (classification and regression trees) base learner is at the heart of the gradient-boosting augmentation technique. With gradient boosting, each new predictor is

trained using the residual errors from the previous predictor's predictions as labels. To put it another way, the algorithm uses past errors to try to make better decisions in the future. The term *boosting* describes this strategy, which is where "Gradient boosting" comes from. Gradient boosting relies heavily on a regularization parameter known as *shrinkage* Model performance may also be greatly enhanced by adjusting the values of *n estimators* and *Max-depth* [12].

- *Catboost*: Yandex created category boosting. Catboost builds on Gradient boosting and refines it. Catboost is the only boosting algorithm with a low prediction time. It predicts 8x quicker than XGBoost because of its symmetrical tree architectures. Catboost handles categories automatically. Instead of one-hot encoding categorical data, it employs ordered target statistics. It's employed in regression analysis, classification, searches, recommendation systems, etc. [13].

### 2.2 Datasets

The dataset that was utilized in this experiment was collected from the replies of students ranging from a variety of age groups and having completed a variety of educational institutions. The data included a total of 23 features, and there were around 950 records that each had their unique characteristics. Each of the properties was assigned a specific type, either an integer, a float, or a category variable. Students at a variety of educational levels, from elementary school to doctoral programmers, are represented in this dataset. It also includes some of the most important characteristics, such as Internet facility, Interaction in online mode, Performance online, and other characteristics that are necessary for assessing whether a student is satisfied. Student level of satisfaction in online mode is the target variable that will be used to measure student satisfaction. This variable will have three levels: Bad, Average, and Good. The dataset includes enough samples from both male (%55) and female (%45) students, showing that there is no imbalance in terms of gender [14].

### 2.3 Data Sampling and Analysis

In this work, we have performed further pre-processing on the selected dataset by using the methods shown in Fig. 2. We evaluated the percentage of student satisfaction in online mode during the emergency to be 87.5% for *Good*, 8.5% for *average*, and 4.0% for *Bad* responses. The following are some inferences that may be drawn from the data:

- Students who perform well in traditional classroom settings are often unimpressed by the online learning environment, whereas those who perform poorly are at ease there.
- Students who are averagely content with online mode have given themselves ratings between 6 and 8, students who are dissatisfied with online mode have given themselves ratings usually below 4, and students who are good with online mode have given themselves ratings above 8.

The correlation between features gives us a few important details regarding the relationship between the features and their effect on the target variable. Figure 1 illustrates a heatmap depicting the correlation between individuals. The heatmap demonstrates explicitly that there is a positive association between interaction in online mode, resolving doubts with faculties, and performance in online mode together with pleasure in online mode. This implies that students' engagement while they are online encourages them to do better in online mode. It has been shown that a student's level of happiness with the online method of education is closely correlated to how well they perform in the online mode of education. This suggests that students who do well in the online mode of education are satisfied with the online style of education [15].

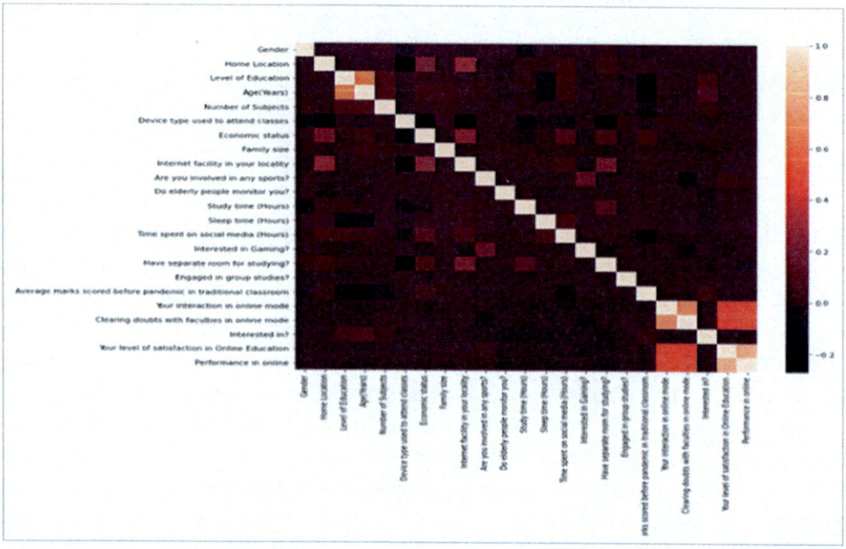


**Fig. 1.** Heatmap histogram showing the correlation between individuals.

## 3 The Proposed Model

The widespread use of online education during times of emergency has become the norm. These emergencies include natural disasters, the spread of disease, and other catastrophic events. As a result of the epidemic, countries have moved their educational systems online rather than relying on the more conventional offline methods. Meanwhile, the level of satisfaction among students with their online education is the primary focus of attention. If students are unhappy and uneasy with this approach to learning, then the knowledge gained via it will be subpar [16].

Therefore, our research proposes a model, represented in Fig. 2, that is divided into four distinct parts. Phase one involves gathering and cleaning up data provided by students from a wide range of educational institutions. In the second round of feature engineering, we encoded, found outliers, and normalized the data with the *MinMax* scaler. There are both ordinal and categorical variables in the data collection. Label encoding was performed on columns with ordinal characteristics to convert them to numbers, as this is a requirement of the ML model. Target encoding was conducted, where the average target value was substituted for the categories in the remaining category columns. For working with interval values, a unique function was developed, which takes an interval and returns an average of its lower and upper bounds. The aim variable was a satisfaction scale from 0 to 2 with the categories *Bad*, *Average*, and *Good* represented. As a result of their strong performance and resistance to collinearity, numerous Tree-based models were trained in the third modeling phase. Following data cleaning and feature engineering, the dataset was divided 75:25 between a train set and a test set [17].

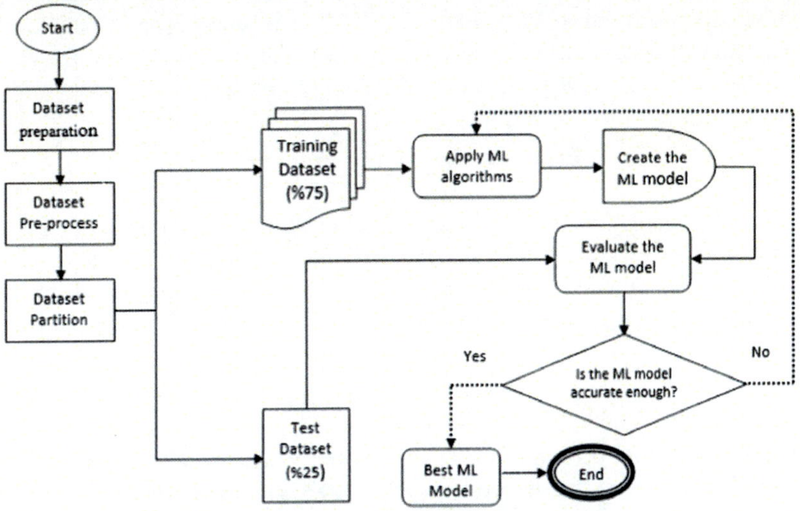


**Fig. 2.** The proposed model for prediction

Using the K-Fold function in the sklearn package, we cross-validate our model by dividing the training dataset into five equally sized parts. Initially, Xgboost, Random Forest, Catboost, and Gradient boosting were each trained on the training data. Model stacking is used to better enhance predictions on the available data. Model stacking is a strategy for enhancing model predictions by combining the output of numerous models and giving that combined output as input to a final model that generates the predictions. This last model is typically referred to as a Meta-learner. Model stacking is preferable to using a single model because it compensates for the shortcomings of the underlying models while amplifying their strengths. The final model from model stacking is reliable and applies well to novel data. The stacking of models is shown in detail in Phase 3 of the accompanying diagram [18]. In the fourth phase, the data are analyzed and visualized so that the predictions may be comprehended in greater precision.

## 4 Results and Discussions

The proposed model's key objective is to predict, with high accuracy, the learner's degree of satisfaction with their online learning experience considering a wide range of variables, including but not limited to their access to the internet, the amount of time they devote to studying, and the like. This means that, besides the F1 score, precision is an important criterion to consider. As a result, we used four metrics—accuracy, precision, recall, and F1 score—to assess how well the final model performed. In the equations, we use the symbols TP for a *True Positive*, TN for a True Negative, FP for a *False Positive*, and FN for a *Negative False*.

## 4.1 Accuracy, Precision, Recall, and F1 Score

- The accuracy of a prediction is measured as the proportion of correct predictions (TP + TN) to all observations (TP + FP + FN + TN). Accuracy is an essential metric since it shows how accurate our projected value is to the trustworthy one. To evaluate the efficacy of our Accuracy, we use the following formula:

$$\text{Accuracy} = \frac{TP + TN}{TP + TN + FP + FN}$$

- The term "precision" refers to the ratio of "True Positive" (TP) points to the total number of "True Positive" and "False Positive" points. When we wish to determine the total number of points in the dataset that have been correctly categorized, precision becomes an important factor. The following formula is used to calculate precision:

$$\text{Precision} = \frac{TP}{TP + FP}$$

- The proportion of properly predicted positive observations relative to the sum of all correctly predicted positive observations (TP + FN) is referred to as the recall. When we wish to determine the number of points that were properly categorized out of all the points that might have been accurately predicted in an unbalanced dataset, recall becomes essential because we want to know how many points we got right. The recall is determined using the following formula:

$$\text{Recall} = \frac{TP}{TP + FN}$$

- When making a categorization, Precision and Recall are constantly at odds with one another. So, we need the best possible metric by which to judge a model's efficacy. The F1 Score is the number that is the harmonic mean of the Recall and Precision values, and it is often regarded as the ideal assessment measure to employ when there is an asymmetry in the dataset. The formula for the F1 Score is as follows:

$$\text{F1 Score} = 2 \times \frac{Precision \times Recall}{Precision + Recall}$$

All the separate models were trained using the preprocessed data and the new data we provided. There should be more metrics besides accuracy used to evaluate the model's efficacy given that the dataset utilized has an imbalance in classes. As a result, the F1 score has been our primary focus. It is possible to see how Single models do in terms of performance in Fig. 3.

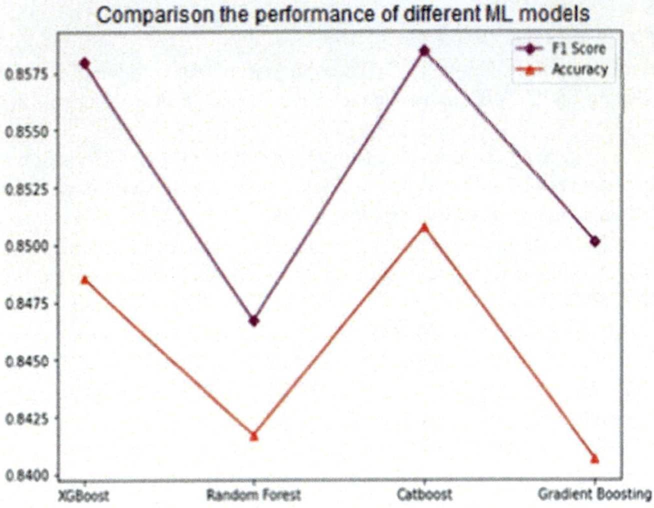


**Fig. 3.** The ML techniques evaluation

All the individual models were trained by applying the preprocessed data to them, in addition to the fresh data that we supplied. Given that the dataset that was utilized contains an uneven distribution of classes, there ought to be more criteria besides accuracy that are used to evaluate the usefulness of the model. Because of this, the F1 score has been the major concentration of our efforts. Figure 3 demonstrates how the proposed model fare in terms of performance, and it is feasible to examine this information.

## 5 Conclusion

The applications of machine learning as well as the amount of data are quite helpful in any industry. The data that is retrieved from the educational sector because of the online mode may be utilized for analysis and the discovery of insights into the performance of students, as well as for the improvement of students' performances. In this article, we presented a model stacking system that received an F1 score of 0.864 on the dataset that was gathered. According to the findings of this research, a large percentage of students take online classes and report feeling neither completely happy nor unsatisfied with their experience. Therefore, a significant number of the students who are impacted are between the ages of 19 and 23, and as a result, university sessions may be held offline, and if students need further information, they can choose to study online material. This creates the ideal balance. As was said before, future work will concentrate on gathering data from a variety of areas and educational backgrounds under a variety of conditions, analyzing students' levels of satisfaction and performance in hybrid modes of instruction (both online and offline), and developing fully-fledged pipelines for modeling.

## References


1. Chakraborty, P., Mittal, P., Gupta, M.S., Yadav, S., Arora, A.: Opinion of students on online education during the COVID-19 pandemic. Hum. Behav. Emerg. Technol. **3**(3), 357–365 (2021)
2. Zhang, N., Liu, Q., Zheng, X., Luo, L., Cheng, Y.: Analysis of social interaction and behavior patterns in the process of online to offline lesson study: a case study of chemistry teaching design based on augmented reality. Asia Pac. J. Educ. 1–22 (2021)
3. Nartiningrum, N., Nugroho, A.: English teachers' perspectives on challenges, suggestions, and materials of online teaching amidst the global pandemic. IJEE (Indones. J. Engl. Educ.) **1**(1), 108–126 (2021)
4. Moise, D., Diaconu, A., Negescu, M.D.O., Gombos, C.C.: Online education during pandemic times: advantages and disadvantages. Eur. J. Sustain. Dev. **10**(4), 63 (2021)
5. Rietveld, J.R., Hiemstra, D., Brouwer, A.E., Waalkens, J.: Motivation and productivity of employees in higher education during the first lockdown. Adm. Sci. **12**(1), 1 (2022)
6. Saleem, F., Ullah, Z., Fakieh, B., Kateb, F.: Intelligent decision support system for predicting student's E-learning performance using ensemble machine learning. Mathematics **9**(17), 2078 (2021)
7. Chen, T., Guestrin, C.: Xgboost: a scalable tree boosting system. In: Proceedings of the 22nd ACM SIGKDD International Conference on Knowledge Discovery and Data Mining, pp. 785–794 (2016)
8. Tang, Q., Xia, G., Zhang, X., Long, F.: A customer Churn prediction model based on XGBoost and MLP. In: 2020 International Conference on Computer Engineering and Application (ICCEA), pp. 608–612. IEEE (2020)
9. Massaro, A., Panarese, A., Giannone, D., Galiano, A.: Augmented data and XGBoost improvement for sales forecasting in the large-scale retail sector. Appl. Sci. **11**(17), 7793 (2021)
10. Bentéjac, C., Csörgő, A., Martínez-Muñoz, G.: A comparative analysis of gradient boosting algorithms. Artif. Intell. Rev. **54**(3), 1937–1967 (2021)
11. Breiman, L.: Random forests. Mach. Learn. **45**(1), 5–32 (2001)
12. Friedman, J., Hastie, T., Tibshirani, R.: Additive logistic regression: a statistical view of boosting (with discussion and a rejoinder by the authors). Ann. Stat. **28**(2), 337–407 (2000)
13. Pathak, M., Jain, A.: Solving Fashion Recommendation—The Farfetch Challenge (2021). arXiv:2108.01314
14. Sujatha, R.: Online Education System – Review, Ver. 1, December 2021. https://doi.org/10.17632/bzk9zbyvv7.1. https://data.mendeley.com/datasets/bzk9zbyvv7/1
15. Iwendi, C., et al.: COVID-19 patient health prediction using boosted random forest algorithm. Front. Public Health **8**, 357 (2020)
16. Pokhrel, S., Chhetri, R.: A literature review on impact of COVID-19 pandemic on teaching and learning. High. Educ. Future **8**(1), 133–141 (2021)
17. Pargent, F., Pfisterer, F., Thomas, J., Bischl, B.: Regularized target encoding outperforms traditional methods in supervised machine learning with high cardinality features. Comput. Stat. 1–22 (2022)
18. Fayaz, M., Khan, A., Rahman, J.U., Alharbi, A., Uddin, M.I., Alouffi, B.: Ensemble machine learning model for classification of spam product reviews. Complexity (2020)

# Integrating Artificial Intelligence and ChatGPT into Higher Engineering Education

Horia Alexandru Modran[1](✉), Tinashe Chamunorwa[1], Doru Ursuțiu[1,2], and Cornel Samoilă[1,3]

[1] Transilvania University of Brașov, Brașov, Romania
{horia.modran,chamunorwa.tinashe,udoru,csam}@unitbv.ro
[2] Romanian Academy of Scientists, Bucharest, Romania
[3] Romanian Academy of Technical Sciences, Bucharest, Romania

**Abstract.** Artificial Intelligence (AI) has the potential to revolutionize the field of engineering education by providing personalized and on-demand support to teachers and students. The current paper analyses the ways that AI can be used in engineering education, pointing out both the benefits it can bring to students and teachers and the potential outcomes of its ethical usage. By reviewing the literature on the use of AI in education, with a focus on engineering education, and providing case studies and practical experiments to illustrate its practical applications in the educational process, the goal is to provide insight into the benefits of using AI in engineering education. Chat GPT can be a valuable tool in higher engineering education, providing students with personalized support and enhancing the learning experience. Limitations and cases of misuse of ChatGPT were also identified, like bias, cheating, plagiarism, and inaccuracies. It is also very important to emphasize that ChatGPT's responses should be used as supplementary information and not a substitute for physical courses or hands-on laboratories. A few experiments were performed to demonstrate the possible usage and capabilities of Artificial Intelligence for engineers. This paper describes one of them, carried out using only graphical programming environments (LabVIEW with DeepLTK from Ngene).

**Keywords:** AI · ChatGPT · PSoC6 · LabVIEW · DeepLTK · Ngene

## 1 Introduction

Artificial Intelligence (AI) has the potential to revolutionize the field of engineering education by providing personalized and on-demand support to learners. With the increasing complexity of engineering concepts and the need for a skilled workforce, incorporating AI into engineering education can help address the field's challenges.

A comprehensive review of relevant research articles was conducted to explore the potential use of AI in engineering education. The case studies of AI applications in engineering education were also considered, to illustrate the practical applications of AI in the educational process. The review is focused on the various ways AI can be used in engineering education, such as providing personalized support, generating learning materials, providing feedback on student work, and supporting collaborative learning.


M. E. Auer et al. (Eds.): ICL 2023, LNNS 899, pp. 499–510, 2024.
https://doi.org/10.1007/978-3-031-51979-6_52

The application of Artificial Intelligence in education is an emerging interdisciplinary field that applies AI algorithms in education to improve the instructional and learning environment [1]. AI-aided education includes intelligent education, innovative virtual learning, and data analysis and prediction. In AI learning system, learner model is critical for improving independent learning capabilities. It is established based on behavior data of learners generated from the learning process [2]. Intelligent education systems provide timely and personalized instruction and feedback for both instructors and learners. They are designed to improve learning efficiency and value by multiple computing technologies, especially machine learning (ML) related technologies [3].

An AI educational platform usually consists of intelligent (machine learning) algorithm and teaching contents. With the help of personalized learning, the students choose the topics they are interested in, while the teachers can use those answers to adjust the teaching materials and courses [4]. In another study, Sharma et al. [5] observed that AI in education has taken the form of intelligent tutoring systems, adaptive learning systems, and other systems that improve the quality of the educational process.

Achieving a good quality STEM education requires careful consideration of several factors (educational, environmental, social), rather than simply applying AI technologies in education [6]. AI can contribute to revolutionizing the ongoing educational paradigm. As AI educational solutions continue to evolve and improve, they may help solve some challenges in the educational processes and adapt to every student's learning capacity [7]. Each student has a different pace of learning and AI provides a possibility to avoid this problem through personalized learning [8].

The same research team already proposed in a previous study a teaching practice to teach engineers all the necessary steps for developing, validating, and deploying machine learning-based systems [9]. The approach was based on learning by examples and by comparisons with human intelligence, but, however, since then, the use of Artificial Intelligence in education has seen a great advance using generative pre-trained machine learning models, like ChatGPT [10].

While the research on the use of AI technologies in education is a topic addressed in several studies, the application of ChatGPT specifically in educational contexts is relatively new. ChatGPT, developed by OpenAI, has shown remarkable capabilities in generating human-like responses and engaging in natural language conversations. As there have multiple possible uses of ChatGPT in educational scenarios, including tutoring, content generation, and student support, there has a huge potential for ChatGPT to enhance the learning experience by providing personalized assistance, answering questions, and promoting a deep understanding of complex topics.

The literature reviewed demonstrates the potential benefits of integrating AI technologies, such as ChatGPT, in educational settings. This paper analyses the ways that AI can be used in engineering education, pointing out both the benefits it can bring to students and teachers and the potential outcomes of its ethical usage. A few experiments were performed to demonstrate the possible usage and capabilities of Artificial Intelligence. This paper describes one of them, carried out using only graphical programming environments (LabVIEW with DeepLTK from Ngene).

## 2 Using AI in Education

### 2.1 Teachers' Perspective

In recent years, the integration of Artificial Intelligence (AI) in Engineering Education has emerged as a promising field to enhance student learning and engagement. AI technologies offer unique opportunities to address the challenges faced by engineering teachers, such as the constantly changing learning curriculum, diverse student needs, evaluation methods, and the demand for practical hands-on experiences. This section explores the potential applications of AI and ChatGPT in improving the course and teaching, being exemplified and applied in a case study - the Virtual Instrumentation course.

Virtual Instrumentation refers to the use of software and hardware tools to simulate and replicate real-world engineering instruments and systems [11]. It provides a hands-on experience for students to acquire and analyze data, and to gain practical skills. By integrating AI into Virtual Instrumentation, engineering educators can augment learning and provide a more interactive, personalized, and adaptive educational experience.

First, the teachers can use ChatGPT to create a learning plan in accordance with the latest technological developments.

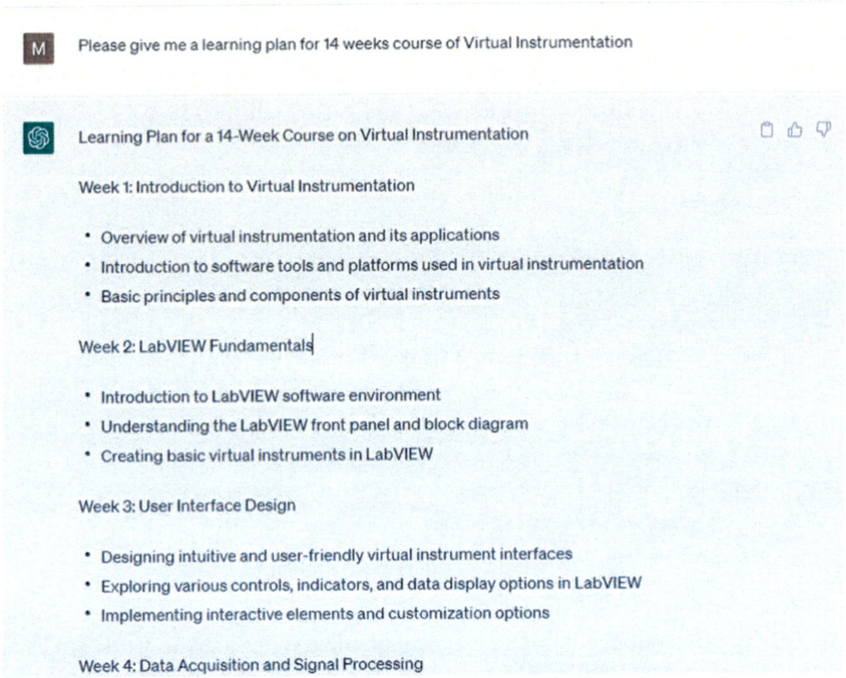


**Fig. 1.** Learning plan for the virtual instrumentation course.

Figure 1 illustrates an extract from the learning plan for a 14-week course in Virtual Instrumentation. As ChatGPT mentions at the end of the proposed solution, the learning plan can be customized based on the specific needs and objectives of the course and can be adjusted to accommodate additional topics or hands-on sessions, if required. The proposed plan includes most of the basic notions necessary for engineers to be able to gain the skills required by companies, such as basic principles of Virtual Instrumentation, designing User Interfaces, data acquisition and simulation, Real-Time and FPGA Systems, etc. Therefore, this can be used by both teachers for the course they teach, as well as by students for individual study.

Another way that ChatGPT can be used by teachers is to compose possible examination subjects for both the written exam and laboratory (practical) exam. ChatGPT has the advantage of remembering the previous answers and, therefore, it can provide evaluation subjects based on the content from the previous step. Figure 2 shows a sample exam for hands-on usage of Virtual Instrumentation using the LabVIEW programming environment, while in Fig. 3 a possible topic for the written examination.

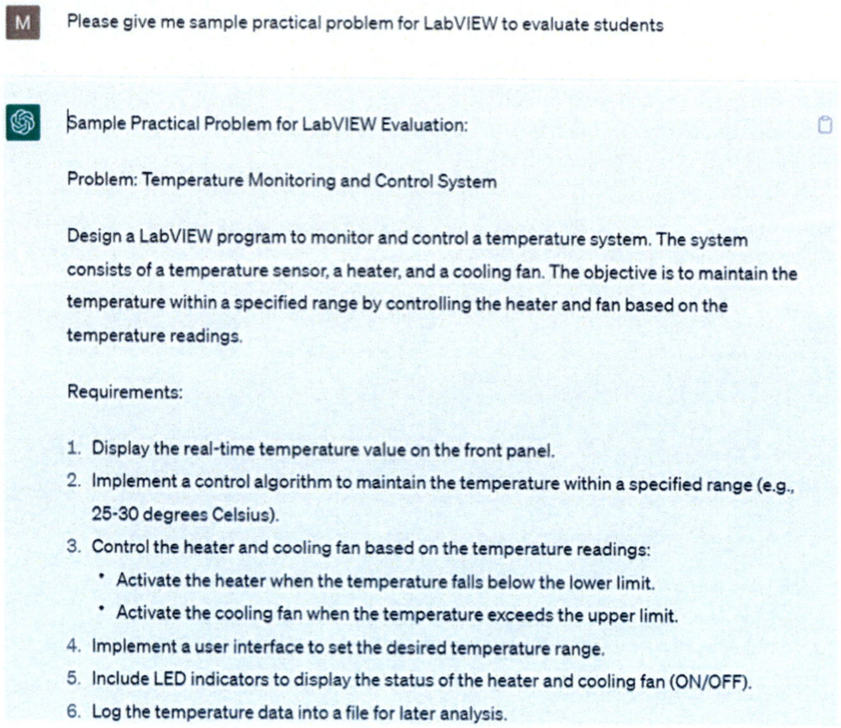


**Fig. 2.** Sample practical exam problem.

As it can be observed in Fig. 2, this practical problem is designed to assess students' understanding of several LabVIEW basic programming concepts, like data acquisition and control, user interface design, and file I/O functions. However, based on the teaching experience, this evaluation can be further customized based on the instructor on their specific learning objectives of the course, as well as depending on the assimilation capacity of the students observed throughout the whole semester.

These sample questions generated by ChatGPT for the written exam (Fig. 3), cover almost all aspects of virtual instrumentation and LabVIEW programming that were already proposed in the learning plan. The complexity and number of the questions that will be included in the final examination remain at the discretion of the teacher. It is also strongly recommended that the grading and the conditions for promoting the exam will be clearly and explicitly defined by the teacher and communicated to the students in advance.

M Thank you for previous response, please give me as well some sample questions for written examination for same course.

Sample Written Examination Questions for Virtual Instrumentation Course:

1. Define virtual instrumentation and explain its significance in the field of engineering.
2. Compare and contrast virtual instruments with traditional hardware-based instruments. Discuss the advantages and limitations of virtual instrumentation.
3. Describe the components of the LabVIEW development environment and their functions.
4. Explain the concept of data acquisition in virtual instrumentation. Discuss the types of sensors and interfaces commonly used for data acquisition.
5. Discuss the importance of user interface design in virtual instrumentation. Explain the principles and guidelines for designing effective user interfaces.
6. Describe the steps involved in developing a virtual instrument using LabVIEW. Provide a brief explanation of each step.
7. Discuss the various methods of instrument control in virtual instrumentation. Explain the advantages and limitations of each method.
8. Explain the role of LabVIEW in real-time and FPGA-based virtual instrumentation. Discuss the applications and advantages of real-time and FPGA platforms.
9. Discuss the concept of calibration in virtual instrumentation. Explain the importance of instrument calibration and the techniques used for calibration.
10. Describe the process of logging and analyzing data in virtual instrumentation. Explain the methods and tools used for data logging and analysis.

**Fig. 3.** Written examination sample questions.

## 2.2 Students' Perspective

This subchapter investigates the potential of utilizing ChatGPT as a helpful resource to support students in the Virtual Instrumentation course.

While Virtual Instrumentation offers numerous advantages in terms of flexibility, scalability, and cost-effectiveness, students often encounter challenges in comprehending complex concepts and troubleshooting issues that arise during their learning and practice. These challenges can hinder their learning experience and limit their ability to apply virtual instrumentation effectively in practical scenarios. Therefore, there is a need to explore innovative approaches to enhance the learning process and support students in overcoming these difficulties.

By integrating ChatGPT into the learning environment, the aim is to enhance students' understanding of virtual instrumentation concepts, provide conceptual clarification, and assist them with troubleshooting issues. The goal is to empower students with an efficient and interactive tool that can aid their learning and problem-solving skills throughout the course.

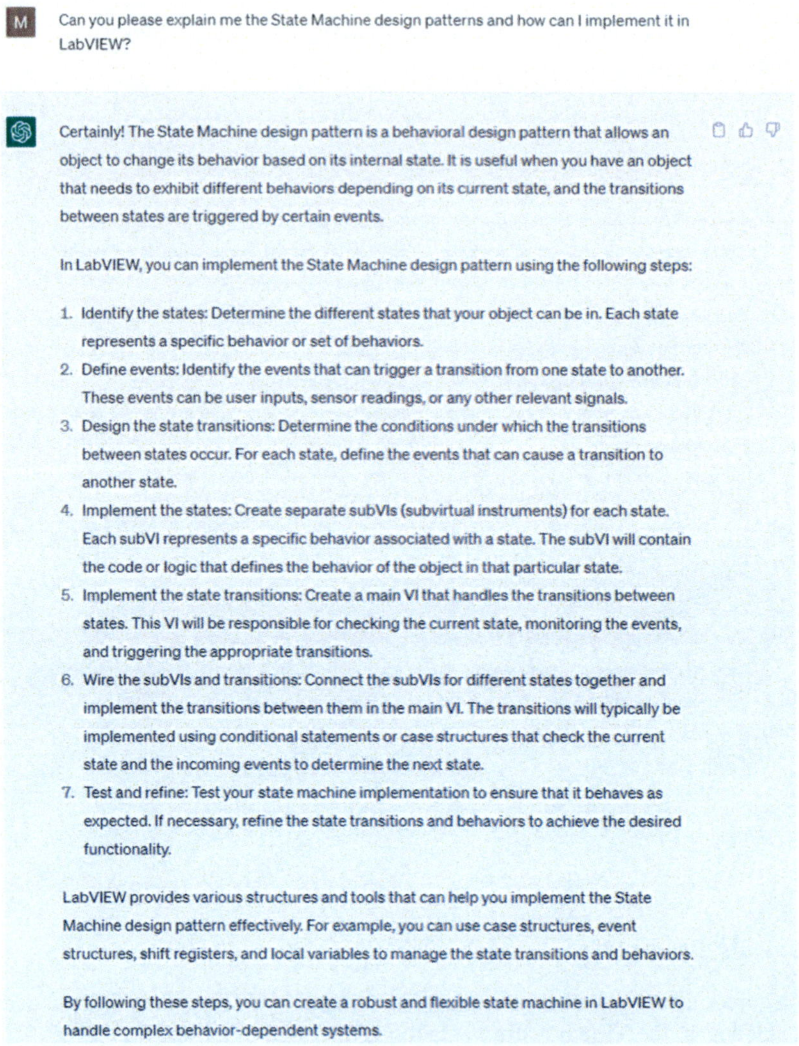


**Fig. 4.** Using ChatGPT to explain concepts.

Students can use ChatGPT to access information from a vast knowledge base, including textbooks, lecture materials, and online resources. They can ask questions about specific concepts, definitions, or examples related to Virtual Instrumentation and the LabVIEW Programming Environment. Figure 4 shows the detailed explanations given by ChatGPT to a sample question, helping students to grasp complex topics more effectively. Considering that the answer is very comprehensive and explains in detail each step for implementing the State Machine design pattern, ChatGPT can be used by the students as a knowledge resource to reinforce their understanding and retain new information for future use.

During the Virtual Instrumentation practical laboratories, students may encounter technical issues, bugs, or errors that require troubleshooting. ChatGPT can assist the students in diagnosing problems, identifying potential solutions, and guiding them through the debugging process. By describing the issue to ChatGPT, students can receive step-by-step instructions and suggestions for troubleshooting approaches. As can be seen in Fig. 5, ChatGPT can also offer insights into common causes for an error and suggest debugging strategies, empowering students to effectively resolve issues and enhance their problem-solving skills.

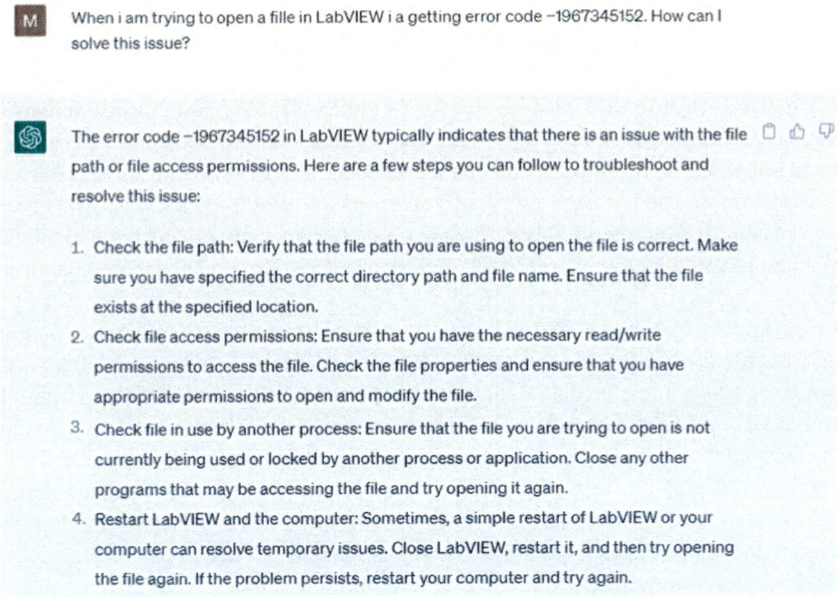


**Fig. 5.** Debugging in LabVIEW with ChatGPT.

Furthermore, ChatGPT has the potential to support adaptive learning and personalization in the Virtual Instrumentation course. By analyzing student interactions, ChatGPT can tailor its responses and learning materials to meet individual learning capacities and preferences. It can also help the student to track his own progress, identify knowledge gaps or even to evaluate himself. This personalized approach enhances the student's engagement, motivation, and overall learning outcomes.

Figure 6 illustrates a sample problem proposed by ChatGPT for self-evaluation of the LabVIEW basic concepts. However, it cannot evaluate the Virtual Instrument developed in LabVIEW, but it can provide answers and tips if the student has any questions.

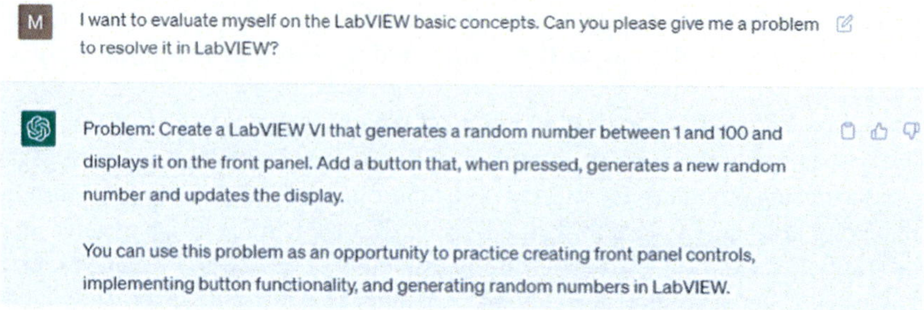


**Fig. 6.** LabVIEW problem for self-evaluation.

## 3 Limitations and Unethical Usage of ChatGPT

While ChatGPT can be a valuable resource for the students in the Virtual Instrumentation course and laboratory, it is very important that teachers should be aware of potential unethical uses and find ways to prevent them. Students should use ChatGPT in the Virtual Instrumentation course and laboratory only in an ethical and integrity way. Therefore, instructors and institutions should emphasize the importance of responsible and ethical use of AI tools, providing guidelines and support to ensure a fair and transparent learning process.

As the answers given by ChatGPT are difficult to be detected by traditional plagiarism software, students may attempt to cheat by asking ChatGPT to generate lab reports, experimental procedures, or data analysis without conducting the actual experiments themselves. They might misuse ChatGPT by requesting detailed solutions to lab assignments. This practice inhibits individual learning and prevents students from developing their problem-solving and analytical skills. However, the instructor can detect and prevent this kind of situation by asking students to explain and demonstrate all the steps for conducting the actual experiment or for solving the lab assignment.

Students may rely only on ChatGPT to interpret experimental data they acquired, bypassing their own understanding and analysis. This can lead to incorrect conclusions or misinterpretations, affecting the integrity of the experimental results. To avoid this, teachers can structure assignments and projects in a way that encourages students to progress through different stages and, by providing incremental feedback for each stage, teachers can assess the authenticity and originality of students' work.

Taking these aspects into consideration, the instructors should design assessments that require students to apply their knowledge, skills, and critical thinking abilities, making it difficult for them to rely solely on ChatGPT for solutions or answers. By incorporating practical and real-world scenarios, students are encouraged to demonstrate their understanding and problem-solving skills.

Nevertheless, ChatGPT has certain limitations that should be considered for this case study, like the inability to perform physical experiments, lack of domain-specific knowledge, limited context understanding, and potential for incorrect or biased responses. By being aware of these limitations, students and teachers can leverage the strengths of ChatGPT while supplementing it with hands-on experimentation, in-depth domain knowledge, and critical thinking to enhance the overall learning experience.

An example of incorrect information given by ChatGPT is given when it is asked to recommend articles or bibliography for a specific topic – in this case, most of the articles and references suggested don't even exist (Fig. 7). Since ChatGPT cannot generate a proper bibliography, the reports or articles generated by it can be easily identified.

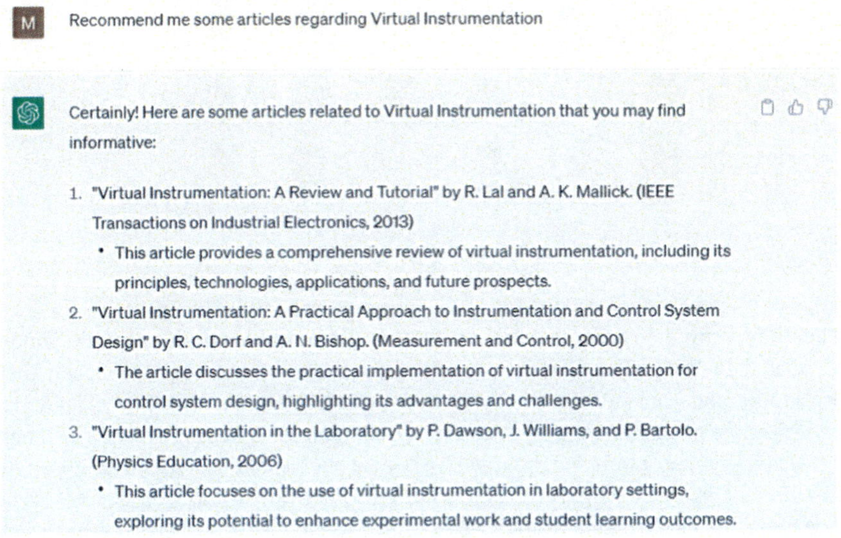


**Fig. 7.** Articles recommended by ChatGPT that don't exist.

## 4 Developing Graphical Machine Learning Models

As a teacher, introducing artificial intelligence (AI) concepts and applications to engineering students can greatly enhance their problem-solving skills and prepare them for the evolving technological landscape. Deep Learning Toolkit (DeepLTK) from Ngene provides a powerful framework for teaching AI and Deep Learning principles for the engineering domain. Being integrated into LabVIEW's graphical programming environment, it can be used by students without the need to write code. This section explores how DeepLTK can be effectively utilized to teach AI to engineering students, covering key steps and considerations.

First of all, with the help of tutorials and examples provided by Ngene [12], students can be introduced to the fundamental concepts of Deep Learning, such as neural networks, activation functions, layers, and training processes. DeepLTK is a comprehensive framework that simplifies the creation, configuration, and deployment of deep neural networks in LabVIEW by providing support for various layers (such as input, convolutional, fully connected, pool, dropout, SoftMax, etc.). Figure 8 illustrates the function palette developed by Ngene that contains all the functions needed for training and evaluating deep neural networks.

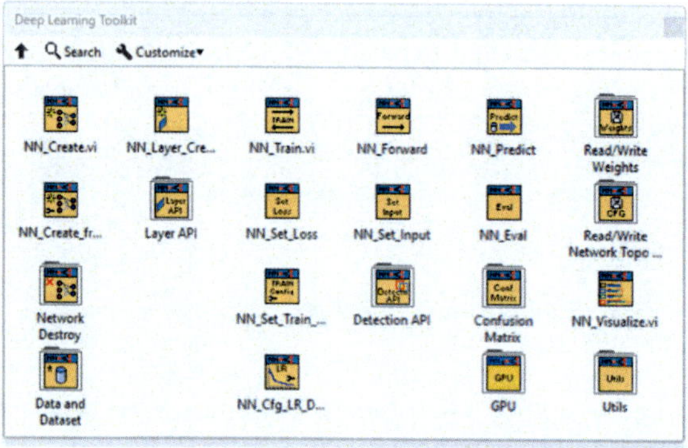


**Fig. 8.** DeepLTK function palette.

There are various learning resources, including examples, tutorials and documentation related to DeepLTK. Students should explore these resources to deepen their understanding and gain practical insights into using the toolkit effectively. They can start with simple examples to grasp the basics of deep learning using DeepLTK. There are a few step-by-step examples provided by Ngene such as image classification on the MNIST dataset, speech recognition, and simple Boolean operations.

The transition can be made from those examples to engineering-specific applications of Deep Learning using DeepLTK. Engineering students can apply Artificial Intelligence to known areas such as signal processing, image recognition, anomaly detection, predictive maintenance or systems control. Various case studies and real-world examples can be illustrated to highlight the relevance of AI in engineering domains. Hands-on projects should be assigned to students to solve engineering problems using DeepLTK. Through these projects, they learn and practice how to design, train, and evaluate deep learning models tailored to specific applications within their discipline.

By leveraging the Deep Learning Toolkit from Ngene as a teaching tool, engineering educators can empower students with Artificial Intelligence skills applicable to their future careers.

## 5 Discussions and Conclusions

To evaluate the effectiveness of integrating ChatGPT in the Virtual Instrumentation course, student feedback is being collected through online anonymous survey on Google Forms (https://forms.gle/Km8jMrow88aNaPYt5). The feedback aims to gauge the overall perception of ChatGPT as a learning tool and its impact on students' learning experience. Students expressed positive responses so far, highlighting the usefulness of ChatGPT in accessing information, clarifying concepts, and obtaining timely assistance. Therefore, they recommended the use of ChatGPT in the future iterations of the Virtual Instrumentation course (Fig. 9).

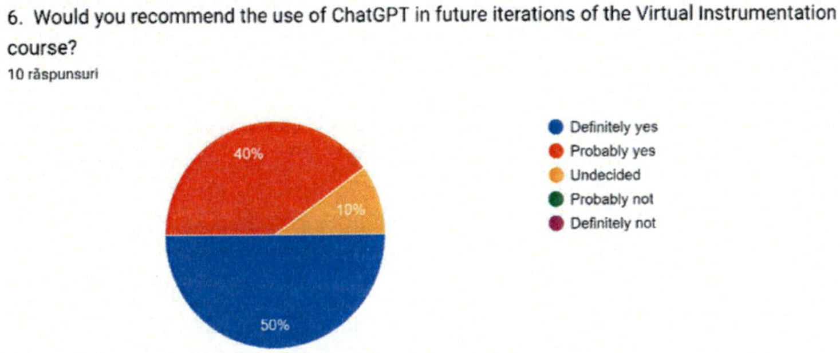


**Fig. 9.** Survey question.

The integration of ChatGPT in engineering education demonstrated promising results in knowledge transfer. Students reported that ChatGPT provided accurate and relevant responses to their questions, effectively addressing their queries, and promoting a deeper understanding of Virtual Instrumentation concepts.

ChatGPT proved to be a valuable resource for troubleshooting and problem-solving. Students found the model's assistance in diagnosing technical issues and guiding them through the debugging process to be beneficial. By following ChatGPT's instructions and suggestions, students reported an improvement in their problem-solving skills and a greater ability to overcome challenges independently. The interactive nature of the system allowed for a dynamic learning experience, encouraging students to think critically and apply their knowledge to practical scenarios.

Despite the positive outcomes, several limitations and challenges were identified in the integration of ChatGPT in the Virtual Instrumentation course. Firstly, the model's responses were sometimes verbose or overly technical, requiring further simplification or contextualization. The model's inability to provide real-time feedback during hands-on lab sessions was also noted as a limitation. Additionally, there were instances where ChatGPT struggled to handle ambiguous or context-dependent queries, leading to less accurate responses. It was crucial to emphasize to students that ChatGPT's responses should be used as supplementary information and not a substitute for physical courses and hands-on laboratories. There are also other chatbots that are using generative AI; therefore, the current study can be improved in the future by testing and comparing several platforms.

Furthermore, by leveraging DeepLTK as a teaching tool, engineering educators can empower students with AI skills applicable to their future careers. Through a combination of foundational knowledge, hands-on projects and collaborative learning, students can gain a solid understanding of deep learning principles and their practical implementation. By embracing DeepLTK in the classroom, teachers can empower engineering students with the necessary skills to tackle complex engineering challenges using AI technologies.

**Acknowledgments.** We would like to express our great appreciation to NI Romania for providing of free LabVIEW license. Our gratitude also goes to Ngene from Armenia for providing us with a free trial license for the Deep Learning Toolkit for LabVIEW. Their collaboration and generosity are highly appreciated.

## References

1. Xu, W., Ouyang, F.: The application of AI technologies in STEM education: a systematic review from 2011 to 2021. Int. J. STEM Educ. **9**, 59 (2022). https://doi.org/10.1186/s40594-022-00377-5
2. Chen, L., Chen, P., Lin, Z.: Artificial intelligence in engineering education: a review. IEEE Access **8**, 75264–75278 (2020). https://doi.org/10.1109/ACCESS.2020.2988510
3. Kahraman, H.T., Sagiroglu, S., Colak, I.: Development of adaptive and intelligent Web-based educational systems. In: Proceeding of the 4th International Conference on Applied Sciences, Information and Communication Technology, pp. 1–5 (2010). https://doi.org/10.1109/ICAICT.2010.5612054
4. Kim, Y., Soyata, T., Behnagh, R.F.: Towards emotionally aware AI smart classroom: current issues and directions for engineering and education. IEEE Access **6**, 5308–5331 (2018). https://doi.org/10.1109/ACCESS.2018.2791861
5. Sharma, R.C., Kawachi, P., Bozkurt, A.: The landscape of artificial intelligence in open, online and distance education: promises and concerns. Asian J. Distance Educ. **14**(2), 1–2 (2019). https://doi.org/10.5281/zenodo.3730631
6. Xu, W., Ouyang, F.: A systematic review of AI's role in the educational system based on a proposed conceptual framework. Educ. Inf. Technol. **27**, 4195–4223 (2022). https://doi.org/10.1007/s10639-021-10774-y
7. Hwang, G.J., Xie, H., Wah, B.W., Gasevic, D.: Vision, challenges, roles and research issues of artificial intelligence in education. Comput. Educ.: Artif. Intell. **1**, 100001 (2020). https://doi.org/10.1016/j.caeai.2020.100001
8. Chassignol, M., Khoroshavin, A., Klimova, A., Bilyatdinova, A.: Artificial intelligence trends in education: a narrative overview. Procedia Comput. Sci. **136**, 16–24 (2018). https://doi.org/10.1016/j.procs.2018.08.233
9. Modran, H.A., Ursutiu, D., Samoila, C., Chamunorwa, T.: Learning methods based on artificial intelligence in educating engineers for the new jobs of the 5th industrial revolution. In: Educating Engineers for Future Industrial Revolutions. ICL 2020. Advances in Intelligent Systems and Computing, vol. 1329 (2021). Springer, Cham. https://doi.org/10.1007/978-3-030-68201-9_55
10. ChatGPT Homepage. https://openai.com/blog/chatgpt. Last accessed 18 April 2023
11. Jennings, R., De La Cueva, R.: LabVIEW Graphical Programming, 5th edn. McGraw-Hill Publishing House (2020). ISBN 9781260135268
12. Deep Learning Toolkit Documentation. https://www.ngene.co/deep-learning-toolkit-for-labview. Last accessed 10 May 2023

# Successful Practices of Artificial Intelligence Technologies in Educational Activities

Olga Kharina(✉)

HSE University, Malaya Ordynka 17, 119017 Moscow, Russia
olaishra@mail.ru

**Abstract.** The article examines and compares the results of the use of artificial intelligence in the educational environment in different countries. Among the leaders in the introduction of artificial intelligence in public life and education is India, as the state that is fastest in the transition processes to an innovative component in all spheres of life. Successful cases of using artificial intelligence in Turkey and China are also considered. The purpose of the article is to choose the most adaptive practices for Russia, their consideration and forecast of application in the domestic educational environment. The result of the work was the systematization of the obtained results and the definition of the practical importance of artificial intelligence in the life of Russian citizens.

**Keywords:** Artificial intelligence · Education · High technology · India · Turkey · China · Innovation

## 1 Introduction

Technological advances such as artificial intelligence are part of the processes shaping the 21st century. However, the speed, scale and ways in which technology is developed and deployed can exacerbate existing digital and economic contradictions. The adoption of artificial intelligence (AI) accelerated during the COVID-19 pandemic worldwide and across sectors. Artificial Intelligence is now poised to impact the lives of citizens in all aspects of their lives. The urgent need to educate the next generation of workers and citizens has led countries to focus on national AI strategies and educational planning concepts.

States now need not only to ensure that their populations have access to the educational resources they need to acquire new skills, but also to become a driving force in addressing the pressing challenges they face.

It is important to note that the term "artificial intelligence" is a popular science, while science uses another formulation, "artificial cognitive system," which is "a hardware and information-software system whose action is similar to that of human mechanisms" [7]. In this paper, artificial intelligence is understood as a hardware and information-software complex, which has broad capabilities and action algorithms that are similar to the human mind. The purpose of this paper is to select the most adaptive practices that are available in different countries, for Russia, their consideration for possible application in the domestic educational environment.


M. E. Auer et al. (Eds.): ICL 2023, LNNS 899, pp. 511–519, 2024.
https://doi.org/10.1007/978-3-031-51979-6_53

## 2 Materials and Methods

The article presents an analytical review of the use of advances in the field of artificial intelligence and its algorithms. A brief description of the practices of application of AI is formed, which allows us to identify the main areas of its application, where the main emphasis is placed on education.

Educational environment refers to one of the most conservative spheres of society, but the accelerated introduction of digital educational resources (including those based on artificial intelligence) is becoming increasingly popular: the number of online courses is growing, visual practices are introduced, various interactive webinars are held, forms of control are moving to a remote format with automatic scoring [9]. In order to assess the problems and prospects of using AI technologies in the modern educational space, it seems appropriate to consider the practices of AI application, its main methods and products, and to understand how foreign experience can be applied in Russia in the framework of educational and pedagogical processes.

## 3 Results

### 3.1 Technovation Experiment

The foreign practice of transforming traditional forms of learning into technological educational formats is as follows: organization of educational processes using educational distance platforms; broadcasting educational content through distance networks and channels; use of social networks, messengers and other resources for educational purposes; transition to online media and transfer of "material" educational materials in the digital plane; introduction of various digital products based on artificial intelligence [16].

In 2020, Technovation, a member of the UNESCO Global Education Coalition, launched a 5-week pilot program aimed at engaging girls in inclusive online learning [21]. Technovation's Ideas Lab was a free online artificial intelligence program open to girls ages 10–18 in Brazil, India, Kenya, Mexico, Nigeria, and Pakistan. The students learned about artificial intelligence and created their own machine learning models as prototype tools to solve problems in their countries. 1,500 female students from 6 countries developed 250 ideas to solve problems ranging from climate change to domestic violence (see Fig. 1).

Based on the results, Technovation offered the following tips: provide regular feedback, get support from mentors, and discuss the results for quality control. After participating in Technovation's Idea Lab, a fully online, low-sensory artificial intelligence program, at least 60% of students showed improvements in their performance, high-tech skills, and systematization when expressing their ideas. For example, a student from Brazil decided to develop an artificial intelligence-based tool to support students with hearing impairments, actively interacting with a mentor from Ericsson. Sharing ideas, engaging artificial intelligence, and incorporating the results into education is a chain that can work productively in all areas of human life in the future.

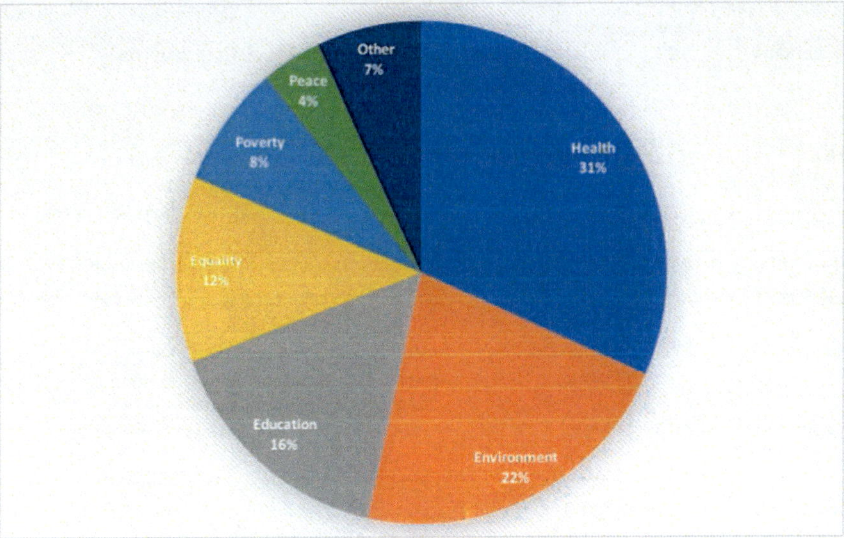


**Fig. 1.** Problems addressed by Technovation girls in 2021.

### 3.2 Artificial Intelligence in India

Artificial Intelligence has revolutionized daily life in India and spurred the development of Indian society. India's ambitious mission in artificial intelligence will become a reality as the Ministry of Electronics and Information Technology (MeitY) launches a five-year program. Through it, the government aims to provide students with a platform for mastering artificial intelligence skills and access to relevant artificial intelligence tools so they can be prepared for the digital revolution of the future [14].

The government has launched the Responsible Artificial Intelligence for Youth program, which will help bridge the skills gap in students and prepare them to use artificial intelligence. This program is for students in grades 8 to 12 in public schools in the states [19].

Many schools in India are already using artificial intelligence in education. Among the technologies using artificial intelligence are the following:

**Chatbots**. They are increasingly appearing in education to help students understand specific topics in different lessons, as well as to repeat material in preparation for grading. An example is Yugasa Bot [6]. It shortens the cycle of tasks assigned to teachers and is used in classrooms to replace email communication between teachers, students, and parents. In addition, India has a well-developed system of AI Startups that have become successful in education over the years (see Fig. 2).

**Virtual Reality (VR)**. One recent innovation in education is virtual reality, which is being used for everything from teaching history to helping students learn math. Virtual reality is a three-dimensional computer environment that people can explore and interact with. Skugal VR is a great way to help students feel connected to a process or subject [13]. When they are in different classrooms but using the same virtual reality program,

| Name | Founded Year | Location | Mission |
|---|---|---|---|
| HackerRank | 2009 | Mountain View (United States), Bengaluru (India) | Technical recruitment platform and code-challenge community |
| Miko | 2015 | Mumbai (India) | AI-powered companion robot for children that can talk, respond, educate, provide entertainment, and understand the child's needs, emotions, likes, and dislikes |
| Financepeer | 2017 | Mumbai (India) | An online platform offering education loans |
| Practically | 2018 | Hyderabad (India) | An AR-based STEM learning app for students |
| iNurture | 2005 | Bengaluru (India) | An education solution provider enabling universities to develop and run job-oriented undergraduate and postgraduate courses |
| Square Panda | 2018 | Mumbai (India) | Multi-sensory phonics learning system |
| Saarthi Pedagogy | 2018 | Ahmedabad (India) | ERP software and solutions for schools |
| Verzeo | 2018 | Bengaluru (India) | Multi-disciplinary learning programs |
| Thinkster | 2010 | Bengaluru (India) | It provides students with a personalized one-on-one session with math tutors, analyzes the performance of students |
| EduGorilla | 2016 | Lucknow (India) | Exam preparation solutions |

**Fig. 2.** AI in education startups in India.

they can safely communicate while remaining separated by distance. With virtual reality, students can explore things they may never have the opportunity to see or learn about in real life. The same is true for teachers. Teachers can find much more engaging ways to teach their students. Anyone who has tried virtual reality knows that it is a much more immersive experience than sitting in front of a screen or being inside a computer environment. Increased engagement and deep understanding are just two of the benefits to students and teachers.

**Learning Management System.** It provides a centralized, intuitive system for managing all of the school's online activities. These tools can be used for a variety of purposes, but they are often used to upload coursework, teachers communicate with students and parents, and track progress. These systems allow course syllabi to be placed in one space, as well as course assignments [2]. This means that teachers can provide feedback on any assignment or grade at any time. Students have instant access to their grades without having to wait until the end of the semester. Many topics can be explored through movies and online lectures using artificial intelligence. A student can get help from an artificial intelligence-powered digital tutor who helps them with homework problems and gives them hints on how to complete their assignments. Artificial intelligence can even be used to create a learning management system that can understand how students think and help them learn better. Currently, there are systems that can help teachers create content, help parents monitor their children's progress in the system, and assess them using artificial intelligence.

**Robotics**. Robots are becoming a great teaching resource in India for students and teachers alike. For example, they have become teachers at the International School in Hyderabad (in southern India) [20]. For teachers, this means that robots can provide opportunities to spend more time with students who need extra help. They allow for experimentation with new ways of teaching. Teachers can also use robotics as a teaching tool to teach lessons about current events or even math concepts such as fractions. As technology advances, they will play an increasingly important role in people's lives.

It's worth noting that only about 47% of students in India take advantage of AI to varying degrees. The growth in the use of high technology in education is planned only by 2035 [1].

### 3.3 Analysis of Artificial Intelligence in the Context of Education in Turkey

In 2019, Turkey launched a national artificial intelligence development plan that has generated international interest. The action plan aims to accelerate the use of AI in teaching, strengthen education management and create an artificial network infrastructure [4]. Turkish educational institutions have launched 50 new programs related to artificial intelligence. Learning is being aligned to be based on a block of subjects defined as "TUBAI" [3]: combining disciplines to focus on an applied field, including "mathematics, computer science, physics, biology, sociology and psychology."

The Turkish government has launched national "talent programs" to attract artificial intelligence researchers to the country and has supported numerous programs to develop Turkish researchers locally [18].

Intelligent learning systems, seen as the second generation of machine learning, are one of the most popular applications of artificial intelligence in education in Turkey. LESs, which are computer-based learning systems with individual databases or structured blocks of learning material (specifying what is to be taught) and teaching methodologies that make inferences based on the student's subject competence, are being introduced in education to provide a dynamic education [15]. The system generates a step-by-step route for the student using appropriate learning materials and activities based on the student's successes and failures. This route is continually updated with respect to the level of difficulty, provides prompts and explanations based on student feedback, and adjusts to meet their specific requirements. The goal is to make it easier for the student to master the topic at hand.

Computer Based Education (CBE) is also an important part of education in Turkey [8]. It also includes e-learning systems. The increasing use of artificial intelligence technologies has led to the development of innovative adaptive and intelligent educational systems. Numerous well-known CBE systems such as ITS, learning management system, adaptive hypermedia and multimedia system, and test and quiz system are also AIEd (artificial intelligence based learning) systems, hence there are parallels between CBE and AIEd.

It is important to note that Turkey's artificial intelligence strategy is characterized by a contrast between state and market approaches: tensions of this kind define and legitimize the educational space. Statements about advanced applications of artificial intelligence focus on a pedagogical strategy for commercial growth rather than any specific educational goal. Turkey needs to develop a comprehensive program on artificial intelligence, a phased plan that will optimize educational processes at all levels with existing successful initiatives in this area.

### 3.4 Features of AI Application in Education in China

Over the past decade, China has made rapid progress in the technology of the Fourth Industrial Revolution and has become a leader in many high-tech areas, such as artificial intelligence.

An example of an AI program in China is Squirrel AI. Squirrel AI specializes in "intelligent adaptive education" [10]. The company invests in artificial intelligence scientists so that they can invest more research in the field. Squirrel AI uses an algorithm that allows students to get 70% of their learning suggestions from artificial intelligence and the other 30% from traditional teachers. This allows for child-centered learning while retaining some human control to manage the machine learning process.

Artificial Intelligence Education in China also aims to shift the focus of China's education system from a mass conveyorized education approach to a higher quality education for the masses. Artificial intelligence programs are tailored to the needs and abilities of each child. The size of the artificial intelligence (AI) online education market in China had its evolution. In 2020, the AI online education market in China reached a value of 368 billion yuan. It was expected that the market would continue to expand based on predictions. However, the Chinese government has recently imposed significant regulations on the education sector with the aim of gaining control over the extensive extra-curricular education industry (see Fig. 3).

In 2017, the State Council published an Artificial Intelligence Development Plan in which China is to become a "major global center of innovation in artificial intelligence" by 2030 [11]. One of the key areas of this plan is to increase people's awareness and use of artificial intelligence.

To implement this policy, in September 2018, the Ministry of Education initiated a primary and secondary artificial intelligence education program to work with city science academies, schools, and private sector partners to develop high-level artificial intelligence courses for students [12].

The cities and schools chosen to launch pilot artificial intelligence courses have been given significant flexibility and autonomy in curriculum development and textbook selection. Since 2018, several versions of AI textbooks have been released - most co-authored by learning experts paired with AI industry practitioners - usually differing in their target audience, content focus, and relationship to broader information technology teachings.

The private sector also plays an active and instrumental role in complementing China's public policy on artificial intelligence education. Among the pioneers in the field was Hong Kong-based artificial intelligence company SenseTime, which collaborated with the Massive Open Online Course Center (MOOC Center) at East China Normal University to publish "Fundamentals of Artificial Intelligence" in 2018, which is considered the world's first textbook on artificial intelligence. The company later

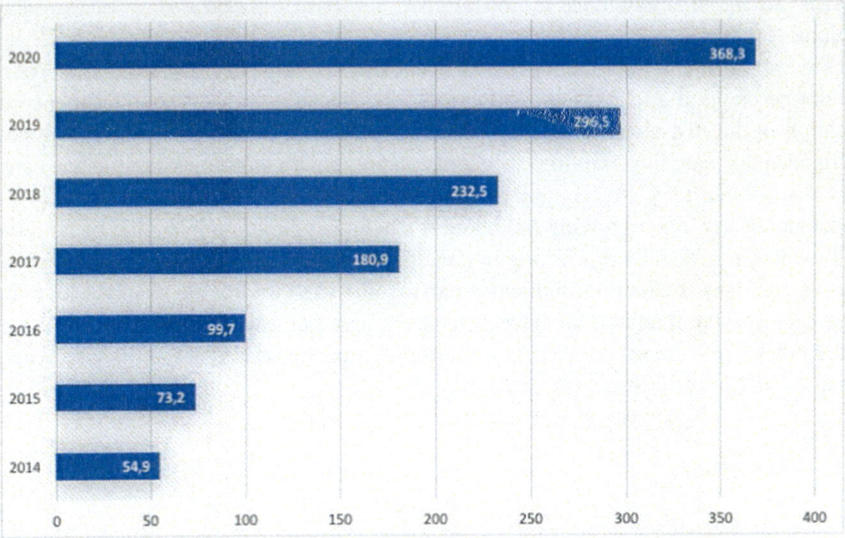


**Fig. 3.** Size of AI online education market in China from 2014 to 2020 (in billion yuan).

founded a separate subsidiary, SenseTime Edu, to support the offering of an integrated "full suite of educational solutions using artificial intelligence." It includes initiatives such as providing learning platforms, labs, and learning robots using its proprietary technologies, as well as compiling a series of AI textbooks for all ages [17].

iFlytek, a Chinese leader in intelligent speech and artificial intelligence in 2018 partnered with Northwestern Pedagogical University and the National Center for Educational Technology to create the first Artificial Intelligence textbook for junior high schools. The textbook uses an exclusive online workshop developed by iFlytek. The company also organized teacher trainings and robotics competitions [5].

East China Normal University has prioritized work at the intersection of artificial intelligence and education, founding the Shanghai Institute of Artificial Intelligence Education (iAIE) in 2020. Its current focus is on using artificial intelligence technology to empower education.

## 4 Discussion

Artificial Intelligence is replacing humans in more fields, including education. It's not just teaching, but also grading, writing essays, and making recommendations to students about what they should study next.

Scientists disagree on whether artificial intelligence should be used to teach students. There is a risk that it will remove the human element from education. However, artificial intelligence in education has many advantages. Artificial intelligence can grade papers and essays much faster than humans. This will give teachers more time to work with students on critical thinking and critical analysis skills. Artificial intelligence can also help teachers build feedback to students who need more practice on specific topics or skills. Artificial Intelligence is not fatigued and has no mood swings or everyday problems.

However, artificial intelligence in education also has some negative aspects. A robot may not be as good a teacher as a human. A disadvantage of artificial intelligence in education is that the technology may not always be successful in teaching. Artificial intelligence does not feel emotions. Students do not feel that artificial intelligence cares about them when they are lectured or when they have a question and when they do not get an answer. It's a growing field, and it's being studied in universities around the world, where professors are working to develop artificial intelligence technologies that improve our lives. Artificial intelligence can also be used to provide adaptive learning for students, where it adjusts the pace of learning based on each student's performance. On the other hand, some people worry about the impact of artificial intelligence where human interaction is waning.

## 5 Conclusion

Artificial Intelligence technology helps students by facilitating certain learning tasks. For example, real-time language translation makes information more accessible to students around the world. Artificial intelligence can increase efficiency, personalization, and simplify administrative tasks, giving instructors more time and freedom to understand and adapt.

In Russia, given the successful practices and innovations, government agencies, schools, universities and private companies together can create a unified system of interaction in education, following common goals in a technologically developing world. Artificial intelligence can be used as an educational tool that guides students toward their goals by providing personalized feedback based on artificial intelligence algorithms. It has the potential to make life easier for everyone through automation. AI is a critical driver of change in education. Every student will have equal access regardless of his or her learning ability or disability due to physical characteristics, which will be the basis for the harmonious development of Russia's future society.

## References

1. Analytics Draft. https://analyticsdrift.com/does-ai-have-potential-to-transform-the-indian-education-system/. Last accessed 25 July 2022
2. Ashri, D., Sahoo, B.P.: Open book examination and higher education during COVID-19: case of University of Delhi. J. Educ. Technol. Syst. **50**(1), 73–86 (2021)
3. Cifci M.A.: Optimizing WSNs for CPS using machine learning techniques. In: Luhach, A.K., Elçi, A. (eds.) Artificial Intelligence Paradigms for Smart Cyber-Physical Systems. IGI Global, pp. 204–228 (2021)
4. Dalal, R.J., Gupta, S., Mishra, A.P.: Artificial intelligence in assisted reproductive technology: present and future. Int. J. Infertil. Fetal. Med. **11**(3), 61–64 (2020)
5. Ecnu.edu.cn 29.12.2020. http://english.ecnu.edu.cn/02/a2/c1703a262818/page.htm. Last accessed 25 March 2023
6. Hello Yubo. https://helloyubo.com/chatbot/educational-chatbots-five-use-cases-in-india-2022/. Last accessed 25 July 2022

7. Kuteynikova, D.L., Izhaev, O.A., Zenin, S.S., Lebedev, V.A.: Cyberphysical, cyberbiological and artificial cognitive systems: essence and legal properties. Russ. Law: Educ. Pract. Sci. **3**(111), 75–79 (2019)
8. Lee, Y.: An analysis of the influence of block-type programming language-based artificial intelligence education on the learner's attitude in artificial intelligence. J. Kor. Assoc. Inf. Educ. **23**(2), 189–196 (2019)
9. Luchsheva, L.V.: Social problems of using artificial intelligence in higher education: tasks and prospects. Sci. Tatarstan **4**, 84–89 (2020)
10. Medium.com. https://edtechchina.medium.com/the-worlds-most-valuable-ai-unicorn-is-implementing-education-initiatives-in-china-e53995dda504. Last accessed 25 March 2023
11. Medium.com. https://medium.com/@jiahe/the-next-generation-ai-development-plan-whats-inside-72824a9bcc3. Last accessed 25 March 2023
12. Moe.gov.cn. http://www.moe.gov.cn/srcsite/A16/s3342/201804/t20180425_334188.html. Last accessed 25 March 2023
13. Official Website of Skugal VR for Education. https://skugal.com/products/skugal-vr. Last accessed 25 March 2023
14. Official Website of the Government of India. https://indiaai.gov.in/article/artificial-intelligence-towards-a-new-dawn-for-new-india. Last accessed 24 March 2023
15. Orman, A., Sebetci, Ö.: Artificial Intelligence (AI) studies in the TR index: a systematic review. Düzce Univ. J. Sci. Technol. **10**(1), 465–475 (2022)
16. Pavlyuk, E.S.: Analysis of foreign experience of the influence of artificial intelligence on the educational process in a higher educational institution. Mod. Pedag. Educ. **1**, 65–72 (2020)
17. Sensetime. https://www.sensetime.com/en/news-detail/54844?categoryId=1072#:~:text=SenseTime%20believes%20education%20is%20the%20engine%20for%20driving%20future%20innovation.&text=The%20SenseTime%20Edu%20brand%20features,primary%20and%20secondary%20school%20students. Last accessed 25 March 2023
18. Tamer, H.Y., Övgün, B.: Yapay zeka bağlamında dijital dönüşüm ofisi. Ank. Üniv. SBF Derg. **75**(2), 775–803 (2020)
19. The Official Website of the National Program Responsible AI for Youth. https://responsibleaiforyouth.negd.in. Last accessed 25 March 2023
20. Times of India. https://timesofindia.indiatimes.com/city/hyderabad/robots-turn-teachers-in-this-school-in-city/articleshow/93172286.cms. Last accessed 25 March 2023
21. UNESCO. https://www.unesco.org/en/articles/education-sustainable-development. Last accessed 25 July 2022

# Inappropriate Benefits and Identification of ChatGPT Misuse in Programming Tests: A Controlled Experiment

Hapnes Toba[1](✉), Oscar Karnalim[1], Meliana Christianti Johan[2], Terutoshi Tada[3], Yenni Merlin Djajalaksana[4], and Tristan Vivaldy[2]

[1] Master of Computer Science Study Program, Faculty of Information Technology, Maranatha Christian University, Bandung, Indonesia
{hapnestoba,oscar.karnalim}@it.maranatha.edu

[2] Bachelor of Informatics Study Program, Faculty of Information Technology, Maranatha Christian University, Bandung, Indonesia
meliana.christianti@it.maranatha.edu, 2172032@maranatha.ac.id

[3] Media Culture Department, Faculty of Information Sciences and Arts, Toyo University, Kawagoe, Japan
t_tada@toyo.jp

[4] EC-Council, Tampa, FL, USA
yenni.djajalaksana@eccouncil.org

**Abstract.** While ChatGPT may help students to learn to program, it can be misused to do plagiarism, a breach of academic integrity. Students can ask ChatGPT to complete a programming task, generating a solution from other people's work without proper acknowledgment of the source(s). To help address this new kind of plagiarism, we performed a controlled experiment measuring the inappropriate benefits of using ChatGPT in terms of completion time and programming performance. We also reported how to manually identify programs aided with ChatGPT (via student behavior while using ChatGPT) and student perspective of ChatGPT (via a survey). Seventeen students participated in the experiment. They were asked to complete two programming tests. They were divided into two groups per the test: one group should complete the test without help while the other group should complete it with ChatGPT. Our study shows that students with ChatGPT complete programming tests two times faster than those without ChatGPT, though their programming performance is comparable. The generated code is highly efficient and uses complex data structures like lists and dictionaries. Based on the survey results, ChatGPT is recommended to be used as an assistant to complete programming tasks and other general assignments. ChatGPT will be beneficial as a reference as other search engines do. Logical and critical thinking are needed to validate the result presented by ChatGPT.

**Keywords:** ChatGPT · Controlled experiments · Plagiarism · Programming · Topic modeling


M. E. Auer et al. (Eds.): ICL 2023, LNNS 899, pp. 520–531, 2024.
https://doi.org/10.1007/978-3-031-51979-6_54

## 1 Introduction

In programming education, plagiarism is a common breach of academic integrity [1]. It is about the reuse of one's program (or even part of it) with insufficient acknowledgment to the owner [2]. While code reuse is somewhat encouraged in programming, there is a need to cite the source in the program comments [3].

Plagiarism can happen due to either pressure, opportunity, or misrationalization [4]. Students are tempted to plagiarize if they are pressured by the circumstances, they misrationalise the act with incorrect justifications, and there are opportunities to cheat (e.g., limited plagiarism detection mechanism). To deal with pressure, instructors can promote early submissions via incentives [5] or replace large assessments with many small assessments [6]. To deal with misrationalization, instructors are expected to inform students about the matter in their courses [7]. Clear explanations about academic integrity can reduce the number of cases of plagiarism [8].

To deal with opportunity, instructors need to avoid reusing assessments [9], encourage the use of student case studies [10], or impose additional grading methods such as oral presentation [11]. There is also a need to check student programs for plagiarism and penalize the perpetrators. Typically, instructors are aided with automated tools to identify similar programs. Some of the tools are MOSS [12], Sherlock [13], Plaggie [14], and SSTRANGE [15] (more publicly available automated tools can be seen in [16]). It is worth noting that instructors need to investigate similar programs and find sufficient evidence for plagiarism (burden of proof [13]). The high similarity is not always a result of plagiarism [17], and some similarities are coincidental [18].

Most of the tools mentioned in the previous paragraph are based on artificial intelligence (AI) which provides machine learning models to predict the occurrences of some events, including plagiarism. Beyond that, AI has also shown its possibilities in education such as supporting student engagements [19], personal tutoring system [20], information searching [21], and even generating exam questions [22]. Recently, another possibility is coming and interrupting the traditional ways of information gathering in the form of ChatGPT which is based on a large language model [23]. By using ChatGPT people can search beyond factoid-based information. Generated-based content such as story-telling and programming codes are ready to be presented by ChatGPT.

Misuse of ChatGPT introduces a new way of plagiarism in programming: students can ask the tool to write the solution (or at least part of it) at which the solution is generated from other people's work without citing the source(s), and thus lack of accountability [23]. Lecturers need to talk about the ethics of ChatGPT with their students. Due to the recency of ChatGPT misuse, to the best of our knowledge, there are no studies formally measuring the inappropriate benefits of using ChatGPT in programming education and reporting how to manually identify programs aided with ChatGPT.

In response to the aforementioned gap, we present a controlled experiment about inappropriate benefits and identification of ChatGPT misuse in programming education. The experiment involves 17 computing undergraduates who appear to know how their colleagues can breach academic integrity. These students are nominated based on academic values and attitudes that will guarantee their integrity and not misuse the tool. We also asked about their perspective on ChatGPT. Our study has thus the following research questions: 1) RQ1: How substantial are the inappropriate benefits of ChatGPT misuse in programming tests? 2) RQ2: What programming features can be used to identify ChatGPT-aided programs?

## 2 Method

A controlled experiment was conducted to address the research questions. RQ1 was addressed by comparing completion time and programming performance between tests with and without the help of ChatGPT. Students with ChatGPT appeared to have unfair benefits if they completed the test faster or they got higher marks in a statistically significant manner. RQ2 was addressed by asking students which programming features could be useful to identify ChatGPT-aided programs after they had used the tool for a test. We also asked several additional questions to understand more about the student perspective regarding ChatGPT.

Twenty computing undergraduate students were invited to participate with 17 of them accepting the invitation. They are never involved in plagiarism, and they have good impressions from instructors at our faculty. We are unable to consider more students as some of them might learn to misuse ChatGPT via our experiment and our current education environment is not fully ready to handle such a situation. While our selected computing undergraduates are never involved in plagiarism, they know how their colleagues do plagiarism as most of them are former tutors and/or are actively engaged with other students (as a member of the student senate or as laboratories staff). Each student was incentivized with Rp. 72.000 e-money (around 5 USD, sufficient to cover up to two-day meals for undergraduates).

The students were asked to complete the mid-test and final test from the previous offering of introductory programming (second semester of the 2021/2022 academic year). The course covers basic concepts of programming in Python: input, output, branching, looping, function, array, matrices, searching, and sorting. The first five concepts were covered in the mid-test while all of them were covered in the final test. Per the test, four programming tasks were introduced with 25 of 100 marks each. Details of the tasks in both tests can be seen in Table 1. The tasks were given in more detail with input-output examples. A test was expected to be completed in two hours. All students passed the course with A or B+ marks.

The students were split into two groups. The first group (nine students) completed the mid-test by themselves and the final test with the help of ChatGPT. The second group (eight students) completed the mid-test with ChatGPT and the final test by themselves. They were asked to record their completion time per programming task. In their corresponding ChatGPT session, they were also asked to report their step-by-step interactions with ChatGPT and which programming factors that could be useful for identifying ChatGPT-aided works. We also asked them to answer several additional questions to understand more about their perspective of ChatGPT. The complete set of survey questions can be seen in Table 2.

The first nine questions were introduced to address RQ1; Q01 was used to determine the group the participant is in; Q02–Q05 and Q06–Q09 were used to capture the completion time of each task. Question Q18 was introduced to address RQ2. The rest of the questions were introduced to understand the student perspective of ChatGPT. Questions

**Table 1.** Task details.

| ID | Task | Test | Difficulty |
|---|---|---|---|
| T1 | Money conversion | Mid | Easy |
| T2 | Statistics of students' GPAs | Mid | Medium |
| T3 | Shop invoice | Mid | Medium |
| T4 | 2D pattern of a combination of symbols and numbers | Mid | Challenging |
| T5 | Calculator | Final | Easy |
| T6 | Searching emoticons in a string | Final | Medium |
| T7 | The closest and the farthest 2D coordinates from a particular coordinate | Final | Challenging |
| T8 | Advanced searching and sorting of items | Final | Challenging |

Q14–Q17, Q19, Q21, and Q23 should be responded to on a 5-point Likert scale where 1 represents strongly disagree and 5 represents strongly agree. Questions Q10–Q13, Q18, Q20, Q22, and Q24 are open-ended questions.

RQ1 was addressed via completion time in seconds and programming performance, both per task and overall. Students with ChatGPT were compared with students without ChatGPT and they were favored with the use of the tool if their average completion time is shorter or their programming performance is better (i.e., higher mark) after being validated with an unpaired t-test with a 95% confidence rate. One of our authors is the instructor of introductory programming and they were in charge of marking the tests for measuring programming performance.

To identify the programming features that can be used to identify ChatGPT-aided programs in RQ2, some open-ended questions were asked to the participants. These questions are intended to extract the characteristics of how the students deal with the keywords when they use ChatGPT. We are interested to analyze how the core keywords, apart from the textual description in the programming assignments, would be explored by the students.

To achieve that objective, we propose a topic relation graph descriptive analyses which are based on the topic modeling [24]. The analysis consists of the following four main steps. In the first steps, the textual answers from each survey form are extracted. For each survey question, a bigram-based Latent Dirichlet Allocation (LDA) Topic Modeling is formed [25]. A bigram model is a sequence of two adjacent elements from a string of words that are expected to form relevant words with stronger semantic relations in the textual description.

The coherence scores are used to limit the number of effective topics, which are tuned for the number of topics from 2 to 10 with 10 words in each topic. The coherence scores measure how likely the words in a topic are semantically related. The higher the coherence score is, the more that the words in a topic are considered semantically related [26]. After the number of topics is determined, a topic relation graph is constructed to determine how one word is related to the other words in a topic and how they are related to other topics. The topic relation graph would also be useful to control the topic centrality. It

**Table 2.** Complete set of the survey questions.

| ID | Question |
|---|---|
| Q01 | Which test was done with the ChatGPT? |
| Q02–Q05 | Write down the time required to complete mid test task 1 to task 4 |
| Q06–Q09 | Write down the time required to complete final test task 1 to task 4 |
| Q10–Q13 | Elaborate how ChatGPT was used to complete task 1 to task 4 of given test (depend on the group the participant is in) |
| Q14 | ChatGPT can help students who are not proficient in programming to complete programming tasks |
| Q15 | ChatGPT-generated code can be easily understood by students who are not proficient in programming |
| Q16 | Students who are not proficient in programming are likely to use ChatGPT to complete programming tasks |
| Q17 | If the participant is a tutor, they are able to differentiate ChatGPT-generated code from original submissions |
| Q18 | Which factors that might be useful to differentiate ChatGPT-generated code from original submissions |
| Q19 | The participant believes with the correctness of ChatGPT-generated code |
| Q20 | Explain the reason for Q19 response |
| Q21 | The participant recommends the use of ChatGPT for completing programming tasks so long as it is properly cited in the comments |
| Q22 | Explain the reason for Q21 response |
| Q23 | The participant recommends the use of ChatGPT for completing general tasks so long as it is properly cited in the comments |
| Q24 | Explain the reason for Q23 response |

would also be beneficial to extract core keywords for constructing hypothetical concepts for the open-ended questions in the surveys.

In short, the following steps are executed: first, all unique words in the optimized number of topics as nodes are extracted. Following that, the edges between the highest weighted word in a topic with the overlapping words in other topics are created. In this way, associated concepts within the window boundary, i.e., a path of associated keywords in the bigram model will be determined [27].

## 3 Result and Discussion

### 3.1 Inappropriate Benefits of ChatGPT

Table 3 shows that in the mid-test, students with ChatGPT generally had shorter completion time (around half than that of students without ChatGPT) and the difference is statistically significant according to a two-tailed unpaired t-test with a 95% confidence rate ($p < 0.001$). While observing per task, the difference is only statistically significant on T2 ($p < 0.01$) and T3 ($p < 0.01$). Calculating student GPAs and generating shop invoices are quite common and ChatGPT could help students well. T1 (money conversion) is also a common task but since it is quite easy to solve, the completion time is not significantly affected. In terms of programming performance, students with ChatGPT had comparable performance to those without ChatGPT. All differences are insignificant according to a two-tailed unpaired t-test with a 95% confidence rate.

**Table 3.** Averaged mid-test result.

| Task ID | Non-ChatGPT | | ChatGPT | |
|---|---|---|---|---|
| | Mark | Time (min) | Mark | Time (minutes) |
| T1 | 21.56 | 12.44 | 21.63 | 9.13 |
| T2 | 21.67 | 34.89 | 22.88 | 10.38 |
| T3 | 22.75 | 30.67 | 20.89 | 16.50 |
| T4 | 10.50 | 29.56 | 13.78 | 20.00 |
| T1–T4 | 77.75 | 107.56 | 77.80 | 56.01 |

For the final test, Table 4 depicts that the findings are somewhat similar to those of the mid-test. Students with ChatGPT completed the tasks two times faster than those without ChatGPT ($p < 0.001$) but they had comparable programming performance. Further observation shows that differences in the completion time are also statistically significant for most tasks: T5 with $p < 0.01$, T6 with $p < 0.001$, and T8 with $p < 0.01$. T7 is the only task at which students with ChatGPT had comparable completion times with their counterparts. Informal post-discussion with the students notes that for T7, ChatGPT-generated code had a slightly different purpose and needed further alignment to the task.

**Table 4.** Averaged final-test result.

| Task ID | Non-ChatGPT | | ChatGPT | |
|---|---|---|---|---|
| | Mark | Time (min) | Mark | Time (min) |
| T5 | 22.38 | 20.38 | 22.33 | 11.89 |
| T6 | 19.75 | 24.88 | 17.67 | 5.67 |
| T7 | 15.13 | 29.25 | 19.33 | 16.33 |
| T8 | 12.38 | 29.69 | 13.67 | 12.78 |
| T5–T8 | 69.63 | 104.19 | 73.00 | 46.67 |

### 3.2 Programming Features to Identify ChatGPT-Aided Programs

In the open-ended surveys Q10–Q13, the participants are requested to explain their experiences in detail to complete the programming assignments. Since we are interested in the core keywords that the participants used during the task completion in general, during the analyses we mixed all the answers of Q10–Q13. Another reason for this approach is that we hypothesize that each participant would have a particular specific strategy to interact with ChatGPT.

By performing the LDA topic modeling, the number of effective topics for Q10–Q13 is 8 with a coherence score of 0.749. The topic relation graph of these questions can be followed in Fig. 1. In this figure can be inferred that the participants are mainly using a copy-and-paste strategy from the textual description of the programming assignments. This can be seen in the core keywords as suggested by the topic relation graph. Some keywords, such as GPA, rupiah, rectangle, coordinate, box, and discount are initially coming from the textual descriptions of the assignments.

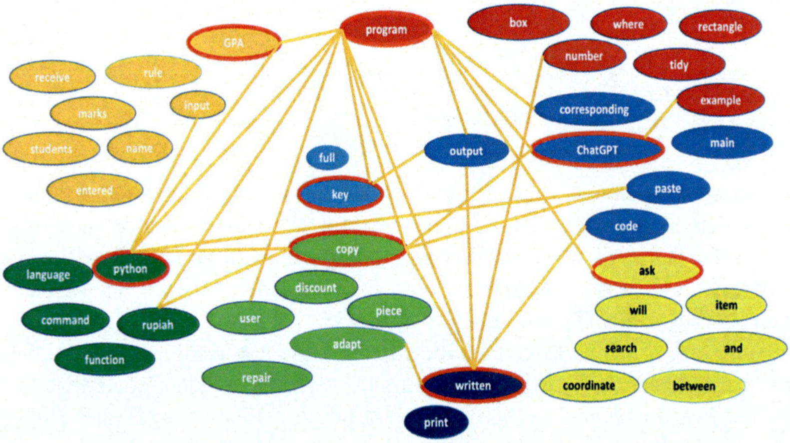


**Fig. 1.** Topic relation graph for Q10–Q13 (general strategy to use ChatGPT).

Several participants made special copy-and-paste strategies by using variations of the input and output formats in the program and also considering the programming language as described in the assignments. After they found the suitable retrieved code example from ChatGPT, in the subsequent steps they will try to adapt the code and repair it as completely as possible according to the assignment. Some also try to translate the assignment into English and enter it into ChatGPT. They made subsequently some adjustments to functions and variable naming.

It is interesting that although the participants have in general almost the same copy-and-paste strategy, they have the confidence to differentiate the ChatGPT-generated codes and those which were written by relatively new programmers. The respondents were confirming that ChatGPT will be very helpful for students who have programming skills limitations. Around 41% of the respondents strongly agree, and around 35% agree about this fact. But, on the other hand, there are some critical characteristics that ChatGPT has, in comparison to the standard coding style. This is confirmed by the results of Q17, which state that around 30% of the respondents strongly agree that they can differentiate codes generated by ChatGPT, and around 35% agree about the fact.

In further analysis of Q18, we explored which factors that might be useful to differentiate ChatGPT-generated code from original submissions based on the participants' point-of-view. The topic relation graph of Q18 can be seen in Fig. 2, which consists of 7 topics and a coherence score of 0.692. The participants describe that computer codes generated by ChatGPT have structural and syntactical advances in comparison to those written by beginners.

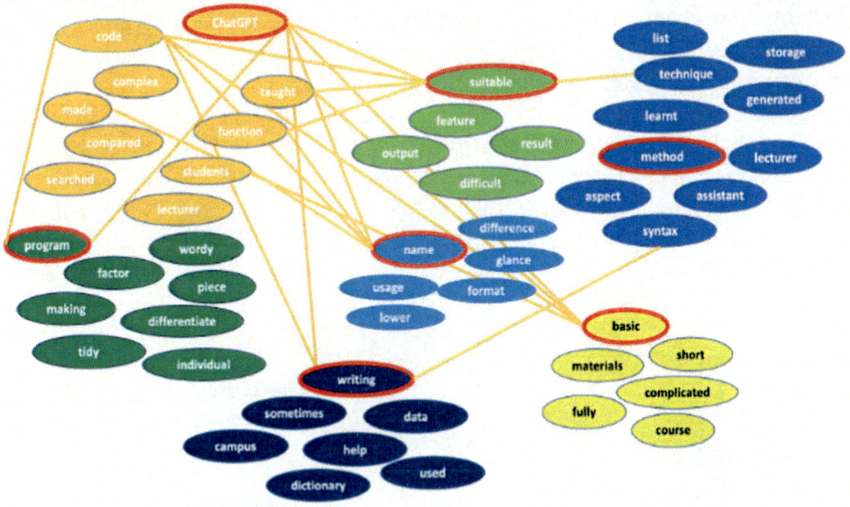


**Fig. 2.** Topic relation graph for Q18 (factors to differentiate ChatGPT).

Some aspects, such as the strategy to efficiently usage of memory (storage) and function arrangement are rather advanced. This is by using, for instance, a list or a dictionary data structure. Further, the generated ChatGPT code itself is mostly tidier and shorter (efficient), than those developed by students. In some cases, it is obvious to observe that the generated codes by ChatGPT are beyond the typical course materials for first-year students.

### 3.3 Student Perspective of ChatGPT

Further analyses from the student perspective regarding the use of ChatGPT are also important as a means to identify the tendency of ChatGPT usage during course assignments. In Q15, Q16, and Q19, we explored the general opinion of the participants on whether ChatGPT would be beneficial for them. In this sense, it is not only that they could answer the assignment correctly, but also emphasize ideas about the knowledge-containing material in the course. Commonly, a teaching assistant or a lecturer will provide some activities to confirm the course learning objectives, such as presentations or small quizzes.

From the answers to Q15, most of the participants (47%) do not agree and strongly do not agree (18%) that generated code of ChatGPT will be easily understood by newbies or those who are not proficient in programming. However, as teaching assistants, they have majority agreements that most non-proficient students would use ChatGPT-generated codes to fulfill their assignments (Q16), i.e., 29% strongly agree and 41% agree. These facts suggest that reconfirmation of submitted assignments would be important to guarantee the authenticity of the codes and to ensure the achievement of learning objectives as well. These facts are also supported by the results of Q19. The participants have a high appreciation for the generated codes by ChatGPT, and they consider the correctness of the generated codes (6% strongly agree, 35% agree, and 41% neutral) with some arguable opinions.

In the open-question survey of Q20, the participants were asked to express their opinion about their belief in the correctness of ChatGPT-generated code. The topic relation graph can be followed in Fig. 3. Four topics are considered effective with a coherence score of 0.654. We can infer that the participants have critical opinions regarding ChatGPT. To be considered correct, the code from ChatGPT should fulfill the expected output and results as described in the assignments. Further, the generated code needs to be written properly as taught in the course, such as how the classes and functions are defined with acceptable factoring and suitable data structures.

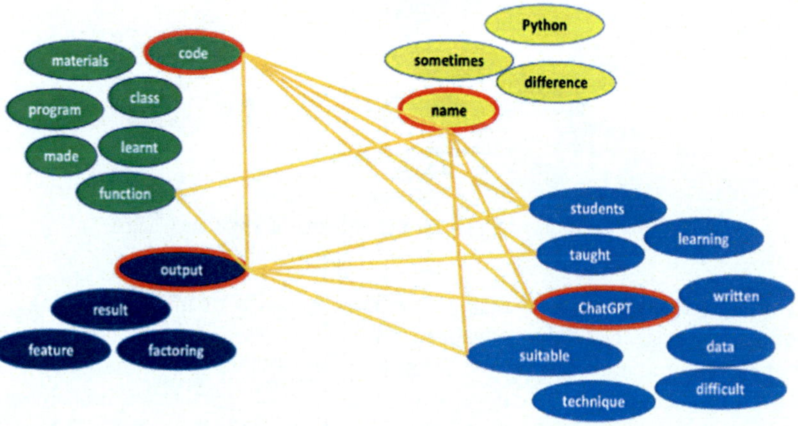


**Fig. 3.** Topic relation graph for Q20 (participants' beliefs in the correctness of ChatGPT).

Some questions in Q21 and Q23 are being asked to investigate the participants' recommendation of using ChatGPT in programming and general tasks. Almost half of the participants strongly agree (12%) and agree (35%) to recommend the utilization of ChatGPT in programming tasks (Q21). In this case, it would be necessary to refer credits and citations to ChatGPT, although only part of the code uses the generated form. A comparable result can also be seen for Q23 which states that the majority of the respondents strongly agree (12%) and agree (47%) to use ChatGPT in general tasks beyond programming tasks.

Specific reasons for using ChatGPT in programming tasks are asked in Q22, and analyzed in a topic relation graph, with 3 effective topics and a coherence score of 0.612. ChatGPT would be useful as a means to help with programming tasks. Especially for first-year students, ChatGPT will be beneficial to help students understand the code and learn to program from code examples retrieved by ChatGPT. Thus, ChatGPT can be considered a web search engine such as Stack Overflow but gives more information and directly focuses the result on highly probable right answers. Nevertheless, ChatGPT can be used to validate computer program codes, and this will be useful for students to learn how to write an acceptable computer program. Comparable comments are also applied for general assignment tasks beyond programming in Q24. Our survey participants stressed the important aspect of logical and critical thinking when using ChatGPT. ChatGPT is indeed a trend, but it must be used wisely, especially in academics.

## 4 Conclusion and Future Work

In this research, we have identified ChatGPT misuse in programming assignments for first-year computer science students in a controlled experimental environment. The main strategy when using ChatGPT in programming tasks as suggested by our respondents is by copy-and-paste approach. The most used part to be copied to ChatGPT is the task descriptions, followed by the input and output structures.

The retrieved and generated code by ChatGPT is subsequently adapted to variable and function naming. Although it seems easy to retrieve the code, the ChatGPT codes are sometimes rather difficult to understand by those who just learn to program. The generated code is highly efficient and uses complex data structures like lists and dictionaries. Based on the survey results, ChatGPT is recommended to be used as an assistant to complete programming tasks and other general assignments. ChatGPT will be beneficial as a reference as other search engines do. Logical and critical thinking are needed to validate the result presented by ChatGPT.

For further research, we plan to identify the differences in the semantics and code structures of ChatGPT and human-generated code. It will be important to examine the control flow graphs of the codes. And thus, the ChatGPT will be beneficial as a means to extract important features to be (machine)-learned to code efficiently, and at the same time to avoid plagiarism.

**Acknowledgment.** The research presented in this paper was supported by the Research Institute and Community Service (LPPM) at Maranatha Christian University, Bandung, Indonesia.

## References

1. Simon, Cook, B., Sheard, J., Carbone, A., Johnson, C.: Academic integrity: differences between computing assessments and essays. In: Proceedings of the 13th Koli Calling International Conference on Computing Education Research, pp. 23–32. Association for Computing Machinery, New York, NY (2013). https://doi.org/10.1145/2526968.2526971
2. Fraser, R.: Collaboration, collusion and plagiarism in computer science coursework. Inform. Educ. **13**, 179–195 (2014). https://doi.org/10.15388/infedu.2014.10

3. Pangestu, M.A., Simon: Work in progress: an automated management system for references in programming code. In: 2021 IEEE Global Engineering Education Conference (EDUCON), pp. 1301–1305 (2021). https://doi.org/10.1109/EDUCON46332.2021.9454112
4. Albluwi, I.: Plagiarism in programming assessments: a systematic review. ACM Trans. Comput. Educ. **20**, 6:1–6:28 (2019). https://doi.org/10.1145/3371156
5. Spacco, J., Fossati, D., Stamper, J., Rivers, K.: Towards improving programming habits to create better computer science course outcomes. In: Proceedings of the 18th ACM Conference on Innovation and Technology in Computer Science Education, pp. 243–248. Association for Computing Machinery, New York, NY (2013). https://doi.org/10.1145/2462476.2465594
6. Allen, J.M., Vahid, F., Downey, K., Edgcomb, A.D.: Weekly programs in a CS1 class: experiences with auto-graded many-small programs (MSP). Presented at the 2018 ASEE Annual Conference & Exposition, June 23 (2018)
7. Simon, Sheard, J., Morgan, M., Petersen, A., Settle, A., Sinclair, J.: Informing students about academic integrity in programming. In: Proceedings of the 20th Australasian Computing Education Conference, pp. 113–122. Association for Computing Machinery, New York, NY (2018). https://doi.org/10.1145/3160489.3160502
8. Mason, T., Gavrilovska, A., Joyner, D.A.: Collaboration versus cheating: reducing code plagiarism in an online MS computer science program. In: Proceedings of the 50th ACM Technical Symposium on Computer Science Education, pp. 1004–1010. Association for Computing Machinery, New York, NY (2019). https://doi.org/10.1145/3287324.3287443
9. Simon: Designing programming assignments to reduce the likelihood of cheating. In: Proceedings of the Nineteenth Australasian Computing Education Conference, pp. 42–47. Association for Computing Machinery, New York, NY (2017). https://doi.org/10.1145/3013499.3013507
10. Bradley, S.: Creative assessment in programming: diversity and divergence. In: Proceedings of the 4th Conference on Computing Education Practice, pp. 1–4. Association for Computing Machinery, New York, NY (2020). https://doi.org/10.1145/3372356.3372369
11. Halak, B., El-Hajjar, M.: Plagiarism detection and prevention techniques in engineering education. In: 2016 11th European Workshop on Microelectronics Education (EWME), pp. 1–3 (2016). https://doi.org/10.1109/EWME.2016.7496465
12. Schleimer, S., Wilkerson, D.S., Aiken, A.: Winnowing: local algorithms for document fingerprinting. In: Proceedings of the 2003 ACM SIGMOD International Conference on Management of Data, pp. 76–85. Association for Computing Machinery, New York, NY (2003). https://doi.org/10.1145/872757.872770
13. Joy, M., Luck, M.: Plagiarism in programming assignments. IEEE Trans. Educ. **42**, 129–133 (1999). https://doi.org/10.1109/13.762946
14. Ahtiainen, A., Surakka, S., Rahikainen, M.: Plaggie: GNU-licensed source code plagiarism detection engine for Java exercises. In: Proceedings of the 6th Baltic Sea Conference on Computing Education Research: Koli Calling 2006, pp. 141–142. Association for Computing Machinery, New York, NY (2006). https://doi.org/10.1145/1315803.1315831
15. Karnalim, O.: Maintaining academic integrity in programming: locality-sensitive hashing and recommendations. Educ. Sci. **13**, 54 (2023). https://doi.org/10.3390/educsci13010054
16. Blanchard, J., et al.: Stop reinventing the wheel! Promoting community software in computing education. In: Proceedings of the 2022 Working Group Reports on Innovation and Technology in Computer Science Education, pp. 261–292. Association for Computing Machinery, New York, NY (2022). https://doi.org/10.1145/3571785.3574129
17. Mann, S., Frew, Z.: Similarity and originality in code: plagiarism and normal variation in student assignments. In: Proceedings of the 8th Australasian Conference on Computing Education, vol. 52, pp. 143–150. Australian Computer Society, Inc., AUS (2006)

18. Simon, Karnalim, O., Sheard, J., Dema, I., Karkare, A., Leinonen, J., Liut, M., McCauley, R.: Choosing code segments to exclude from code similarity detection. In: Proceedings of the Working Group Reports on Innovation and Technology in Computer Science Education, pp. 1–19. Association for Computing Machinery, New York, NY (2020). https://doi.org/10.1145/3437800.3439201
19. Ayub, M., Toba, H., Wijanto, M.C., Parsaoran, R., Sani, A., Hernita, Y.T.: The impact of developing a blended learning sub-system on students online learning engagement. JOTSE: J. Technol. Sci. Educ. **11**, 556–568 (2021). https://doi.org/10.3926/jotse.1196
20. Zawacki-Richter, O., Marín, V.I., Bond, M., Gouverneur, F.: Systematic review of research on artificial intelligence applications in higher education – where are the educators? Int. J. Educ. Technol. Higher Educ. **16**, 39 (2019). https://doi.org/10.1186/s41239-019-0171-0
21. Shukla, L., Singh, J.N., Johri, P., Kumar, A.: Artificial intelligence in information retrieval. In: 2022 4th International Conference on Advances in Computing, Communication Control and Networking (ICAC3N), pp. 1–5 (2022). https://doi.org/10.1109/ICAC3N56670.2022.10074291
22. Wijanarko, B.D., Heryadi, Y., Toba, H., Budiharto, W.: Question generation model based on key-phrase, context-free grammar, and Bloom's taxonomy. Educ. Inf. Technol. **26**, 2207–2223 (2021). https://doi.org/10.1007/s10639-020-10356-4
23. van Dis, E.A.M., Bollen, J., Zuidema, W., van Rooij, R., Bockting, C.L.: ChatGPT: five priorities for research. Nature **614**, 224–226 (2023). https://doi.org/10.1038/d41586-023-00288-7
24. Karnalim, O., Ayub, M., Toba, H.: Mitigating plagiarism and collusion in K-12 as initial knowledge for programming instructors in higher education. World Trans. Eng. Technol. Educ. (WTE&TE) **21**, 44–49 (2023)
25. Blei, D.M., Ng, A.Y., Jordan, M.I.: Latent Dirichlet allocation. J. Mach. Learn. Res. **3**, 993–1022 (2003)
26. Chang, J., Gerrish, S., Wang, C., Boyd-graber, J., Blei, D.: Reading tea leaves: how humans interpret topic models. In: Advances in Neural Information Processing Systems. Curran Associates, Inc. (2009)
27. Yang, P.: KeypartX: Graph-Based Perception (Text) Representation (2022). http://arxiv.org/abs/2209.11844. https://doi.org/10.48550/arXiv.2209.11844

# Accelerating Higher Education Transformation: Simulation-Based Training and AI Coaching for Educators-in-Training

Jasmin Cowin[1], Birgit Oberer[2], James Lipuma[3], Cristo Leon[3], and Alptekin Erkollar[2(✉)]

[1] Touro University, New York, NY, NY 10036, USA
jasmin.cowin@touro.edu
[2] ETCOP Institute for Interdisciplinary Research, Klagenfurt, Austria
{oberer,erkollar}@etcop.at
[3] New Jersey Institute of Technology, Newark, NJ 07102, USA
{lipuma,leonc}@njit.edu

**Abstract.** As the world undergoes remarkable transformations powered by Artificial Intelligence, the challenge arises for educational systems and institutions to adapt. How can we adequately equip educators-in-training to flourish in unprecedented change? The emergence of flexible, hybrid, and socially engaged learning environments has created a need for effective training methodologies that empower educators-in-training to thrive in this new paradigm. Higher education institutions need to expand aspiring educators' human and professional potential amidst accelerating change, in line with the clarion call of the Sustainable Development Goal 4 Quality Education "By 2030, substantially increase the supply of qualified teachers, including through international cooperation for teacher training in developing countries, especially least developed countries and small island developing States" [1]. Simulation-based training coupled with Artificial Intelligence offers a solution to equip educators with the necessary skills and competencies to navigate complex real-world educational settings to succeed in classrooms of the 21st century. Simulation-based training allows educators-in-training to develop their skills and build confidence in their abilities to effectively engage with students in multifaceted classroom environments by providing a safe and controlled space for experimentation and practice. In conclusion, this paper and presentation explore the shifting teaching paradigms in higher education using simSchool and Mursion simulation platforms as examples and examine inclusive and dynamic practices that promote sustainable systems change in line with SDG 4. Quality Education, supporting educators-in-training by identifying strengths and encouraging personal and professional growth through AI feedback loops and faculty coaching.

**Keywords:** Artificial Intelligence · Educators-in-training · Simulation-based training · Sustainable systems change


M. E. Auer et al. (Eds.): ICL 2023, LNNS 899, pp. 532–541, 2024.
https://doi.org/10.1007/978-3-031-51979-6_55

## 1 Introduction

To prepare educators-in-training for AI-driven changes in education, this paper explores simulation-based training and mentoring as effective methods [2–5]. The shift toward flexible, hybrid, and socially engaged learning requires educators to adapt and excel in their roles. The paper focuses on a new generation of instructional simulators with AI-driven algorithms that reduce bias, promote social-emotional learning, and improve instructional effectiveness for preK-12 educators. The integration of theory and practice is critical in teacher preparation programs [6–8]. AI and other technologies need to be integrated into education given SDG 4's call for qualified teachers, especially in developing countries. AI-based observation simulations provide teachers-in-training with personalized feedback to improve their strategies, teaching performance, and decision-making, thereby promoting effective teaching practices [10–15]. Targeted simulations enable educators to embrace diversity and tailor interventions to students' needs, making them more effective teachers [16, 17].

## 2 What Are SimSchool and Mursion

simSchool and Mursion are simulation platforms used in teacher education. Winn (1993) describes immersive VR as providing first-person, non-symbolic experiences designed to facilitate learning [18]. Figure 1 compares both simSchool and Mursion in terms of simulation-based learning environments for practice-based teacher education. Mursion, also known as TeachLivE™ [19], is a mixed-reality simulated classroom technology integrated into teacher education programs at 80 universities worldwide. It provides experiential learning scenarios for preservice teachers. Trained improvisational actors act as puppeteers, controlling live avatars in virtual classrooms. They manipulate the avatars' actions and interactions to create authentic classroom experiences [21]. On the other hand, simSchool operates as an asynchronous self-service arena, providing simulations with up to 10 trillion learner profiles. It complements face-to-face experiences and can serve as a stand-alone solution for teacher training. Unlike Mursion, simSchool is a non-immersive desktop VR system accessed from a stationary position using a personal computer [20]. Both simSchool and Mursion aim to bridge the gap between coursework and field experiences in educator preparation programs (EPPs) by providing unique and effective training opportunities for teacher candidates. Research-validated outcomes include improved confidence in teaching, amplified technology self-efficacy, expanded knowledge of instructional strategies, increased knowledge of classroom management techniques, amplified understanding of students' cognitive, emotional, and cultural differences, improved ability of faculty to monitor candidate progress, and streamlined data capture and analysis [22]. Simulation-based learning (SBL) with emotionally intelligent student avatars is now feasible. This review focuses on simSchool, a web-based online virtual classroom with emotionally intelligent sim students. Numerous studies have explored the potential of simulations in teaching and learning [23–26]. Effective simulations depend on accuracy and contextual relevance. They have been widely used in fields such as medicine, nursing education, aviation, corporate work, safety training, and the military to enable trainees to make informed decisions based on industry best practices in virtual environments [27].

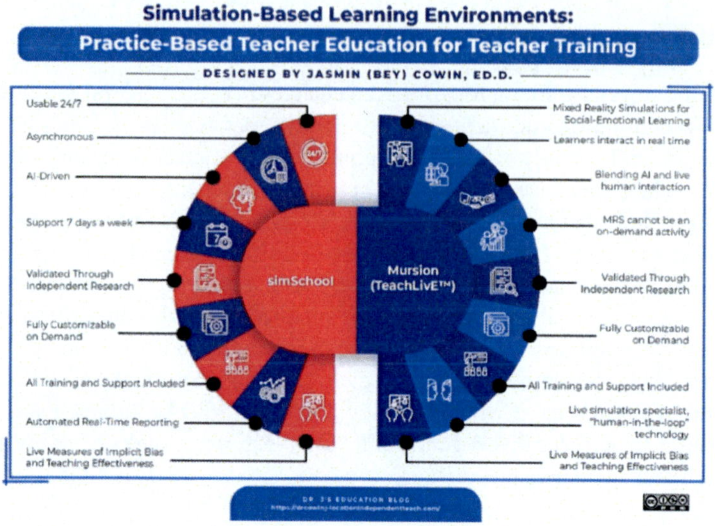


**Fig. 1.** Simulation-based learning environments, practice-based teacher education for teacher training

## 3 Practice-Based Teacher Education and AI Coaching for Educators-in-Training

SBL environments, in the era of ChatHPT and AI, bridge the gap between theoretical frameworks, observations of teacher candidates, and practice teaching through practice-oriented simulations [28]. To address the limited opportunities for pre-service teachers (PSTs) to authentically practice and improve their teaching skills, simulations can be incorporated as an innovative form of micro-teaching [29, 30]. Traditional micro-teaching using actual faculty and students or peers as students has limitations in providing objective feedback and accurately representing diverse student populations [31]. Simulations offer a valuable method in global education settings to connect educators-in-training with real-life situations and challenges. In foreign language instruction, simulations can enhance teachers' communication skills through high-leverage instructional tasks and activities. In content-based courses, simulations are used to develop students' expertise in their chosen fields of study. Global simulations take this approach further by transforming entire courses into simulated worlds where students assume authentic roles and responsibilities. This immersive learning environment allows educators to simultaneously develop their communication and professional skills [32].

## 4 Literature Review

With the remarkable changes being driven by artificial intelligence (AI) in education, higher education institutions must prepare educators-in-training to excel in this evolving paradigm. Simulation-based training combined with AI has emerged as a potential solution to address this challenge. This literature review will examine articles relevant to the integration of technology in educational settings and the complexities of implementing AI in educational assessment. In their article "Engaging training simulations for socially demanding roles" [33], the authors discuss and analyze the role of educational technology in promoting literacy. They emphasize the need to find a balance between traditional teaching methods and the integration of technology in educational settings. Fluck and Fox address the ongoing debate about the impact of technology on literacy development, acknowledging conflicting views about its effectiveness. They argue for the integration of technology as a tool to actively engage students and facilitate their knowledge construction. The article highlights various educational technology options, including multimedia resources, interactive software, and online platforms, to support literacy development. Gardner et al. in "Artificial Intelligence in Educational Assessment: 'Breakthrough? Or Buncombe and Ballyhoo?'" [34] explore the complexities of implementing AI in educational assessment. They critically examine the potential benefits and challenges of using AI-driven assessment tools and ask whether they represent a true breakthrough or transitional hype. The article "Toward a New Generation of Personality Theories: Theoretical contexts for the five-factor model" [35] provides a comprehensive overview of the five-factor model (FFM) used by simSchool to create avatar personalities. The authors analyze the historical background of personality theories, identify the limitations of previous frameworks, and present the FFM as an alternative. The FFM includes five broad dimensions of personality: Neuroticism, Extraversion, Openness to Experience, Agreeableness, and Conscientiousness. These dimensions are embedded in a theoretical framework that considers biological, environmental, and cultural factors that influence personality development.

## 5 21st-Century Simulation-Based Learning (SBL)

The benefits of simulation-based learning (SBL) were identified by Badiee and Kaufman [36] as (a) decision-making in the classroom, (b) practice with feedback and advice, (c) improved self-efficacy in the classroom, and (d) improved collaboration and social interaction. Dieker et al. [24] emphasized that successful simulations create a sense of realism, so that users perceive the simulated world as real. Comparing a new teacher entering the classroom to a novice pilot flying an airplane for the first time, simulations offer focused training opportunities similar to flight simulators for pilots [37]. Kai et al. [38] explore the application of decision tree modeling and educational data mining. They investigate the concepts of wheel-spinning and productive persistence, key behavioral patterns in skill-building exercises, and examine the underlying factors that influence these behaviors (Table 1).

## 6 Teaching Simulations, Intelligent Tutoring Systems and Teacher Self-efficacy

Researchers at the University of Virginia explored ways to supplement traditional methods of teacher preparation, such as student teaching, with digitally mediated simulations [39]. The research concludes that simulations provide pre-service teachers with

**Table 1.** Comparative analysis of learning behaviors: wheel-spinning vs productive persistence

| Behavioral patterns | Wheel-spinning | Productive persistence |
|---|---|---|
| Approach | Repetitive and unproductive practice | Sustained effort and strategic problem-solving |
| Goal orientation | Lack of clear goals or direction | Clear, specific, and challenging goals |
| Problem-solving | Ineffective problem-solving, repeating same actions | Actively seeking new solutions and adjusting strategies |
| Response to failure | Fear of failure, avoidance of risks | Embracing challenges as opportunities for growth |
| Learning foundation | Inadequate knowledge or skills | Building a strong foundation of understanding |
| Mindset | Fixed mindset, resistance to change | Growth mindset, belief in development through effort |
| Metacognition | Limited self-reflection and adjustment | Utilizing metacognitive strategies for monitoring and improvement |
| Learning environment | Lack of support or resources | Supportive learning environment with guidance and feedback |

opportunities to engage with content they might not otherwise encounter. In addition, simulations allow for repeated practice and immediate, targeted coaching and feedback. Teacher self-efficacy refers to teachers' beliefs about their ability to effectively manage professional tasks and challenges. Studies show that teachers with high levels of self-efficacy experience higher job satisfaction, lower job-related stress, and less difficulty managing student misbehavior. This concept is rooted in social cognitive theory and focuses on judgments about one's ability to cope with ambiguous, unpredictable, and stressful situations [40]. Roll et al. in "Improving students' help-seeking skills using metacognitive feedback in an intelligent tutoring system" [41] conducted a study to improve students' help-seeking skills using metacognitive feedback in an intelligent tutoring system (ITS). In the experiment, students were provided with feedback that emphasized the importance of asking for help when needed. The results showed that such feedback improved students' help-seeking behavior and led to better learning outcomes. This study contributes to the understanding of how intelligent tutoring systems can effectively support students in developing essential metacognitive skills. Dieker et al. [24] emphasized the importance of a cyclical process of action, feedback, debriefing, and adjustment in simulation-based learning, known in military contexts as the Action Review Cycle (ARC) [42]. This iterative approach is critical for effective learning outcomes. A meta-analysis by Gegenfurtner et al. [43] supported the importance of feedback in simulation activities. The study showed that receiving feedback not only improved self-efficacy, but also facilitated the transfer of acquired skills to real-world situations, thereby enhancing the practical application of skills. As artificial intelligence becomes more integrated into the classroom, teachers are challenged to adapt to a new paradigm and technological landscape in order to effectively engage and support their students. They must also upskill themselves to engage in forward-thinking high-leverage practices (HLPs) while leveraging AI-driven feedback.

## 7 Discussion and Implications

The U.S. Department of Education's report, "Artificial Intelligence and the Future of Teaching and Learning" [44], expresses interest in using AI and simulation projects to improve the nation's educational landscape. The department aims to improve the learning experiences of pre-service educators, address student weaknesses, and develop effective teaching methods. Their interests in educational innovations include artificial intelligence, virtual worlds, simulation systems, data analysis, human performance evaluation, and human behavior modeling to gain a comprehensive understanding of the field.

Simulation-based approaches offer a cost-effective solution for teacher education, especially in regions with limited resources for traditional programs [45]. Virtual classrooms and realistic scenarios provide valuable hands-on experience in a safe environment [46–48]. Such interactive training develops essential pedagogical skills, classroom management techniques, and the ability to adapt to diverse student needs. Simulation training prepares teachers-in-training for the real-world challenges they may encounter in the classroom, such as language barriers and diverse cultural backgrounds. Reflective learning through AI and instructor debriefing enhances the integration of knowledge and experience gained in simulations, aligning with the demands of the 21st century workforce. Simulation technology could play a critical role in providing educators-in-training in Africa and small nation states with the opportunity to practice high-leverage teaching skills along with AI-driven feedback. This aligns with Sustainable Development Goal 4.c, which aims to increase the availability of qualified teachers worldwide. The integration of simulation technology complements existing teacher training experiences and focuses on helping teacher candidates change their habits, rather than simply acquiring new knowledge. UNESCO data highlights the challenges of recruiting and training enough teachers for universal access to education, particularly in sub-Saharan Africa, where teacher availability and quality are pressing issues [49–51]. In addition, simulation training at universities can have a significant financial impact by reducing the costs associated with traditional face-to-face methods that require time-consuming on-site visits by teachers.

## 8 Conclusion

The global teacher shortage has detrimental effects on students, educators in training, and public education systems. Lack of qualified teachers and high teacher turnover affect student learning and waste valuable resources. This shortage also hinders the professionalization of the teaching profession and disproportionately affects high-poverty schools, hindering the goal of ensuring inclusive and equitable quality education for all under SDG 4 [9]. AI technologies, including classroom simulations, hold great promise for technology-based tools and strategies in education. AI-powered instructional simulations can create new self-paced, on-demand learning systems for teachers and educators. These simulations with AI-driven algorithms can detect and address implicit biases in a simulated classroom environment and improve the teaching effectiveness of educators in training. The combination of simulation-based training (SBT) and AI equips pre-service educators with the skills and competencies they need to excel in complex, real-world educational settings and succeed in 21st century classrooms. SBT provides a safe and controlled environment for educators-in-training to develop their skills and build confidence through experimentation and practice to effectively engage with students in diverse classroom learning environments. AI research aims to create lifelike entities for XR platforms to effectively replace humans. This advance could improve scalability without relying on paid actors. However, it raises ethical concerns in higher education and requires the protection of staff and student rights. Responsible use of AI, privacy, and mitigation of displacement are critical to harnessing the benefits of AI in XR while maintaining ethical standards. Simulation-based learning offers several opportunities to practice complex skills in higher education and to implement effective scaffolding techniques [52]. Advances in computer hardware and software enable innovative methodologies using simulation-based education (SBE) tools to enhance the learning experience. In addition, the globalization of e-learning practices allows students from different regions and universities to access these educational experiences, promoting the widespread availability of simulations for educators in training and fostering international and inter-university collaboration in education [53].

SDG 4 emphasizes transcending physical boundaries through digital connectivity to foster interdisciplinary skills, teamwork, and multicultural learning through simulation training. Shared virtual environments could enable educators-in-training to collaboratively address complex challenges, draw on their diverse expertise, and explore global teaching perspectives. This will enhance their problem-solving skills and promote the application of high-leverage teaching practices in their unique educational settings.

## References

1. SDSN: 4.c By 2030, substantially increase the supply of qualified teachers, including through international cooperation for teacher training in developing countries, especially least developed countries and small island developing States—indicators and a monitoring framework, August 1, 2012. https://indicators.report/targets/4-c/. Accessed 24 Jul. 2023
2. Adinda, D., Mohib, N.: Teaching and instructional design approaches to enhance students' self-directed learning in blended learning environments. Electron. J. e-Learn. **18**(2), 162–174 (2020)
3. Blitz, C.L.: Can online learning communities achieve the goals of traditional professional learning communities? What the literature says. REL 2013-003. U.S. Department of Education, Institute of Education Sciences, National Center for Education Evaluation and Regional Assistance, Regional Educational Laboratory Mid-Atlantic, Washington, DC (2013). https://eric.ed.gov/?id=ED544210. Accessed 5 Oct. 2020
4. Koller, V., Hervey, S., Magnotta, M.: Technology-Based Learning Strategies. Social Policy Research Associates (2005). https://www.doleta.gov/reports/papers/tbl_paper_final.pdf

5. León, C., Lipuma, J.: Transforming traditional professional development into blended learning communities. In: Ciências Socialmente Aplicáveis: Integrando Saberes e Abrindo Caminhos, vol. VIII, pp. 219–229. Editora Artemis (2023). https://doi.org/10.37572/EdArt_30052381118
6. NCES: Press Release—U.S. Schools Report Increased Teacher Vacancies Due to COVID-19 Pandemic, New NCES Data Show. IES, March 3, 2022. https://nces.ed.gov/whatsnew/press_releases/3_3_2022.asp. Accessed 24 Jul. 2023
7. Querolo, N., Rockerman, O., Ceron, E.: The system pushed me out of public education': US faces the great teacher resignation. Bloomberg.com, September 2, 2022. https://www.bloomberg.com/features/2022-america-teachers-great-resignation/. Accessed 24 Jul. 2023
8. O'Hagan, C.: World Teachers' Day: UNESCO sounds the alarm on the global teacher shortage crisis|UNESCO. UNESCO, October 4, 2022. https://www.unesco.org/en/articles/world-teachers-day-unesco-sounds-alarm-global-teacher-shortage-crisis. Accessed 24 Jul. 2023
9. United Nations: Transforming our world: the 2030 Agenda for Sustainable Development. United Nations, Resolution adopted by the General Assembly on 25 September 2015. 70/1. A/RES/70/1, October 2015. https://documents-dds-ny.un.org/doc/UNDOC/GEN/N15/291/89/PDF/N1529189.pdf?OpenElement
10. Vilalta, J.M., Betts, A., Gómez, V.: Higher education's role in the 2030 agenda: The Why and How of GUNi's Commitment to the SDGs. In: Sustainable Development Goals: Actors and Implementation. A Report from the International Conference, 1st ed., pp. 10–14. Global University Network for Innovation (GUNi), Barcelona España (2018). http://www.acup.cat/sites/default/files/2018-06/Higher%20Education%27s%20Role.pdf
11. UN Global Compact: UN Global Compact 20th-Anniversary Progress Report: Uniting Business in the Decade of Action (2020). https://www.unglobalcompact.org/library/5747. Accessed 8 Jul. 2021
12. Vázquez-Montilla, E., Just, M., Triscari, R.: Teachers' dispositions and beliefs about cultural and linguistic diversity. UJER **2**(8), 577–587. https://doi.org/10.13189/ujer.2014.020806
13. Davies, P.: What is evidence-based education? Br. J. Educ. Stud. **47**(2), pp. 108–121. https://doi.org/10.1111/1467-8527.00106
14. Chen, D.W., Nimmo, J., Fraser, H.: Becoming a culturally responsive early childhood educator: a tool to support reflection by teachers embarking on the anti-bias journey. Multicult. Perspect. **11**(2), 101–106 (2009). https://doi.org/10.1080/15210960903028784
15. Danielson, C.: Enhancing Professional Practice: A Framework for Teaching, 2nd ed. Association for Supervision & Curriculum Development, Alexandria, VA (2007)
16. Hook, V., Cheri, W.: Preservice teachers' perceived barriers to the implementation of a multicultural curriculum. J. Instr. Psychol. December 2002. https://www.semanticscholar.org/paper/Preservice-Teachers%27-Perceived-Barriers-to-the-of-a-Hook-Cheri/706441d43e7284283d74f4ffbcb358df4bb49cf0. Accessed 24 Jul. 2023
17. Wasson, D.H., Jackson, M.H.: Assessing cross-cultural sensitivity awareness: a basis for curriculum change. J. Instr. Psychol. **29**(4), 265–276 (2002)
18. Winn, W.: A conceptual basis for educational applications. Human Interface Technology Laboratory, August 1993. https://www.hitl.washington.edu/projects/learning_center/winn/winn-paper.html. Accessed 24 Jul. 2023
19. Ersozlu, Z., Ledger, S., Ersozlu, A., Mayne, F., Wildy, H.: Mixed-reality learning environments in teacher education: an analysis of TeachLivE$^{TM}$ research. Sage Open **11**(3), 21582440211032156 (2021). https://doi.org/10.1177/21582440211032155
20. Christensen, R., Knezek, G., Tyler-Wood, T., Gibson, D.: SimSchool: an online dynamic simulator for enhancing teacher preparation. IJLT **6**, 201–220 (2011). https://doi.org/10.1504/IJLT.2011.042649

21. Fischetti, J., Ledger, S., Lynch, D., Donnelly, D.: Practice before practicum: simulation in initial teacher education. Teach. Educ. Q. **57**(2), 155–174 (2022). https://doi.org/10.1080/08878730.2021.1973167
22. Zach, S., Ophir, M.: Using simulation to develop divergent and reflective thinking in teacher education. Sustainability **12**(7), Article no. 7, January 2020. https://doi.org/10.3390/su12072879
23. Dalgarno, B., Gregory, S., Knox, V., Reiners, T.: Practising teaching using virtual classroom role plays. Austr. J. Teach. Educ. **41**(1), January 2016. https://doi.org/10.14221/ajte.2016v41n1.8
24. Dieker, L.A., Rodriguez, J.A., Lignugaris/Kraft, B., Hynes, M.C., Hughes, C.E.: The potential of simulated environments in teacher education: current and future possibilities. Teach. Educ. Spec. Educ. **37**(1), 21–33, February 2014. https://doi.org/10.1177/0888406413512683
25. Lateef, F.: Simulation-based learning: just like the real thing. J. Emerg. Trauma Shock **3**(4), 348–352 (2010). https://doi.org/10.4103/0974-2700.70743
26. McGarr, O.: The use of virtual simulations in teacher education to develop pre-service teachers' behaviour and classroom management skills: implications for reflective practice. J. Educ. Teach. **47**, 1–13 (2020). https://doi.org/10.1080/02607476.2020.1733398
27. Milgram, P., Colquhoun, H.A.: In mixed reality: merging real and virtual worlds. In: Mixed Reality: Merging Real and Virtual Worlds, 1st ed., pp. 5–30. Springer (1999)
28. O'Callaghan, C., Piro, J.S.: Virtual simulations in a practice based teacher education. Field Exp. J. **17**, 94–118 (2016)
29. Lindqvist, H., Weurlander, M., Wernerson, A., Thornberg, R.: Conflicts viewed through the micro-political lens: beginning teachers' coping strategies for emotionally challenging situations. Res. Pap. Educ. **35**(6), 746–765 (2020). https://doi.org/10.1080/02671522.2019.1633559
30. Kaufman, D., Ireland, A.: Enhancing teacher education with simulations. TechTrends **60**(3), 260–267 (2016). https://doi.org/10.1007/s11528-016-0049-0
31. Grossman, P., Hammerness, K., McDonald, M.: Redefining teaching, re-imagining teacher education. Teach. Teach. **15**(2), 273–289 (2009). https://doi.org/10.1080/13540600902875340
32. Perez, Y., Poole, P.: Making language real: developing communicative and professional competences through global simulation. Simul. Gaming **50**(6), 725–753 (2019). https://doi.org/10.1177/1046878119869756
33. Fluck, A.E., Fox, A.: Engaging training simulations for socially demanding roles. In: ASCILITE, pp. 398–406. The University of Tasmania: The University of Tasmania, Hobart (2011). https://www.ascilite.org/conferences/hobart11/downloads/ProceedingsV3.pdf
34. Gardner, J., O'Leary, M., Yuan, L.: Artificial intelligence in educational assessment: 'Breakthrough? Or buncombe and ballyhoo?' J. Comput. Assist. Learn. **37**(5), 1207–1216 (2021). https://doi.org/10.1111/jcal.12577
35. McCrae, R.R., Costa Jr. P.T.: Toward a new generation of personality theories: theoretical contexts for the five-factor model. In: The Five-Factor Model of Personality: Theoretical Perspectives, pp. 51–87. Guilford Press, New York, NY, USA (1996)
36. Badiee, F., Kaufman, D.: Design evaluation of a simulation for teacher education. Sage Open **5**(2), 2158244015592454 (2015). https://doi.org/10.1177/2158244015592454
37. Zibit, M., Gibson, D.: simSchool: The game of teaching. Innov.: J. Online Educ. **1**(6) (2005)
38. Kai, S., Almeda, V., Baker, R.S., Heffernan, C., Heffernan, N.: Decision tree modeling of wheel- spinning and productive persistence in skill builders. J. Educ. Data Min. **10**(1), 36–71 (2018). https://doi.org/10.5281/zenodo.3344873
39. Kaskamanidis, Z.: Classroom simulators to support teacher training. Teach. Mag. (2022). https://www.teachermagazine.com/sea_en/articles/classroom-simulators-to-support-teacher-training. Accessed 24 Jul. 2023

40. Bandura, A., Schunk, D.H.: Cultivating competence, self-efficacy, and intrinsic interest through proximal self-motivation. J. Pers. Soc. Psychol. **41**(3), 586–598 (1981). https://doi.org/10.1037/0022-3514.41.3.586
41. Roll, I., Aleven, V., McLaren, B.M., Koedinger, K.R.: Improving students' help-seeking skills using metacognitive feedback in an intelligent tutoring system. Learn. Instr. **21**(2), 267–280 (2011). https://doi.org/10.1016/j.learninstruc.2010.07.004
42. Holman, P., Devane, T., Cady, S., Adams, W.A.: The Change Handbook: Group Methods for Shaping the Future, 2nd edn. Berrett-Koehler Publishers, San Francisco (2007)
43. Gegenfurtner, A., Quesada-Pallarès, C., Knogler, M.: Digital simulation-based training: a meta-analysis. Br. J. Edu. Technol. **45**(6), 1097–1114 (2014). https://doi.org/10.1111/bjet.12188
44. USDOE: Artificial intelligence and the future of teaching and learning: insights and recommendations. Office of Educational Technology, May 2023. https://tech.ed.gov/ai-future-of-teaching-and-learning/. Accessed 24 Jul. 2023
45. Dotger, B., Dotger, S., Masingila, J., Rozelle, J., Bearkland, M., Binnert, A.: The right 'Fit': exploring science teacher candidates' approaches to natural selection within a clinical simulation. Res. Sci. Educ. **48**(3), 637–661 (2018). https://doi.org/10.1007/s11165-016-9582-2
46. Cohen, J., Wong, V.: Using classroom simulators to transform teacher preparation. November 17, 2021. https://www.brookings.edu/articles/using-classroom-simulators-to-transform-teacher-preparation/
47. Ronfeldt, M., Brockman, S.L., Campbell, S.L.: Does cooperating teachers' instructional effectiveness improve preservice teachers' future performance? Educ. Res. **47**(7), 405–418 (2018). https://doi.org/10.3102/0013189X18782906
48. Cohen, J., Wong, V., Krishnamachari, A., Berlin, R.: Teacher coaching in a simulated environment. Educ. Eval. Policy Anal. **42**(2), 208–231 (2020). https://doi.org/10.3102/0162373720906217
49. UNESCO: Teacher education and learning outcomes|Unesco IIEP Learning Portal. IIEP Learning Portal, March 1, 2023. http://learningportal.iiep.unesco.org/en/issue-briefs/improve-learning/teacher-education-and-learning-outcomes. Accessed 24 Jul. 2023
50. Bayley, S., Wole Meshesha, D., Ramchandani, P., Rose, P., Woldehanna, T., Yorke, L.: Socio-emotional and academic learning before and after COVID-19 school closures: evidence from Ethiopia. Research on Improving Systems of Education (RISE), November 2021. https://doi.org/10.35489/BSG-RISE-WP_2021/082
51. OECD and The World Bank: How learning continued during the COVID-19 pandemic: global lessons from initiatives to support learners and teachers. OECD (2022). https://doi.org/10.1787/bbeca162-en
52. Chernikova, O., Heitzmann, N., Stadler, M., Holzberger, D., Seidel, T., Fischer, F.: Simulation-based learning in higher education: a meta-analysis. Rev. Educ. Res. **90**(4), 499–541 (2020). https://doi.org/10.3102/0034654320933544
53. Campos, N., Nogal, M., Caliz, C., Juan, A.A.: Simulation-based education involving online and on-campus models in different European universities. Int. J. Educ. Technol. High. Educ. **17**(1), 8 (2020). https://doi.org/10.1186/s41239-020-0181-y

# The Role of Discussion Forum and Assignment Choice for Students with Different Educational Backgrounds

Maryam Ghalkhani(✉) and Moein Mehrtash

W Booth School of Engineering Practice and Technology, McMaster Automotive Resource Center (MARC), MARC-273, 200 Longwood Road South, Hamilton, ON L8P 0A6, Canada
ghalkham@mcmaster.ca

**Abstract.** Educators regularly struggle with how to engage students in an environment in which students are increasingly distracted especially with online courses. Courses delivered online found a rapidly growing component of higher education since online delivery makes colleges and universities accessible to a more significant number of students at a fraction of the cost of in-person delivery. Therefore, The challenge for active student engagement increases considerably in online teaching and learning. The paper proposes an active-learning strategy to improve class engagement and enhance student learning by adapting the online discussion board for use during and out of the class session to keep students connected and engaged with the weekly topics. Student feedback shows that combining the technologies, synchronous teaching, and active learning activities in both online classes is effective for interactive learning and using student–content interactions, student-instructor interactions, and student-student interactions in online teaching and learning.

**Keywords:** Discussions forums · Students engagement · Student participation · Engineering-pedagogical approaches · Active student engagement

## 1 Introduction

The focus of this paper is on demonstrating the benefits of active-learning pedagogy using discussion forums and the support they may provide for learning in online courses and big classrooms with over 200 students from different educational backgrounds.

The classroom teaching and delivering instruction can be divided into three groups: (1) a blended learning method which is more traditional while the classes are only in-person and digital resources are used to extend the classroom, and (2) a hybrid format where the class meets in-person several times over the semester, but an online learning management system such as D2L, Canvas, Blackboard, ANGEL, WebCT or similar environment is used for the content delivery, learning management and assessment, and (3) the online learning style which is an internet-based learning environment where the class never meet in-person and the institution will use a learning management system, to enable online learning. Online learning could be in the form of synchronous learning


M. E. Auer et al. (Eds.): ICL 2023, LNNS 899, pp. 542–550, 2024.
https://doi.org/10.1007/978-3-031-51979-6_56

where the class must be online in a virtual platform at the same time, or in the form of asynchronous learning where students are not required to be online at the same time and students access materials at their own pace.

Over the past decade developing and maintaining student engagement, and increasing opportunities for students have been studied extensively. In traditional learning environments with lecture-style practices, student activities were limited to listening and taking notes or asking occasional questions during lecture time. Therefore, the lack of student engagement was improved by introducing alternative teaching practices to enable deep learning in several experiential learning activities [1–3].

It is common sense among researchers and educators that learning only through lectures is not enough for improved learning outcomes in the regular classrooms" for undergraduate science, technology, engineering, and mathematics (STEM) courses [4, 5].

Although there is a wide range of acceptable descriptions for Active Learning (AL), generally AL seeks to enhance student participation and provide opportunities for students to reflect on the content, ideas, and concerns involved in the course. Therefore, any classroom activity in which students are asked to concentrate and participate instead of just watching, listening, and taking notes can be considered an AL approach [6, 7].

As shown in Table 1 and Fig. 1, Emma R. Wester et al., studied how the shifting from in-person to online classes, due to the COVID-19 pandemic, impacted three concepts of student engagement such as behavioural, cognitive, and emotional [8].

Behavioural engagement refers to the student's contribution to academic or social activities such as students' incidence of participating in class discussions, meeting with instructors, and studying with peers outside of class. The study indicates that overall behavioural engagement did not change, although students participated less frequently in class discussions but met with professors more often outside of class. Cognitive engagement is defined by a student's ability to understand new ideas and use thinking skills and implement learning strategies such as reviewing the content and participating in question-and-answer sessions; this study used the student effect, consisting of self-efficacy and sense of belonging, to assess a student's cognitive engagement with class material. The study shows although students' sense of belonging and self-efficacy increases over the semester, no significant transformation in cognitive engagement was reported. Emotional engagement has been synonymously linked to terms such as feelings, moods and reactions, including students' approaches toward science, their perceived value of the course, and their stress. The analysis of emotional parameter concepts focused on the student's value of the course, stress level, and attitude. The study found a significant decrease between the start of the transition and the end of the semester since Student's positive attitude toward science was significantly lower at the end of the semester [8].

Therefore, the investigation explores the emotional engagement in students' relationships with the instructor and the peer group contributing to students' cognitive engagement in academic activities and learning outcomes surprisingly. For the same reason, we can better construct mechanisms for supporting student learning during disruptions to education caused by emergencies, and to identify multiple environmental and individual factors that contribute to students' emotional and cognitive involvement in academic activities and build the capacity for students to successfully cope with various academic

**Table 1.** Overview of surveys and questions administered using the Qualtrics survey system [8].

| Engagement construct | Student parameter measured | Survey or Question Used | No. of Questions | Format | Example question |
|---|---|---|---|---|---|
| Behavior | Meetings with professor | "How often did you meet with professors outside of class time?" | 1 | 5-point likert scale: 0 = Never, 1 = Once a semester, 2 = 1-2 times a mo, 3 = Once a wk, 4 = More than once a wk | See questions used |
| | Participation in class | "How often did you participate in class discussions?" | 1 | | |
| | Study group | "How often did you study with other students outside of class time?" | 1 | | |
| Cognitive | Self-Efficacy | General Self-efficacy Scale (GSE) (62) | 10 | 5-point Likert Scale: 1 = Strongly disagree, 2 = Somewhat disagree, 3 = Neither agree nor disagree, 4 = Somewhat agree, 5 = Strongly agree | "I can solve most problems if I invest the necessary effort" |
| | Sense of Belonging | Ingram Sense of Belonging survey (63) | 20 | | "I would find it easy to join study groups with other students if I wanted to" |
| Emotional | Attitude | Colorado Learning Attitudes about Science (CLASS) (64) | 14 | | "I think about the science I experience in everyday life" |
| | Value | Papanastasiou survey (65) | 8 | | "The skills I have acquired in this class will be helpful to me in the future" |
| | Stress | Papanastasiou survey (65) | 8 | | "This class is stressful" |

All surveys were combined into one for ease of student participation

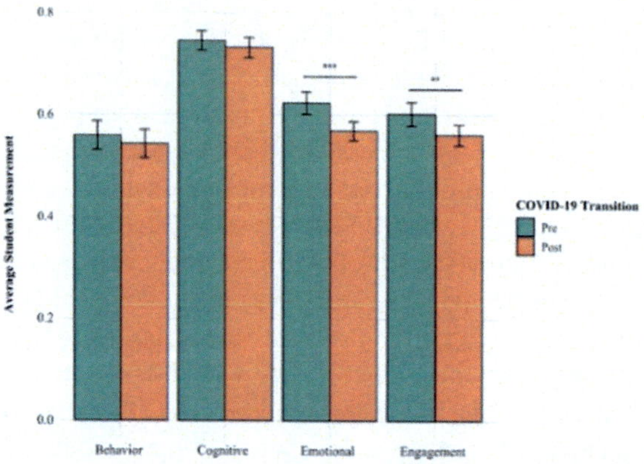


**Fig. 1.** Engagement of students before (Pre) and after (Post) the COVID-19 transition to online learning [8].

tasks and participate in academic activities provided by the instructor at any situation [7].

For engineering graduates, emotional intelligence is a necessary consideration for employment and later for career growth as it has a direct impact on confidence, motivation, self-discipline, self-regulation and leadership skills [9]. Therefore, one common approach in engineering is solving more practice problems during class time or replacing in-person lectures with video lectures since integrating art into engineering education can improve critical thinking using innovative pathways and their capacity to solve problems [10].

Since engineering education should provide essential training for students to develop problem-solving skills, studying the challenges of the large lecture-focused courses in higher education will improve the critical thinking indicators that students need in higher institutions.

One of the methods that have been applied to integrate emotions in multime-dia-based learning environments is the Theory of Emotional Design which is created on cognitive science theories and product design. This Theory classifies interaction with a product based on three levels; visceral which is the appreciation of beauty and is used to impact learners through a good user interface by promoting better information ability and online communication, behavioural activities which is the physical interactions with the system, and reflective which is reflective thoughts [9, 11].

This study accomplishes the impact of emotional design (EI) in improving students' learning outcomes as the learner's characteristics were represented as their emotional intelligence which is critical in students' adaption of design and technology. Additionally, in this study, the importance of individuality and adaptability were assessed using EI as an important aspect of online learning since those with low levels of emotional intelligence were more pleased with the neutral design (NeuD) of the courseware in comparison to emotionally designed courseware as displayed in Figs. 2 and 3.

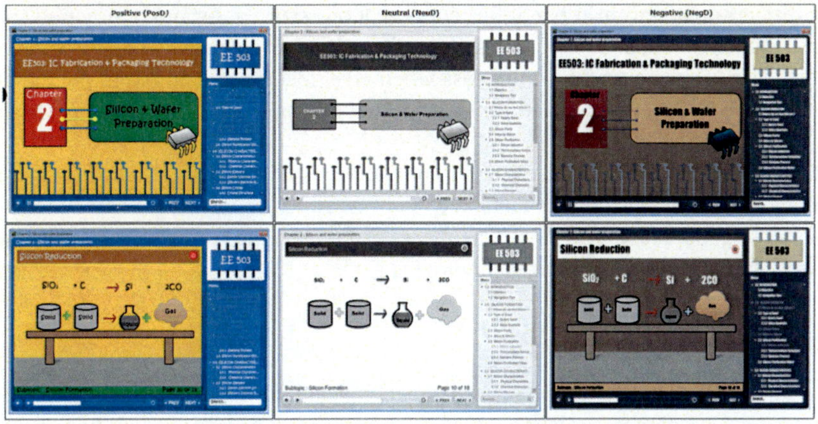


**Fig. 2.** Screenshots of the courseware [11].

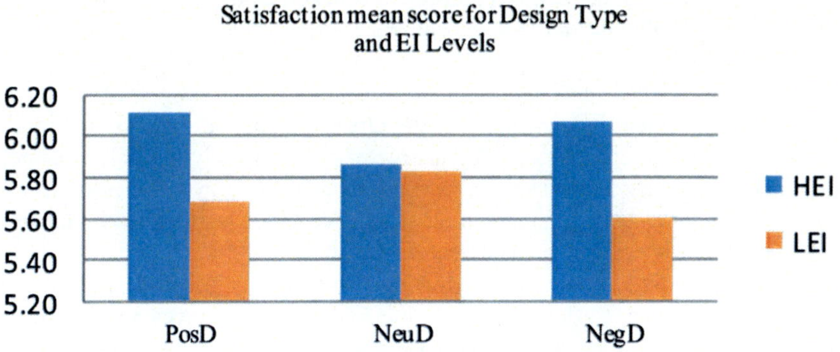


**Fig. 3.** Graphical representation of the difference of each design group against different levels of EI [11].

The effects of emotional design and aesthetics are not the only two components that affect the cognitive and affective learning outcomes of students in the same class with different backgrounds. Thus, the literature suggests the potential to optimize student engagement in hybrid and fully online environments, which combine synchronous and asynchronous activities using social interaction [12].

Therefore, As more instructors shift toward active learning, research needs to identify how different types of activities affect students' cognitive engagement with concepts in the classroom. While the use of technology helps with the safe delivery of teaching, it also makes it challenging to engage an entire class of students during the session and make sure students are actively following the class materials. This issue has become more important as instructors have been forced to adopt a variety of online learning tools to manage instruction.

Therefore, this study explores new pedagogical approaches to enhance some of the current challenges such as class sizes, limited funding support, and difficulties in active involvement and learning with a diverse cohort of students. The adaption of online discussion boards in either online courses or big-size class-rooms communication between students creates a lot of noise and difficulties to manage the classroom deal with the issue of engaging the entire class when all students are required to participate. Moreover, in any classroom, some students might be comfortable speaking and sharing their feedback with others [13].

Morover, other studies have reported confusion about the navigation of the online environment by nursing students that have overshadowed the learning, or tendencies of accounting students to overlook critical aspects of the online environment as they are not particularly fond of computer technologies [14].

## 2 Methodology and Results

Now a day, the use of technology to support teaching and learning is an essential skill for instructors and learners since technology has a strong impact on students' studying habits, accessibility of information, communication and collaboration skill and inspiration. Moreover, students appreciate the freedom of independent practice and online/offline learning using the technologies since appropriate educational technologies rises accessibility to learning materials and various learning methods to achieve the requirement of various students.

The approach is to design an online course with activities and assignments that help the students deepen their understanding of the already discussed material and reflect on their knowledge. In this approach, the instructor can frequently request students to respond to specific questions and interact with peer students. Moreover, the instructor can simply screen the students' work, share immediate feedback on learners' understanding, and discuss with the specific students whenever needed.

The course materials are documented using PowerPoint slides and the lecture is pre-recorded and posted on the D2L platform. Additional, resources such as recent publications, web links, customized handouts, announcements related to the course, recorded sessions of synchronous and asynchronous teaching and learning, and video-recorded laboratory experiments will be provided to students as an extra reference in the same e-learning platform. One of the weekly assignments is answering a question from the weekly materials using a discussion forum as Discussion Post which will remain available to students for one week.

Additional assignments on discussion forums as Tutorial Discussions are used for assessments online, and communication with the learners during the live classes which will be available to students only during class time to share their feedback and ask their questions about the other students' presentations since students in this class are expected to prepare a group research project about the different applications of AI based on their interests. In addition, they are motivated to prepare a 10–15-min group presentation based on the same topic.

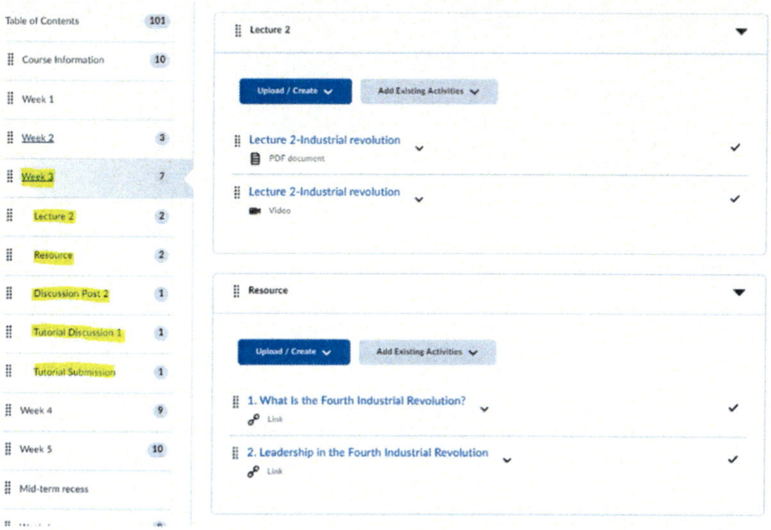


**Fig. 4.** Weekly course structure and provided resources.

Figures 4 and 5 show a screenshot view of the course shell created in the e-learning platform that includes the activities on the course site.

The Online discussion forum is a web-based application included in D2L that helps students in the same class to post messages to the discussion threads, interact and receive feedback from other students and instructors at the same time which is beneficial to create a deeper understanding of the subject matter being discussed during and after the class time [15, 16].

This teaching strategy has been used in teaching artificial intelligence (AI) courses to over 200 students in the same class from different Engineering programs, Business schools, Law schools, schools of Medicine, and Science and Management schools. The outcome illustrates an increase in critical thinking skills for students and provides them with the opportunity to learn the different applications of the same AI technics and motivates them to find more applications of AI in their program and share them with students from other programs. Moreover, this created a strong social and communication link between students.

This innovation was initially tried in a synchronous online class. The pre-recorded lectures and extra online sources were provided to students to read/watch and share their feedback about the given question during the week using a discussion forum, and during

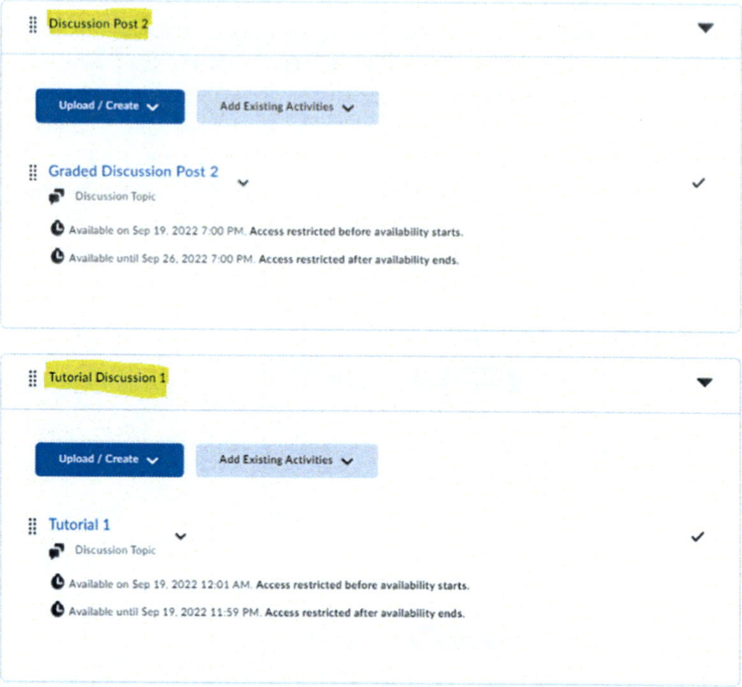


**Fig. 5.** The weekly assignments and students' interaction using the Discussion forum.

the 3 h of live sessions offered a short group presentation based on the assigned topics. The student engagement is shown in Fig. 6. In this assignment, students were expected to discuss the pros and cons of the application of artificial intelligence in the automotive industry. This result proves completing the weekly assignment using a discussion board significantly improves students' interaction with the instructor and with each other since from 200 students in the class, 190 students responded to the initial post by creating a new thread. Later the student interaction increased by an additional 199 feedback which means in total 389 responses were exchanged between the students.

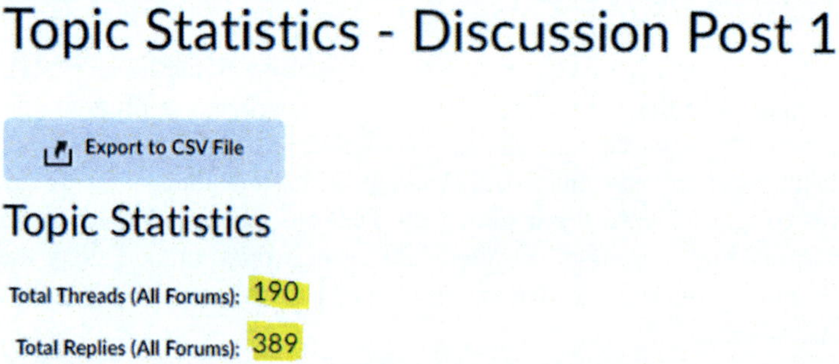


**Fig. 6.** Students' participation in an offline assignment in a class with 200 students.

Although students were enjoying the weekly discussion they were showing less motivation to join live classes since the recording was available to them to watch on their own time. Additionally, it was difficult for them to maintain their concentration during the entire class time and just listen to the weekly presentations done by their class-fellow. Therefore, to inspire students to participate in live sessions and motivate them to bring their attention to the class and be engaged with the presenters and instructor, the study was expanded and more forums were involved in the course to be available only during class while students can share their understanding of the class presentations with others. The rubrics wad added to each discussion forum to assist in grading as well.

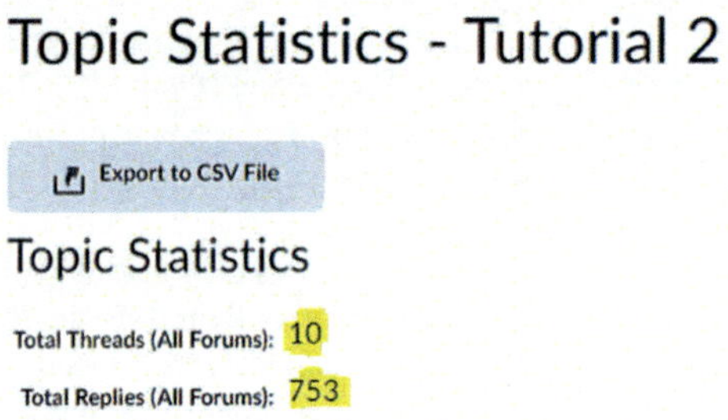


**Fig. 7.** Student participation in a live class with 200 students.

In this in-class assignment, 10 different questions from the in-class presentations were asked from students and the total replies from students are 753 as shown in Fig. 7.

## 3 Conclusions

This result in Figs. 1 and 2 confirms how much student engagement can impact their performance and motivation in an educational environment. More studies on the influence of students' communication skills, and the development of argumentation opportunities for students using synchronous and asynchronous communications tools would be beneficial in designing a more effective learning approach. The proposed framework can be used in hybrid and blended course delivery methods as well. Though, it is challenging to implement multiple student presentations for large-size face-face courses because the implementation performance depends not only on designing activities but also on the technology used to exercise the activities during course conduction.

## 4 Future Work

The future work will focus on how Online Discussion Forums can be used as an instrument to develop review-based learning and teaching resources which can be accessible through mobile technologies on E-learning.

Moreover, the use of discussion forums will be evaluated in the laboratory courses to evaluate how sharing the observation, participation, and experimentation results of the students can improve their analyzing skills.

## References


1. Centea, D., Apostolou, K., Mehrtash, M.: Development of a Classroom Response System: A Web-Based Approach Used in SEPT, pp. 91–101 (2020)
2. Cosbey, R., Wusterbarth, A., Hutchinson, B.: Deep learning for classroom activity detection from audio. In: ICASSP 2019–2019 IEEE International Conference on Acoustics, Speech and Signal Processing (ICASSP), pp. 3727–3731. IEEE (2019)
3. Mehrtash, M., Ghalkhani, K., Singh, I.: IoT-based experiential E-Learning Platform (EELP) for online and blended courses. In: 2021 International Symposium on Educational Technology (ISET), pp. 252–255. IEEE (2021)
4. Shoufan, A.: Lecture-free classroom: fully active learning on moodle. IEEE Trans. Educ. **63**(4), 314–321 (2020)
5. Wu, S.P., Van Veen, B., Rau, M.A.: How drawing prompts can increase cognitive engagement in an active learning engineering course. J. Eng. Educ. **109**(4), 723–742 (2020)
6. Guimarães, L.M., Lima, R.D.S.: Active learning application in engineering education: effect on student performance using repeated measures experimental design. Eur. J. Eng. Educ. **46**(5), 813–833 (2021)
7. Pietarinen, J., Soini, T., Pyhältö, K.: Students' emotional and cognitive engagement as the determinants of well-being and achievement in school. Int. J. Educ. Res. **67**, 40–51 (2014)
8. Wester, E.R., Walsh, L.L., Arango-Caro, S., Callis-Duehl, K.L.: Student engagement declines in STEM undergraduates during COVID-19–driven remote learning. J. Microbiol. Biol. Educ. 22(1) (2021). https://doi.org/10.1128/jmbe.v22i1.385
9. Kumar, J.A., Muniandy, B., Wan Yahaya, W.A.J.: Exploring the effects of emotional design and emotional intelligence in multimedia-based learning: an engineering educational perspective. New Rev. Hypermedia Multimedia **25**(1–2), 57–86 (2019)
10. Alharthi, M., Zhang, K.: Faculty's use of social media in flipped classrooms: a mixed-method investigation. Int. J. Technol. Educ. Sci. (IJTES) **5**(3), 394–410 (2021)
11. Fishwick, M.: Emotional design: why we love (or hate) everyday things. J. Am. Cult. **27**(2), 234 (2004)
12. Heilporn, G., Lakhal, S., Bélisle, M.: An examination of teachers' strategies to foster student engagement in blended learning in higher education. Int. J. Educ. Technol. High. Educ. **18**, 1–25 (2021)
13. Cooper, A.D.: Using the discussion board during your online synchronous class to engage students. Mark. Educ. Rev. 1–4 (2022)
14. Seethamraju, R.: Effectiveness of using online discussion forum for case study analysis. Educ. Res. Int. **2014** (2014)
15. Onyema, E.M., Deborah, E.C., Alsayed, A.O., Noorulhasan, Q., Sanober, S.: Online discussion forum as a tool for interactive learning and communication. Int. J. Recent Technol. Eng. **8**(4), 4852–4859 (2019)
16. Nor, N.F.M., Razak, N.A., Aziz, J.: E-learning: analysis of online discussion forums in promoting knowledge construction through collaborative learning. WSEAS Trans. Commun. **9**(1), 53–62 (2010)

# Author Index





M. E. Auer et al. (Eds.): ICL 2023, LNNS 899, pp. 551–553, 2024.
https://doi.org/10.1007/978-3-031-51979-6